Springer Series on Environmental Management

Robert S. DeSanto, Series Editor

Springer Series on Environmental Management
Robert S. DeSanto, Series Editor

Disaster Planning:
The Preservation of Life and Property
Harold D. Foster
1980/275 pp./48 illus./cloth
ISBN 0-387-90498-0

Enviromental Effects
of Off-Road Vehicles:
Impacts and Management
in Arid Regions
R. H. Webb
H. G. Wilshire (Editors)
1983/560 pp./149 illus./cloth
ISBN 0-387-90737-8

Natural Hazard Risk Assessment
and Public Policy:
Anticipating the Unexpected
William J. Petak
Arthur A. Atkisson
1982/489 pp./89 illus./cloth
ISBN 0-387-90645-2

Global Fisheries:
Perspectives for the '80s
B. J. Rothschild (Editor)
1983/approx. 224 pp./11 illus./cloth
ISBN 0-387-90772-6

Heavy Metals in Natural Waters:
Applied Monitoring and Impact
Assessment
James W. Moore
S. Ramamoorthy
1984/256 pp./48 illus./cloth
ISBN 0-387-90885-4

Landscape Ecology:
Theory and Applications
Zev Naveh
Arthur S. Lieberman
1984/376 pp./81 illus./cloth
ISBN 0-387-90849-8

Organic Chemicals in Natural Waters:
Applied Monitoring and Impact
Assessment
James W. Moore
S. Ramamoorthy
1984/282 pp./81 illus./cloth
ISBN 0-387-96034-1

The Hudson River Ecosystem
Karin E. Limburg
Mary Ann Moran
William H. McDowell
1986/344 pp./44 illus./cloth
ISBN 0-387-96220-4

Human System Responses to Disaster:
An Inventory of Sociological Findings
Thomas E. Drabek
1986/512 pp./cloth
ISBN 0-387-96323-5

The Changing Environment
James W. Moore
1986/256 pp./40 illus./cloth
ISBN 0-387-96314-6

Balancing the Needs
of Water Use
James W. Moore
1989/280 pp./39 illus./cloth
ISBN 0-387-96709-5

Air Pollution and Forests:
Interactions between Air Contaminants
and Forest Ecosystems, Second Edition
William H. Smith
1990/640 pp./illus./cloth
ISBN 0-387-97084-3

William H. Smith

Air Pollution and Forests

Interactions between Air Contaminants
and Forest Ecosystems

Second Edition

Springer-Verlag
New York Berlin Heidelberg
London Paris Tokyo

William H. Smith
Clifton R. Musser Professor of Forest Biology
Yale University
School of Forestry and Environmental Studies
New Haven, CT

On the cover: Looking east from the town of Woodstock to the White Mountain region of north-central New Hampshire.

Library of Congress Cataloging-in-Publicaton Data
Smith, William H., 1939-
    Air pollution and forests: interaction between air contaminants
and forest ecosystems / William H. Smith. — 2nd ed.
        p.   cm. — (Springer series on environmental management)
    Includes bibliographical references.
    ISBN 0-387-97084-3
    ✓1. Air—Pollution—Environmental aspects. ✓2. Forest ecology.
✓3. Plants, Effect of air pollution on.   I. Title.   II. Series.
    QH545.A3S64   1990
    574.5′2642—dc20   89-28540
                                                               CIP

Printed on acid-free paper

Typeset by Printworks, Madison, Connecticut
Printed and bound by R. R. Donnelley & Sons, Harrisonburg, Virginia
Printed in the United States of America.

9 8 7 6 5 4 3 2 1

ISBN 0-387-97084-3 Springer-Verlag New York Berlin Heidelberg
ISBN 3-540-97084-3 Springer Verlag Berlin Heidelberg New York

7091

To Fox, Scott, Philip, Tyler, friends, and students,
for their encouragement and inspiration.

Series Preface

This series is dedicated to serving the growing community of scholars and practitioners concerned with the principles and applications of environmental management. Each volume will be a thorough treatment of a specific topic of importance for proper management practices. A fundamental objective of these books is to help the reader discern and implement human's stewardship of our environment and the world's renewable resources. For we must strive to understand the relationship between humankind and nature, act to bring harmony to it, and nurture an environment that is both stable and productive.

These objectives have often eluded us because the pursuit of other individual and societal goals has diverted us from a course of living in balance with the environment. At times, therefore, the environmental manager may have to exert restrictive control, which is usually best applied to humans, not nature. Attempts to alter or harness nature have often failed or backfired, as exemplified by the results of imprudent use of herbicides, fertilizers, water, and other agents.

Each book in this series will shed light on the fundamental and applied aspects of environmental management. It is hoped that each will help solve a practical and serious environmental problem.

Robert S. DeSanto
East Lyme, Connecticut

Preface

This book was made possible by the research performed by a very large number of scientists. Their productivity and publication have allowed the author to construct this review. This volume provides a compendium of the most significant relationships between forests and air pollution under low-, intermediate-, and high-exposure conditions. Under conditions of a low dose, the vegetation and soils of forest ecosystems function as important sources and sinks for air pollution. When exposed to an intermediate dose, individual tree species or individual members of a given species may be subtly and adversely affected by nutrient stress, impaired metabolism, predisposition to entomological or pathological stress, or direct disease induction. Exposure to a high dose may induce acute morbidity or mortality of specific trees.

At the ecosystem level, the impact of these various interactions would be very variable. In the low-dose relationship, pollutants are exchanged between the atmospheric compartment, the available nutrient compartment, other soil compartments, and various elements of the biota. Depending on the nature of the pollutant, this transfer can be innocuous or stimulatory to the forest. Forest exposure to intermediate dose conditions may result in an inimical influence. The ecosystem impact in this instance may include reduced productivity and biomass, alterations in species composition or community structure, increased insect outbreaks or microbial disease epidemics, and increased morbidity. Under conditions of a high dose and concentrated mortality, ecosystem impacts may include gross simplification, impaired energy flow, and biogeochemical cycling, changes in hydrology and erosion, climate alteration, and major impacts on associated ecosystems as well as forest destruction.

The author hopes that this book will provide a strategy for a comprehensive introduction to the complex relationship between forest systems and air contaminants. An understanding of this relationship is essential for the protection and wise management of our forest resources.

William H. Smith
Yale University
New Haven, Connecticut

Contents

1
Introduction

Forest systems have enormous variability. Forests may differ in soil type, climate, aspect, elevation, species composition, age, and health. Forests may be uneven aged, even aged, all aged or overmature. Forests may be reproduced by seed, by coppice, or by planting. Some forests have their structure completely shaped by natural forces, some may be influenced by human forces as well as natural forces, while other forests may be completely artificial in design and establishment. Forest trees may be arrayed along a continuum of human management efforts ranging from no management to intensive management (Table 1-1). All of this variability makes generalizations about forest tree systems and their stresses very difficult to formulate, but it also makes them essential to attempt.

A. Perspective

Forest stress may be viewed from a variety of perspectives: an individual tree, a population of trees (members of the same species in a given area), a stand of trees (all tree species in a given area), or a forest ecosystem (Figure 1-1). Stress considerations at the individual tree level typically involve considerations of abnormal physiology — how stress alters metabolic processes, for example, photosynthesis, respiration, translocation, or flowering. At the population level, stress impacts on rates of reproduction, morbidity, and mortality are especially significant. From the perspective of forest stands, the concepts of competition, vegetative interaction, and the spatial arrangement of trees are central to stress intensity and significance. Stress alteration of forest system functions, such as bio-

Table 1-1. United States forest tree management continuum.

Type	Description	Apprixunate Size (ha)	Management Intensity
Unmanaged natural forest	areas of low tree density and/or slow growth	120×10^6	none
Managed natural forest	commercial forest land recreation/ wilderness areas	190×10^6	medium
Plantation forest	row pattern planting with regular spacing	20×10^6	high
Christmas tree plantation	short rotation, holiday market	20×10^4	high
Suburban forest	specimen, shade ornamental trees	80×10^6	high
Urban forest	park, street trees	40×10^6	high
Arboreta	trees established for study, enjoyment	4×10^3	very high
Forest tree seed orchard	seed production areas	4×10^3	very high
Forest tree nursery	seedling production areas	6×10^3	very high

geochemical cycling and energy storage, and forest system structure, such as succession and food webs, are encompassed in the forest ecosystem perspective. Unfortunately, our understanding of the relationship between air pollution and forest systems is not equivalent across these perspectives. We know most about the relationship between individual trees, less about the relationship at the population and stand levels, and least about ecosystem level responses. This difference in understanding is due to our relatively deficient understanding of complex systems compared to individual trees and to the relative differences in cost, time, and complexity of research efforts on single plants versus systems of plants. Due to the holistic, integrative, and management utility of the forest ecosystem perspective, it is the goal of contemporary research, and of this book, to describe the interaction of atmospheric contaminants with forest systems at this level.

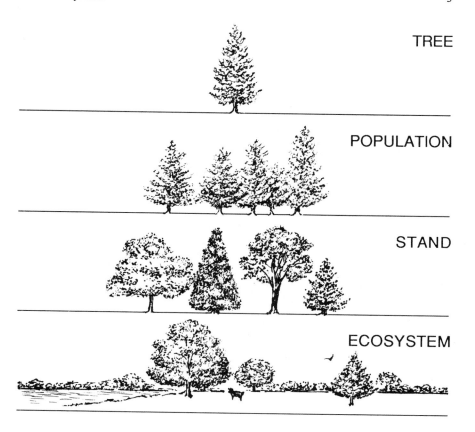

TREE

POPULATION

STAND

ECOSYSTEM

Figure 1-1. Forest stress may be viewed from the perspective of individual trees, populations of trees, stands of trees or forest ecosystems.

B. Forest Ecosystems

Maintenance of life over extended time requires a flow of energy through a living system and a complete cycling of all chemical elements required for life. Individuals, populations, and even communities of populations are not sufficient for this maintenance. The smallest natural systems that permit energy flow and complete chemical cycling consist of at least several interacting populations of various species and their nonliving environments. These minimal systems are termed *ecosystems* (Botkin and Keller, 1982). An ecosystem is appropriately defined as any unit of nature that includes all of the organisms in a given area interacting with the physical environment so that a flow of energy leads to a clearly defined trophic structure, biotic diversity, and biogeochemical cycles (Odum, 1971a). All ecosystems have common components organized into structural patterns and united by functional processes. The components include inorganic substances, organic substances, climate, autotrophs, heterotrophs, and de-

Table 1-2. Ecosystem elements. All units of nature with energy flow and complete chemical cycles, or ecosystems, have common components organized in structural patterns and united by functional processes.

Components	Structural Patterns	Functional Processes
inorganic substances[a]	food webs[g]	energy flow[j]
organic substances[b]	diversity: species	production = energy storage,
climate[c]	variation in space[h]	biomass
autotrophs (producers)[d]	succession: species[i]	biogeochemical Cycling[k]
heterotrophs(consumers)[e]	variation in time	
decomposers[f]		

[a]*Inorganic substances* - Elements and compounds required for metabolism and other materials involved in biogeochemical cycles. Examples include carbon, nitrogen, carbon dioxide, and water.

[b]*Organic substances* - Carbon compounds that join living and nonliving components of ecosystems. Examples include proteins, carbohydrates, lipids, and humic materials.

[c]*Climate* - The meteorologic "context" of the troposphere in which the ecosystem is located. Elements involved include temperature, precipitation, solar radiation, and wind.

[d]*Autotrophs (producers)* - Organisms capable of synthesizing their own food materials. Green plants able to make their own food from simple substances are the dominant producers of ecosystems. Trees, of course, are the dominant producers in forests.

[e]*Heterotrophs (consumers)* - Organisms requiring preformed food materials for metabolism. They are extremely variable in size and morphology, and range from microorganisms to mammals. Heterotrophs are commonly characterized as herbivores (plant eating) or carnivores (meat eating).

[f]*Decomposers* - Heterotrophic organisms capable of degrading complex organic debris from decomposing plants and animals (detritus), utilizing some of the decomposition products for their own use, and releasing organic residues and inorganic substances for use by other organisms. Decomposers are primarily microorganisms, including bacteria, fungi, and protozoa.

[g]*Food webs* - Organisms within ecosystems feed on one another. Energy, elements, and compounds are transferred from organism to organism through food webs consisting of integrated systems of interacting food chains. Organisms are grouped in food webs into trophic levels. All organisms in a food web that are the same number of steps away from the original source of energy are at the same trophic level.

[h]*Diversity* - The number of species in a given ecosystem is variable. Desert ecosystems have fewer species than do tropical rain forests. Natural ecosystems generally have more species than do agricultural ecosystems.

[i]*Succession* - The number of species may change as an ecosystem matures and succession proceeds. Succession involves recognizable, repeated, and predictable changes in species composition over time. Primary succession occurs when an ecosystem is initially established, whereas secondary succession involves the reestablishment of an ecosystem following disturbance.

[j]*Energy flow* - Energy must be continually added to an ecosystem in a usable form. Generally it flows in from outside the ecosystem, mainly through photosynthesis. Within the ecosystem energy is passed from one organism to another, and some is released as heat. Production is stored energy and is termed *biomass* or *organic* matter.

[k]*Biogeochemical cycling* - Chemical elements move into and out of ecosystems. This cycling connects biological cycles to geologic cycles. Chemical elements cycle within an ecosystem from organism to organism through water and air and soils and rocks.

Source: Adapted from Botkin and Keller (1982), and Odum (1971b).

composers. Structural elements include food webs and patterns in time and space. Functional elements include energy flow and storage, and biogeochemical cycling (Table 1-2).

The earth is covered by a mosaic of ecosystems. These ecosystems are connected and influence one another in a variety of ways. Temperate forest ecosystems occupy a position of prominence among all ecosystems. Temperate forests (1.8 billion ha) are second only to tropical forest ecosystems (2 billion ha) in size. Temperate forest ecosystems (200–400 tons ha^{-1} yr^{-1}) are second only to rain forest ecosystems (400–500 tons ha-1 yr-1) in biomass. In terms of primary productivity, temperate forest ecosystems (5–20 tons ha^{-1} yr^{-1}) rank third behind only tidal zone (20–40 tons ha^{-1} yr^{-1}) and rain forest (10–30 tons ha^{-1} yr^{-1}) ecosystems. Unfortunately, temperate forest ecosystems are also located in the zone of maximum air pollution because of their extensive distribution throughout the zone of primary urbanization and industrialization of the earth.

In general, forest ecosystems develop through a rapid growth stage that leads to some manner of maturity or steady state, typically an oscillating steady state. Steady state is the condition in which there is no net change in biomass over time. The early successional growth stage is characterized by a high production to respiration ratio, high yields (net production), short food webs, low diversity, small size of organisms, open nutrient cycles, and a lack of relative stability. Mature forest stages, on the other hand, are characterized by a low biomass to respiration ratio, complex food webs, low net production, high diversity, and relatively high stability. In summary, major energy flow shifts from production to maintenance as the system ages (Odum, 1971b). Species diversity declines along the latitudinal continuum from the wet tropical forests through temperate zone forests to boreal forests. Temperate zone forests, the systems of interest for this book, combine characteristics of relatively low diversity with relatively high stability (Langford and Buell, 1969). In mesic, temperate forest ecosystems a peak in community diversity can be expected 100–200 years after the initiation of the secondary successional sequence, that is, when elements of both the pioneer and stable communities are present. A decrease in both diversity and primary production takes place when the entire community is made up of shade-tolerant mature species (Loucks, 1970).

In natural forest ecosystems, disturbance or perturbation caused by environmental or biotic forces (stress factors) may be necessary to maintain maximum diversity and productivity. The natural tendency in forest ecosystems toward periodic perturbation at intervals of 50–200 years recycles the system and maintains a periodic wave of peak diversity and a corresponding wave of peak primary production (Loucks, 1970). There is substantial evidence to support the hypothesis that insect outbreaks, fire, and wind storms, even those events that cause massive destruction in the short run, may play beneficial and essential roles in forest ecosystems in the long term. These roles are presumed to regulate tree species competition, species composition and succession, primary production, and nutrient cycling. It is equally true that forest stress factors can severely impact management objectives imposed on forest systems by human managers. Stress impacts are not absolute, however, and the goals and objectives of management interests (Table 1-1) determine the significance of stress events.

Table 1-3. Biotic and abiotic stress factors that cause disease and injury of forest trees.

	Biotic	Abiotic
Cause Disease	fungi	air pollution
	bacteria	drought
	viruses	salt
	mistletoes	adverse soil chemistry
	nematodes	or physical structure
	mycoplasma	nutrient deficiencies
	insects	and excesses
	mites	competition
Cause Injury	insects	wind
	mites	fire
	rodents	temperature extremes
	mammals	moisture extremes
	birds	lightning
	humans	volcanic eruptions
		landslides
		avalanches
		harvesting
		pesticides
		ice
		snow
		salt, other chemicals
		radiation

C. Forest Stress

Factors capable of causing injury, disease and mortality in forest systems are termed *stresses*. Stress factors are very varied and include climatic, pathologic, entomologic, anthropogenic, wildlife, fire and stand dynamic elements. Stresses recognized to have widespread and general importance in forest systems are listed in Table 1-3. Despite the diversity of important forest stress factors, several general forest stress principles can be distilled from our understanding of forest systems. These principles are presented in Chapter 20 of this book.

Forest systems are dynamic, not static. They are characterized by variability rather than constancy and continually change in time and space. Adaptation, adjustment, and evolution take place with time as the biotic and abiotic components interact. The dominants of a community are replaced by new species as energy flow and nutrient cycling are altered. As the sequential changes in species composition and community structure, known as *succession* , occur, ecosystems originate, develop, mature, and ultimately decline and disaggregate. Through succession, ecosystems evolve toward the most stable state possible within the constraints of the environment.

The evidence is overwhelming that the activities of human beings are presently altering this environment in numerous ways at both regional and global scales. Forests will respond!

References

Botkin, D.B. and E.A. Keller. 1982. Environmental Studies. Merril Publishing Co., Columbus, OH., 506 pp.

Langford, A. N. and M.F. Buell. 1969. Integration, identity and stability in the plant association. Adv. Ecol. Res. 6: 83-135.

Loucks, O.L. 1970. Evolution of diversity, efficiency and community stability. Am. Zool. 10: 17-25.

Odum, E.P. 1971b. Ecosystem structure and function. In J.A. Wiens, ed., Proc. 31st Annual Biol. Colloquium. Oregon State Univ. Press, Corvallis, OR, PP. 11-24.

Odum, J.A. 1971b. Ecosystem structure and function. In J.A. Wiens, ed., Proc. 31st Annual Biol. Colloquium. Oregon State Univ. Press, Corvallis, OR, PP. 11-24.

2
Air Pollution

The atmosphere that surrounds the earth is a mixture of gases and suspended solids whose density decreases rapidly with elevation until space void is reached at hundreds of kilometers above the surface of the planet. The lowest portion of the atmosphere is designated the *troposphere* and it contains approximately 70% of the mass of the atmosphere. At the poles of the earth, the troposphere is about 8 km thick, while at the equator the thickness approximates 16 km. The troposphere contains essentially all storm systems and weather events and is subject to the input of a large number of compounds resulting from both natural processes and human activities at the surface of the earth. The *stratosphere* extends from the troposphere to very approximately 50 km above the earth. This layer contains the critically important natural ozone band, which is significant for attenuating short wave radiation received from the sun. Persistent chemicals released to the troposphere, for example, carbon dioxide, nitrous oxide, and halocarbons, may become reactive in the higher energy stratosphere (Chapter 19). The highest energy environment of the atmosphere is designated the *mesosphere* and extends from approximately 50 to 100 km above the surface of the earth.

The study of air pollutants, and the effects of air pollutants, is multidisciplinary. It very importantly involves the disciplines of atmospheric chemistry and physics, meteorology and chimatology, hydrology and soils, and the biological and life sciences. Comprehensive study of the air pollution topic encompasses investigation of the source and release of pollutants, their transport and transformations, and the deposition and influence on a variety of receptors. Most of these topics are well beyond the scope of this book. We will explore only the "effects" component of the *air pollution system* in this volume (Figure 2-1).

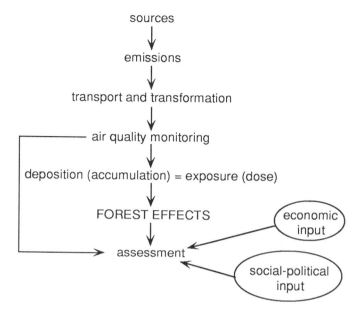

Fig. 2-1. The air pollution system from source through forest effects assessment. Forest effects are the focus of this book.
Source: Modified from Heck (1984) and Heck and Heagle (1985).

A. Pollutant Materials

For the purpose of this book, an appropriate definition of air pollutants is *materials that occur in the troposphere in quantities in excess of normal amounts.* These materials may be solid, liquid, and gaseous in character and they may result from both natural and human (anthropogenic) processes. Natural sources of air pollution are diverse and include volcanic and other geothermal eruptions, forest fires, gases released from vegetation, wind- blown soil and other debris, pollen, spores, and sea spray particles. Anthropogenic sources are also diverse, and include a variety of combustion and industrial activities. The specific materials that contaminate the troposphere are as varied as the sources and have differential importance depending on their ability to influence natural processes or elements of the biota and human health or activities. Air pollutants of particular importance, or potential importance, to forest systems are presented in Table 2-1.

Particulate or gaseous materials released directly into the troposphere in large amounts, by natural or anthropogenic processes, are termed *primary pollutants.* Particulate or gaseous materials formed in the atmosphere from precursors released in large amounts to the troposphere, by natural or anthropogenic processes, are designated *secondary pollutants.* The chemistry of the lower atmosphere is very dynamic and particles are formed from gases, large particles are

Table 2-1. Materials of Importance*, or Potential Importance, to Forests that Pollute the Troposhere[a].

I. Particulate pollutants[b]

   A. Primary[c]
     1. Inorganic
      *a. heavy metal and other
        elements[d]
      *b. salts
        i. sulfate
        ii. nitrate
        iii. ammonium
        iv. chloride
        v. fluoride
      c. acids
      d. bases
      e. variable (dust, soil)

     2. Organic
      a. spores
      b. pollen
      c. soot (pure carbon)

   B. Secondary[e]
     1. Inorganic
      a. hydronium (H^+)[f]
      *b. sulfate $(SO_4^=)$
      *c. nitrate (NO_3^-)
      d. chloride (Cl^-)
      e. ammonium (NH_4^+)
      f. other ions[g]

     2. Organic
      a. condensed
       hydrocarbons
      b. carboxylic acids
      c. dicarboxylic acids
      d. aliphatic nitrates
      e. other[h]

II. Gaseous Pollutants[i]
   A. Primary[c]
     1. Inorganic
      *a. oxides
        i. carbon (CO, CO_2)
        ii. sulfur (SO_2)
        iii. nitrogen $(NO, NO_2,$
         $N_2O)$
      b. halogens
        i. bromine (Br_2)
        ii. chlorine (Cl_2)
        iii. hydrogen bromide (HBr)
        iv. hydrogen chloride (HCl)
        *v. hydrogen fluoride (HF)
      c. other
        i. mercury (Hg_2)
        ii. ammonia (NH_3)
        iii. hydrogen sulfide (H_2S)

     2. Organic
      a. hydrocarbons[j]
      b. oxygenated
       hydrocarbons[k]
      c. halocarbons
      d. mercaptans
      e. sulfides

   B. Secondary[e]
     1. Inorganic
      a. ozone (O_3)
      b. hydrogen peroxide (H_2O_2)
      c. nitric acid (HNO_3)
      d. nitrous acid (HONO)

     2. Organic
      a. peroxyacetylnitrate
       and homologues
      b. oxygenated
       hydrocarbons[k]
      c. N-nitroso
       compounds
      d. peracetic acid

---

[a]References employed in the preparation of this Table included; Gaffney et al. (1987), Hanst et al. (1977), McCune (1986), and Urone (1976).
[b]Solid or liquid material with typical dimension between 10 nm and 1 mm and formed by grinding or atomization processes, condensation of supersaturated vapors, or from gas reactions that form nonvolatile solids.
[c]Materials released directly into the troposphere by natural or anthropogenic processes.

[b]Solid or liquid material with typical dimension between 10 nm and 1 mm and formed by grinding or atomization processes, condensation of supersaturated vapors, or from gas reactions that form nonvolatile solids.
[c]Materials released directly into the troposphere by natural or anthropogenic processes.
[d]Major species of atmospheric particles: Ca, Zn, Cu, Ni, Mn, Sn, Cd, V, Sb, Fe, Mg, Ti, K, Si, Al, Pb, As, and Se.
[e]Materials formed in the atmosphere from precursors released into the troposhere by natural or anthropogenic processes.
[f]Samples of rain or melted snow generally have a pH in the range of 3.0–8.0. Pure water in equilibrium with atmospheric carbon dioxide, carbonic acid formed, has a pH of 5.6. As a result, the term acid precipitation is generally applied to precipitation samples with a pH $\leq$ 5.6. Natural contamination of precipitation with nitrogen and sulfur compounds, organic acids and alkaline particles can cause the pH of precipitation to vary locally in the range of 5–7.
[g]Additional ions commonly found in precipitation include calcium (Ca^+), magnesim (Mg^+), potassium (K^+), and sodium (Na^+).
[h]A large number of organic molecules released to or formed in the troposphere typically increase in polarity and aqueous solubility as they are oxidized and become dissolved in rain, mist or fog particles.
[i]Compounds that have boiling points below approximately 200°C.
[j]Hydrocarbons include alkanes, alkenes, acetylenes, and aromatics.
[k]Oxygenated hydrocarbons include alcohols, phenols, ethers, aldehydes, ketones, peroxides, and organic acids.
[l]Halocarbons(chloro fluorocarbons) include trichlorofluoromethane, dichlorofluoromethane, trichlorotrifluoroethane, and dichlorotetrafluoroethane.

formed from small particles, and oxidized materials are formed from reduced materials. The troposphere is an oxidizing medium and reduced species released to it are commonly oxidized to oxidants or acids. Both inorganic and organic molecules are oxidized, frequently with associated increases in polarity and aqueous solubility (Gaffrey et al. 1987). Trace gases in the atmosphere may become highly soluble in precipitation. As a result, precipitation transfer is a critical process for both particulate and gaseous pollutants.

Human beings have been polluting the atmosphere for millennia, and it is quite impossible to know the exact ranges of normalcy for tropospheric materials. In concept, however, unpolluted tropospheric air contained a variety of naturally generated particles plus gases in the approximate concentrations presented in Table 2-2. Gas concentrations in Table 2-2, and gas and particulate concentrations throughout this book, will generally be expressed as micrograms per cubic meter in air ($\mu g\ m^{-3}$) (for gases, standard conditions of 25°C and 760 mm mercury are assumed). Gaseous pollutant concentrations will also be presented in parts per billion (ppb) units. While it is impossible to specify trace gas concentrations of unpolluted air, global sources of air pollution vary greatly, and as a result, various portions of the troposphere may be designated " relatively clean" and "relatively polluted," respectively. A comparison of trace gas concentrations for air environments so designated is presented in Table 2-3.

Due to increasing awareness of the adverse consequences of air pollution, numerous countries of the temperate latitudes have promulgated air quality standards. In the United States, the Clean Air Act of 1970 directed the Environmental Protection Agency to establish air quality standards for particulates, sulfur dioxide, nitrogen dioxide, carbon monoxide, hydrocarbons, and ozone. A stan-

Table 2-2. Approximate Gaseous Composition of Unpolluted Tropospheric Air on a Wet Basis.

Gas	$\mu g\ m^{-3}$
Nitrogen	9×10^8
Oxygen	3×10^8
Water	2×10^7
Argon	1×10^7
Carbon dioxide	5×10^5
Neon	1×10^4
Helium	8×10^2
Methane	$6\text{--}8 \times 10^2$
Krypton	3×10^3
Nitrous oxide	9×10^2
Xenon	4×10^2
Hydrogen	4×10^1

dard for lead was added in 1978. Primary and secondary standards were established. Primary standards are intended to protect human health, while secondary standards are intended to protect public welfare. The latter includes consideration of the air pollution influence on vegetative health, materials weathering, and visibility. The National Ambient Air Quality Standards of the United States as amended through February 1979 are presented in Table 2-4. It is of constant interest to compare various dose-response relationships reviewed in this book with ambient air quality measurements and standards. Unfortunately ambient air quality monitoring has been concentrated in urban areas due to the overriding concern with human health effects. Only recently have we begun to monitor rural air quality in the environments of agricultural, forest, and other wildland systems.

Table 2-3. Trace Gas Concentrations for Relatively "Clean" and "Polluted" Atmosphere.

Gas	$\mu g\ m^{-3}$	
	Clean air	Polluted air
Carbon dioxide	57.6×10^4	72.0×10^4
Carbon monoxide	115	$46\text{-}80.5 \times 10^3$
Methane	920	1533
Nitrous oxide	450	?
Nitrogen dioxide	1.9	376
Ozone	39	980
Sulfur dioxide	0.5	524
Ammonia	7.0	14.0

Source: Urone (1976).

Table 2-4. National Ambient Air Quality Standards of the United States (as amended through Febrary 1979)[a].

Pollutant	Standards (μg m^{-3})	
	Primary	Secondary
Particulates (total suspended)		
annual	75	60
24-hr	260	150
Sulfur dioxide		
annual	80	
24-hr	365	
3-hr		1,300
Nitrogen dioxide		
annual	100	100
Carbon monoxide		
8-hr	10,000	10,000
1-hr	40,000	40,000
Hydrocarbons		
3-hr	160	160
Ozone		
1-hr	240	240
Lead		
3-month	1.5	

[a]Short-term standards (24 hr and less) are not to be exceeded more than once a year. Long-term standards are maximum permissible concentrations never to be exceeded.

B. Pollutant Deposition

The transfer of air pollutants from the troposphere to components of forest systems is accomplished by a variety of deposition mechanisms. These mechanisms may be conveniently divided into wet and dry deposition. Wet deposition involves the movement of dissolved gases and large particles, generally with diameters greater than 20 μm, via incident precipitation. The precipitation may be direct in the form of rain, snow, sleet or hail, or indirect in the form of cloudwater, fog, mist, or rime ice. Acid precipitation is defined as rain or snow having a pH of less than 5.6. The pH parameter is a measurement of the difference in hydrogen ion activity between an unknown solution and a standard buffer of assigned pH value. Upon ionization, water yields hydrogen and hydroxyl ions. When the activity of these ions is equal, water is neutral and the pH recorded will be 7. At pH values below 7 water becomes increasingly acid, and above 7 it becomes increasingly alkaline. In the absence of air pollutants, the pH of precipitation is presumed to be dominated by carbonic acid formed from ambient carbon dioxide, which produces a pH of approximately 5.6–6.0. The pH of precipitation presently falling in North and Central Europe and in the northeastern United States and adjacent portions of Canada is commonly in the range of 3–5.5. Individual storm events have been recorded with pH values between 2.0 and 3.0.

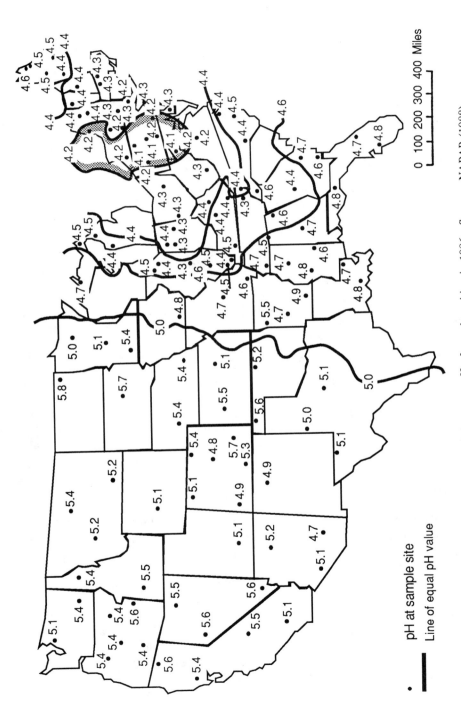

Fig. 2-2. Precipitation-weighted annual average pH of wet deposition in 1986. Source: NAPAP (1988).

• pH at sample site

▬ Line of equal pH value

Table 2-5. Droplet diameter and settling velocity of various precipitation forms.

Precipitation	droplet diameter μm	settling velocity cm sec[-1]
cloud or fog	< 100	1–25[a] 1–80[b]
drizzle	100 – 500	25–200
rain	> 500	>200[a]

Source: [a]Daube et al. (1987), [b] Lovett (1984).

The precursors of acid precipitation are presumed to be gaseous sulfur and nitrogen compounds of the atmosphere. The oxidation of sulfur dioxide and nitrogen oxides leads to the formation of sulfuric and nitric acids. Analyses of more than 1500 precipitation samples, with a median pH of 4.0, from New York and New Hampshire, revealed that in 80–100% of the cases, low pH was attributable to sulfuric and nitric acid (Galloway et al., 1976). In central New Hampshire, Likens (1975) observed that precipitation hydrogen-ion content was 60% due to sulfuric acid, 34% due to nitric acid, and 6% due to various organic acids. In theory it is possible that acid precipitation precursors may have natural or anthropogenic sources. In the European (Oden, 1976) and United States (Likens, 1976; Hitchcock, 1976) locations of greatest precipitation acidity, however, anthropogenic sources are judged to play a dominant role, as air mass trajectories pass over high human emission sources prior to reaching these most seriously impacted regions (Figure 2-2). Additional consideration of the general nature of acid precipitation is beyond the scope of this book. Excellent and comprehensive reviews are available (see Table 2-9).

In order to estimate the quantity of air pollution deposited by wet deposition, it is generally calculated as the concentration of chemical contained in the precipitation times the amount of precipitation recorded. Pollutant concentrations in rain generally decrease as rainfall amount, intensity, and duration increase. Amount of rain recorded, however, exerts the most significant influence on pollutant transfer (Lindberg et al., 1982). Wet deposition amounts will also vary with the season of the year, frequently being higher during the growing season (Lindberg, 1982), and with the form of precipitation, frequently being higher in fog and cloud water relative to other forms of precipitation (Weathers et al., 1988, Schemenauer 1986, Barrie and Schemenauer 1986, Daube et al. 1987, Weathers et al. 1986a, 1986b) (Table 2-5). The hydrogen ion activity of precipitation represents a useful index of contamination and forest stress potential, and varies with the position of collection in the forest (Table 2-6) and with the form of precipitation (Table 2-7).

Dry deposition involves the transfer of gases, fine particles (1-20 μm), and very fine particles (<1 μm) to forest and soil surfaces. The mechanism of dry

Table 2-9. Reviews of air pollution interaction with forest systems provided after the first edition of *Air Pollution and Forests* was published (1981).

General	
Acid Rain Foundation 1985	Kozlowski and Constantinidou 1986
American Forestry Association 1987	Krause 1987a
Appalachian SAF 1986	Krause 1987b
Bedford 1986	Last 1983
Bicknell 1987	Last 1982
Bormann 1987	Legge, A. H. and S. V. Krupa 1986
Bormann 1986	Mackenzie and El-Ashry 1988
Bormann 1985	McLean 1983
Davis et al. 1983	McLaughlin 1985
Fernandez 1986	McLaughlin et al. 1983
Georgii 1986	Postel 1984
Goldstein and Legge 1986	Skelly et al. 1987
Grodzinski et al. 1984	Smith 1987
Halbwachs 1983	Smith 1986
Heck et al. 1988	Smith 1985
Heck et al. 1986	Smith 1984
Hutchinson and Meema 1987	Tomlinson 1983
Huttunen 1981	Treshow 1984
Jäger et al. 1986	Ulrich 1981
Keller 1983	Ulrich and Pankrath 1983
Knabe 1983	Unsworth and Ormrod 1982
Kozlowski 1986	Woodman and Cowling 1987
Kozlowski 1985	

Pollutant focus	

Acid deposition:

Backiel and Hunt 1986	Lefohn and Brocksen 1984
Bennett et al. 1985	Legge and Crowther 1987
Breece and Hasbrouck 1984	Linthurst 1984a
Bruck 1987	Linthurst 1984b
Cowling and Linthurst 1981	Martin 1986
Electric Power Research Institute 1983	Mayo 1987
Evans 1982	Morrison 1984
Evans 1984	National Council of the Paper Industry
Fernandez 1983	for Air and Stream Improvement 1981
Fuhrer and Fuhrer–Fries 1982	National Research Council 1986
Garner et al. 1988	Pinkerton 1984
Hasbrouck 1984	Prinz and Brandt 1986
Huckabee et al 1988	Reuss and Johnson 1986
Jacobson and Troiano 1983	Shepard 1985
Johnson and Siccama 1983	Society of American Forsters 1984
Kulp 1987	Ulrich 1983

Lead (other heavy metals):

Backhaus and Backhaus 1986	Johnson et al. 1982
Baes and McLaughlin 1986	Lindberg et al. 1982
Baes and McLaughlin 1984	Smith et al. 1988
Friedland et al. 1984a	U. S. Environ. Protection Agency 1986a
Friedland et al. 1984b	Zöttl 1985b

Table 2-9. (continued)

Oxidants:

Adams and Taylor 1987	Shriner et al. 1982
Asmore 1984	Skelly 1987
Harkov and brennan 1982	Skelly et al. 1984
Karnosky et al. 1986	Taylor and Norby 1986
McBride and Miller 1987	Taylor and Norby 1985
McClenahen 1984	Temple and Taylor 1983
Miller et al. 1982	U. S. Environ. Protection Agency 1986b
Moskowitz et al. 1982	

Particulates:

U. S. Environ. Protection Agency 1982

Sulfur dioxide:

Bewley and Parkinson 1984	U.S. Environ. Protection Agency 1982
Winner et al. 1985	

Regional focus

Canada:

Addison and Rennie 1987	Linzon 1986
Fraser et al. 1985	Rennie 1987

Central Europe:

Ashmore et al. 1985	Prinz 1987
Bell 1986	Prinz et al 1987
Bell 1984	Prinz et al. 1982a
Binns and Redfern 1983	Prinz et al. 1982b
Federal Minister Food, Agric., Forestry (FRG) 1982	Rehfuess 1987
	Rehfuess 1985
Flückiger et al. 1986	Rehfuess 1983
Kandler 1983	Roberts 1983
Knabe 1985	Schöpfer and Hradetzky 1984
Krause 1985	Schütt and Cowling 1985
Krause et al. 1986	Steinbeck 1984
Nilsson and Duinker 1987	Ulrich 1984
Plochmann 1984	Zöttl 1987

Norway:

Overrein 1980	Overrein et al. 1981

Sweden:

Abrahamsen 1984	Andersson and Olsson 1985
Abrahsen and Tveite 1983	Falkengren–Grerup 1986
Andersson 1987	Kvist 1985
Andersson 1986	Tamm et al. 1984

United States:

Armentano and Loucks 1983	Lindberg and Harriss 1981
Bartuska and Medlarz 1986	Louchs 1981
Bennett et al. 1986	Miller 1987
Bruck 1987	Miller et al. 1982
Burgess 1984	Molliton and Raynal 1983
Esserlieu and Olson 1986	Raynal et al. 1983
Heck 1984	Smith 1984
Johnson and Siccama 1984	Smith 1982

Table 2-6. Hydrogen ion activity (mean) of precipitation collections made in a forest canopy opening and under forest canopies at the Huntington Forest, Newcomb, New York, between May 1979 and May 1980.

Collector location	pH
Collector in forest opening (bulk precipitation)	4.06
Collector under hardwood canopy (throughfall)	4.18
Collector under conifer canopy (throughfall)	4.02

Source: Mollitor and Raynal (1983).

deposition actually involves a series of processes. In the initial stage of dry deposition, gas molecules and small particles are entrained from the airstream by turbulent eddies erected by the friction of air mass movement over the forest canopy. Once in the canopy, pollutants must penetrate the boundary layer around tree surfaces. Boundary layers are small and have thicknesses of the order of a millimeter or less. For gases and ultra fine particles (<0.1 μm), movement through the boundary layer may be accomplished by diffusion. Larger particles, 2-3μm diameter, propelled by wind may be able to penetrate the boundary layer by impaction. Particles with diameters approximately 0.1–2 μm have very lim-

Table 2-7. Hydrogen ion activity (mean) of cloud water samples.

Site	Sample number	Sample year	pH
Mt. Mitchell, NC, USA	231	1986	3.42[a]
Shenandoah National Park, VA, USA	38	1986	3.67[a]
Whiteface Mt. NY, USA	329	1986	3.68[a]
Whitetop Mt, VA, USA	129	1986	3.49[a]
Mt. Tremblant, Ontario, Canada	20	1985	3.82[b]
Roundtop Mt, Ontario, Canada	4	1985	3.71[b]

Sources: [a]Mohnen 1987, [b]Schemenauer 1986.

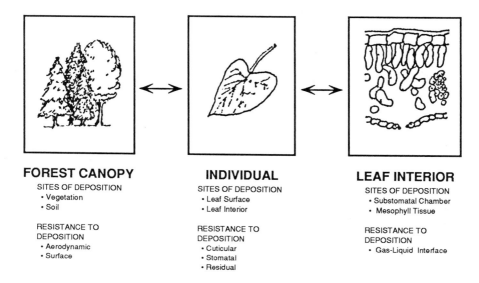

FOREST CANOPY	**INDIVIDUAL**	**LEAF INTERIOR**
SITES OF DEPOSITION	SITES OF DEPOSITION	SITES OF DEPOSITION
• Vegetation	• Leaf Surface	• Substomatal Chamber
• Soil	• Leaf Interior	• Mesophyll Tissue
RESISTANCE TO DEPOSITION	RESISTANCE TO DEPOSITION	RESISTANCE TO DEPOSITION
• Aerodynamic	• Cuticular	• Gas-Liquid Interface
• Surface	• Stomatal	
	• Residual	

Fig. 2-3. Deposition of pollutant gases to forest trees; sites of, and resistances to, deposition.
Source: modified from Taylor et al. (1987).

ited means to penetrate boundary layers (Lovett, 1987). Gases that move through the boundary layer surrounding a leaf will either adsorb to the surface of the leaf or enter the leaf through a stomatal opening (Figure 2-3).

Unfortunately, unlike wet deposition, dry deposition is presently very difficult to quantify. For field measurement, techniques in current use include: micrometeorological procedures to estimate deposition as a function of pollutant depletion in the air above the canopy, accumulation inventories that measure the accumulation of deposited material on natural or artificial surfaces, and inferential methods that do not directly record deposition but estimate it employing a flux for-mula (Lovett, 1987). An example of the latter is the simple dry deposition model:

pollutant flux (dry deposition rate) = deposition velocity (vd) x pollutant concentration
typical units: $\mu g\ cm^{-1}\ sec^{-1}$ $cm\ sec^{-1}$ $\mu g\ cm^{-3}$

Deposition velocity (vd) is the ratio of the rate of deposition divided by the concentration of the pollutant in the boundary layer. It is variable depending on the chemistry of the pollutant, the surface characteristics of the receptor, and the turbulence of the air environment. Sehmel (1980) has listed 16 micrometeorological variables, 14 particle characteristics, 4 gas characteristics, and 10 receptor variables that influence deposition velocity. In general, deposition velocity increases with solubility of the pollutant, particle diameter and density, wetness and roughness of the surface, and turbulence and wind speed of the air. The deposition velocity of numerous gases significant for forest systems, for example,

Table 2-8. Deposition velocities to leaf surfaces for several pollutants as determined under managed environmental conditions.

Pollutant	cm sec^{-1}
O_3	0.2–0.7
NO	0.01–0.1
NO_2	0.1–0.8
HNO_3	0.5–5.0
NH_3	0.2–0.6
PAN	0.1–0.6
SO_2	0.2–3.0
H_2S	0.2–0.4

Source: Taylor et al. (1987).

sulfur dioxide, ozone, hydrogen fluoride, nitrogen dioxide, ammonia, and peroxyacetynitrate, is very approximately 0.1 cm sec^{-1} for stable air conditions and roughly 10 cm sec^{-1} for highly turbulent conditions. It is not uncommon to assume a deposition velocity value of 1 cm sec^{-1} for the above gases under general conditions (Peterson, 1977). If the deposition velocity of a pollutant is 1.0 cm sec^{-1}, it suggests that the surface is completely removing the pollutant from a layer of air 1.0 cm thick each second, with the "clean" layer immediately replaced by a "new" contaminated layer. Deposition velocities of several pollutant gases to leaf surfaces, as determined under managed environmental conditions, are provided in Table 2-8.

For particles deposited by impaction, for example, cloud moisture with a droplet size of approximately 10 μm, deposition velocities were estimated at 6 cm sec^{-1} at 2 m sec^{-1} wind speed and 70 cm sec^{-1} at 10 m sec^{-1} wind speed (Reiners et al., 1987). Deposition velocities of cloud droplets are also a function of droplet size and density, wind speed, and surface characteristics. It is common practice to assume a deposition velocity of 1 cm sec^{-1} for particles as well as gases (Peterson, 1977).

Excellent reviews of the critically important wet and dry deposition processes have been provided by Albritton et al. (1987), Barrie and Schemenauer (1986), Chamberlain (1986), Davidson et al. (1982), Fowler (1980), Garland (1978), Hales et al. (1987), Hicks (1986), Hosker (1986), Hutchinson and Hicks (1983), Lovett (1987), Parker (1987), Sehmel (1980), Taylor et al. (1987), United States Environmental Protection Agency (1983), Wesely and Hicks (1986).

C. Air Pollution Interaction with Forests

1. Pollution Scales

The history of air pollution interaction with forest systems has been one concerned with forest responses over increasingly larger areas. During the first two thirds of the twentieth century, research and regulatory efforts focused on local air pollutants and acute effects on vegetation. Pollutants of primary concern were sulfur dioxide, particulate and gaseous fluoride compounds, and numerous heavy metals (such as lead, copper, and zinc). Occasional interest was expressed in other inorganic gases including ammonia, hydrogen sulfide, hydrogen chloride, and chlorine.

The sources of these pollutants continue to be typically discrete and stationary facilities for energy production (such as fossil-fuel electric-generating plants and gas-purification plants), metal-related industries (such as copper, nickel, lead, zinc, or iron smelters, and aluminum production plants), and diverse other industries (such as cement plants, chemical plants, fertilizer plants, and pulp mills).

Forest areas directly affected by these facilities are typically confined to a radius of a few kilometers immediately surrounding the plant, and for a distance up to tens of kilometers downwind. The dimensions of the surrounding and downwind zones of influence are variable, and are primarily controlled by the strength of the effluent at its source, local meteorology, regional topography, and the susceptibility of the vegetation. In any case, the forest influence is confined to a region of generally less than a thousand hectares.

During the past three decades researchers have become increasingly aware of regional-scale air pollutants. The regional designation is applied because these contaminants may affect forests tens, hundreds, or even thousands of kilometers from their origin. The regional air pollutants of greatest documented or potential influence for forests include: oxidants, most importantly, ozone; trace metals, most importantly, heavy metals (such as cadmium, cobalt, copper, lead, mercury, molybdenum, nickel, vanadium, and zinc); and acid deposition, most importantly, sulfuric and nitric acids. Ozone and sulfuric and nitric acids are termed *secondary* air pollutants because they are synthesized in the atmosphere rather than released directly into it. The precursor chemicals, released directly into the atmosphere and causing secondary pollutants to form, include hydrocarbons and nitrogen oxides in the case of ozone, and sulfur dioxide and nitrogen oxides in the case of sulfuric and nitric acid. Because secondary air pollutants may form over tens or hundreds of kilometers from the site of precursor release, and because small particles may remain airborne for days or weeks, these pollutants may be transported 100 kilometers or more than 1000 km from their origin. Eventual wet and dry deposition of the pollutants onto lakes, or forests may occur over large areas.

Evidence is available for long-range transport of regional-scale pollutants from numerous sources, including satellites, surface deposition monitoring, and paleolimnological studies. The long-distance transport of regional pollutants means they may have interstate, international, and even intercontinental significance. It means further that the forests subject to their deposition exceed tens of thousands of square kilometers.

In the past 25 years, people have become concerned with a third scale of air pollution — global. The global atmosphere is changing; a variety of trace gases, for example, methane, nitrous oxide and carbon dioxide, are increasing in concentration, and novel gases, for example, halocarbons, are being added. What is the significance of these changes for global radiation balances? What will be the impact on forest systems of global atmospheric change?

Fortunately, societal interest in the relationship between air contaminants and natural resources has dramatically increased in the past decade. This interest has stimulated more research to understand mechanisms of air pollution stress on forest systems, more monitoring of atmospheric deposition in forested areas, and more thinking on the extraordinarily difficult task of assessing the impact of subtle, long term stress on complex forest systems. Since the first edition of this book was published in 1981, a large number of reviews concerning air pollution and forests have been provided (Table 2-9).

2. Forest Exposure

There is much uncertainty concerning the complex interactions of air pollutants and forest systems. Despite this uncertainty, we recognize that the fundamental "dose response" relationship can be applied to this interaction, as it can be applied to any interaction between a toxic material and a biological system. The dose-response concept recognizes that as the exposure to a toxin is altered, the reaction of the receiving organism will change.

The dose-response concept itself is very complex (Krupa and Kickert, 1987). In the traditional toxicological context, dose represents the amount of material administered (ingested, injected) into a test organism. Lefohn and Benedict (1984) currently emphasize that air pollution studies commonly use ambient exposure (pollutant concentration x duration of pollutant episode) interchangeably with dose. The actual amount of any pollutant taken up by vegetation, however, is highly variable and regulated by numerous biological, chemical, and physical factors (Taylor et al., 1982). In addition, much uncertainty exists regarding the appropriate exposure statistic to employ in an effort to describe dose. Is the long term mean, the short term maximum, or the repeated exposure to varying ambient levels above a threshold concentration the most meaningful parameter (Larsen and Heck, 1985)? In the case of ozone influence on vegetation, considerable evidence stresses the importance of episodic maximum concentrations (Lefohn et al., 1986, Lefohn and Runeckles, 1987). Also are mixtures of pollutants, rather than single contaminants, most significant for vegetative health? With regard to major phytotoxic gaseous pollutants, for example, sulfur dioxide, ozone, and nitrogen dioxide, evidence suggests that cooccurrence of episodic peaks are rare, of short duration, and are separated by long time intervals (Lefohn and Tingey, 1984, Lefohn et al., 1987a,b).

Despite all of this complexity, it remains fair to suggest that the response of individual trees, or of forests, will generally become more adverse and obvious as the exposure to pollutants (an index of potential dose) is increased. In an effort to provide reference points for evaluation of ambient exposure information and experimental exposure detail, Tables 2-10 through 2-23 have been provided. These tables provide ambient information on gaseous pollutant concentrations [parts per billion (ppb)] and ambient flux rates of particulate pollutants in kilo-grams per hectare per year (kg ha^{-1} yr^{-1}).

D. Book History

In 1972, the author was asked to review the relationship between air pollution and forests for the annual meeting of the American Association for the Advancement of Science. This formidable assignment was approached by divid-ing the interactions between air contaminants and forest ecosystems into three classes (Smith, 1974). Under conditions of a low dose (Class I relationship), the vegetation and soils of forest ecosystems function as important sources and sinks for air pollutants. When exposed to an intermediate dose (Class II relationship), individual tree species or individual members of a given species may be subtly and adversely affected by nutrient stress, impaired metabolism, predisposition to entomological or pathological stress, or direct disease induction. Exposure to high dose (Class III relationship) may induce acute morbidity or mortality of specific trees. At the ecosystem level, the impact of these various interactions would be very variable. In the Class I relationship, pollutants would be ex-changed between the atmospheric compartment, the available nutrient compart-ment, other soil compartments, and various elements of the biota. Depending on the nature of the pollutant, the ecosystem impact in this case could be unde-tectable (innocuous effect) or stimulatory (fertilizing effect). If the effect of the air pollution dose on some component of the biota is inimical then a Class II re-lationship is established. The ecosystem impact in this case could include re-duced productivity or biomass, alterations in species composition or community structure, increased insect outbreaks or microbial disease epidemics, and increased morbidity. Under conditions of a high dose and a Class III relationship, ecosys-tem impacts may include gross simplification, impaired energy flow and biogeo-chemical cycling, changes in hydrology and erosion, climate alteration, and ma-jor impacts on associated ecosystems.

While these classes of interaction are conceptual and artificial, and not neces-sarily discrete entities in time or space, they have proved useful in teaching, in the preparation of this book, and in organizing and establishing research priorities.

The information reviewed in this second edition again concerns the temperate latitudes of the Northern Hemisphere, primarily the area between latitude 40° and 60°N, because the primary release of anthropogenic air pollutants is within this zone, and as a result the research on effects is concentrated in this area. The in-formation reviewed is principally from North America. This reflects the relative availability and accessibility of the literature to the author. The literature review was completed in November 1988 and the book contains only a small amount of material published after that date. The author sincerely hopes the information presented will provide assistance to the reader in understanding the extremely complex relationships between air pollutants and forest systems.

Table 2-10. Ozone (O_3) ambient concentrations.

	ppb	
Urone 1976	-9	($18\ \mu g\ m^{-3}$) clean air baseline
Lefohn & Jones 1986	-10	($20\ \mu g\ m^{-3}$), USA sites not influenced by urban sources maximum hourly means do not exceed (generally)
Taylor & Norby 1986	-10–20	(20-$39\ \mu g\ m^{-3}$) North American, remote regions, annual mean range
NCASI 1988	-10–20	(20-$39\ \mu g\ m^{-3}$) USA rural region natural background during growing season
Evans 1985	-12	($24\ \mu g\ m^{-3}$) USA National Forest sites, five forests exceeded this hourly mean, 1979
Lefohn & Jones 1986	-15	($29\ \mu g\ m^{-3}$) USA sites influenced by urban sources, 50–70% of hourly means equal or exceed this concentration
Mohnen & Bradow 1986	-15–40	(29–$78\ \mu g\ m^{-3}$) Whiteface Mt., NY, winter range
Prinz 1984	-20–27	(39–$53\ \mu g\ m^{-3}$) FRG, southern region @ 740 m, annual mean 1977–1982
Taylor & Norby 1986	-20–60	(39–$118\ \mu g\ m^{-3}$) Shenandoah National Park, VA,USA, monthly mean range, 1980
Skelly 1987	-21–50	(41–$98\mu g\ m^{-3}$) USA, rural areas, 7 hr (daily) growing season mean range
Prinz 1984	-22–62	(43–$122\mu g\ m^{-3}$) FRG, Black Forest region, monthly mean range, 1980–83
Taylor & Norby 1986	-25	($49\ \mu g\ m^{-3}$) USA, eastern region, agricultural areas, typical growing season
Mohnen & Bradow 1986	-25–60	(49–$118\ \mu g\ m^{-3}$) Whiteface Mt., NY, growing season range
Linzon et al. 1984	-30	($59\ \mu g\ m^{-3}$) agricultural crop, toxicity threshold, growing season mean
Prinz 1984	-32–39	(63–$76\ \mu g\ m^{-3}$) FRG, southern region @1000m, 1977–1982 annual mean
Pratt et al. 1983	-33	($65\ \mu g\ m^{-3}$) USA upper midwest, rural region, annual mean 1977–1981

Table 2-10. (continued)

	ppb	
Pinkerton & Lefohn 1987	−36	(71 µg m⁻³) USA, northern New England, NY, daily growing season mean 1978–1983
U.S.E.P.A. 1988	−36–40	(70–80 µg m⁻³) USA, England, annual mean
Pinkerton & Lefohn 1987	−37	(73 µg m⁻³) USA upper Great Lakes region, annual mean
Prinz 1984	−38–49	(74–96 µg m⁻³) FRG, southern region @ 2964 m, annual mean 1977–1982
Pinkerton & Lefohn 1987	−39	(76 µg m⁻³) USA, NY, PA, MD daily (7 hr) growing season mean, 1978–1983
Linzon et al. 1984	−40-50	(78–98 µg m⁻³) threshold range, acute foliar symptoms, white ash, eastern white pine, southern Ontario, Canada, daily (7 hr) growing season mean
Taylor & Norby 1986	−40–60	(78–118 µg m⁻³) USA, eastern forest regions, daily (7 hr) growing season mean range
Pinkerton & Lefohn 1987	−43	(84 µg m⁻³) USA, southern coastal plain region, daily (7 hr) growing season mean, 1978–1983
Mohnen & Bradow 1986	−44	(86 µg m⁻³) Whiteface Mt., NY, long-term mean, 1978–1985
Schemenauer & Anlauf 1987	−44.5	(87 µg m⁻³) Roundtop Mt., Quebec, Canada, highest monthly mean, May–Sept 1986
Linzon et al. 1984	−45	(88 µg m⁻³) southern Ontario, Canada, annual mean 1979–1981
Mohnen & Bradow 1986	−47	(92 µg m⁻³) Mt. Mitchell, NC, USA, monthly mean June–Nov 1986
Mohnen & Bradow 1986	−49	(96 µg m⁻³) Whitetop Mt., VA, USA, monthly mean 1985–1986
Skelly et al. 1983	−49	(96 µg m⁻³) Big Meadows area, Shenandoah National Park, VA @ 1050 m, monthly (8 hr) mean May–Sept. 1979–1982
Linzon et al. 1984	−50–52	(98–102 µg m⁻³) southern Ontario, Canada, annual mean range 1974–1978
Pinkerton & Lefohn 1987	−51	(100 µg m⁻³) USA southern Piedmont, mountain-ridge-valley

Table 2-10. (continued)

	ppb	
		region, daily (7 hr) growing season mean
Lefohn & Mohney 1986	−51	(100 μg m⁻³) Brotjacklriegel, Bavaria, FRG, daily mean (7 hr), growingseason mean
Lefohn & Mohnen 1986	−51	(100 μg m⁻³) Schauinsland, Baden-Wurttemberg, FRG, daily mean (7 hr) growing season 1983
Pinkerton & Lefohn 1987	−53	(104 μg m⁻³) USA, pine-land region, southern NJ, daily mean (7 hr) growing season 1978–1983
Evans et al. 1983	−54	(106 μg m⁻³) USA, eight remote National Forest sites, growing season mean
NCASI 1987	−55–85	(108–167μg m⁻³) threshold range for forest productivity impact, USA
Vogler 1982	−57–85	(112–167 μg m⁻³) USA, southern Sierra Nevada Mts, CA @ 1654–1860 m, daily mean 1977–1981
Lefohn & Mohnen 1986	−58	(114 μg m⁻³) Waldhof, Niedersachen, FRG, daily mean (7 hr), growing season, 1983
Ashmore 1984	−60	(118 μg m⁻³) threshold impact on radish, hourly mean
Gilliam et al. 1988	−67	(131 μg m⁻³) Shenandoah, National Park, VA, USA, peak daily conc., Feb–March 1987
Hayes & Skelly 1977	−80	(157 μg m⁻³) threshold, visible symptoms, eastern white pine, rural VA, USA
Mann et al. 1980	−80	(157 μg m⁻³) threshold, visible symptoms, eastern white pine
McBride & Miller 1987	−80–120	(157–235 μg m⁻³) San Bernardino National Forest, CA, USA, 33% ponderosa pine mortality, 50–99 yr age class, hourly mean 1974–1983
Gilliam et al. 1988	−84	(165 μg m⁻³) Shenandoah National Park, VA, USA, daytime peak, May–June 1987
Schemenauer & Anlauf 1987	−95	(187 μg m⁻³) Montmorency, Quebec, Canada, maximum 15 min concentration,1986

Table 2-10. (continued)

	ppb	
		growing season
Kohut et al. 1988	−100	(196 µg m⁻³) Ithaca, NY, USA, peak episodic concentration
Lefohn & Mohnen 1986	−113	(221 µg m⁻³) Mt. Mitchell, NC, USA, hourly peak, June–Nov 1986
Lefohn & Mohnen 1986	−120	(235 µg m⁻³) Whiteface Mt., NY, USA, hourly peak, 1973–1983
Lefohn & Mohney 1986	−120	(235 µg m⁻³) Whitetop Mt., VA, USA, hourly peak, 1985-86
National Ambient Standard (USA)	−122	(240 µg m⁻³) 1 hr maximum permissible air quality concentration, not to be exceeded more than once per year
Evans et al. 1983	−125	(245 µg m⁻³) USA eight National Forest sites, maximum hourly mean 1979
Russo–Savage 1989	−125	(245 µg m⁻³) Sayre Peak Meteoroligical Station, Mt. Moosilauke, N.H., maximum hourly mean (June), 1988 growing season
Pratt et al. 1983	−150	(294 µg m⁻³) USA, rural upper midwest region, peak hourly mean 1977–1981
Department of Environmental Protection, CT, USA	−200	(392 µg m⁻³) Stratford CT, USA, urban concentration 1987
McBride & Miller 1987	−200–250	(392–490 µg m⁻³) San Bernardino National Forest, CA, USA, daily range of episodic peaks (typical)
McBride & Miller 1987	−300–400	(588–784 µg m⁻³) San Bernardino National Forest, CA, USA, range of episodic peaks (extreme)

Table 2-11. Sulfur dioxide (SO_2) ambient concentrations.

	ppb	
Urone 1976	-0.19	(0.5 μg m^{-3}) clean air baseline
Mohnen & Bradow 1986	$-1-2$	(2.6–5.2 μg m^{-3}) Whiteface Mt., NY, growing season monthly mean
Mohnen & Bradow 1986	$-1-3$	(2.6–7.9 μg m^{-3}) Whiteface Mt., NY, winter season monthly mean
U.S.E.P.A. 1985	$-3.8-5.3$	(10–14 μg m^{-3}) rural eastern USA, typical range 1970s
Lefohn & Mohnen 1986	-10	(26 μg m^{-3}) Schauinsland, Baden-Württemberg, FRG, growing season 24 hr mean, 1984
Lefohn & Mohnen 1986	-20	(52. 4 μg m^{-3}) Whiteface Mt., NY, episodic peaks
Lefohn & Mohnen 1986	-23	(60 μg m^{-3}) Brotjacklriegel, Bavaria, FRG, growing season, 24 hr mean, 1984
National Ambient Air Quality Standard (USA)	-30.6	(80 μg m^{-3}) annual maximum permissible concentration not to be exceeded
Lefohn & Mohney 1986	-40	(105 μg m^{-3}) Whiteface Mt., NY, long-term summer/winter maximum hourly mean
Lefohn & Mohney 1986	-91	(238 μg m^{-3}) Bodenmais, Bavaria, FRG, maximum hourly mean 1984
National Ambient Air Quality Standard (USA)	-139	(365 μg m^{-3}) 24 hr maximum permissible concentration, not to be exceeded more than once per year
Lefohn & Mohnen 1986	-312	(817 μg m^{-3}) Arzberg, Bavaria, FRG, maximum hourly mean, 1984
Lefohn & Mohnen 1986	-541	(1417 μg m^{-3}) Hof, Bavaria, FRG, maximum hourly mean, 1984

Table 2-12. Nitrogen oxides. (NO_x = NO and NO_2) ambient concentrations.

	ppb	
Pratt et al. 1983	–0.1-7.2	(0.16–11 µg m^{-3}) rural areas, USA, annual mean
Urone 1976	–1.0	(1.9 µg m^{-3}) clean air baseline
Lefohn and Mohnen 1986	–1-2	(2-3 µg m^{-3}) Whiteface Mt., NY, typical range
U.S.E.P.A. 1985	–1.1	(2 µg m^{-3}) rural western USA, typical 1970's
Pratt et al. 1983	–< 2	(3.1 µg m^{-3}) (NO_2) rural upper midwest USA, 500 km from urban area, annual mean 1977-1981
U.S.E.P.A. 1985	–2.7-10.6	(5–20 µg m^{-3}) rural eastern USA, typical summer range 1970s
Pratt et al. 1983	–3-4	(3.7–5.0 µg m^{-3}) (NO) rural upper midwest USA, 500 km from urban area, annual mean 1977-1981
Lefohn & Mohnen 1986	–6	(11 µg m^{-3}) Schauinsland, Baden-Württemberg, FRG, growing season 24 hr mean
Pratt et al. 1983	–6	(11 µg m^{-3}) (NO_2) rural upper midwest, USA, 50 km from urban area, annual mean 1977-1981
Lefohn & Mohnen 1986	–9	(17 µg m^{-3}) Brotjackbriegel, Bavaria, FRG, growing season 24 hr mean
National Ambient Air Quality Standard (USA)	–53	(100 µg m^{-3}) (NO_2) annual maximum permissible concentration, not to be exceeded

Table 2-13. Hydrogen chloride (HCl), ammonia (NH$_3$), and peroxyacetylnitrate (PAN) ambient concentrations.

	ppb	HCl
U.S.E.P.A. 1985	−0.7-7	(1–10 µg m^{-3}) (dry) rural western USA, typical range 1970s
		NH$_3$
Urone 1976	−0.08	(7.0 µg m^{-3}) clean air baseline
U.S.E.P.A. 1985	−0.7–2.8	(0.5–2µg mz^{-3}) (dry) rural eastern and western USA typical range 1970s
		PAN
U.S.E.P.A. 1985	−0.02–0.2	(0.1-1 µg m^{-3}) rural western USA, typical range 1970s
U.S.E.P.A. 1985	−0.1–0.6	(0.5–3 µg m^{-3}) rural eastern USA, typical range 1970s
Lewis et al. 1983	−1.17	(5.8 µg m^{-3}) New Brunswick, NJ, peak daily mean, 9/78–5/80
Temple & Taylor 1982	−9.4	(47 µg m^{-3}) Riverside CA, peak monthly mean, Sept. 1980
Lewis et al. 1983	−10.6	(49 µg m^{-3}) New Brunswick, NJ, peak hourly mean, 9/78-5/80
Temple & Taylor 1982	−25.1	(124 µg m^{-3}) Riverside CA, peak daily mean, Oct. 1980
Temple & Taylor 1982	−41.6	(206 µg m^{-3}) Riverside CA, peak hourly mean, 1980

Table 2-14. Hydrogen peroxide (H_2O_2) and nitric acid vapor (HNO_3) ambient concentrations.

	ppb	H_2O_2
Mohnen & Bradow 1986	−0.2–2.8	(0.28–3.89 µg m^{-3}) (dry) Whiteface Mt., NY, summer 1986 range
Mohnen & Bradow 1986	−0.1–2.6	(0.14–3.6 µg m^{-3}) (dry) Whitetop Mt., Virginia, summer 1986 range
Mohnen & Bradow 1986	−200	(278 µg m^{-3}) (wet-cloud water) Whiteface Mt., NY, summer 1986, typical
Mohnen & Bradow 1986	−200–2000	(278–28 X 10^5) (wet-cloud water), WhitefaceMt., NY, summer 1983, range
		HNO_3
U.S.E.P.A. 1985	−0.1–1.2	(0.3–3 µg m^{-3}) (dry) rural eastern USA, typical range 1970s
U.S.E.P.A. 1985	−≤ 1.0	(≤ 2.6 µg m^{-3}) (dry) rural western USA, typical range 1970s
Mohnen & Bradow 1986	−0.1–1.0	(0.3–2.6 µg m^{-3}) (dry) Whiteface Mt., NY, typical range, early 1980s
Mohnen & Bradow 1986	−10	(26 µg m^{-3}) (dry) Whiteface Mt., NY, episodic peaks, early 1980s

Table 2-15. Sulfur (S) deposition.

	kg ha^{-1} yr^{-1}	
U.S.E.P.A. 1985	–0.7–3.3 S	(2-10 kg SO$_4^=$) (wet) USA west of Mississippi River
Kociuba 1984	–1.6 S	(4.8 kg SO$_4^=$) (dry) Beaverlodge, 1982, Alberta, Canada
Lau 1985	–2.2S	(6.8 kg SO$_4^=$)(wet) Beaverlodge, 1978–1982, Alberta, Canada
Sullivan et al. 1988	–3.3 S	(≤10 kg SO$_4^=$)(wet) no acidification risk, North American lakes > 4 ha, due to atmospheric deposition
WHS	–3.3–5.0S	(10–15 kg SO$_4^=$) (wet) acidification potential, high risk North American lakes
Rennie 1987	–3.3–16.5 S	(10–50 kg SO$_4^=$) (wet) eastern Canada range, 1981
Lau 1985	–3.5 S	(10.5 kg SO$_4^=$) (wet) Whitecourt, 1978-1982, Alberta, Canada
U.S.E.P.A. 1985	–5.0–15 S	(15–45 kg SO$_4^=$) (wet) USA east of Mississippi River
Fernandez et al. 1986	–5.1 S	(15.4 kg SO$_4^=$) (wet) Greenville, ME, 1979–1982
Rennie 1987	–5.6–7.6 S	(17–23 kg SO$_4^=$) (wet) southern Nova Scotia, Canada
Rennie 1987	–5.9–7.3 S	(18–22 kg SO$_4^=$) (wet) north, central Ontario, 1985, Canada
U.S.E.P.A. 1988	–6.6 S	(20 kg SO$_4^=$) (wet) New Hampshire
Kociuba 1984	–7.2 S	(21.9 kg SO$_4^=$) (dry) Whitecourt, 1982, Alberta, Canada
Mollitor & Raynal 1983	–8.7 S	(26 kg SO$_4^=$) (wet) Huntington Forest, May 1979–1980, Newcomb, NY
Rennie 1987	–8.9–10.1 S	(27–33 kg SO$_4^=$) (wet) southern Quebec, Canada
Hedin et al. 1987	–9.5 S	(28.5 kg SO$_4^=$) (wet) 1983 Hubbard Brook Forest, New Hampshire
Rennie 1987	–9.6–11.6 S	(29–35 kg SO$_4^=$) (wet) south central Ontario, 1985, Canada
Lau 1985	–9.9 S	(30 kg SO$_4^=$) (wet) Truro, 1978-1982, Nova Scotia, Canada
U.S.E.P.A. 1988	–9.9 S	(30 kg SO$_4^=$) (wet) New York
Erb 1987	–10–15.5 S	(29–47 kg SO$_4^=$) Pennsylvania range, 1986
NCASI 1988	–12.2 S	(37 kg SO$_4^=$) (wet) Pennsylvania range, 1981–1986

Table 2–15. (continued)

	kg ha^{-1} yr^{-1}	
Lau 1985	−18.9 S	(57.2 kg SO$_4^=$) (wet) Quebec City, 1978–1982, Quebec, Canada
Lovett et al. 1982	−21 S	(65 kg SO$_4^=$) (wet- precipitation) balsam fir forest, 1220 m, Mt. Moosilauke, NH
Frink and Voigt 1976	−26.8 S	(81 kg SO$_4^=$) (wet) Windsor, CT, annual mean, 1929–1948
Mayer & Ulrich 1978	−47-51 S	(51–155 kg SO$_4^=$) beech forest, 1968–74, 40 km NW Göttingen, FRG
Mayer & Ulrich 1978	−80-86 S	(242–260 kg SO$_4^=$) spruce forest, 1968–1974, 40 km NW Göttingen, FRG
Lovett et al. 1982	−91 S	(276 kg SO$_4^=$) (wet-cloud deposition) balsam fir forest, 1220 m, Mt. Moosilauke, NH
Rennie 1987	−99 S	(300 kg SO$_4^=$) (wet) highly stressed site, eastern region, FRG

Table 2-16. Nitrogen (N) deposition.

kg ha⁻¹ yr⁻¹		
U.S.E.P.A. 1985	$-0.5-2.3$	(NO_3^-–N) (2-10 kg NO_3^-) (wet) USA west of Mississippi River
Lau 1985	-0.7	(NO_3^-–N) (3.2 kg NO_3^-) (wet) Beaverlodge, 1978–1982, Alberta, Canada
Fernandez et al. 1986	-0.8	(NH_4^+ –N) (1.08 kg NH_4^+) (wet) Greenville, 1979–1982, ME, USA
U.S.E.P.A. 1985	$-0.8-3.1$	(NH_4^+ –N) (1-4 kg NH_4^+) (wet) USA range, 1978–1984
Lau 1985	-1.0	(NO_3^-–N) (4.3 kg NO_3^-) (wet) Whitecourt, 1978–1982, Alberta, Canada
Frink and Voight 1976	-1.7	(NH_4^+ –N) (2.2 kg NO_4^+) (wet) Windsor, CT, annual mean 1929–1948
U.S.E.P.A. 1985	$-1.8-4.1$	(NO_3^-–N) (8-18 kg NO_3^-) (wet) USA range 1978–1984
Fernandez et al. 1986	-2.1	(NO_3^-–N) (9.1 kg NO_3^-) (wet) Greenville, ME, USA,1979–1982
Rennie 1987	$-2.3-6.9$	(NO_3^-–N) (10-30 kg NO_3^-) (wet) s. Ontario, 1981, Canada
Rennie 1987	$-2.3-6.9$	(NO_3^-–N) (10-30 kg NO_3^-) (wet) eastern Canada range 1981
U.S.E.P.A. 1985	$-2.3-6.9$	(NO_3^-–N) (10-30 kg NO_3^-) (wet) USA east of Mississippi River
Lau 1985	-2.5	(NO_3^-–N) (10-30 kg NO_3^-) (wet) Truro, 1978–1982, Nova Scotia, Canada
Fernandez 1986	$-3-18$	(N) (wet) low elevation forest, range, eastern USA
Lovett et al. 1982	-3.3	(NH_4^+ –N) (4.2 kg NH_4^+) (wet-bulk precipitation), balsam fir forest, Mt. Moosilauke, 1220 m, NH, USA
U.S.E.P.A. 1988	-3.5	(NO_3^-–N) (15 kg NO_3^-) (wet) NH (mean), USA
Mollitor & Raynal 1983	-3.5	(NO_3^-–N) (15.2 kg NO_3^-) (wet) Huntington Forest, Newcomb, NY, USA
Frink and Voight 1976	-3.8	(NO_3^-–N) (16.5 kg NO_3^-) (wet) Windsor, CT, annual mean 1929–1948
U.S.E.P.A. 1988	-4.6	(NO_3^-–N) (20 kg NO_3^-) (wet) western NY, USA
N.C.A.S.I. 1988	-5.3	(NO_3^-–N)(23 kg NO_3^-) (wet) PA mean 1981–1986

Table 2-16.   (continued)

	kg ha^{-1} yr^{-1}	
Lovett et al. 1982	−5.4	(NO$_3^-$–N) (23.4 kg NO$_3^-$) (wet- bulk precipitation) balsam fir forest, Mt. Moosilauke, 1220 m, NH, USA
Lau 1985	−5.9	(NO$_3^-$–N) (25.8 kg NO$_3^-$) (wet) Quebec City, 1978–1982, Quebec, Canada
Likens et al. 1977	−6.5	(NO$_3^-$–N) (28 kg NO$_3^-$) (wet) Hubbard Brook, White Mt. National Forest, NH, USA
Derwent 1986	−9.0	(N) (wet + dry) maximum deposition, southern England
Kelly & Meagher 1986	−11.3	(N) (wet, dry - NO, NO$_2$ HNO$_3$) Camp Branch Forest, 1982–1984, e. TN, USA
Kelly & Meagher 1986	−11.6	(N) (wet, dry, NO, NO$_2$; HNO$_3$) Cross Creek Forest, 1982–1984, e. TN, USA
Lovett et al. 1982	−12.7	(NH$_4^+$ –N) (16.3 kg NH$_4^+$)(wet-cloud deposition) balsam fir forest, Mt. Moosilauke, 1220 m, NH, USA
Kelly & Meagher 1986	−13.2	(N) (wet, dry - NO, NO$_2$,HNO$_3$) Walker Branch forest, 1982–1984, e. TN, USA
Lovett et al. 1982	−23.3	(NO$_3$ - N) (101.5 kg NO$_3^-$) (wet-cloud deposition) balsam fir forest, Mt. Moosilauke,1220 m, NH, USA
Cline et al. 1988	−35	(N) maximum deposition rate, southeastern USA
WHS	−42	(N) (91 kg – 200 pounds) commercial urea fertilization rate, Douglas-fir forests, Pacific Northwest, USA
Lovett et al. 1982	−44.7	(N) (wet-precipitation and cloud deposition total, NH$_4^+$, NO$_3^-$) balsam fir forest, Mt. Moosilauke, 1220 m, NH, USA

Table 2-17. Hydrogen (H^+) deposition.

	kg ha^{-1} yr^{-1}	
U.S.E.P.A. 1985	−0.0005− 0.1	(wet) USA west of Mississippi River, rain pH 5.0-6.0
Lau 1985	−0.02	(wet) Whitecourt, 1978–1982 mean, Alberta, Canada
Lau 1985	−0.03	(wet) Beaverlodge, 1978–1982 mean, Alberta Canada
U.S.E.P.A. 1985	−0.2-0.8	(wet) USA east of Mississippi River, rain pH 4.0-4.8
Lau 1985	−0.43	(wet) Truro, 1978–1982 mean, Nova Scotia, Canada
Hedin et al. 1987	−0.55	(wet) Hubbard Brook, 1983, White Mt. National Forest, NH, USA
Lau 1985	−0.57	(wet) Quebec City, 1978–1982 mean, Quebec, Canada
U.S.E.P.A. 1985	−0.60	(wet) western PA, 1978–1984 mean, USA
Mollitor & Raynal 1983	−0.7	(wet) Huntington Forest, 1979–1980, Newcomb, NY, USA
Ulrich 1980	−0.7	(wet) Solling, FRG
Likens et al. 1980	−0.75	(wet) Hubbard Brook, 1964–1965, White Mt. National Forest, NH, USA
Likens et al. 1980	−0.85	(wet) Hubbard Brook, 1976--1977, White Mt. National Forest, NH, USA
Matzner and Ulrich 1985	−1.0	(bulk throughfall) beech forest, Göttingen, 1981–1982, FRG
Lovett et al. 1982	−1.5	(wet-bulk precipitation) balsam fir forest, 1220 m, Mt. Moosilauke, NH, USA
Binkley and Richter 1986	−1.6	(wet 0.7 + dry 0.9) chestnut oak forest, TN, USA
Matzner and Ulrich 1985	−2.0	(bulk-throughfall) beech forest, 1969-81, Solling, FRG
Lovett et al. 1982	−2.4	(wet-cloud deposition) balsam fir forest, 1220 m, Mt. Moosilauke, NH, USA
Metzner and Ulrich 1985	−4.0	(bulk-throughfall) spruce forest, 1969–1981, Solling, FRG

Table 2-18. Lead deposition.

	kg ha^{-1} yr^{-1}	
		Forest, FRG
Kahl et al. 1982	−0.099	maximum input to Dream Lake (elevation 792 m), NH, USA prior to 1980
Zöttl 1985	−0.128	(bulk) spruce forest, Black Forest, FRG
Tuüby & Zöttl 1984	−0.132	(wet) pine forest, upper Rhine Plain, FRG
Turner et al. 1984	−0.140	(bulk) pine barrens, 1980−1982 s. NJ, USA
Tyler 1981	−0.150	spruce forest, Sweden
Lindberg & Harriss 1981	−0.150	Walker Branch forest, e. TN, USA
Lindberg et al. 1982	−0.15	chestnut oak forest, e. TN, USA
Friedland et al. 1984	−0.175	boreal forest, 1966−1980, Camels Hump Mt., VT, USA
Smith et al. 1988	−0.190	(bulk) Hubbard Brook forest,1975−1984 mean,White Mt. National Forest, NH, USA
Reiners et al 1975	−0.200	montane forests, NH, USA
Groet 1976	−0.238	White Mts., NH, USA
Getz et al. 1977	−0.24	(wet) watershed ecosystem, USA
Smith & Siccama 1981	−0.266	(bulk) Hubbard Brook forest, 1975−1978 mean,White Mt. National Forest, NH, USA
Zöttl 1985	−0.285	(bulk) spruce forest canopy opening, Solling region, FRG
Schlesinger & Reiners 1974	−0.300	(bulk) White Mts., NH, USA
Siccama & Smith 1978	−0.317	(bulk) Hubbard Brook forest, 1975, White Mt. National Forest, NH, USA
Wiersma 1979	−0.329	Great Smoky Mts. National Park, NC-TN, USA
Turner et al. 1984	−0.350	pine barrens, 1978−1979, s. NJ, USA
Heinrichs & Mayer 1977	−0.405	spruce-beech forest, Solling region, FRG
Friedland et al. 1984	−0.521	boreal forest, 1966−1980, Camels Hump Mt., VT, USA
Schlesinger & Reiners 1974	−1.3	artifical foliage collectors, White Mt. National Forest, NH, USA
Lazrus et al. 1970	−1.32	urban area, NJ, USA
Jackson & Watson 1977	−1000.	immediate vicinity, metal smelter, southeastern MI, USA

Table 2-19.   Cadmium deposition.

	kg ha^{-1} yr^{-1}	
Trüby & Zöttl 1984	−0.002	(wet) spruce forest, Black Forest, FRG
Zöttl 1985	−0.004	(bulk) spruce forest canopy opening, Black Forest, FRG
Trüby & Zöttl 1984	−0.005	(wet) pine forest, upper Rhine Plain, FRG
Smith et al. 1988	−0.008	(bulk) Hubbard Brook forest, 1975 −1984 mean,White Mt. National Forest, NH, USA
Zöttl 1985	−0.016	(bulk) (throughfall) spruce forest, Black Forest, FRG
Zöttl 1985	−0.016	(bulk) spruce forest canopy opening, Solling region, FRG
Smith & Siccama (unpublished)	−0.02	(bulk) Hubbard Brook forest, 1975 −1978 (unpublished) mean,White Mt. National Forest, NH, USA
Zöttl 1985	−0.02	(throughfall) spruce forest, Solling region, FRG

Table 2-20.   Copper deposition.

	kg ha^{-1} yr^{-1}	
Smith et al. 1988	−0.016	(bulk) Hubbard Brook forest,1975– 1984 mean, White Mt. National Forest, NH, USA
Trüby & Zöttl 1984	−0.02	(wet) spruce forest, Black Forest, FRG
Groet 1976	−0.02	northeastern USA
Friedland et al. 1984	−0.021	northern hardwood forest, 1966– 1980, Camels Hump Mt., VT, USA
Zöttl 1985	−0.023	(bulk) spruce forest canopy opening, Black Forest, FRG
Zöttl 1985	−0.025	(throughfall) spruce forest canopy opening, Black Forest, FRG
Friedland et al. 1984	−0.028	boreal forest, 1966-80, Camels Hump Mt., VT, USA
Trüby & Zöttl 1984	−0.03	(wet) pine forest, upper Rhine Plain, FRG
Zöttl 1985	−0.23	(throughfall) spruce forest, Solling region, FRG
Zöttl 1985	−0.24	(bulk) spruce forest canopy opening, Solling region, FRG

Table 2–21. Zinc deposition.

	kg ha⁻¹ yr⁻¹	
Kahl et al. 1982	-0.012	maximum input to Dream Lake (elevation 792 m), NH, USA prior to 1980
Kahl et al. 1982	-0.013	maximum input to Unnamed Pond (elevation 141 m), ME, USA prior to 1933
Kahl et al. 1982	-0.027	maximum input to Little Long Pond, (elevation 72 m), ME, USA prior to 1950
Friedland et al. 1984	-0.086	boreal forest, 1966–1980, Camels Hump Mt., VT, USA
Groet 1976	-0.135	northeastern USA
Smith et al. 1988	-0.139	(bulk) Hubbard Brook forest, 1975–1984 mean, White Mt. National Forest, NH, USA
Smith & Siccama (unpublished)	-0.15	(bulk) Hubbard Brook forest, 1975–1978 (unpublished) mean White Mt. National Forest, NH, USA
Friedland et al. 1984	-0.178	northern hardwood forest, 1966–1980, Camels Hump Mt. VT, USA

Table 2-22.   Manganese deposition.

kg ha^{-1} yr^{-1}		
Trüby & Zöttl 1984	−0.08	(wet) spruce forest, Black Forest, FRG
Zöttl 1985	−0.096	(bulk) spruce forest canopy opening, Black Forest, FRG
Smith et al. 1988	−0.098	(bulk) Hubbard Brook forest, 1975–1984 mean, White Mt. National Forest, NH, USA
Trüby & Zöttl 1984	−0.10	(wet) pine forest, upper Rhine Plain, FRG
Smith & Siccama (unpublished)	−0.10	(bulk) Hubbard Brook forest, 1975–1978 mean, White Mt. National Forest, NH, USA
Zöttl 1985	−0.39	(bulk) spruce forest canopy opening, region, FRG
Zöttl 1985	−1.23	(throughfall) spruce forest, Black Forest, FRG
Zöttl 1985	−6.48	(throughfall) spruce forest, Solling region, FRG

Table 2-23.   Cobalt and nickel deposition.

kg ha^{-1} yr^{-1}		Cobalt
Trüby & Zöttl 1984	−0.005	(wet) spruce forest, Black Forest, FRG
Trüby & Zöttl 1984	−0.01	(wet) pine forest, upper Rhine plain, FRG
		Nickel
Trüby & Zöttl 1984	−0.008	(wet) spruce forest, Black Forest, FRG
Smith et al. 1988	−0.014	(bulk) Hubbard Brook forest, 1975–1984 mean, White Mt. National Forest, NH, USA
Smith & Siccama (unpublished)	−0.018	(bulk) Hubbard Brook forest, 1975–1978 mean, White Mt. National Forest, NH, USA
Trüby & Zöttl 1984	−0.042	(wet) pine forest, upper Rhine plain, FRG

References

Abrahamsen, G. 1984. Effects of acidic deposition on forest soil and vegetation. Phil. Trans. Royal Soc., London (B) 305: 369-382.

Abrahamsen, G. and B. Tveite. 1983. Effects of air pollutants on forest growth. Report No. PM 1636. National Swedish Environment Protection Bd., Solna, Sweden, pp. 199-219.

Acid Rain Foundation. 1985. Proceedings Air Pollutants Effects on Forest Ecosystems. Acid Rain Foundation, St. Paul, MN, 439 pp.

Adams, M.B. and G.E. Taylor Jr. 1987. Effects of ozone on forests in the Northeastern United States. Publica. No. ORNL-53, Oak Ridge National Laboratory, Oak Ridge, TN.

Addison, P.A. and P.J. Rennie. 1987. Influence of the Long-Range Transport of Air Pollution on the Canadian Forest (annotated bibliography). Publ. No. 428, Canadian Forestry Service, Otawa, Canada.

Albritton, D.F. Fehsenfeld, B. Hicks, J. Miller, S. Liu, J. Hales, J. Shannon, J. Durham, and A. Patrinos. 1987. Atmospheric processes. Chap. 4. in Vol. III. Atmospheric Processes and Deposition. Interim Assessment. National Acid Precipitation Assessment Program, Washington, DC pp. 37-59.

American Forestry Association. 1987. White Paper on the Forest Effects of Air Pollution. American Forestry Assoc., Washington, DC, 17 pp.

Andersson, F. 1986. Acidic deposition and its effects on the forests of Nordic Europe. Water Air Soil Pollu. 30: 17-29.

Andersson, F. 1987. Air pollution impact on Swedish forests. Present evidence and future development. In proceedings Fourth Annual Wilderness Conference, Sept. 1987, Ester Park, CO, 10 pp.

Andersson, F. and B. Olsson. 1985. Lake Gårdsjön. An Acid Forest Lake and its Catch-ment. Ecological Bulletins, Vol. 37, Publishing House of the Swedish Research Councils, Stockholm, Sweden, 336 pp.

Appalachian Society of American Foresters. 1986. Proceedings Atmospheric Deposition and Forest Productivity. Appalachian SAF, Clemson, SC, 119 pp.

Armentano, T.V. and O.L. Loucks. 1983. Air pollution threats to U. S. national parks of the Great Lakes region. Environ. Conserva. 10: 303-313.

Ashmore, M.R. 1984a. Effects of ozone on vegetation in the United Kingdom. Proceedings Workshop on Ozone. Swedish Environmental Research Institute, Göteborg, Sweden.

Ashmore, M.R. 1984b. Effect of ozone on vegetation in the United Kingdom. Proc. Workshop on the Evaluation and Assessment of the Nature, Magnitude and Geographical Extent of Effects of Phototochemical Oxidants on Human Health, Agricultural Crops, Forestry Materials and Visibility. Swedish Environmental Research Institute, Goteborg, Sweden.

Ashmore, M R., J.N. B. Bell,and A.J. Rutter. 1985. The role of ozone in forest damage in West Germany. Ambio 14: 81-87.

Backhaus, B. and R. Backhaus. 1986. Is atmospheric lead contributing to mid-European forest decline. Sci. Tot. Environ. 50: 223-225.

Backiel, A. and F.A. Hunt. 1986. Acid rain and forests. An attempt to clear the air. Amer. Forests. 92: 42-48.

Baes, C.F. and S.B. McLaughlin. 1986. Multielemental Analysis of Tree Rings: A Survey of Coniferous Trees in the Great Smoky Mountains National Park. Publ. No. ORNL-6155kj, Oak Ridge National Laboratory, Oak Ridge, TN. 48 pp.

Baes, C.F. and S.B. McLaughlin. 1984. Trace elements in tree rings: Evidence of recent and historical air pollution. Science 224: 494-497.

Barrie, L.A. and R.S. Schemenauer. 1986. Pollutant wet deposition mechanisms in precipitation and fog water. Water Air Soil Pollu. 30: 91-104.

Bartuska, A.M. and S.A. Medlarz. 1986. Spruce-fir decline – Air pollution related? Proceedings Atmospheric Deposition and Forest Productivity, Appalachian Soc. of American Foresters. Raleigh, NC, pp. 55-73.

Bedford, B.L. (ed.). 1986. Proceedings Modification of Plant-Pest Interactions by Air Pollutants. Publ. No. ERC-117, Ecosystems Research Center, Cornell Univ., Ithaca, NY, 128 pp.

Bell, J.N.B. 1984. Air pollution problems in Western Europe. In M. J. Kozol and F. R. Whatley, eds., Gaseous Air Pollutants and Plant Metabolism. Butterworths, London, pp. 3-24.

Bell, J.N B. 1986. Effects of acid deposition on crops and forests. Experientia 42: 363-371.

Bennett, D.A., R.L. Goble and R.A. Linthurst. 1985. The Acidic Deposition Phenomenon and Its Effects: Critical Assessment Document. U.S. Environmental Protection Agency, Publ. No. 600/8-85-001, U.S.E.P.A., Washington, DC.

Bennett, J.P., M.K. Esserlieu, and R. J. Olson. 1986. Ranking Wilderness Areas for Sensitivities and Risks to Air Pollution. U.S.D.A. Forest Service, Genl. Tech. Report No. INT-212, Intermountain Forest Range Exp. Sta., Ft. Collins, CO, pp. 73-84.

Bewley, R.J.F. and D.Parkinson. 1984. Effects on sulfur dioxide pollution on forest soil microorganisms. Can J. Microbiol. 30: 179-185.

Bicknell, S.H. (ed.). 1987. Proceedings California Forest Response Program Planning Conference. Humboldt State Univ. Foundation, Arcata, CA.

Binkley, D. and D. Richter. 1986. Nutrient cycles and H^+ budgets of forest ecosystems. Adv. Ecol. Res. (in press).

Binns, W.O. and D.B. Redfern. 1983. Acid Rain and Forest Decline in West Germany. Research and Development Paper No. 131, Forestry Commission, Edinburgh, Scotland.

Bormann, F.H. 1987. Landscape ecology and air pollution. In M. G. Turner, ed., Landscape Heterogeneity and Disturbance. Ecological Studies: Vol. 64. Springer–Verlag, New York, pp. 37-57.

Bormann, F.H. 1986. Air pollution and temperate forests: Creeping degradation? In G. M. Woodwell, ed., Proc. Biotic Impoverishment. Cambridge Univ. Press, New York.

Bormann, F.H. 1985. Air pollution and forests: An ecosystem perspective. BioScience 35: 434-441.

Breece, L. and S. Hasbrouck. 1984. Forest Responses to Acidic Deposition. Land and Water Resources Center. Univ. Maine, Orono, ME, 117 pp.

Bruck, R.I. 1987. Forests and Acid Deposition: Status of Knowledge (in press).

Bruck, R.I. 1987. Recent Advances in Understanding the Etiology and Epidemiology of Forest Decline in the Southern United States. World Resources Institute, Washington, DC, 133 pp.

Burgess, R.L. (ed.). 1984. Effects of Acid Deposition on Forest Ecosystems in the Northeastern United States: An Evaluation of Current Evidence. State Univ. of NY, Syracuse, NY, 148 pp.

Chamberlain, A.C. 1986. Deposition of gases and particles on vegetation and soils. In A.H. Legge and S.V. Krupa, eds., Air Pollutants and Their Effects on the Terrestrial Ecosystem. John Wiley and Sons, New York, pp. 189-209.

Cline, M.L., R.J. Stephans, and D H. Marx. 1988. Influence of atmospherically deposited nitrogen on mycorrhizae: A critical literature review. Annu. Rev. Phytopath (in press).

Cowling, E.B. and R.A. Linthurst. 1981. The acid precipitation phenomenon and its ecological consequences. BioScience 31: 649-654.

Daube, B. Jr., K D. Kimball, P.A. Lamar, K.C. Weathers. 1987. Two new ground-level cloud water sampler designs which reduce rain contamination. Atmos. Environ. 21: 893-900.

Davidson, C.I., J.M. Miller and M.A. Pleskow. 1982. The influence of surface structure on predicted particle dry deposition to natural grass canopies. Water Air Soil Pollu. 18: 25-43.

Davis, D.D., A.A. Millen, and L. Dochinger (eds.). 1983. Proceedings Air Pollution and the Productivity of the Forest. Izaak Walton League of America, Arlington, VA, 344 pp.

Derwent, R.G. 1986. Anthropogenic dry and wet deposition from UK and European sources. ETSU Report No. 37, AERE, Harwell, Oxfordshire, UK.

Electric Power Research Institute. 1983. Acid Rain Research: A Special Report. Palo Alto, CA, EPRI J. 8: 1-64.

Elias, R.W., T.K. Hinkley, Y. Hirao, and C.C. Patterson. 1976. Improved techniques for studies of mass balances and fractionations among families of metals within terrestrial ecosystems. Geochim. Cosmochim. Acta 40: 583-587.

Esserlieu, M.K. and R.J. Olson. 1986. Biological Vulnerabilities of National Park Service Class I Areas to Atmospheric Pollutants. Oak Ridge National Laboratory, Publ. No. ORNL/TM-9818. Oak Ridge National Laboratory, Oak Ridge, TN.

Evans, G.F. 1985. The National Air Pollution Background Network Final Project Report. Report No. EPA-600-S4-85-038. U.S. Environmental Protection Agency, Washington, DC.

Evans, G.P. Finkelstein, B. Martin, N. Possiel, and M. Graves. 1983. Ozone measurements from a network of remote sites. J. Air Pollu. Control Assoc. 33: 291-296.

Evans, L.S. 1982. Biological effects of acidity in precipitation on vegetation: A review. Environ. Exper. Bot. 22: 155-169.

Evans, L.S. 1984. Acidic precipitation effects on terrestrial vegetation. Ann. Rev. Phytopathol. 22: 397-420.

Falkengren-Grerup, V. 1986. Soil acidification and vegetation changes in deciduous forest in southern Sweden. Oecologia 70: 339-347.

Federal Minister of Food, Agriculture and Forestry (FRG). 1982. Forest Damage due to Air Pollution. Bonn, Federal Republic of Germany, 63 pp.

Fernandez, I.J. 1986. Air pollution: Synthesis of the role of major air pollutants in determining forest health and productivity. In T.C. Hennessey, P.M. Dougherty, S.V. Kossuth, and J.D. Johnson, eds., Stress Physiology and Forest Productivity. Martinus Nijhoff Publishers, New York, pp. 217-239.

Fernandez, I.J. 1983. Acidic Deposition and its Effects on Forest Productivity: A Review of the Present State of Knowledge, Research Activities and Information Needs. Tech. Bull. No. 392. National Council of the Paper Industry for Air and Stream Improvement, New York, 104 pp.

Fernandez, I.J., L. Wortmann, and S.A. Norton. 1986. Composition of Precipitation at the National Atmospheric Deposition Program/National Trends Network (NADP/NTN) Site in Greenville, Maine. Tech. Bull. No. 118, Maine Agricul. Exper. Sta., Orono, ME, 24 pp.

Flückiger, W., S. Braun, S. Leonardi, N. Asche, and H. Flückiger-Keller. 1986. Factors contributing to forest decline in northwestern Switzerland. Tree Physiol. 1: 177-184.

Fowler, D. 1980. Removal of sulphur and nitrogen compounds from the atmosphere in rain and by dry deposition. In D. Drablos and A. Tollan, eds., Proceedings Ecological Impact of Acid Precipitation. SNSF Project, Oslo, Norway, pp. 22-32.

Fraser, G.A., E.E. Phillips, G.W. Lamble, G.D. Hogan, and A.G. Teskey. 1985. The Potential Impact of the Long Range Transport of Air Pollutants on Canadian

Forests. Information Report No. E-X-36. Canadian Forestry Service, Quebec, Canada, 43 pp.

Friedland, A.J., A H. Johnson, and T.G. Siccama. 1984a. Trace metal content of the forest floor in the Green Mountains of Vermont: Spatial and temporal patterns. Water Air Soil Pollu. 21: 161-170.

Friedland, A.J., A.H. Johnson, and T.G. Siccama. 1984b. Trace metal content of the forest floor in the Green Mountains of Vermont: Spatial and temporal trends. Water Air Soil Pollu. 21: 161-170.

Friedland, A.J., A.H. Johnson, T.G. Siccama, and D.L. Mader. 1984c. Trace metal profiles in the forest floor of New England. Soil Sci. Soc. Am. 48: 422-425.

Frink, C.R. and G.K. Voigt. 1976. Potential effects of acid precipitation on soils in the humid temperate region. In U.S.D.A. Forest Service, Genl. Tech. Rep. No. NE-23, Broamall, PA, pp. 685-709.

Führer, J. and C. Führer-Fries. 1982. Interactions between acidic deposition and forest ecosystem processes. Eur. J. For. Pathol. 12: 377-390.

Gaffrey, J.S., G.E. Streit, W.D. Spall, and J.H. Hall. 1987. Beyond acid rain. Eviron. Sci. Technol. 21: 519-524.

Galloway, J.N., G.E. Likens, and E.S. Edgerton. 1976. Acid precipitation in the Northwestern United States: pH and acidity. Science 194: 722-724.

Garland, J.A. 1978. Dry and wet removal of sulphur from the atmosphere. Atmos. Environ. 12: 349-362.

Garner, J.H.B., T. Pagano, K. Joyner, and E.B. Cowling. 1988. A Critical Assessment of the Role of Acid Deposition and Other Airborne Sulfur-and Nitrogen-Derived Pollutants in the Forests of Eastern North America. U.S. Environmental Protection Agency, Publ. No. EPA-88-620, U.S.E.P.A., Research Triangle Park, NC.

Georgii, H.W. 1986. Atmospheric Pollutants in Forest Areas. D. Reidel Publishing Co., Boston, MA, 387 pp.

Getz, L.L., A.W. Haney, R.W. Larimore, J.W. McNurney, H.V. Leland, P.W. Price, G.L. Rolfe, R L. Wortman, J L. Hudson, R.L. Solmon, and K.A. Reinbold. 1977. Transport and distribution in a watershed ecosystem. In W.R. Boggess, ed., Lead in the Environment. Report No. NSF-RA-770214. National Science Foundation, Washington, DC, pp. 105-133.

Gilliam, F.S., J.T. Sigmon, M.A. Reiter, and D.O. Krovetz. 1988. Elevational and spatial variation in daytime ozone concentrations in the Virginia Blue Ridge Mountains: Implications for forest exposure. Can. J. For. Res (in press).

Goldstein, R.A. and A.H. Legge. 1986. Ecosystem analysis of air pollution effects. In A.H. Legge and S.V. Krupa, eds., Air Pollutants and Their Effects on the Terrestrial Ecosystem. Vol. 18. Advan. Environ. Sci. Technol., John Wiley and sons, New York, pp. 631-636.

Grodzinski, W., J. Weiner, and P.F. Maycock. 1984. Forest Ecosystems in In-dustrial Regions. Springer–Verlag, New York, 277 pp.

Groet, S.S. 1976. Regional and local variations in heavy metal concentrations of bryophytes in the northeastern United States. Oikos 27: 445-456.

Halbwachs, G. 1983. Effects of air pollution on vegetation. In W. Holzner, M.J.A. Werger, and I. Ikusima, eds., Man's Impact on Vegetation. D.W. Junk Publishers, Boston, MA, pp. 55-67.

Hales, J.M., B.B. Hicks, and J.M. Miller. 1987. The role of research measurement networks as contributors to federal assessments of acid deposition. Bull. Amer. Meteor. Soc. 68: 216-225.

Hanst, P.L., J.W. Spence, and M. Miller. 1977. Atmospheric chemistry of N-nitroso dimethylamine. Environ. Sci. Technol. 11: 403-405.

Harkov, R.S. and E. Brennan. 1982. An ecophysiological analysis of woody and herbaceous plants to oxidant injury. J. Environ. Mamt. 15: 251-261.

Hasbrouck, S. 1984. Acid Deposition and Forest Decline. Land and Water Resources Center, Univ. Maine, Orono, ME, 11 pp.

Hayes, E.M. and J.M. Shelly. 1977. Transport of ozone from the northeast U.S. into Virginia and its effect on eastern white pines. Pl. Dis. Rept. 61: 778-780.

Heck, W.W. 1984. Defining gaseous pollution problems in North America. In M.J. Koziol and F.R. Whatley, eds., Gaseous Air Pollutants and Plant Metabolism. Butterworths, London, pp. 35-48.

Heck, W.W. and A.S. Heagle. 1985. SO₂ effects on agricultural systems: A regional outlook. In W.E. Winner, H.A. Mooney, and R.A. Goldstein, eds., Sulfur Dioxide and Vegetation. Physiology, Ecology, and Policy Issues. Stanford Univ. Press, Stanford, CA, pp. 418-430.

Heck, W.W., A.S. Heagle, and D.S. Shriner. 1986. Effects on vegetation: Native, crops, forests. In A.S. Stern, ed., Air Pollution, Academic Press, New York, pp. 247-350.

Heinrichs, H. and R. Mayer. 1977. Distribution and cycling of major and trace elements in two Central European forests ecosystems. J. Environ. Qual. 6: 402–407.

Hicks, B.B. 1986. Measuring dry deposition: A re-assessment of the state of the art. Water Air Soil Pollu. 30: 75-90.

Hitchcock, D.R. 1976. Atmospheric sulfates from biological sources. Jour. Air Pollu, Control. Assoc. 26: 210-215.

Hosker, R.P. Jr. 1986. Practical application of air pollutant deposition models—Current status, data requirements, and researh needs. In A.H. Legge and S.V. Krupa, eds., Air Pollutants and Their Effects on the Terrestrial Ecosystem. John Wiley and Sons, New York, pp. 505-567.

Huckabee, J.W. and J.S. Mattice, L.F. Pitelka, D.B. Porcella, and R.A. Goldstein. 1988. An assessment of the ecological effects of acidic deposition. Arch. Environ. Contam. Toxicol. (in press).

Hutchison, B.A. and B.B. Hicks (eds.). 1983. The forest-atmosphere interaction. Proceedings Forest Measurements Conference. D. Reidel Publishing Com., Boston, MA, 684 pp.

Hutchinson, T.C. and K.M. Meema (eds). 1987. Effects of Atmospheric Pollutants on Forests, Wetlands and Agricultural Ecosystems. NATO ASI Series. Vol. G16. Springer–Verlag, New York.

Huttunen, S. (ed.). 1981. Proceedings air pollutants as additional stress factors under northern conditions. Silva Fennica 15: 382-504.

Jackson, D.R. and A.P. Watson. 1977. Disruption of nutrient pools and transport of heavy metals in a forested watershed near a lead smelter. J. Environ. Qual. 6: 331-338.

Jacobson, J.S. and J.J. Troiano. 1983. Dose-response functions for effects of acidic precipitation on vegetation. Water Qual. Bull. (Canada) 8: 67-71, 109.

Jäger, V.H.J., H.J. Weigel and L. Grünhage. 1986. Physiological and biochemical aspects of the impact of air pollutants on forest trees. Eur. J. For. Pathol. 16: 98-109.

Johnson, A.H. and T.G. Siccama. 1983. Acid deposition and forest decline. Environ. Sci. Technol. 17: 294-306.

Johnson, A.H. and T.G. Siccama. 1984. Decline of red spruce in the northern Appalachians: Assessing the possible role of acid deposition. TAPPI J. 67: 68-72.

Johnson, A.H., T.G. Siccama, and A.J. Friedland. 1982. Spatial and temporal patterns of lead accumulation in the forest floor in the northeastern U.S. J. Environ. Qual. 11: 577-580.

Kahl, J.S., S.A. Norton, and J.S. Williams. 1982. Chronology, magnitude, and paleolimnological record of changing metal fluxes related to atmospheric deposition of acids and metals in New England. Proc. Division of Environmental Chemistry meeting, April 1982, Las Vegas, NV. American Chemical Society, Washington, DC.

Kandler, O. 1983. Waldsterben: Emissions order epidemiehypothese? Naturwiss. Rdsch. 11: 488-490.

Karnosky D.F., P. Berrang, and R. Mickler. 1986. A Genecological Evaluation of Air Pollution Tolerances in Hardwood Trees in Eastern Forest Parks. Contract No. CX-001-4-0057 Report, National Park Service, Air Quality Division, Denver, CO, 93 pp.

Kelly, J.M. and J. F. Meagher. 1986. Nitrogen input/output relationships for three forested watersheds in eastern Tennessee. In D.L. Correll, ed., Watershed Research Perspectives. Smithsonian Institution Press, Washington, DC., pp. 360-391.

Keller, T. 1983. Air pollutant deposition and effects on plants. In B. Ulrich and J. Pankrath, eds., Effects on Accumulation of Air Pollutants in Forest Ecosystems. D. Reidel Publishing Co., New York, pp. 285-294.

Knabe, W. 1983. Chemical events in the atmosphere and their impact on the environment. Pontificial Academiae Scientiarum Scripta Varia 56: 553-606.

Knabe, W. 1985. Das Waldsterben aus immissionsäkologischer Sicht. Geogra-phische Rdsch. 5: 249-256.

Kociuba, P.J. 1984. Estimate of sulphate deposition in precipitation for Alberta. Report of Atmospheric Environment Service, Environment Canada, Edmonton, Alberta, Canada, 6 pp.

Kohut, R., J.A. Laurence, and R.G. Amundson. 1988. Comparison of the responses of seedling and sapling red spruce exposed to ozone and acidic precipitation under field conditions. In Vol. I. Project Status Reports, Forest Response Program Annual Meting, 22–26 Feb. 1988, Corpus Christi, TX, pp. 228-229.

Kozlowski, T.T. 1985. Measurement of effects of environmental and industrial chemicals on terrestrial plants. In V.B. Vouk, G.C. Butler, D.G. Hoel, and D.B. Peakall, eds., Methods for Estimating Risk of Chemical Injury: Human and Non-Human Biota and Ecosystems. SCOPE, Vienna, Austria, pp. 573-609.

Kozlowski, T.T. 1986. The impact of environmental pollution on shade trees. J. Arbor. 12: 29-37.

Kozlowski, T.T. and H.A. Constantinidou. 1986. Responses of woody plants to environmental pollution. For. Abst. 47: 5-132.

Krause, G.H.M. 1987a. Impact of air pollutants on above-ground plant parts of forest trees. In Proc. Effects of Air Pollution on Terrestrial and Aquatic Ecosystems. Commission of the European Communities, Grenoble, France.

Krause, G.H.M. 1987b. Forest decline and the role of air pollutants. In Acid Rain: Scientific and Technical Advances. Selper Ltd., London, UK, pp. 621-632.

Krause, G.H.M. 1985. Forest effects in West Germany. In Proc. Air Pollution and the Productivity of the Forest. Izaak Walton League of America, Arlington, VA.

Krause, G.H.M., U. Arndt, C.J. Brandt, J. Bucher, G. Kenk, and E. Matzner. 1986. Forest decline in Europe: Development and possible causes. Water Air Soil Pollu. 31: 647-668.

Krupa, S. and R.N. Kickert. 1987. An analysis of numerical models of air pollutant exposure and vegetation response. Environ. Pollu. 44: 127-158.

Kulp, J. L. 1987. Effects on forests. Chap. 7. Effects on Acidic Deposition. Interim Assessment. In Vol. IV. National Acid Precipitation Assessment Program, Washington, DC, 59 pp.

Kvist, K. 1985. Luftföroreningar effekter i lanlbruket. Värtskyddscrapporter. Jordbruk 32: 115-127.

Larsen, R.I. and W.W. Heck. 1985. An air quality data analysis system for interrelating effects, standards, and needed source reductions: Part 9. Calculating effective ambient air quality parameters. J. Air Pollu. Control Assoc. 35: 1274-1279.

Last, F.T. 1982. Effects of atmospheric sulphur compounds on natural and man-made terrestrial and aquatic ecosystems. Agric. Environ. 7: 299-387.

Last, F.T. 1983. Direct effects of air pollutants, singly and in mixtures, on plants and plant asemblages. In H. Ott and H. Stangl, eds., Proceedings Acid De-position — A Challenge for Europe. Commission of European Communities, Brussels, Belgium, pp. 105-126.

Lau, K.Y. 1985. A 5-year (1978–1982) summary of precipitation chemistry measurements in Alberta. Report of Pollution Control Division, Alberta En-vironment, Edmonton, Alberta, Canada.

Lazrus, A.L., E. Lorange, and J.P. Lodge Jr. 1970. Lead and other metal ions in United States precipitation. Environ. Sci. Technol. 4: 55-58.

Lefohn, A.S. and H.M. Benedict. 1984. Exposure considerations associated with characterizing ozone ambient air quality monitoring data. In S.D. Lee, ed., Transactions Evaluation of the Scientific Basis for Ozone/Oxidants Standards. Air Pollu. Control Assoc., Specialty Conf., Nov 1984, Houston, TX, pp. 17-27.

Lefohn, A.S. and R.W. Brocksen. 1984. Acid rain effects research — A status report. J. Air Pollu. Control. Assoc. 34: 1005-1013.

Lefohn, A.S. and C.K. Jones. 1986. The characterizaion of ozone and sulfur dioxide air quality data for assessing possible vegetation effects. J. Air Pollu. Control Assoc. 36: 1123-1129.

Lefohn, A.S. and V.A. Mohnen. 1986. The characterization of ozone, sulfur dioxide, and nitrogen dioxide for selected monitoring sites in the Federal Republic of Germany. J. Air Pollu. Contol Assoc. 36: 1329-1337.

Lefohn, A.S. and V.C. Runeckles. 1987. Establishing standards to potect vegetation — ozone exposure/dose considerations. Atmos. Environ. 21:561-568.

Lefohn, A.S. and D.T. Tingey. 1984. The co-occurrence of potentially phytotoxic concentrations of various gaseous air pollutants. Atmos. Environ. 18: 2521-2526.

Lefohn, A.S., W.E. Hogsett, and D.T. Tingey. 1986. A method for developing ozone exposures that mimic ambient conditions in agricultural areas. Atmos. Environ. 20: 361-366.

Lefohn, A.S., C.E. Davis, C.K. Jones, D.T. Tingey, and W.E. Hogsett. 1987a. Co-occurrence patterns of gaseous air pollutant pairs at different minimum concentrations in the United States. Atmos. Environ. 21: 2435-2444.

Lefohn, A.S., W.E. Hogsett, and D.T. Tingey. 1987b. The development of sulfur dioxode and ozone exposure profiles that mimic ambient conditions in the rural southeastern United States. Atmos. Environ. 21: 659-669.

Legge, A.H. and R.A. Crowther. 1987. Acidic Deposition and the Environment: A Literature Overview. Acid Deposition Research Program, Kananaskis Centre for Environmental Research, University of Calgary, and Aquatic Resource Man-agement Limited, Calgary, Alberta Canada, 235 pp.

Legge, A.H. and S.V. Krupa. 1986. Air Pollutants and Their Effects on the Terrestrial Ecosystem. John Wiley and Sons, New York, 662 pp.

Lewis, T.E., E. Prennan, and W.A. Lonneman. 1983. PAN concentrations in ambient air in New Jersey. J. Air Pollu. Cont. Assoc. 33: 885-887.

Likens, G.E. 1975. Acid precipitation: Our understanding of the phenomenon. Proc. Conf. Emerging Environmental Problems: Acid Precipitation. May 1975, Renssalaeville, New York, EPA-902/9-75-001. U. S. Environmental Protection Agency, New York, New York, 115 pp.

Likens, G. E. 1976. Acid Precipitation. Chem. Eng. News 54: 29-44.

Likens, G.E., F.H. Bormann, and J.S. Eaton. 1980. Variations in precipitation and streamwater chemistry at the Hubbard Brook Experimental Forest during 1964 to 1977. In T.C. Hutchinson and M. Havas, eds., Effects of Acid Precipitation on Terrestrial Ecosystems. Plenum Publishing Corp., New York, pp. 443-464.

Lindberg, S.E. and R.C. Harriss. 1981. The role of atmospheric deposition in an eastern U.S. deciduous forest. Water Air Soil Pollu. 16: 13-31.

Lindberg, S.E., R.C. Harriss, and R.R. Turner. 1982. Atmospheric deposition of metals to forest vegetation. Science 215: 1609-1611.

Lindberg, S.E. and R.R. Turner. 1983. Trace metals in rain at forested sites in the eastern United States. In Heavy Metals in the Environment. Vol. I. CEP Con-sultants Ltd, Edinburg, UK, pp. 107-114.

Linthurst, R.A. (ed). 1984a. Direct and Indirect Effects of Acidic Deposition on Vegetation. Vol. 5. Acid Precipitation Series. Butterworth Publishers, Boston, MA, 117 pp.

Linthurst, R.A. 1984b. The Acidic Deposition Phenomenon and its Effects: Critical Assessment Review Papers. Vol. II. Effects Sciences. U.S. Environmental Pro-tection Agency, Publ. No 600/8-83-016BF, U.S.E.P.A., Washington, DC.

Linzon, S.N. 1986. Effects of gaseous pollutants on forests in eastern North America. Water Air Soil Pollu. 31: 537-550.

Linzon, S.N. R.G. Pearson, J.A. Donnan, and F.N. Durham. 1984. Ozone effects on crops in Ontario and related monetary values. Ontario Ministry of the Environment, Publ. No. ARB-13-84-Phyto, Ontario, Canada, 60 pp.

Loucks, O.L. 1981. A Scenario for Research on the Effects of Stress in the Eastern Deciduous Forest. Institute of Ecology Report, Indianapolis, IN, 36 pp.

Lovett, G.M. 1984. Rates and mechanisms of cloud water deposition to a subalpine balsam fir forest. Atmos. Environ. 18: 361-371.

Lovett, G.M. 1987. Atmospheric deposition: Processes and measurement methods. In S.H. Bicknell, ed., Proceedings California Forest Response Program Planning Conference, Feb. 22-24, 1987, Humboldt State Univ., Arcata, CA, pp. 7-23.

Lovett, G.M., W.A. Reiners, and R.K. Olson. 1982. Cloud droplet deposition in subalpine balsam fir forests: Hydrological and chamical imputs. Science 218: 1303-1304.

MacKenzie, J.J. and M.T. El-Ashry. 1988. Ill Winds: Airborne Pollution's Toll on Trees and Crops. World Resources Institute, Washington, DC, 71 pp.

MacLean, D.C. 1983. Air pollution and horticulture: An overview. Hort Science 18: 674-675.

Mann, L.K., S.B. McLaughlin, and D.S. Shriner. 1980. Seasonal physiological responses of white pine under chronic air pollution stress. Environ. Exp. Bot. 20: 99-104.

Martin, H.C. (ed). 1986. Proceedings International Symposium on Acidic Pre-cipitation. Water Air Soil Pollu. 30: 1-1045, 31: 1-517.

Matzner, E. and B. Ulrich. 1985. Implications of the chemical soil conditions for forest decline. Experientia 41: 578-584.

Mayer, R. and B. Ulrich. 1978. Input of atmospheric sulfur by dry and wet depositon to two central European forest ecosystems. Atmos. Environ. 12: 375-377.

Mayo, J.M. 1987. The Effects of Acid Deposition on Forests. Publ. No. ADRP-B 09/87. Acid Deposition Research Program, Dept. Biology, Emporia State Univ., Emporia, KS, 80 pp.

McBride, J.R. and P.R. Miller. 1987. Responses of American forests to photochemical oxidants. In T.C. Hutchinson and K.M. Meema, eds., Effects of Atmospheric Pollutants on Forests, Wetlands and Agricultural Ecosystems. Springer–Verlag. New York, pp 217-228.

McClenahen, J.R. 1984. Air pollutant effects on forest communities. In D.D. Davis, A.A. Millen and L. Dochinger, eds, Proceedings Air Pollution and the Productivity of the Forest. Izaak Walton League of America, Washington DC, pp. 83-94.

McCune, D.C. 1986. Characterization of air pollutants in the plant's environment. In B.L. Bedford, ed, Modification of Plant-Pest Interactions by Air Pollutants,

Publica. No. ERC-117 Ecosystems Research Center, Cornell Univ., Ithaca, NY, pp. 53-70.

McLaughlin, S.B. 1985. Effects of air pollution on forests. A critical review. J. Air Pollu. Control Assoc. 35: 512-534.

McLaughlin, S.B., T.J. Blasing, L.K. Mann, and D.D. Duvick. 1983. Effects of acid rain and gaseous pollutants on forest productivity.

Miller, P.R. 1987. Concept of Forest Decline in Relation to Western Forests. World Resource Institute, Washington, D.C., 67 pp.

Miller, P.R., O.C. Taylor, and R.G. Wilhour. 1982. Oxidant Air Pollution Effects on a Western Coniferous Ecosystem. U. S. Environmental Protection Agency, Publ. No. 600/D-82-276, Corvallis, OR, 10 pp.

Mohnen, V.A. 1987. Exposure of Forests to Gaseous Air Pollutants and Clouds. Report to U. S. Environmental Protection Agency, Contract No. 813-934010. Chemistry Project. U.S.E.P.A. Research Triangle Park, NC.

Mohnen, V.A. and R.L Bradow. 1986. MCCP Management Center Report on Hydrogen Peroxide and Other Gas/Aqueous Pollutants. Dec. 1, 1986. U.S.E.P.A., Washington, DC.

Mollitor, A.V. and D.J. Raynal. 1983. Atmospheric deposition and ionic input in Adirondack forests. J. Air Pollu. Control Assoc. 33: 1032-1036.

Morrison, I.K. 1984. Acid rain. A review of literature on acid deposition effects in forest ecosystems. For. Abst. 45: 483-506.

Moskowitz, P.D., E.A. Coveney, and W.H. Morris. 1982. Oxidant air pollution: A model for estimating effects on U. S. vegetation. J. Air Pollu. Control Assoc. 32: 155-160.

NAPAP (National Acid Precipitation Assessment Program). 1988. Annual Report— 1987. National Acid Precipitation Assessment Program, 722 Jackson Place NW, Washington, DC, 76 pp.

National Council of the Paper Industry for Air and Stream Improvement. 1981. Acidic Deposition and its Effects on Forest Productivity — A Review of the Present State of Knowledge, Research Activities, and Information Needs. National Council for Air and Stream Improvement Tech. Bull. No. 110, New York, 66 pp.

National Council of the Paper Industry for Air and Stream Improvement. 1987. Air Quality/Forest Health Program News. Vol. 2. NCASI, 260 Madison, Ave, New York.

National Council of the Paper Industry for Air and Stream Improvement. 1988. Air Quality/Forest Health Program News. Vol 3. 260 Madison Ave, New York.

National Research Council. 1986. Acid Deposition. Long Term Trends. National Academy Press, Washington, DC, 506 pp.

Nilsson, S. and P. Duinker. 1987. The extent of forest decline in Europe. A synthesis of survey results. Environment 29: 4-9, 31.

Oden, S. 1976. The acidity problem—An outline of concepts. In LS. Dochinger and T.A. Seliga, ed., Proc. 1st Intl. Symp. Acid Precipitation and the Forest Ecosystem. U.S.D.A. Forest Service, Gen. Tech. Rep. No. NE-23, Upper Darb, PA, pp. 1-36.

Odum, J.A. 1971. Ecosystem structure and function In J.A. Wiens, (ed.), Proc.31st Annual Biol. Colloquium, Oregon State Univ. Press, Corvallis, OR.

Overrein, L.N. 1980 Acid precipitation impact on terrestrial and aquatic systems in Norway. In P.R. Miller, ed., Proceedings Effects of Air Pollutants on Mediterranean and Temperate Forest Ecosystems. U.S.D.A. Forest Service Gen. Tech, Report No. PSW-43, U.S.D.A. Forest Service, Berkeley, CA, pp. 145-151.

Overrein, L.N., H.M. Seip, and A. Tollan. 1981. Acid Precipitation. Effects on Forest and Fish. Final Report of the SNSF Project 1972–1980. Norwegian Council for Scientific and Industrial Research, Oslo, Norway.

Parker, G.G. 1987. Uptake and release of inorganic and organic ions by foliage: Evaluation of dry deposition, pollutant damage, and forest health with throughfall studies. Tech. Bull. No. 532, National Council of the Paper Industry for Air and Stream Improvement, New York.

Peterson, E.W. 1977. Wet and Dry Deposition — A Synopsis Containing Estimates of Deposition Velocities for Some Pollutant and Trace Gases in the Atmosphere. Publ. No. CERL-037. National Oceanic and Atmospheric Administration, Washington, D.C.

Pinkerton, J.E. 1984. Acidic deposition and its relationship to forest productivity. TAPPI J. 67: 36-39.

Plochmann, R. 1984 Air pollution and the dying forests of Europe. Am. For. 90: 17-21,. 56.

Postel, S. 1984. Air Pollution, Acid Rain, and the Future of Forests. Paper No. 58. Worldwatch Institute, Washington, DC, 54 pp.

Pratt, G.C., R.C. Henrickson, B.I. Chevone, D. A. Christopherson, M.V. O'Brien, and S.V. Krupa. 1983. Ozone and oxides of nitrogen in the rural upper-midwestern U.S.A. Atmos. Environ. 17: 2013-2023.

Prinz, B. 1984. Recent forest decline in the Federal Republic of Germany and contribution of the LIS for its explanation. In Excursion Guide, German/American Information Exchange on Forest Dieback, May 16, 1984. Landsanstalt für immissionsschutz des landes Nordrhein-Westfalen, Essen. FRG, pp. 1-35.

Prinz, B. 1987. Causes of forest damage in Europe. Major hypotheses and factors. Environment 29: 11-15, 32-37.

Prinz, B. and C.J. Brand (eds). 1986. Acidic Precipitation. Formation and Impact on Terrestrial Ecosystems. VDI Commission for Air Pollution Prevention, Düsseldorf, FRG, 281 pp.

Prinz, B., G.H.M. Krause, and K.D. Jung. 1987. Development and causes of novel forest decline in Germany. In T. C. Hutchinson and K. M. Meema, eds., Effects of Atmospheric Pollutants on Forests, Wetlands and Agricultural Ecosystems. Springer–Verlag, New York, pp. 1-24.

Prinz, B., G.H.M. Krause, and H. Stratmann. 1982. Forest Damage in the Federal Republic of Germany. Publ. No. 28, Land Institute for Pollution Control, Essen, FRG, 140 pp.

Raynal, D.J., F.S. Raleigh, and A.V. Mollitor. 1983. Characterization of Atmospheric Deposition and Ionic Input at Huntington Forest, Adirondack Mountains, NY. Publica. No. 83-003, State Univ. of New York, Syracuse, New York, 85 pp.

Rehfuss, K.E. 1983. Walderkrankungen und Immissionen — eine Zwischenbilanz. Allg. Forsztg. 24: 601-610.

Rehfuess, K.E. 1985. On the causes of decline of Norway spruce (*Picea abies* Karst.) in Central Europe. Soil Use Mamt. 1: 30-33.

Rehfuess, K.E. 1987. Perceptions of forest diseases in Central Europe. Forestry (UK) 60 (in press).

Reiners, WA., R.H. Marks, and P.M. Vitousek. 1975. Heavy metals in subalpine and alpine soils of New Hampshire. Oikos 26: 264-275.

Reiners, W.A., G.M. Lovett, and R.K. Olson. 1987. Chemical interactions of a forest canopy with the atmospherre. In H. Moses, V.A. Mohmen, W.E. Reifsnyder, and D.H. Slade, eds., Proceedings Forest-Atmosphere Interaction Workshop, Publ. No. CONF-85-10250, U.S. Dept. of Energy, Washington, DC, pp. 111-147.

Rennie, P.J. 1987. Air pollution and the forestry sector: Challenges and requirements. In Proc. Woody Plant Growth in a Changing Physical and Chemical Environment, International Union Forestry Research Organizations Workshop, Univ. British Columbia, Vancouver, BC, Canada, July 27–31, 1987.

Reuss, J.O. and D.W. Johnson. 1986. Acid Deposition and the Acidification of Soils and Waters. Ecological Studies No. 59. Springer–Verlag, New York, 119 pp.

Roberts, L. 1983. Is acid deposition killing West German forests? BioScience 33: 302-305.

Russo–Savage, S. F. 1989. Personal Communication. Environmental Studies Program. Dartmouth College, Hanover, NH.

Schemenauer, R.S. 1986. Acidic deposition to forests: The 1985 chemistry of high elevation fog (CHEF) project. Atmos. Ocean 24: 303-328.

Schemenauer, R.S. and K.G. Aulauf. 1987. Geographic variation of ozone concentrations at high and low elevtion rural sites in Quebec. In Proc. North American Oxidant Symposium, 25–27 Feb. 1987, Quebec City, Canada, pp. 412-429.

Schlesinger, W.H. and W.A. Reiners. 1974. Deposition of water and cations on artifical foliar collectors in fir krummholz of New England mountains. Ecology 55: 378-386.

Schöpfer, V.W. and J. Hradetzky. 1984. Circumstantial evidence: Air pollution is the determinative factor causing the forest decline. Forstwissenschaftliches centralblatt 103: 1-18.

Schütt, P. and E.B. Cowling. 1985. Waldsterben, a general decline of forests in Central Europe: Symptoms, development and possible causes. Pl. Dis. 69: 548-558.

Sehmel, G.A. 1980. Particle and gas dry deposition: A review. Atmos. Environ. 14: 983-1011.

Shepard, M. 1985. Forest stress and acid rain. Electric Power Research Institute J. (Palo Alto, CA) Sept 1985: 16-25.

Siccama, T.G. and W.H. Smith. 1978. Lead accumulation in a northern hardwood forest. Environ. Sci. Technol. 12: 593-594.

Skelly, J.M. 1987. Oxidant induced effects to forest ecosystems: Hidden injury versus devastating forest declines: Reality versus conjecture. In Proc. North American Oxidant Symposium, Quebec, Canada, pp. 21-41.

Skelly, J.M., D.D. Davis, W. Merrill, E.A. Cameron, H.D. Brown, D.B. Drummond, and L.S. Dochinger (eds.). 1987. Diagnosing Injury to Eastern Forest Trees. Pennsylvania State Univ., University Park, PA, 122 pp.

Skelly, J.M., Y.S. Yang, B I. Chevone, S.J. Long, J.E. Nellssen, and W.I. Winner. 1983. Ozone concentrations and their influence on forest species in the Blue Ridge Mountains of Virginia. In D.D. Davis, A.A. Millen, and L. Dochinger, eds., Proc. Air Pollution and the Productivity of the Forest. Izaac Walton League of America, Washington, DC, 344 pp.

Skelly, J.M., Y.S. Yang, B.J. Chevone, and S.J. Long. 1984. Ozone concentrations and their influence on forest species in the Blue Ridge Mountains of Virginia. In D. D. Davis, A. A. Millen, and L. Dochinger, eds. Proceedings Air Pollution and the Productivity of the Forest Izaak Walton League of America, Washington, DC, pp 143-159.

Smith, W.H. 1974. Air pollution — Effects on the structure and function of the temperate forest ecosystem. Environ. Pollu. 6: 111-129.

Smith, W.H. 1982. The Influence of Acid Deposition on the Forests of the Tug Hill (New York) Study Area: A General Assessment. Temporary State Commission on Tug Hill, Watertown, New York, 21 pp.

Smith, W.H. 1984a. Ecosystem pathology: A new perspective for phytopathology. For. Ecol. Mgmt. 9: 193-219.

Smith, W.H. 1984b. Effects of regional air pollutants on forests in the USA. Forst. Centralblatt. 103: 48-61.

Smith, W.H. 1985. Forest quality and air quality. J. For. 83: 82-92.

Smith, W.H. 1986. Fossil fuel combustion and forest ecosystem health. In R. Markuszewski and B.D. Blaustein, eds., Fossil Fuels Utilization: Environmental Concerns. Symposium Series No. 319. American Chemical Society, Washington, DC, pp. 254-266.

Smith, W.H. 1987. Energy production and forest ecosystem health. In S.K. Majumdar, F.J. Brenner, and E.W. Miller, eds., Environmental Consequences of Energy Production. Pennsylvania Academy of Science, Harrisburg, PA, pp. 432-444.

Smith, W.H. and T.G. Siccama. 1981. The Hubbard Brook ecosystem study: Biogeochemistry of lead in the northern hardwood forest. J. Environ. Qual. 10: 323-333.

Smith, W.H., T.G. Siccama, and S.L. Clark. 1988. Atmospheric Deposition of Heavy Metals and Forest Health: An Overview and a Ten-Year Budget for the Input/Output of Seven Heavy Metals to a Northern Hardwood Forest. Publ. No. FWS-8702. School of Forestry and Wildlife Resources, Virginia Polytechnic Institute and State University, Blacksburg, VA.

Society of American Foresters. 1984. Acidic Deposition and Forests. SAF, Bethesda, MD, 48 pp.

Steinbeck, K. 1984. West Germany's Waldsterben. Forestry 82: 719-720.

Sullivan, T.J., J.M. Eilers, M.R. Church, D.J. Bick, K.N. Eshleman, D.H. Landers, and M.S. DeHaan. 1988. Atmospheric wet sulphate deposition and lakewater chemistry. Nature 331: 607-609.

Tamm, C.O., F. Andersson, H. Eriksson, J. Hällgren, K. Kvist, K. Lundkvist, N. Nykvist, J. Persson, T. Troedsson, and B. Ronne. 1984. Effects of Air Pollution on Forest Land. Outline for a Research Program. Report of Sweedish University of Agricultural Sciences. Uppsala, Sweden, 197 pp.

Taylor, G.E. and R.J. Norby. 1985. The significance of elevated levels of ozone on natural ecosystems of North America. In Proceedings Evaluation of the Scientific Basis for an Ozone/Oxidant Standard. J. Air Pollu. Control Assoc. (in press).

Taylor, G.E. Jr. and R.J. Norby. 1986. The significance of elevated levels of ozone on natural ecosystems of North America. In S.D. Lee, ed., Evaluation of the Scientific Basis for Ozone/Oxidants Standards. Air Pollu. Control Assoc., Speciality Conference Proc., Pottsburgh, PA, pp. 152-175.

Taylor, G.E. Jr., S.B. McLaughlin, and D.S. Shriner. 1982. Effective pollutant dose. In M.H. Unsworth and D. P. Ormrod, eds, Effects of Gaseous Air Pollution in Agriculture and Horticulture. Butterworths, London.

Taylor, G.E. Jr., D.D. Baldocchi, and P.J. Hanson. 1987. Pollutant deposition to individual leaves and plant canopies: Sites of regulation and relationship to injury. Oak Ridge National Laboratory Report, Oak Ridge, TN.

Temple, P.J. and O.C. Taylor. 1982. World-wide measurements of peroxyacetyl nitrate (PAN) and implications for plant injury. Atmos. Environ. 29: 302-306.

Temple, P.J. and O.C. Taylor. 1983. World-wide ambient measurements of peroxyacetyl nitrate (PAN) and implications for plant injury. Atmos. Environ. 17: 1583-1587.

Tomlinson, G.H. 1983. Air pollutants and forest decline. Environ. Sci. Technol. 17: 246A-256A.

Treshow, M. 1984. Air Pollution and Plant Life. John Wiley and Sons, New York, 486 pp.

Trüby, P. and H. W. Zöttl. 1984. Heavy metal turnover in a spruce ecosystem of the higher Black Forest (Bärhalde) and a pine ecosystem of the Upper Rhine Plain (Hartheim). Angew. Botanik 58: 39-45.

Turner, R.S., A.H. Johnson, and D. Wang. 1984. Biogeochemistry of lead in McDonalds Branch Watershed, New Jersey pine barrens. J. Environ. Qual. 14: 305-310.

Tyler, G. 1981. Leaching of metals from the A-horizon of a spruce forest soil. Water Air Soil Pollu. 15: 353-369.

Ulrich, B. 1980. Production and consumption of hydrogen ions in the ecosphere. In T. C. Hutchinson and M. Havas, eds., Effects of Acid Precipitation on Terrestrial Ecosystems. Plenum Press, New York, pp. 255-282.

Ulrich, B. 1981. The destabilization of forest ecosystems by the accumulation of air contaminants. Der Forst. Holzwirt. 36: 525-532.

Ulrich, B. 1983. Effects of accumulation of air pollutants in forest ecosystems. In H. Ott and H. Stangl, eds., Proceedings Acid Deposition — A Challenge for Europe. Commission of European Communities, Brussels, Belgium, pp. 127-146.

Ulrich, B. and J. Pankrath (eds.). 1983. Effects of Accumulation of Air Pollutants in Forest Ecosystems. D. Reidel Publishing Co., Boston, MA, 389 pp.

Ulrich, V. 1984. Effects of air pollution on forest ecosystems and waters — The principles demonstrated at a case study in central Europe. Atmos. Environ. 18: 621-628.

Unsworth, M.H. and D.P. Ormrod (eds.). 1982. Effects of Gaseous Air Pollution in Agriculture and Horticulture. Butterworths, London, 552 pp.

Urone, P. 1976. The primary air pollutants - gaseous. Their occurrence, sources and effects. In A. C. Stern, ed., Air Pollution. Vol. I. Academic Press, New York, pp. 23-75.

U. S. Environmental Protection Agency. 1982. Air Quality Criteria for Particulate Matter and Sulfur Oxides. Vol. III. U.S.E.P.A. Publ. No. 600/8-82-029c Research Triangle Park, NC.

U. S. Environmental Protection Agency. 1983. The Acidic Deposition Phenomenon and its Effects. Critical Assessment Review Papers. Vol. I. Atmospheric Sciences. U.S. E. P. A. Publ. No. 600/8-83-016AF, Washington, DC.

U. S. Environmental Protection Agency. 1985. The Acidic Deposition Phenomenon and Its Effects. Publ. No. EPA-600-8-85-001. U.S.E.P.A., Washington, DC, 159 pp.

U. S. Environmental Protection Agency. 1986a. Air Quality Criteria for Lead. Vols. I and II. U.S.E.P.A. Publ. No. 600/8-83-028aF, Research Triangle Park, NC.

U. S. Environmental Protection Agency. 1986b. Air Quality Criteria for Ozone and Other Photochemical Oxidants. Vol. III. U.S.E.P.A. Publ. No. 600/8-84-020cF, Research Triangle Park, NC.

Volger, D. 1982. Ozone monitoring in the southern Sierra Nevada. U.S.D.A. Forest Service. Forest Pest Management Report No. 82-17, Berkeley, CA, 43 pp.

Weathers, K.C., G.E. Likens, F.H. Bormann, J.S. Eaton, W.B. Bowden, J.L. Andersen, D.A. Cass, J.N. Galloway, W.C. Keene, K D. Kimball, P. Huth, and D. Smiley. 1986a. A regional acidic cloud/flog water event in the eastern United States. Nature 319: 657-658.

Weathers, K.C., G.E. Likens, F.H. Bormann, J.S. Eaton, K.D. Kimball, J. N. Galloway, T.G. Siccama, and D. Smiley. 1986b. Chemical concentrations in cloud water from four sites in the eastern United States. In Proc. Acid Deposition Processes at High Elevation Sites. NATO Advanced Research Workshop. D. Reidel, New York.

Weathers, K.C., G.E. Likens, F.H. Bormann, S.H. Bicknell, B.T. Bormann, B. Daube Jr., J.S. Eaton, J.N. Galloway, W.C. Keene, K.D. Kimball, W.H. McDowell, T.G. Siccaina, D. Smiley, and R.A. Tarrant. 1988. Cloud water chemistry from ten sites in North America. Environ. Sci. Technol. (in press).

Wesely, M.L. and B.B. Hicks. 1986. Practical aspects of measuring dry deposition. In Methods for Acid Deposition Measurements. Publ. No. EA-4663, Electric Power Research Institute, Palo Alto, CA.

Wiersma, G.B. 1979. Kinetic and exposure commitment analyses of lead behavior in a biosphere reserve. Report No. 15. Monitoring and Assessment Research Center, Chelsea College, University of London, London, U.K..

Winner, W.E., H.A. Mooney, and R.A. Goldstein (eds). 1985. Sulfur Dioxide a Vegetation. Physiology, Ecology and Policy Issues. Stanford University Press, Stanford, CA, 593 pp.

Woodman, J.N. and E.B. Cowling. 1987. Airborne chemicals and forest health. Environ. Sci. Technol. 21: 120-126,

Zöttl, H.W. 1985a. Heavy metal levels and cycling in forest ecosystems. Experientia 41: 1104-1113.

Zöettl, H.W. 1985b. Role of heavy metals in forest ecosystems. T.D. Lekkas, ed., Proc. Heavy Metals in the Environment. CEP Consultants, Edinburgh, Scotland, pp. 8-15.

Zöttl, H.W. 1987. Element turnover in ecosystems of the Black Forest. Forstwissenschaftliches Centralblatt 106: 105-114.

SECTION I

FORESTS FUNCTION AS SOURCES AND SINKS FOR AIR CONTAMINANTS — CLASS I INTERACTIONS

3
Role of Forests in Major Element Cycles: Carbon, Sulfur, and Nitrogen

Despite the fundamental and enormous importance of carbon, sulfur, and nitrogen to the biota, our appreciation of the cycling, residence, and flux of these elements among various ecosystems, oceans, and the atmosphere is incomplete. Despite our deficiencies, efforts are being made to refine our understanding of major element cycles and to more accurately estimate global budgets. This is especially important for those interested in air pollution, as compounds containing these elements are extremely significant air contaminants under certain circumstances. Forest ecosystems, because of their extensive distributions and their varied functions, play important roles in global element cycles. The imprecision of our estimates of global nutrient budgets is large and inconclusive conclusive, and predictions based on them must be qualified and cautious.

Carbon, sulfur, and nitrogen have analagous biogeochemical cycles. All three cycles contain a biological reduction step with a reduced residence period in the biota, where these elements exist prior to reintroduction to the atmosphere (Deevey, 1973; Hitchcock and Wechsler, 1972). Since oxides and oxidized compounds of these three elements can be important atmospheric contaminants and since forest ecosystems are involved in the global release of these compounds, it is important to review the significance of forests as sources of carbon, sulfur, and nitrogen oxides, and oxidized end products.

A. Carbon Pollutants

Carbon monoxide and carbon dioxide are both very important atmospheric contaminants. Human activities are responsible for the introduction of increasing quantities of these gases to the atmosphere. Carbon monoxide is particularly important because of its potent mammalian toxicity, while carbon dioxide is most significant because of its ability to regulate global temperature. Neither gas is thought to cause direct damage to vegetation at ambient concentrations presently monitored. Carbon monoxide has not been shown to produce acute effects on plants at concentrations below 100 ppm (11.5 $\times$ 10^4 μg m$^{3)}$ for exposures from 1 to 3 weeks (U.S. Environmental Protection Agency, 1976). The threshold of carbon dioxide toxicity to plants is in such excess of ambient conditions as to be completely unimportant. The hypotheses that the increasing concentration of carbon dioxide in the atmosphere will result in elevated global temperatures, and for forest fertilization however, have enormous implications for the health of forest ecosystems and will be discussed in more detail later.

1. Atmospheric Increase of Carbon Dioxide

The carbon dioxide concentration of the global atmosphere has been estimated to have been approximately 270 $\pm$ 10 ppm (4.8 $\times$ 10^5 μg m^{-3}) in the middle of the nineteenth century. Today the carbon dioxide concentration approximates 350 ppm (6.3 $\times$ 10^5 μg m^{-3}). Evidence for this increase has come from careful carbon dioxide monitoring carried out over the last 30 years in Hawaii, Alaska, New York, Sweden, Australia, and the South Pole. The current rate of increase is about 1.5 ppm (2.7 $\times$ 10^3 m^{-3}) annually. In the early to mid portion of the next century, if the increasing rate continues, the carbon dioxide amount in the the global atmosphere may be nearly two times the present value (Trabalka et al. 1986).

2. Carbon Cycle

Pool sizes, transfer patterns, and flux rates of the carbon cycle are not fully nor very accurately appreciated. Bolin (1986) has presented estimates of the major carbon pools and flux rates. The atmosphere contains 720 $\times$ 10^9 tons of carbon, the biota 830 $\times$ 10^9 tons, and below-ground pools include 60 $\times$ 10^9, 500 $\times$ 10^9, 1400 $\times$ 10^9 and 5000 $\times$ 10^9 tons in surface detritus, peat, soil and fossil fuels, respectively. The largest carbon pool is the global ocean system with a carbon content of approximately 20 $\times$ 10^{15} tons.

A dynamic and critically important flux of carbon, as carbon dioxide, occurs between the global biota and the surface waters of the ocean. A biogeochemical cycle for carbon is presented in Figure 3-1.

Forests occupy approximately one-third of the area of global terrestrial ecosystems (Reichle et al., 1973; Whittaker, 1975) and constitute from approximately 60% (Reichle et al., 1973) to 90% (Woodwell, 1978) of the total terrestrial carbon pool. Global net primary production and total mass of carbon for various ecosystems are listed in Table 3-1.

The natural net input of carbon dioxide to the atmosphere from vegetative systems is close to zero (equilibrium) when the systems exist in natural, undisturbed states. Gross input by forest fires, organic matter decomposition, and

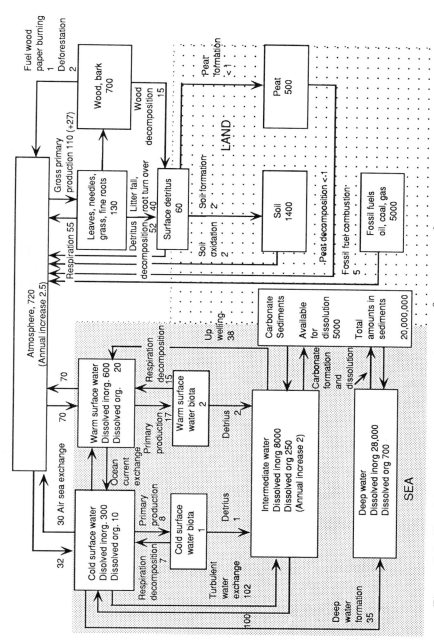

Figure 3-1. The global carbon cycle. Units: inventories 10^9 ton, fluxes 10^9 tons yr^{-1}. Source: Bolin (1986).

Table 3-1. Major Ecosystems of the Earth: Area, Net Primary Production, and Mass of Carbon.

Plant community	Area (10^6 km^2)	Total net primary production of carbon $(10^{15} \text{ g yr}^{-1})$	Total plant mass of carbon (10^{15} g)
Tropical rain forest	17.0	16.8	344
Tropical seasonal forest	7.5	5.4	117
Temperate evergreen forest	5.0	2.9	79
Temperate deciduous forest	7.0	3.8	95
Boreal forest	12.0	4.3	108
Woodland and shrubland	8.5	2.7	22
Savanna	15.0	6.1	27
Temperate grassland	9.0	2.4	6.3
Tundra and alpine meadow	8.0	0.5	2.3
Desert scrub	18.0	0.7	5.9
Rock, ice, and sand	24.0	0.03	0.2
Cultivated land	14.0	4.1	6.3
Swamp and marsh	2.0	2.7	13.5
Lake and stream	2.0	0.4	0.02
Total continental	149	52.8	827
Open ocean	332	18.7	0.45
Upswelling zones	0.4	0.1	0.004
Continental shelf	26.6	4.3	0.12
Algal bed and reef	0.6	0.7	0.54
Estuaries	1.4	1.0	0.63
Total marine	361	24.8	1.74
Full total	510	77.6	828

Source: Woodwell (1978).

plant respiration is balanced by carbon dioxide uptake for photosynthesis in existing and newly established vegetation. Lugo and Brown (1986) argue that frequent and minor disturbances to forests are common and that forests slowly sequester atmospheric carbon. Under conditions of widespread and rapid forest destruction, however, equilibrium conditions, or near equilibrium conditions may be lost.

Table 3-2. Estimates of Airborne Fraction of Total Anthropogenic Induced Emmissions of CO_2 into the Atmosphere During the Periods 1860–1980 and 1958–1982.

	Increase in the atmosphere 10^{15} g C	Fossil fuel emissions 10^{15} g C	Terrestrial biosphere emissions 10^{15} g C	Total emissions 10^{15} g C	Airborne fraction
1860–1980					
Maximum	172	180	200	380	0.67
Maximum	129	155	100	255	0.34
1958–1982					
Maximum	56	103	65	168	0.44
Maximum	54	89	35	124	0.32

Source: Bolin (1986).

3. Relationship of Forests to Increasing Atmospheric CO_2 Levels

In 1954, G. Evelyn Hutchinson, presently professor emeritus of zoology at Yale University, suggested in a chapter in *The Earth as a Planet*, edited by G. P. Kuiper that the increasing content of carbon dioxide of the atmosphere is due to the destruction of forests as well as the combustion of fossil fuels (Woodwell, 1978). George M. Woodwell of the Ecosystems Center, Woods Hole, Massachusetts, by providing and summarizing recent evidence, has actively advanced the hypothesis that forest destruction contributes to increasing atmospheric carbon dioxide levels. In addition to forest destruction, conversion of original-natural forests to younger-managed forests, forest fires, and soil management practices may significantly contribute carbon dioxide to the atmosphere. The latter may be the same order of magnitude as the input of fossil fuel carbon dioxide (Bolin et al., 1977, Bolin, 1986) (Table 3–2).

One third of the United States land area or 300 X 10^6 ha (741 million acres) consisted of forest ecosystems in 1979. This area is continually changing in response to clearing of forests to convert to agricultural activity, abandonment of agricultural land to forest development, and loss of forest systems to urban, industrial, or transportation use (Spurr and Vaux, 1976). It has been estimated that over the next 50 years the United States will witness the following net loss of forested land: 8 X 10^6 ha (20 million acres) to crop and pasture use and 3.2 X 10^6 ha (8 million acres) to development pressure (Spurr and Vaux, 1976). Adams et al. (1977) point out, however that the United States forest loss is quite insignificant when compared to the forest destruction in some of the developing countries such as Brazil where the minimum net loss of wood is estimated to be 3.5 tons per capita per year. In Rondonia, the Brazilian Amazon Basin state with the highest rate of forest clearing, LANDSAT inventories suggest that between 3.9 X 10^{12} and 5.5 X 10^{12} g carbon were released in 1981 alone

Table 3-3. Net Average Annual Input of Carbon (as Carbon Dioxide) into the Atmosphere and Accumulated Input since the Early Nineteenth Century due to Human Perturbation of Terrestrial Ecosystems.

	Input (10^9 tons)	
Source	Present average annual	Accumulated
Reduction of forests		
Developed countries	0 ± 0.1	
Developing countries		45 ± 15
Forestation	-0.3 ± 0.1	
Deforestation	0.8 ± 0.4	
Use of fuel wood	0.3 ± 0.2	
Changes of organic matter in soil	0.3 ± 0.2	24 ± 15
Total	1.0 ± 0.6	70 ± 30

Source: Bolin (1977).

(Woodwell et al., 1986). Global forest destruction may be underestimated by overreliance on data from developed countries relative to the developing countries, where forest loss to agriculture and firewood is particularly high (Adams et al., 1977).

Bolin et al. (1977) have concluded that changes in forest area in developed countries do not constitute a significant net flux of carbon to or from the atmosphere. For the developing countries, however, Bolin estimated that approximately 0.12×10^6 km^2 of natural forest is cleared and burned annually and that this loss has significantly added to the atmospheric carbon dioxide load during this century, as indicated in Table 3–3.

Comparison of the ratios of carbon isotopes ^{12}C, ^{13}C, and ^{14}C has been useful to those studying the flux and sources of carbon dioxide in the atmosphere. The procedure is based on the fact that the ratios differ for the major sources of carbon. Fossil fuels do not contain ^{14}C and also have elevated ^{12}C compared to the atmosphere. Burning of fossil fuels liberates carbon deficient in ^{14}C and thus dilutes the ^{14}C of the atmosphere, the so-called Suess effect (Woodwell et al., 1978). Anthropogenic release of CO_2 from both fossil fuels and the biota reduces ^{13}C levels in the atmosphere, but ^{14}C is reduced only by the fossil fuel source. Using the ^{13}C content of woody tissues of trees, Stuiver (1978) has suggested that between 1850 and 1950 the biospheric release of carbon dioxide to the atmosphere was 120×10^9 tons of carbon while the release from fossil fuels combustion was only half this amount (60×10^9 tons). Stuiver cited forest cutting in the Lakes states starting in 1870 and Pacific Northwest logging during the early 1900s as examples of deforestation contributing to the biospheric carbon dioxide release.

It has been suggested that increased atmospheric carbon dioxide may stimulate vegetative growth through enhancement of photosynthesis and that the biota may serve as a sink for carbon dioxide. Woodwell et al. (1978) considered this

Table 3-4. Estimated Net Release of Carbon from Major Global Terrestrial Plant Communities.

Plant community	Net release of carbon ($\times 10^{15}$ g)	% of total
Tropical forests	3.5	44
Temperate zone forests	1.4	18
Boreal forest	0.8	10
Other vegetation (including agriculture)	0.2	3
Total vegetation	5.9	75
Detritus and humus	2.0	25
Total land	7.9	100

Source: Woodwell et al. (1978).

possibility and concluded that the available evidence does not support the notion that the biota is or has been an appreciable sink for carbon. These authors supported the hypothesis that manipulation of terrestrial vegetation is a source rather than a sink for carbon dioxide and that the most probable range for global release from the biota is 4 to 8 $\times 10^{15}$ g of carbon annually. Contributions by various forest types are presented in Table 3–4.

Wong (1978) has estimated the atmospheric input of carbon dioxide from burning wood. He estimated the input from forest fires in boreal and temperate regions resulting from both natural and human-related fires and fires in tropical regions caused by shifting cultivation. As indicated in Table 3–5, these sources

Table 3-5. Area Burned and Net Input of Carbon Dioxide into the Atmosphere from Nonfossil Wood Burning.

Ecosystem	Area burned (10^{10} m^2 yr^{-1})	Net carbon input (10^{15} g yr^{-1})
Boreal forest	1.2	
Boreal nonwooded land	1.3	0
Temperate forest	4.6	
Temperate nonwooded land	4.6	0
Tropical forest: new clearing		
FAO estimate	24.0	1.5
Rural population increase estimate	7.5	(0.7-2.2)
Tropical shifting cultivation: existing	300	0

Source: Wong (1978).

carbon dioxide and deforestation partly compensate one another, and in any resulted in a net input of carbon of approximately 1.5 X 10^{15} g carbon per year and this was solely from new forest clearing.

An examination of global models for the natural carbon dioxide cycle has led Siegenthaler and Oeschger (1978) to conclude that we are deficient in our understanding of the role of the biosphere in the carbon cycle, that fertilization by carbon dioxide and deforestation partly compensate one another, and in any case, that a sink for the large carbon dioxide release proposed by Woodwell and others cannot be found. Broecker et al. (1979) have employed several models to estimate the global carbon budget and have concluded that changes in forest biomass, relative to fossil fuel combustion, have not contributed significant carbon dioxide to the atmosphere.

Clearly expert opinion on the role of forest systems in the global carbon cycle is divided. Comprehensive discussions of key issues in this cycle are provided in Brown (1981) and Trabalka (1985). The latter publication draws the following key conclusions:

1. Uncertainties concerning the adequacy of treatment of the oceans in present models are responsible for a large portion of the uncertainties in the prediction of atmospheric carbon dioxide increases.
2. There is doubt concerning whether the present terrestrial carbon dioxide release estimates represent the net flux of carbon between terrestrial systems and the atmosphere, or whether there are other sources or sinks from changes in currently undisturbed areas, e.g., from climate or carbon dioxide fertilization effects.
3. Model improvement requires more basic understanding of fundamental ecosystem patterns and processes, e.g., natural controls on terrestrial carbon storage and the role of subfossil carbon pools, in addition to improved appreciation of the relationships between ocean, climate, and carbon dioxide.
4. The growth rate of global future carbon dioxide emissions is expected to average approximately 1% per year over the next 75 years.

With specific regard to the role of terrestrial ecosystems in the global carbon cycle, Houghton et al. (1985) have concluded as follows:

1. The natural exchange of carbon between land ecosystems and the atmosphere is within the range of 60–120 X 10^{15} g carbon per year.
2. Human activities have impacted this natural exchange and are currently resulting in a net release into the atmosphere. In 1980 this net release approximated 0.6–2.6 X 10^{15} g carbon.
3. The bulk of the net release is from forest perturbations in tropical latitudes.

The extraordinary rates of tropical deforestation (Table 3–6) provide strong incentive for major research efforts to resolve uncertainties of the global carbon cycle, particularly the role of forest loss in adding to the carbon dioxide loading of the atmosphere.

Table 3-6. Rates of tropical deforestation.

10^6 ha yr^{-1}				
Total Tropics	Tropical America	Tropical Africa	Tropical Asia	Source
24	7.5	8	8.5	FAO (1966) (from Wong 1978)
30				Brunig (1977)
12	6	2	4	Bolin (1977)
7.5	1.3	1.2	5.0	Wong (1978)
9.3–16.6 (5.0–9.1)[a]				Seiler and Crutzen (1980)
11.0–18.5 (6.7–11.0)				Detwiler et al. (1985) (from Seiler and Crutzen 1980)
3.22 (2.84)	2.21	0.07	0.795 (0.465)	Houghton et al. (1983) (from FAO Production Yearbooks)
17.1 (15.15)	5.40	2.85	8.7 (6.8)	(from Meyers 1980)
8.65 (5.76)	2.36 (2.36)	2.86 (1.91)	3.286 (1.39)	(based on population)
(19.95)	(10.05)	(1.73)	(8.18)	Barney (1980)
11.07 (7.27)	5.39 (4.12)	3.68 (1.33)	2.00 (1.82)	FAO (1981)
9.41	5.43	3.09	0.90	FAO Production Yearbook (FAO 1983)
(7.5)[b]	(2.5)[b]	(1.4)[b]	(3.1)[b]	Melillo et al. (1985) (from Meyers 1980)

[a]Values in parentheses refer to closed forests only; open forests or woodlands have been omitted from these values.
[b]Values are for closed, broad-leaved forests only.
Source: Houghton et al. (1985).

4. Carbon Monoxide

Unlike carbon dioxide, the concentration of carbon monoxide appears to be stable or is only slightly increasing in "clean" atmospheres remote from excessive local input of carbon oxides. Carbon monoxide is input to the atmosphere in very approximately equal amounts from anthropogenic sources and from the biota on a global basis (Table 3–7). If the inputs estimated in Table 3–7 approximate the actual fluxes in nature, the role of forest ecosystems in the input of carbon monoxide to the atmosphere may be significant. Since the global atmospheric concentration of carbon monoxide is not increasing dramatically, despite increasing combustion of fossil fuels, it must be assumed that an effective global carbon monoxide sink is operating. It will be suggested in

Table 3-7. Primary Sources of Atmospheric Carbon Monoxide.

	CO (10^6 tons yr^{-1})			
	Total		Northern Hemisphere	
Source	Min.	Max.	Min.	Max.
Methane oxidation	400	4000	200	2000
Anthropogenic	600	1000	540	900
Biota				
Oceans	100	220	40	90
Terrestrial plants	20	200	14	140
Chlorophyll degradation	300	700	200	500

Source: Nozhevnikova and Yurganov (1978).

Chapter 5 that the soils of forest ecosystems may play a particularly important role in this sink function.

B. Sulfur Pollutants

The atmospheric sulfur contaminants of primary interest to those concerned with vegetative and human health effects and environmental quality are currently sulfur dioxide and sulfates. The latter include sulfuric acid, metallic sulfates, and ammonium sulfate. Because of the role of sulfates in precipitation acidity and the increasing geographic area subject to precipitation of lowered pH, considerable interest is focused on the sulfate group. Sulfur containing air contaminants exert a major impact on agricultural and forest ecosystems.

1. Sulfur Cycle

In recent reviews, Graedel (1979) and Ivanov and Freney (1983) have emphasized the uncertainties associated with quantifying and balancing the global sulfur cycle and with identifying the primary means of atmospheric sulfate formation.

Almost without exception, the traditional models of the sulfur cycle have indicated that approximately 50% of the sulfur in the atmosphere results from biological transformations of sulfur in soil and water ecosystems. The microbial activity within these natural ecosystems is presumed to volatilize sulfur in the form of hydrogen sulfide. Additional natural sources include volcanic emissions and sea spray input. The primary elements of the sulfur cycle are presented in Figure 3–2.

Much evidence exists to suggest that microbes produce hydrogen sulfide by two principal routes: sulfate reduction and organic matter decomposition. Sulfate reducers, *Desulfovibrio* and related bacteria, proliferate in swamp, mud, and poorly drained soils and employ sulfate as a terminal electron acceptor. An extremely large and diversified group of microorganisms, including aerobes, anaerobes, thermophiles, psychrophiles, bacteria, actinomycetes, and fungi decompose sulfur containing organic compounds and release hydrogen sulfide

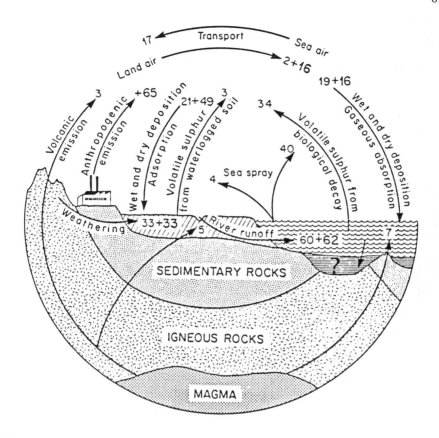

Figure 3-2. The global sulfur cycle. Small type represents natural fluxes. Bold type reflects human impacts on the sulfur cycle. Units are Tg (million metric tons) yr^{-1}. Source: Granat et al. (1976).

(Alexander, 1974; Laishley and Bryant, 1987). While there is little doubt that soil microbes can release hydrogen sulfide by reduction of sulfate and decomposition of organic compounds containing sulfur, there does not appear to be reliable, direct evidence that significant amounts of hydrogen sulfide are released from soils in natural ecosystems (Bremner and Steele, 1978). Microbially generated hydrogen sulfide may be rapidly converted in soil to metallic sulfides, for example, iron sulfide (Ayotade, 1977; Kittrick, 1976). When soils with high levels of sulfate and decomposable organic matter become poorly aerated (poorly drained) and are deficient in cations required to precipitate hydrogen sulfide, the latter gas may be released to the atmosphere. Some evidence exists for direct release of hydrogen sulfide by plants to the atmosphere. Spáleny (1977) recorded that Norway spruce seedlings, provided with 5 g potassium sulfate per day for 7

Table 3-8. Estimates of Global Biogenic Production of Hydrogen Sulfide by Terrestrial Systems.

Reference	Sulfur $(10^6$ tons $yr^{-1})$
Junge (1963)	70
Eriksson (1963)	110
Robinson and Robbins (1968)	68
Friend (1973)	58
Kellogg et al. (1972)	90

Table 3-9. Global Transfer Rates of Sulfur to the Atmosphere.

Sources	Sulfur $(10^6$ tons $yr^{-1})$		
	Terrestrial	Marine	Total
Human activity	50		50
Sea spray sulfate		43	43
Biogenic H_2S	90		90
Total			183

Source: Hill (1973).

days, released 2.2 µg of hydrogen sulfide hr^{-1} kg^{-1} dry weight of needles when exposed to light. No release was recorded in the dark.

The actual quantity of hydrogen sulfide produced by natural systems, while assumed by the traditional sulfur cycle to be substantial, is not directly measured and has been only crudely calculated by balancing the global sulfur cycle to range from 58 to 110 X 10^6 tons of sulfur annually (Table 3–8).

The atmospheric chemistry of the traditional sulfur cycle involves the oxidation of hydrogen sulfide to sulfur dioxide and the oxidation of sulfur dioxide to sulfates. Hydrogen sulfide may be oxidized by reactions involving atomic oxygen, molecular oxygen, and ozone (Kellogg et al., 1972). Sulfur dioxide may be oxidized by reaction with OH, HO^2, and RO^2. The reaction with OH forms HSO_3 and is several orders of magnitude faster than the others (Wolff, 1979). Sulfates are also introduced directly to the atmosphere as sea salt particles from sea spray, as indicated in Table 3–9.

2. New Sulfur Cycle Hypothesis

An alternative and new hypothesis for sulfur input to the atmosphere from natural ecosystems, however, proposes that most of the sulfur volatilized from soils through microbial activity (Zinder and Brock, 1978) is in the form of organic compounds such as carbonyl sulfide, dimethyl sulfide, dimethyl disulfide, carbon disulfide, and methyl mercaptan (Bremmer, 1977; Adams et al., 1979).

The flux of sulfur gases from a variety of soils was given by Adams et al. (1979) to average 72 g sulfur m^{-2} yr^{-1}, with a range of 0.002 to 152 g sulfur m^{-2} yr^{-1}. Most of the soils tested by Adams were tidal or poorly drained and, unfortunately, none was a forest soil. In addition to sulfur dioxide and hydrogen sulfide, recent advances in sulfur gas analysis demonstrate that carbonyl sulfide, carbon disulfide, dimethyl sulfide, and sulfur fluoride are components of tropospheric air (Bremmer and Steele, 1978). Aneja et al. (1979) have provided emission rates of carbon disulfide and carbonyl sulfide from North Carolina salt marshes and have estimated global marsh emissions to approximate 0.07 Tg S yr^{-1}.

Lovelock et al. (1972) found that living foliage of cotton, oak, spruce, and pine trees released from 2 to 43 X 10^{12} g dimethyl sulfide per g (dry wt) hr^{-1} and that decaying leaves released 10–100 times this amount. Soils were found to have dimethyl sulfide emission rates ranging from 21 to 84 X 10^{12} g g^{-1}. Hitchock (1977) estimated the global production of dimethyl sulfide for fresh leaves, senescent leaves and soils to be 0.01, 0.53, and 1.5–4.9 Tg S yr^{-1}, respectively (Table 3–10). Clearly soil is the primary source of terrestrial dimethyl sulfide. Banwart and Bremner (1976) and Bremner and Steele (1978) have thoroughly reviewed sulfur input to the atmosphere and conclude that most of the dimethyl sulfide originates via microbial degradation of methionine or dimethyl-b-propiothetin, most of the carbon disulfide through microbial degradation of cysteine or cystine, and at least some of the carbonyl sulfide by microbial decomposition of thiocyanates and isothiocyanates in plant materials. Dimethyl sulfide is presumed to be directly oxidized in the atmosphere to sulfate without the formation of a sulfur dioxide intermediate (Granat et al., 1976).

It is important to realize that this new hypothesis also has its critics (Maroulis and Bandy, 1977) and that much uncertainty still characterizes our appreciation of the sulfur cycle. Vegetation also serves as a sink for organic sulfur compounds in the atmosphere. In fact, the major global sink for carbonyl sulfide may be plants, accounting for removal of 2–5 Tg yr^{-1} (Brown and Bell, 1986).

3. Input of Sulfur to the Atmosphere: Anthropogenic Sources Relative to Natural, Particularly Forest, Sources

As indicated in Table 3-10, these data suggest a relatively minor input directly from vegetation and a relatively major release from the soil. The relative efficiency of forest soils compared to other soils for dimethyl sulfide release is not clear. It is clear, however, that the maximum estimated dimethyl sulfide release (5.5 X 10^{12} g S yr^{-1}) is modest compared with the anthropogenic release of approximately 65 X 10^{12} g S yr^{-1} (2% sulfate, 98% sulfur dioxide) (Granat et al., 1976). The release of dimethyl sulfide, however, may represent less than 10% of the total biogenic emission of sulfur (Bremner and Steele, 1978).

It is also possible that all models of the sulfur cycle have considerably overestimated the contribution of the biota. Granat et al. (1976) have suggested

Table 3-10. Global Emissions of Dimethyl Sulfide from Natural Sources.

Source	Sulfur (10^{12} g yr^{-1})
Marine algae	0.05
Fresh leaves	0.01
Senescent leaves	0.53
Soils	1.5-4.9
Total	2.1-5.5

Source: Granat et al. (1976).

a total flux of reduced sulfur to the atmosphere of approximately 35–37 X 10 ^{12} g yr^1, less than one-half the estimate of several other authors.

In the Northern Hemisphere, where anthropogenically input sulfur has a residence time of very approximately 1–2 days, human sulfur inputs dominate single source natural fluxes. The current contribution to the sulfur burden in submicron particles in the troposphere of the Northern Hemisphere is 0.17 X 10^{12} g natural relative to 0.23 X 10^{12} g human (Granat et al., 1976).

Ryaboshapko (1983) has provided a mass-balance sulfur budget for a temperate-zone atmosphere contaminated by regional scale sulfur pollution. By his estimation, anthropogenic sulfur inputs represent approximately 42 % of the sulfur content of the atmosphere (Figure 3–3).

The relative importance of human versus natural sulfur sources in a given atmosphere of the developed nations is strongly influenced by distance from major anthropogenic sulfur sources. Regions hundreds of kilometers downwind of major industrial locations are dominated by human-source sulfur. Other locations, however, experience sulfur deposition resulting from inputs from a variety of natural sources, including forest systems.

C. Nitrogen Pollutants

The primary atmospheric contaminants containing nitrogen include ammonia, nitrogen oxides, and nitrates. Direct injury to humans or vegetation from gaseous ammonia is very infrequent and has usually only occurred in localized regions from accidental spills, rather than continuous emissions to the atmosphere. Ammonia is important because of its reactions to form aerosols, especially ammonium sulfate. Nitrogen forms seven oxides that reflect each of its recognized oxidation states. Of these oxides, only nitrous oxide (N_2O), nitric oxide (NO), and nitrogen dioxide (NO_2) exist at measurable concentrations in the atmosphere, and only nitrogen oxide and nitrogen dioxide are important air pollutants (Spedding, 1974). Nitrogen dioxide may directly affect vegetation in regions of high local release. Indirect impact on vegetation is much more important, however, because of the key role of oxides of nitrogen in forming generally phytotoxic oxidants, such as ozone and peroxyacetylnitrates. Since the primary sources of hydrogen ions that cause the phenomenon of acid precipitation are the strong mineral acids, sulfuric and nitric, nitrate is also an important air contaminant (Galloway et al., 1976).

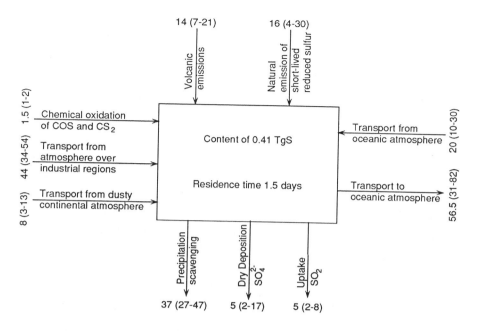

Figure 3-3. Sulfur budget for a temperate zone atmosphere contaminated by regional scale sulfur pollution. Units are Tg S yr⁻¹.
Source: Ryaboshapko (1983).

1. Nitrogen Cycle

The global nitrogen cycle is extremely complex because of the large number and variable nature of the nitrogenous compounds that are involved, the myriad of processes operative, and the large regional and local deviations from generalized patterns. The biota does not use nitrogen (N_2) gas, approximately 80% of the troposphere, directly but rather uses it only in a form combined with hydrogen, oxygen, and other elements. The principal mechanism of nitrogen fixation involves nitrogen-fixing bacteria and fungi that are either free living in the soil or associated in symbiotic relationships with plants. These microbes extract nitrogen from the air and reduce it via enzyme action to ammonia (NH_3). Lighting can combine nitrogen in the air with molecular oxygen (O_2) to form nitrogen oxides (NO). These oxides dissolve in water, forming nitric acid (HNO_3). Nitrifying microorganisms convert soil ammonia to nitrite (NO_2^-) and then to nitrate (NO_3^-). Vegetation absorbs both the reduced and oxidized forms of nitrogen from the soil solution. Decomposer microbes convert bound organic nitrogen back to ammonia. Of the latter, some is absorbed by plants and some is again converted to nitrate by nitrifying bacteria. Denitrification eventually returns nitrogen to the atmosphere. In the absence of oxygen in the soil, certain

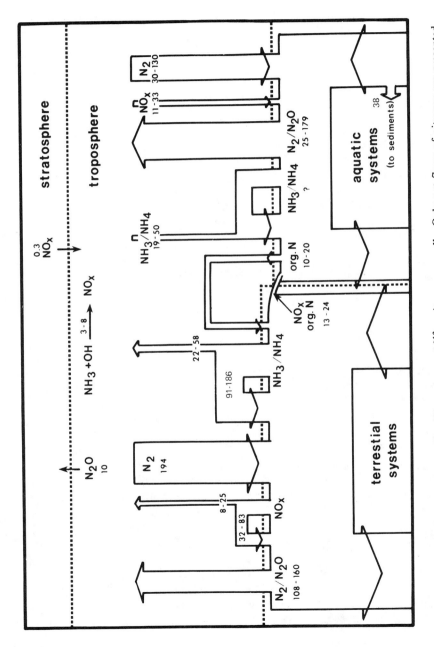

Figure 3-4. The global nitrogen cycle. The units are 10^{12} g nitrogen annually. Only net flows of nitrogen are presented.
Source: Söderlund and Svensson (1976).

microbes convert nitrate to nitrite through enzymatic activity. Other denitrifying bacteria reduce nitrite to nitric oxide (NO), then to nitrous oxide (N_2O), and ultimately to gaseous nitrogen (N_2). The latter two gases diffuse through the soil back to the troposphere. Human inputs of nitrogen to the troposphere are dominated by a wide variety of combustion processes that cause nitrogen (N_2) and oxygen (O_2) to combine and form nitrogen oxides (NO_x).

Söderlund and Svensson (1976) have thoroughly reviewed the various nitrogen cycles and have presented their own (Figure 3–4), which substantially differs in selected aspects from those of earlier efforts. Their model presents a net flow of ammonia and oxides of nitrogen from terrestrial to oceanic systems through the atmosphere. They conclude that natural fluxes of nitrogen oxides to the atmosphere may be smaller than generally suggested. This has the effect of increasing the relative importance of anthropogenic flows to the atmospheric compartment. These latter flows, estimated below, are the most accurately known fluxes of the nitrogen cycle. The greatest uncertainty of the cycle is associated with the quantification of denitrification processes that release molecular nitrogen and nitrous oxide into the atmosphere.

2. Sources of Ammonia

The sources of ammonia input to the atmosphere are varied and include minor release from various industries, coal combustion, fertilizer breakdown, volatilization from wild and domestic animal and human excreta, and the decomposition of organic matter in the soil. With the exception of the industrial release and coal combustion, the above sources are biological processes mediated by a variety of soil microorganisms. Excreta from animals in grasslands may result in ammonia volatilization up to 100 kg N ha^{-1} yr^{-1} depending on grazer density (Bowden, 1986). Alexander (1974) has listed alkaline pH, warm temperatures, high rates of evaporation, and low cation exchange capacity of soils as all favoring microbial ammonia volatilization. Low animal densities, generally low fertilizer application, and typically low pH soils all mitigate against forest soil release of ammonia.

3. Nitrogen Oxides

The high temperature combustion of fossil fuels for transportation, energy generation, and the manufacture of petroleum products all contribute to the input of nitric oxide and nitrogen dioxide to the atmosphere. Söderlund and Svensson (1976) have presented an estimate of 19 X 10^{12} g of nitrogen as the annual anthropogenic input via nitrogen oxide release to the atmosphere. The use of higher combustion temperatures, probable in the future, will increase the atmospheric input of nitrogen oxides from fossil-fuel burning. A less quantifiable, but probably larger amount of nitrogen oxides (exclusive of nitrous oxide) are released via several natural mechanisms including fixation by lightning (Junge, 1958; Ferguson and Libby, 1971), inflow from the stratosphere (Söderlund and

Svensson, 1976), chemical conversion from ammonia in the troposphere (Crutzen, 1974; McConnell, 1973), and loss of gaseous nitric oxide from soils (Robinson and Robbins, 1975).

The global input of nitrogen to the atmosphere in the form of nitrogen oxides from natural sources may range from approximately equal to 4.5 (Söderlund and Svensson, 1976) to 15 (Rasmussen et al., 1975) times greater than the human input.

4. Role of Terrestrial Ecosystems, Particularly Forests, in the Nitrogen Cycle

The single, most important fixation of nitrogen on a global basis occurs in terrestrial ecosystems (Table 3–11). Of the total nitrogen fixed by these ecosystems, approximately equal amounts are fixed by agricultural, grassland, and forest systems (Table 3–12). Tables 3–11 and 3–12 establish the relative importance of forest ecosystems in the global fixation of nitrogen. In light of this importance, what is the significance of forests in the release of ammonia, nitrogen oxides, and nitrates to the atmosphere?

5. Release of Ammonia

Living plants are generally not considered important sources of ammonia to the atmosphere. Farquhar et al. (1979), however, have presented evidence that corn plants release ammonia in normal air as they senesce. The investigators indicated that their observed rate of release was equivalent to 7 g of nitrogen ha^{-1} day $^{-1}$ with a leaf area index of 1. Comparable release of ammonia from tree leaves is undetermined.

Ammonia is generated during the mineralization of humus by a wide variety of bacteria and fungi. The acid pH, cool temperatures, low rates of evaporation, and high cation exchange capacity characteristic of many temperate forest soils would not favor the volatilization of ammonia. Kim (1973), however, monitored the ammonia released from a Korean pine forest, oak forest, and grassland from May through July. His results showed that an average of 3.4, 2.6, and 1.8 kg ammonia ha^{-1} week^{-1} were evolved from the pine and oak forest soil and in the grassland, respectively.

The use of fertilizers containing organic nitrogen compounds can result in substantial volatilization from treated soils. This is particularly true for urea, a common fertilizer, as this compound is hydrolyzed rapidly by cosmopolitan and common urease containing heterotrophic microbes (Alexander, 1974). Urease activity has been reported for numerous forest soils (Wollum and Davey, 1975). The application of urea to forested ecosystems is receiving active research attention. In the United States, experimental applications have been made to Douglas fir (Miller and Reukema, 1974, 1977), western hemlock (DeBell, 1975; Webster et al., 1976) and West Virginia hardwoods (Aubertin et al., 1973) and in Canada to jack pine (Armson, 1972; Morrison et al., 1976). The rate of urea application

Table 3-11. Global Fixation of Nitrogen.

Process	Nitrogen (10^{12} g yr^{-1})
Biotic	
Terrestrial	139
Aquatic, pelagic	20-120
Abiotic	
Industrial	36
Combustion	19
Total	214-314

Source: Söderlund and Svensson (1976).

Table 3-12. Global Nitrogen Fixation by Terrestrial Ecosystems.

Ecosystem	Nitrogen (10^{12} g yr^{-1})
Agricultural	44
Grassland	45
Forest	40
Other	10
Total	139

Source: Burns and Hardy (1975).

utilized in these tests generally ranged from 113 to 555 kg ha^{-1} (100–500 lbs acre^{-1}). The quantity of ammonia lost by volatilization may be substantial and has been estimated to range from 18% to 75% of applied nitrogen (Volk, 1959, 1970). Many of the studies supporting high percentage loss, however, have been performed in laboratory environments. Hargrove and Kissel (1979) have compared ammonia volatilization from surface urea applications in the laboratory and in the field, and found the latter small relative to the former.

While the use of fertilizer in forest management practice is miniscule compared to agricultural practice, it is important to realize that the interest in fertilizer use on forests is growing and that the forest systems receiving the greatest research interest are expansive. Large scale fertilization programs currently exist in Sweden and Finland (Armson et al., 1975).

6. Nitrogen Oxides and Nitrates

The abiotic and biotic processes of denitrification globally release large quantities of nitrogen oxides to the atmosphere. Söderlund and Svensson (1976) have indicated that terrestrial ecosystems may release 1 to 14 X 10^{12} g nitrogen oxide-nitrogen annually. Focht and Verstraete (1977) have estimated that terrestrial denitrification may generate an annual surface flux of approximately 3 kg nitrogen ha^{-1}. The role of terrestrial ecosystems in releasing nitrogen via denitrification is critically important due to the fundamental significance of this process to the global nitrogen cycle, but also because nitrous oxide may play a significant role in the dynamics of stratospheric ozone integrity and in global warming (both topics discussed more fully in Chapter 19).

Denitrification requires anaerobic soil conditions and is stimulated by large amounts of organic matter. The highest rates of denitrification generally occur during wet periods, typically during spring or early summer.

Biological denitrification is a respiratory function that employs nitrate as a terminal electron acceptor. Subsequent nitrogen reductions may release nitrogen dioxide, nitrous oxide, nitric oxide, and dinitrogen. Denitrification is carried out by a diverse group of microbes, including bacteria and fungi, autotrophs and heterotrophs, and aerobes and anaerobes, in a variety of habitats, including poorly drained and well-drained and acid and alkaline soils (Delwiche and Bryan, 1976; Wollum and Davey, 1975; Bremner and Blackmer, 1978). As in the case of ammonia release, fertilization may stimulate denitrification (Blackmer and Bremner, 1979; Hutchinson and Mosier, 1979). In Kim's (1973) study of Korean forest soils, the release of nitrogen dioxide averaged 0.21, 0.12, and 0.19 kg ha^{-1} week^{-1} in a pine forest, oak forest, and grassland, respectively.

Prevailing wisdom has suggested that well-aerated, acid soils of forest regions add very little nitrous oxide to global atmospheres. Recent evidence, however, is not fully consistent with this view (Wollum and Davey, 1975). W.C. Caskey has estimated total nitrogen loss (nitrous oxide plus nitrogen) from two upland North Carolina forests, one undisturbed and one recently harvested (clear cut), to be 9.6 and 25 kg ha^{-1} yr^{-1}, respectively (Iker, 1982). In studies of the northern hardwood forest in central New Hampshire, Bowden and Bormann (1986) have reported that whole-tree harvesting increased the concentration of nitrous oxide in soil water by two orders of magnitude over the concentration anticipated in equilibrium with the atmosphere. In undisturbed stands, the nitrous oxide concentrations were close to expected theoretical values. Presumably, harvesting results in soil warming and increased organic matter, which stimulate decomposition and provide increased soil nitrogen, resulting in enhancement of denitrification.

The most comprehensive review of terrestrial ecosystem nitrogen gas contributions to the global troposphere has been provided by Bowden (1986) (Table 3–13). By this inventory, temperate and tropical forests are suggested to be generally less important contributors than other ecosystems but to have larger than previously estimated roles in nitrogen (N_2) and nitric oxide (N_2O) releases.

D. Summary

Carbon dioxide is increasing in the atmosphere at a rate of approximately 1.5 ppm (2.7 X 10^3 µg m^{-3}) per year. Currently the 700 billion tons of carbon in the atmosphere in the form of carbon dioxide are supplemented with an additional 2.3 billion tons annually, resulting in a 3% increase every decade. Anthropogenic combustion of fossil fuels contributes five billion tons of carbon to the atmosphere annually (Ember, 1978). A comparable or greater amount of carbon may be released to the atmosphere via forest clearance and soil losses.

Table 3-13. Gaseous nitrogen emmissions from undisturbed terrestrial ecosystems. Area units, 10^9 ha; emission units 10^{12} g N yr^{-1}. Symbol "tr" represents $< 10^{12}$ g N yr^{-1}; ND represents no data.

Ecosystem	Area	Global Emission Rate				
		NH$_3$	(N$_2$O + N$_2$)	N$_2$O	NO$_x$	Total[a]
Temperate hardwood	1.55	tr	tr–15.5	tr–1.5	1.5	1.5–15.5
Temperate conifer	0.50	tr	tr	tr–1.0	1.0	1.0–2.0
Tropical forests	2.45	ND	7.1	7.1	ND	7.4–9.8
Prairie/grassland	2.40	0–24	tr	tr–4.8	3–24	4.8–48.0
Arid lands/desert	4.20	4.2	0–80	ND	ND	4.2–84.0
Tundra/boreal	2.00	0	tr	tr	ND	tr
Wetlands/marsh	0.20	ND	6–130	tr–20	ND	6.0–130
Total[a]	13.30[b]	4–28	13–233	7–16	5–26	20–280

[a]Column and row totals may not sum due to rounding errors and differences in source data.
[b]Does not include cultivated land (1.4 x 10^9 ha) or lakes and streams (0.12 x 10^9 ha).
Source: Bowden (1986).

Approximately 75% of carbon introduced to the atmosphere from all sources is removed by effective sink mechanisms. The most important sink is presumed to be the oceans with some potential contribution from new vegetation. The imperfections in our understanding of global carbon pools and cycles and carbon models preclude conclusive assessment of the precise role of forest ecosystems in the increased loading of carbon dioxide in the atmosphere. If forests are ultimately judged to have an important role in the carbon dioxide loading of the atmosphere, it is probable that manipulations of tropical forest ecosystems will prove to be much more important than changes in temperate or boreal systems.

Our appreciation of the sulfur cycle is undergoing major change as new developments in sulfur gas analysis reveal the increasing importance or organic sulfur volatiles from natural systems. Sulfur may be input to the atmosphere in amounts of approximately 50–70 X 10^{12} g yr^{-1} from both human activities and natural systems. It is probable that soil metabolism generates most of the natural contribution and that much of it is in the form of organic sulfur gases, but it is not possible to specifically rank forest soils relative to other soils in this production.

The important nitrogenous gases of air quality interest, ammonia and oxides of nitrogen, are released in important amounts from terrestrial soils. Forest soils do not appear unique nor outstanding in the production of these gases. Recent evidence, however, suggests greater release of nitrogen and nitrous oxide from forest soils, especially those disturbed by harvesting, than previously estimated.The size of forest ecosystems may make their release comparable with agricultural systems. Adoption of broad scale forest fertilization programs would increase ammonia release and enhance denitrification.

Forests clearly have central significance in the global carbon cycle. Forest roles in the sulfur and nitrogen cycles are less dramatic, but important. The vitality and health of worldwide forest ecosystems will significantly influence global biogeochemical cycling.

References

Adams, J.A.S., M.S.M. Mantovani, and L.L. Lundell. 1977. Wood versus fossil fuel as a source of excess carbon dioxide in the atmosphere: A preliminary report. Science 196: 54-56.

Adams, D.F., S.O. Farwell, M.R. Pack, and W.L. Bamesberger. 1979. Preliminary measurements of biogenic sulfur-containing gas emissions from soils. Air Pollu. Control Assoc. 29: 380-383.

Alexander, M. 1974. Microbial formation of environmental pollutants. Adv. Appl. Microbiol. 18: 1-73.

Aneja, V.P., J.H. Oveton, L.T. Cupitt, J.L. Durham, and W.E. Wilson. 1979. Carbon disulphide and carbonyl sulphide from biogenic resources and their contributions to the global sulphur cycle. Nature 282: 493-496.

Armson, K.A. 1972. Fertilizer distribution and sampling techniques in the aerial fertilization of forests. Tech. Report No. 11, University of Toronto, Ontario, Canada, 27 pp.

Armson, K.A., H.H. Krause, and G.F. Weetman, 1975. Fertilization response in the northern coniferous forest. In B. Bernier and C.H. Winget, eds., Forest Soils and Forest Land Management, Proc. Fourth North Amer. Forest Soils Conf., Laval Univ., Quebec. Les Presses de l'Université Laval, Quebec, Canada, pp. 449-466.

Aubertin, G.M., D.E. Smith, and J.H. Patric. 1973. Quantity and quality of streamflow after urea fertilization on a forested watershed: First-year results. In Forest Fertilization Symp. Proc. U.S.D.A. Forest Service, Genl. Tech. Report NE-3, Upper Darby, PA, pp. 88-100.

Ayotade, K.A. 1977. Kinetics and reactions to hydrogen sulfide in solution of flooded rice soils. Plant Soil 46: 381-389.

Banwart, W.L. and J.M. Bremner. 1976. Evolution of volatile sulfur compounds from soils treated with sulfur containing organic materials. Soil Biol. Biochem. 8: 439-443.

Blackmer, A.M. and J.M. Bremner. 1979. Stimulatory effect of nitrate on reduction of N_2O to N_2 by soil microorganisms. Soil Biol. Biochem. 11: 313-315.

Bolin B. 1977. Changes of land biota and their importance for the carbon cycle. Science 196: 613-615.

Bolin, B. 1986. Requirements for a satisfactory model of the global carbon cycle and current status of modeling efforts. In J.R. Trabalka and D.E. Reichle, eds., The Changing Carbon Cycle. A Global Analysis. Springer–Verlag, New York, pp. 403-424.

Bolin, B., E.T. Degens, S. Kempe, and P. Ketner (eds.). 1977. The Global Carbon Cycle. SCOPE Report 13. Wiley, New York, 491 pp.

Bowden, W.B. 1986. Gaseous nitrogen emissions from undisturbed terrestrial ecosystems: An assessment of their impacts on local and global nitrogen budgets. Biogeochemistry 2: 249-279.

Bowden, W.B. and F.H. Bormann. 1986. Transport and loss of nitrous oxide in soil water after forest clear-cutting. Science 233: 867-869.

Bremner, J.M. 1977. Role of organic matter in volatilization of sulfur and nitrogen from soils. In Proceedings of Symposium on Soil Organic Matter Studies, Vol. 11, Braunschweig, Federal Republic of Germany Sept. 6-10, 1976. Internat. Atomic Energy Agency, Vienna, pp. 229-240.

Bremner, J.M. and A.M. Blackmer. 1978. Nitrous oxide: Emission from soils during nitrification of fertilizer nitrogen. Science 199: 295-296.

Bremner, J.M. and C.G. Steele. 1978. Role of microorganisms in the atmospheric sulfur cycle. Adv. Microbial Ecol. 2: 115-2011.

Broecker, W.S., T. Takahashi, H.J. Simpson, and T.H. Peng. 1979. Fate of fossil fuel carbon dioxide and the global carbon budget. Science 206: 409-418.

Brown, S. 1981. Global Dynamics of Biospheric Carbon. Conference Publ. No. CONF-8108131. U.S. Department of Energy, Washington, DC, 194 pp.

Burns, R.S. and R.F.W. Hardy. 1975. Nitrogen Fixation in Bacteria and Higher Plants. Springer–Verlag, New York, 189 pp.

Crutzen, P.J. 1974. Photochemical reactions initiated by and influencing ozone in unpolluted tropospheric air. Tellus 26: 47-49.

DeBell, D.S. 1975. Fertilize western hemlock — yes or no? In Global Forestry and the Western Role. Western Forestry and Conservation Association Proc., Portland, OR, pp. 140-143.

Deevey, E.S. 1973. Sulfur, nitrogen and carbon in the biosphere. In G.M. Woodwell and E.V. Pecan, eds., Carbon and the Biosphere, Proceedings 24th Brookhaven Symposium in Biology, Upton, NY, May 16–18, 1972. Tech. Inform. Center, U.S. Atomic Energy Commission, pp. 182-190.

Delwiche, C.C. and B.A. Bryan. 1976. Denitrification. Annu. Rev. Microbiol. 30: 241-262.

Ember, L.R. 1978. Global environmental problems: Today and tomorrow. Environ. Sci. Technol. 12: 874-876.

Eriksson, E. 1963. The yearly circulation of sulfur in nature. J. Geophys. Res. 68: 4001-4008.

Farquhar, G.D., R. Wetselaar, and P.M. Firth. 1979. Ammonia violatilization from senescing leaves of maize. Science 203: 1257-1258.

Ferguson, E. E. and W.F. Libby. 1971. Mechanism for the fixation of nitrogen by lightning. Nature 229: 37-38.

Focht, D.D. and W. Verstraete. 1977. Biochemical ecology of nitrification and denitrification. Adv. Microbial Ecol. 1: 135-214.

Friend, J.P. 1973. The global sulfur cycle. In:S.I. Rasool, ed., Chemistry of the Lower Atmosphere. Plenum, New York, pp. 177-201.

Galloway, J.N., G.E. Likens, and E.S. Edgerton. 1976. Hydrogen ion speciation in the acid precipitation of the northeastern United States. In L.S. Dochinger and T.A. Seliga, eds, Proc. First Intl. Symp. on Acid Precipitation. U.S.D.A. Forest Service Genl. Tech. Report NE-23, Upper Darby, PA, pp. 383-396.

Graedel, T.E. 1979. The oxidation of atmospheric sulfur compounds. In Proc. MAASS-APCA Technical Conf. on the Question of Sulfates, PA, Pennsysvania, April 13–14, 1978 (in press).

Granat, L., H. Rodhe, and R.O. Hallberg. 1976. The global sulphur cycle. In B.H. Svensson and R. Söderlund, eds., Nitrogen, Phosphorus and Sulphur. SCOPE Report No. 7, Ecological Bulletins (Stockhom) 22: 89-134.

Hargrove, W.L. and D.E. Kissel. 1979. Ammonia volatilization from surface application of urea in the field and laboratory. Soil Sci. Soc. Am. J. 43: 359-363.

Hill, F.B. 1973. Atmospheric sulfur and its link to the biota. In G.M. Woodwell and E.V. Pecan, (eds.), Carbon and the Biosphere, Proc. 24th Brookhaven Symposium in Biology, Upton, NY, May 16–18, 1972. Tech. Inform. Center, U.S. Atomic Energy Commission, pp. 159-181.

Hitchcock, D.R. 1975. Biogenic contributions to atmospheric sulfate levels. Second Annual Conference on Water Reuse, Chicago, IL.

Hitchcock, D.R. 1977. Biogenic contributions to atmospheric sulfate levels: In L.K. Cecil, ed., Proc. Second Nat. Conf. Complete Water Reuse. Water's Interface with Energy Air and Solids. Am. Instit. Chem. Engineers, New York, pp. 291-310.

Hitchcock, D. and A.E. Wechsler. 1972. Biogenic Cycling of Atmospheric Trace Gases. Final Report, NASA Contract HASW-2128, Arthur D. Little, Inc., Cambridge, MA.

Houghton, R.A., W.H. Schlesinger, S. Brown, and J.F. Richards. 1985. Carbon dioxide exchange between the atmosphere and terrestrial ecosystems. In J.R. Trabalka, ed., Atmospheric Carbon Dioxide and the Global Carbon Cycle.U.S. Department of Energy, Publ. No. DOE/ER-0-0239, Washington, DC, pp. 113-140.

Hutchinson, G.L. and A.R. Mosier. 1979. Nitrous oxide emissions from an irrigated cornfield. Science 205: 1125-1127.

Iker, S. 1982. The problem of nitrous oxide. Mosaic 13: 20-25.

Ivanov, M.S. and J.R. Freney. eds., 1983. The Global Biogeochemical Sulfur Cycle. SCOPE Publ. No.19. John Wiley and Sons, New York, 470 pp.

Junge, C.E. 1958. The distribution of ammonia and nitrate in rainwater over the United States. Trans. Am. Geophys. Union 39: 241-248.

Junge, C.E. 1963. Air Chemistry and Radioactivity. Academic Press, New York, pp. 59-74.

Kellogg, W.W., R.D. Cadle, E.R. Allen, A.L. Lazus, and E.A. Martell. 1972. The sulfur cycle. Science 175: 587-596.

Kim, C.M. 1973. Influence of vegetation types on the intensity of ammonia and nitrogen dioxide liberation from soil. Soil Biol. Biochem. 5: 163-166.

Kittrick, J.A. 1976. Control of Zn^{2+} in the soil solution by sphalerite. Soil Sci. Soc. Am. J. 40: 314-317.

Laishley, E.J. and R. Bryant. 1987. Critical Review of Inorganic Sulphur Microbiology with Particular Reference to Alberta Soils. Publ. No. ADRP-B-04/87. Acid Deposition Research Program, Calgary, Alberta, Canada, 50 pp.

Lovelock, J.E, R.J. Maggs, and R.A. Rasmussen. 1972. Atmospheric dimethyl sulphide and the natural sulphur cycle. Nature 237: 452-453.

Lugo, A.E. and S. Brown. 1986. Steady state terrestrial ecosystems and the global carbon cycle. Vegetatio. 68: 83-90.

Marchesani, V.J., T. Towers, and H.C. Wohlers. 1970. Minor sources of air pollutant emissions. J. Air Pollu. Control Assoc. 20: 19-22.

Maroulis, P.J. and A.R. Bandy. 1977. Estimate of the contribution of biologically produced dimethyl sulfide to the global sulfur cycle. Science 196: 647-648.

McConnell, J.C. 1973. Atmospheric ammonia. J. Geophys. Res. 75: 7812-7821.

Miller, R.E. and E.L. Reukema. 1974. Seventy-five-year-old Douglas-fir on high quality site respond to nitrogen fertilizer. U.S.D.A. Forest Service, Res. Note PNW-281, Portland, OR, 8 pp.

Miller, R.E. and D.L. Reukema. 1977. Urea fertilizer increases growth of 20 year-old, thinned Douglas-fir on a por quality site. U.S.D.A. Forest Service, Res. Note PNW-291, Portland, Oregon, 8 pp.

Morrison, I.K., F. Hegye, N.W. Foster, D.A. Winston, and T.L. Tucker. 1976. Fertilizing semimature jack pine (*Pinus banksiana* Lamb.) in northwestern Ontario: Fourth-year results. Report No. 0- X-240, Can. For. Service, Dept. Environment, Sault Ste. Marie, Ontario, Canada, 42 pp.

Nozhevnikova, A.N. and L.N. Yurganov. 1978. Microbiological aspects of regulating the carbon monoxide content in the earth's atmosphere. Adv. Microbial Ecol. 2: 203-244.

Rasmussen, K.H., M. Taheri, and R.L. Kabel. 1975. Global emissions and natural processes for removal of gaseous pollutants. Water Air Soil Pollu. 4: 33-64.

Reichie, DE., B.E. Dinger, W.T. Edwards, W.F. Harris, and P. Sollins. 1973. Carbon flow and storage in a forest ecosystem. In G.M. Woodwell and E.V. Pecan, eds., Carbon and the Biosphere, Proc. 24th Brookhaven Symposium in Biology, Upton, NY, May 16–18, 1972. Tech. Inform. Center, U.S. Atomic Energy Commission, pp. 182-190.

Robinson, E. and R.C. Robbins. 1968. Sources, Abundance and Fate of Gaseous Atmospheric Pollutants. Final Report SRI Project PR-6755. Stanford Research Institute, Menlo Park, CA.

Robinson, E. and R.C. Robbins. 1975. Gaseous atmospheric pollutants from urban and natural sources. In S.F. Singer, ed., The Changing Global Environment. Reidel, Dordrecht, pp. 111-123.

Ryaboshapko, A.G. 1983. The atmospheric sulphur cycle. In M.V. Ivanov and J.R. Freney, eds., The Global Biogeochemical Sulphur Cycle. Scope Publ. No. 19, John Wiley and Sons, New York, pp. 203-296.

Siegenthaler, U. and H. Oeschger. 1978. Predicting future atmospheric carbon dioxide levels. Science 199: 388-395.

Söderlund, R. and B.H. Svensson. 1976. The global nitrogen cycle. In B.H. Svesson and R. Söderlund, eds., Nitrogen, Phosphorus and Sulphur-Global Cycles. SCOPE Report No. 7, Ecological Bulletin No. 22, Royal Swedish Academy of Sciences, Stockholm, pp. 23-73.

Spáleny, J. 1977. Sulphate transformation to hydrogen sulphide in spruce seedlings. Plant Soil 48: 557-563.

Spedding, D.J. 1974. Air Pollution. Clarendon Press, Oxford, 76 pp.

Spurr, S.H. and H.J. Vaux. 1976. Timber: Biological and economic potential. Science 191: 752-756.

Stuiver, M. 1978. Atmospheric carbon dioxide and carbon reservoir changes. Science 199: 253-258.

Trabalka, J.R. 1985. Atmospheric Carbon Dioxide and the Global Carbon Cycle. U.S. Department of Energy, Publica. No. DOE/ER-0-0239, Washington, DC, 315 pp.

Trabalka, J.R., J.A. Edmonds, J.M. Reilly, R.H. Gardner, and D.E. Reichle. 1986. Atmospheric CO_2 projections with globally averaged carbon cycle models. In J.R. Trabalka and D.E. Reichle, eds., The Changing Carbon Cycle of Global Analysis. Springer–Verlag, New York, pp. 534-560.

U. S. Environmental Protection Agency. 1976. Diagnosing Vegetation Injury Caused by Air Pollution. Contract No. 68-02-1344, U.S.E.P.A., Air Pollution Training Institute, Research Triangle Park, NC.

Volk, G.M. 1959. Volatile loss of ammonia following surface application of urea to turf or bare soils. Agron. J. 51: 756-749.

Volk, G.M. 1970. Gaseous loss of ammonia from prilled urea applied to slash pine. Soil Sci. Soc. Am. Proc. 34: 513-516.

Webster, S.R., D.S. deBell, K.N. Wiley, and W.A. Atkinson. 1976. Fertilization of western hemlock. Proc. Western Hemlock Mamt. Conf., Univ. Washington, Seattle, WA, pp. 247-251.

Whittaker, R.H. 1975. Communities and Ecosystems. Macmillan, New York, 385 pp.

Wolff, G.T. 1979. The question of sulfates: A conference summary. J. Air Pollu. Control Assoc. 29: 26-27.

Wollum, A.G. II and C B. Davey. 1975. Nitrogen accumulation, transformation transport. In B. Bernier and C.H. Winget, eds., Forest Soils and Forest Land Management, Proc. Fourth North Amer. Forest Soils Conf., Laval Univ., Quebec. Les Presses de l'Université Laval, Quebec, Canada, pp. 67-106.

Wong. C.S. 1978. Atmospheric input of carbon dioxide from burning wood. Science 200: 197-200.

Woodwell, G.M. 1978. The carbon dioxide question. Sci. Am. 238:34-43.

Woodwell, G M., R.A. Houghton, T.A. Stone, and A.B. Park. 1986. Changes in the area of forests in Rondonia, Amazon Basin, measured by satellite imagery. In J.R. Trabalka and D.E. Reichle, eds., The Changing Carbon Cycle. A Global Analysis. Springer–Verlag, New York, pp. 403-424.

Woodwell, G.M., R.H. Whittaker, W.A. Reiners, G.E. Likens, C.C. Delwiche, and D.B. Botkin. 1978. The biota and the world carbon budget. Science 199: 141-146.

Zinder, S.H. and T D. Brock. 1978. Microbial transformations of sulfur in the environment. Part II. In J.O. Nriagu, ed., Ecological Impacts. Wiley, New York, pp. 445-466.

4
Forests as Sources of Hydrocarbons, Particulates, and Other Contaminants

In addition to whatever contribution forests may make to the atmosphere burden of carbon, sulfur, and nitrogen oxides, they are known to be important natural sources of hydrocarbons and particulates. Volatile hydrocarbons are released by a variety of woody plants during the course of normal metabolism. Pollen, the most significant particulate contaminant released by forests from the standpoint of human health, is also produced, of course, during normal reproductive metabolism. Hydrocarbon aerosols are viewed as an increasingly important particulate emission from forests. Forest burning, whether naturally occurring or artificially ignited, also produces hydrocarbons, and particulates as well as carbon oxides. Even though forest fires may be a natural recurring event in most forest ecosystems, the pollutants generated by this process are not the result of normal metabolism but rather are generated by combustion of forest biomass. As a result, the latter are discussed in Section C.

A. Volatile Hydrocarbons

These organic gases vary greatly in chemical reactivity depending on structure. The paraffins, or aliphatic hydrocarbons, that contain the maximum number of hydrogen atoms (saturated), while not chemically inert are relatively unreactive and of limited importance as atmospheric components (Calvert, 1979). Alkenes (olefins) are open-chain hydrocarbons containing one or more carbon-carbon double bonds, and alkynes contain one or more carbon-carbon triple bonds. Both groups do not contain the maximum number of hydrogen atoms and are termed *unsaturated*. Aromatic hydrocarbons contain ring systems (benzene structure) in which all carbon atoms are linked in a system of conjugated double bonds.

Since all unsaturated organic gases are chemically reactive, many of them are important atmospheric contaminants.

Polluted urban atmospheres contain in excess of 100 different hydrocarbons, the most reactive of which are the olefins. The primary significance of these olefins is their important role in the synthesis of photochemical oxidants. As knowledge of photochemistry has advanced volatile carbon compounds containing oxygen and halogens have also been shown to be important. As a result, *volatile organic compounds* (VOC) is used to describe stable organic compounds that exist in the troposphere and that can participate in the formation of photochemical oxidants. Natural ecosystems, especially forests, are important sources of hydrocarbons.

1. Sources of Hydrocarbons

A variety of anthropogenic and natural processes release hydrocarbons into the atmosphere. The principal human sources are industrial processes, which release a wide variety as chemical solvents (Feldstein 1974, Parsons and Mitzner 1975, Davies et al., 1976), and transportation (Heuss et al. 1974), which includes the emission of hydrocarbons in gasoline vapor as well as in gasoline combustion products. The U.S. Environmental Protection Agency (1986) estimated United States anthropogenic release of VOC in 1980 to include, in millions of metric tons: 8.2 transportation, 2.1 stationary fuel combustion, 8.9 industrial processes, 0.6 solid waste, and 2.9 miscellaneous, for a total release of 22.7 million metric tons. It has been estimated that the total global anthropogenic release of hydrocarbons may approximate 88 million metric tons (Stern et al., 1973). Globally, natural biological processes may release volatile organic gases into the atmosphere in amounts many times the quantity released by the activities of human beings. Natural systems, especially soils (microorganisms) (Taylor, 1984) and vegetation (Gershenzon, 1984, Bufalini 1980, Smith, 1982) release abundant hydrocarbons to the atmosphere.

2. Forests as Sources of Hydrocarbons

In 1955., F. W. Went, then Director of the Missouri Botanical Garden, advanced the hypothesis that vegetation contributes to air pollution by the release of organic gases (Went, 1955). Went was particularly interested in the fate of plant products that were derivatives of isoprene. These compounds included terpenes,

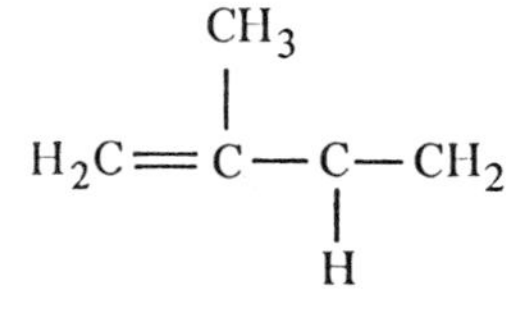
CH3
H2C=C—C—CH2
H

isoprene unit

carotenoids, resin acids, rubber, and phytol. Went estimated that approximately 200 million metric tons of organic volatile plant products were released to the global atmosphere annually (Went, 1960a,b).

Turk and D'Angio (1962), in their efforts to evaluate the sensory quality of "freshness" in natural area atmospheres studied the atmospheric composition along the coast of New Jersey, in a Connecticut state forest, and on Mount Washington in New Hampshire. They detected aromatic and unsaturated organic gases at both the coastal and forest sites and observed that the forest location was the richer of the two. Major et al. (1963) presented evidence that the leaves of several trees were capable of releasing the aldehyde a-hexanal from their leaves when injured at certain times during the growing season.

R.A. Rasmussen, a student of F.W. Went, stressed the importance of terpenes as important organic volatiles released from vegetation (Rasmussen, 1964). Terpenoid compounds are an extremely diverse group of organic materials built of isoprene units. Current wisdom suggests that isopentenyl pyrophosphate, and not isoprene itself, may be the actual building block of the terpenoids (Hess, 1975). Two, three, four, and more isopentenyl residues may be combined to form monoterpenoids (C_{10} compounds), sesquiterpenoids (C_{15} compounds), diterpenoids (C_{20} compounds), and triterpenoids (C_{30} compounds). Hydrocarbon terpenoids are specifically termed *terpenes*. Employing a gas chromatograph mounted in a mobile trailer laboratory, Rasmussen and Went (1965) studied volatile organic release in forest areas in North Carolina, Virginia, Missouri, and Colorado. They found three monoterpenes, α-pinene, β-pinene, and mycrene, along with isoprene, to be particularly abundant. They also observed that concentrations varied with meteorological conditions and density of the vegetation. Terpene release from the vegetation was higher in the summer than the winter. In the fall, leaf litter became a major source of aromatic materials.

pinene

The terpenes characteristic of various tree species are quite variable and are under fairly rigid genetic regulation (Table 4–1). While considerable uncertainty

Table 4-1. Monoterpenes Associated with Various Coniferous Species.

Terpene	Douglas fir[a,b]	Western white pine[c]	Slash pine[d]
α-Pinene	X	X	X
Camphene	X	X	X
β-Pinene	X	X	X
Sabinene	X		
Δ-3-Carene	X	X	X
Myrcene	X	X	X
Limonene	X	X	X
β-Phellandrene	X		
γ-Terpinene	X		
Terpinolene	X		
α-Phellandrene			X
1,8-Cineole	X		
2-Hexenal	X		
Ethyl caproate	X		
cis-Ocimene	X		
Citronellal	X		
Linalool	X		
Fenchyl alcohol	X		
Bornyl acetate	X		
Sesquiterpene HC	X		
Terpinen-4-ol	X		
β-Caryophyllene	X		
Terpene alcohol	X		
Citronellyl acetate	X		
α-Terpineol	X		
Citronellol	X		
Geranylacetate	X		

[a] Radwan and Ellis (1975).
[b] Maarse and Kepner (1970).
[c] Hanover (1966).
[d] Squillace (1971).

has existed concerning the importance of the hemiterpene isoprene, Rasmussen (1970) presented evidence identifying isoprene as an important forest emission to the atmosphere. Numerous angiosperm species are efficient isoprene emitters. Unlike monoterpenes, isoprene is released from foliage only in the light.

In 1972, Rasmussen reviewed his own work and that of several others in an effort to inventory the sources of forest terpenes, to estimate terpene emission rates, and to calculate the significance of tree released hydrocarbons on a global basis. Six monoterpenes, α-pinene, camphene, β-pinene, limonene, myrcene, and β-phellandrene, were judged to be the major terpenic emissions from gymnosperm foliage. Numerous angiosperms, along with a few gymnosperms, were concluded to release isoprene. Table 4–2 indicates the relative importance of

Table 4-2. Percentage of Monoterpene and Hemiterpene Emitters for Various Forest Regions of the United States.

	% total U.S. forest area	% α-pinene emitters	% isoprene emitters
Eastern type groups			
Softwood types			
Loblolly-shortleaf pine	11	∿ 100	Some from oak and sweet-gum associates
Longleaf-slash pine	5	∿ 100	Some from oak and sweet-gum associates
Spruce-fir	4	∿ 75	25 from spruce, which also emits α pinene
White-red jack pine	2	∿ 90	10 from aspen trees
Subtotal	22	∿ 91	∿ 9
Hardwood types			
Oak-hickory	23	∿ 10	70, diluted by hickory, maple, and black walnut
Oak-gum-cypress	7	∿ 50	50 from plurality of oak, cottonwood and willow
Oak-pine	5	∿ 30	60, diluted by black gum and hickory associates
Maple-beech-birch	6	∿ 15	–, terpene foliages are hemlock and white pine
Aspen-birch	5	∿ 20	60, diluted by birch, α-pinene source is balsam fir and balsam poplar
Elm-ash-cottonwood	4	–	30 from cottonwood, sycamore, willow
Subtotal	50	∿ 21	∿ 45
Total	72	–	–
Western type groups			
Softwoods			
Douglas fir	7	∿ 100	–
Ponderosa pine	7	∿ 100	5 from aspen associates
Lodgepole pine	3	∿ 90	10 from Englemann spruce and aspen
Fir-spruce	3	∿ 100	40 from spruce trees
Hemlock-Sitka spruce	2	∿ 100	25 from Sitka spruce
White pine	1	∿ 100	5 from Englemann spruce
Larch	1	∿ 100	–
Redwood	0.5	∿ 100	–
Subtotal	24	∿ 98	∿ 12
Hardwoods	2	–	∿ 100 from aspen trees
Total	26		

Source: Rasmussen (1972).

Table 4-3. Emission rates for total organic gases released by tree foliage.

Sample	Emission rate $\mu g\ g^{-1}\ hr^{-1}$
Raleigh, North Carolina (June, 1977)	
Oaks	26.1
Shortleaf pine	16.3
Loblolly pine	4.85
Red cedar	1.14
Virginia pine	13.6
Pullman, Washington (August–November, 1976)	
Ponderosa pine	2.96
Mugo pine	1.78
Douglas fir	0.86
Juniper	3.25
Spruce	7.26
Tampa, Florida (April–May, 1977)	
Laurel oak	11.2
Turkey oak	26.2
Water oak	27.2
Blue Jack oak	16.5
All oak species (night)	1.20

Source: Zimmerman et al. (1978).

hemiterpene and monoterpene emitters for major United States forest types. Using a generalized release rate of 100 ppb hr^{-1}, Rasmussen estimated the relative importance of forest terpene emissions to the atmosphere on a global basis to approximate 175 million metric tons of reactive materials. This is larger than the amount of hydrocarbons produced by human beings.

In addition to terpenes the biota is a potential source of additional hydrocarbons to the atmosphere. Ethylene is released by numerous angiosperms during flowering and fruit maturation, and in response to injury and infection of certain tissues. Shain and Hills (1972) have provided unique evidence that ethylene is also released by gynosperms under certain conditions. Methane is released from anaerobic ecosystems due to the activity of methanogenic bacteria (Mah et al., 1977; Dacey and Klug, 1979). These organisms are generally presumed unimportant in forest systems, unless the forests contain flooded soils for extended periods. The natural hydrocarbons of significance, however, are judged to be the monoterpenes ($C_{10} H_{16}$) and isoprene ($C_5 H_8$). Methane (CH_4) is presumed unreactive and stress-evolved ethene (C_2H_4) emissions are too small to be significant.

Zimmerman et al. (1978) have employed Teflon enclosures to collect organic compounds from tree foliage. Emission rates for total organic compounds from a variety of tree species at various locations are presented in Table 4–3. Based on these emission rates and appropriate biomass calculations, the total global emissions level of isoprene plus the terpenes may approximate 900 million metric tons yr[-1] (Bufalini 1980).

Total natural hydrocarbon emissions from the state of Pennsylvania were estimated to be 3400 tons day[-1] during August. The majority, approximately 75%, of these emissions were from forests with the balance from agricultural crops, principally corn. The forest emissions were roughly evenly divided among isoprene-emitting and non-isoprene-emitting deciduous and conifer trees (Lamb et al., 1984). For typical northeastern deciduous forests, however, isoprene emissions may account for approximately 78% of the total nonmethane hydrocarbon release (Lamb et al., 1984). Westberg and Lamb (1985) have inventoried the natural hydrocarbon production in the Atlanta, Georgia region. They routinely detected isoprene, α-pinene, β-pinene, Δ^3-carene and myrcene. Enclosure studies provided region-wide mean fluxes of 2500 μg m[-2] hr[-1] for isoprene and 90 μg m[-2] hr[-1] for α-pinene. With appropriate biomass data, these rates suggest 330 tons day[-1] release for isoprene and 24 tons day[-1] release for α-pinene. These natural emission amounts exceed the average anthropogenic release rate in the region, but not the maximum rate of human release.

B. Particulates

1. Pollen

The medical significance of certain tree pollen makes pollen grains one of the most important air contaminants produced by woody vegetation. The time and method of opening of anthers or microstrobili, dehisence, and subsequent release of pollen are variable depending on the tree family. Release in gymnosperms follows a parting of microstrobili sporophylls. Retraction of bract scales caused by dehydration frees pollen to be released by the wind or shaking. All gymnosperm pollen, with the exception of cycads, is distributed following release by wind (anemophilous). Angiosperm pollen release follows one of numerous opening patterns of anther sac walls. Wall rupture is due to shrinkage occasioned by a change in atmospheric humidity. In addition to wind, angiosperm pollen may be distributed following release by birds, bats, and insects (zoophilous/entomophilous) (Stanley and Linskens, 1974).

a. Characteristics

The size of pollen grains among modern flowering plants varies from approximately 5 X 2.4 mm in *Myosotis* to 200 mm or greater in certain Cucurbitaceae and Nyctaginaceae (Erdtman, 1969). Wind pollinated species generally have a grain size in the range of 17–58 mm, while insect-or animal-pollinated species are typically larger or smaller (Stanley and Linskens, 1974). Size, volume, and weight variations of various tree pollens are presented in Table 4–4. The elaborate and variable shape and sculpturing of pollen allows identification to species, and the resistance of the walls of pollen grains, allowing persistence over time,

Table 4-4. Pollen Dimensions, Volume, and Weight for Selected Tree Species.

Species	Dimensions (μm)			Volume (10^{-9} cm^3)	Weight (10^{-9} g)
	Length	Width	Height		
Silver fir	97.8	102.9	62.7	499.4	251.6
Cephalonica fir	97.1	98.6	86.2	422.6	212.2
Norway spruce	85.8	80.5	66.3	278.2	110.8
Scotch pine	41.5	45.9	36.0	35.5	37.0
European larch	76.0	72.0	50.0	180.2	176.3
Douglas-fir	84.8	81.1	54.8	219.2	188.8
Sugar maple	32.5	23.6	24.6	16.5	6.6
Horsechestnut	31.0	16.4	18.2	4.8	0.9
European alder	26.4	22.8	13.7	4.4	1.4
European birch	10.1	10.1	16.8	2.9	0.8
European beech	55.1	40.5	41.1	50.3	26.0
English oak	40.8	26.1	21.5	13.3	5.7
European cut-leaf linden	40.5	40.1	20.6	15.0	6.5
Rock elm	33.4	32.7	17.7	12.8	6.8

Source: Stanley and Linskens (1974).

has made them extremely valuable in ecological, geological, ethnological, and medical studies. Excellent keys and descriptions of tree pollen are provided by Moore and Webb (1978), Wodehouse (1945), and Nilsson et al. (1977).

b. Production

The production and release of tree pollen is very variable and is dependent on species, strategy of pollen transfer, time of year, numbers of anthers, mi

Table 4-5. Pollen Production by Reproductive Parts of Various Trees.

Species	Number of pollen grains anther^{-1}	Number of pollen grains flower^{-1}	Number of pollen grains catkin^{-1}
Norway maple	1,000	8,000	
Apple	1,400-6,250		
European ash	12,500		
Common juniper		400,000	
Scotch pine		160,000	
Norway spruce		600,000	
European white birch			6,000,000
European alder			4,500,000
English oak			1,250,000
European beech			175,000

Source: Erdtman (1969).

crosporophylls and flowers produced, climate, tree age, and health. Entomophilus species (for example, *Acer*, *Malus*) have appreciably lower pollen production per anther than anemophilous species (for example, *Fagus*). Some representative estimates of pollen production for various species are presented in Tables 4–5 and 4–6. The variableness of pollen production may follow a predictable pattern in certain species. *Fraxinus* and *Ulmus* have a high yield approximately every third year, while American beech may give high yields every other year (Stanley and Liskens, 1974).

Hyde (1951) emphasized the importance of the influence of climate on pollen formation by monitoring seven tree genera over a 6 year period in Great Britain (Table 4–7). These data suggest a substantial influence of climatic variation on pollen production.

Of greatest interest to those interested in air contamination by pollen is the amount released per tree per unit time. Some estimations are presented in Table 4–8.

c. Distribution

The movement of pollen is critically important in air quality considerations. While animal disseminated pollen is moved in predictable patterns and for limited distances, anemophilous pollen may be moved considerable distances in complex patterns. The latter has been intensively studied, and even a summary treatment is beyond the scope of this book. Pollen can spread over wide distances. The distance traveled is a function of sedimentation rate, which is dependent on specific gravity, size, form, and degree of clumping (Stanley and Liskens, 1974). For various trees, therefore, dispersion distance may vary considerably (Table 4–9). Tauber (1965, 1967) has proposed a model for pollen dispersion from a forest that has three primary components. The trunk space component is the pollen that falls vertically from the tree and shrub canopies. Since air movement is slower below the canopy than above, most pollen in the trunk space component is deposited in relatively short distances. The canopy component is the pollen that escapes from the tree crowns and enters the faster moving air above the canopy. A portion of this component may be transported to high altitudes and may be carried for considerable distances. The rain component refers to those pollen grains that function as nuclei for water droplet formation. This component may also result in long distance pollen distribution.

Anderson (1967) analyzed the pollen spectra of forest floor moss to study the distribution of tree pollen in a mixed deciduous forest in southern Denmark. For the species studied (*Betula, Quercus, Alnus, Fagus, Tilia,* and *Fraxinus* ; canopy height, 20 m), moss analyses indicated surprisingly short dispersal distances for the majority of the pollen within the forest, that is, less than 20–30 m. Perhaps this largely represented the trunk space component.

Trees near the edge of forest stands or isolated trees (urban/suburban condition), however, may disseminate pollen great distances because they may release into locally turbulent atmospheric conditions. Pollen from trees upslope may tend to concentrate in valleys below (Silen and Copes, 1972).

Table 4-6. Volume of Pollen Yield by Reproductive Parts of Various Trees.

	ml				
Genus	Catkin^{-1}	Inflorescence^{-1}	Strobili^{-1}	Flower^{-1}	Head^{-1}
Alnus	.04				
Betula	.01-.2				
Fagus		.01-.016			
Larix			.01-.014		
Liquidambar				.2-.3	
Pinus			.7-.35		
Platanus					.08
Populus	.5-1				
Pseudotsuga			.007-.037		
Ulmus				.001-.004	

Source: U.S.D.A. Forest Service (1974).

Table 4-7. Annual Variation in Pollen Productivity.

	Average catch	Catch by year as % of 1943-1948 average					
Genus	1943-1948	1943	1944	1945	1946	1947	1948
Pinus	386	92	121	72	127	40	146
Alnus	385	45	226	120	55	34	118
Fraxinus	675	330	25	24	159	54	8
Fagus	273	23	140	11	143	8	275
Betula	620	57	160	42	106	22	214
Ulmus	4579	84	87	180	50	48	150
Quercus	2776	58	90	42	280	10	117

Source: Hyde (1951).

Table 4-8. Annual Pollen Production per Tree for Selected Species.

Species	No. pollen grains ($\times 10^6$)[a]	kg[b]
European beech	409	0.15
Sessile oak	654	—
European hornbeam	3149	—
Norway spruce	5481	0.4
European birch	5562	0.03
European filbert	5603	0.05
Littleleaf linden	5603	—
Scotch pine	6462	0.12
Alder	7239	0.05

[a] From Erdtman (1969).
[b] From Brooks (1971).

Table 4-9. Average Pollen Dispersion Distance at a Windspeed of 10 m sec^{-1}.

Species	km
Norway spruce	22.2
Scotch pine	267.8
English oak	199.0
Alder	546.7
European filbert	267.8

Source: Stanley and Linskens (1974).

Extensive surveys of contemporary pollen distribution have been conducted in numerous parts of North America and Europe (for example, Moore and Webb, 1978). Canadian studies suggest that long-distance transport of pollen from deciduous trees into boreal and tundra regions does not appear to be very significant. Input of *Picea* and *Pinus* pollen into deciduous forest regions, however, is measurable (Lichti-Federovich and Ritchie, 1968). Moore and Webb (1978) have concluded that, under appropriate meteorological conditions, pollen grains may be carried thousands of kilometers from their point of origin.

d. Human Health Aspects

Human allergic response can result from exposure to a wide variety of allergenic materials including fungal spores, animal hair, feathers, dust, anthropod parts, insect and reptile venom, drugs, certain foods, and pollen. Allergic response to the latter, termed *hay fever* or *pollinosis*, is generally recognized as the most prevalent and important of all allergies (Stanley and Linskens, 1974). In sensitive individuals, hay fever symptoms, which include sneezing, watery eyes, nasal obstruction, itchy eyes and nose, and coughing, typically occur within minutes of exposure to allergenic pollen. Clinically hay fever is described as allergic rhinitis or conjunctivitis. Hyde et al. (1978) have suggested that between 10% and 20% of the population of the United States manifest recurrent or persistent allergic rhinitis. The cost to United States industry for lost wages due to airborne pollen allergies is in excess of $500 million annually.

Allergic rhinitis in a given individual may result from exposure to the pollen of any one of several hundred plant species. *Ambrosia* (ragweed) and *Phleum pratense* (timothy grass) pollen, both very widely distributed throughout the north temperate zone, elicit widespread allergic responses and cause approximately half the pollinosis cases in numerous urban areas. In this zone there is considerable temporal variation in disease associated with pollen production and distribution. Ragweed pollinosis is most prevalent from mid August through late September, while grass pollinosis is especially common from late spring through early July.

Table 4-10. Principal Tree Genera with Species that Produce Allergenic Pollen.

Acer	*Juniperus*
Alnus	*Morus*
Betula	*Olea*
Broussonetia	*Platanus*
Carpinus	*Populus*
Carya	*Quercus*
Casuarina	*Salix*
Cupressus	*Ulmus*
Fraxinus	

e. Tree Related Hay Fever

For hay fever sufferers throughout the north temperate zone, pollinosis symptoms are frequently initiated in early spring (actually February through May, depending on location) when a variety of tree species release pollen. A large number of tree species produce allergenic pollen (Table 4–10).

Based on adult skin test evidence, Lewis and Imber (1975) have ranked various tree groups relative to the ability of their pollen to incite rhinitis and asthma in North America. They suggest that box elder, willow, and hickory pollen elicited the greatest allergenic reactivity. Oak, sycamore, poplar, maple, and birch proved moderately allergenic, while elm, cottonwood, and white ash were lowest in producing adult skin reactions.

For any given location in the temperate zone, the most important tree species producing allergenic pollen are highly variable. Numerous regional studies are available (for example, Anderson et al., 1978; Newmark, 1978; Lew and Imber, 1975). Rubin and Weiss (1974) list major tree pollens by state and Wodehouse (1945) by region.

Fortunately gymnosperm pollen is generally not allergenic. Pollen from *Abies* does not appear to cause symptoms, while pollen from *Pinus* only rarely is involved in pollinosis (Newmark, 1978). There is evidence that pollen from certain *Juniperus* and *Cupressus* species is allergenic (Yoo et al., 1975).

2. Hydrocarbon Aerosol

The ability of various hydrocarbons to form particulate organic compounds in the atmosphere is a complex function of their ambient concentration, gas-phase reactivity, and ability to form products with appropriate physical characteristics for gas-to-aerosol conversion. Cyclic olefins (for example, cyclopentene) are probably the most significant class of urban organic aerosol precursors (National Academy of Sciences, 1977). Went (1960a) was the first to suggest that terpenoids from vegetation could serve as precursors for organic particulate formation. He concluded that laboratory and field evidence demonstrated that organic volatiles, specifically α-pinene, could form fine particles when subjected to high light intensities in the presence of nitrogen dioxide as a "light-absorbing catalyst" (Went, 1964). The size of these particles approximated $10^{-7} - 10^{-1}$ cm.

Organic aerosols formed from hydrocarbon precursors released from vegetation were judged to be responsible for the blue haze commonly observed in the Great Smoky Mountains of North Carolina and Tennessee, the sagebrush area in the western United States, the eucalypt forests of Australia, and the forested tropics (Went, 1960a,b). In 1964, Went estimated that organic gases released annually by vegetation may occasion the formation of 500 million tons of submicroscopic particulate matter globally. A mobile laboratory was used to obtain additional evidence for the hypothesis that terpenes were activated to agglomerate into fine particles by a photochemical process (Went et al., 1967).

The fate of gaseous olefins, for example, α-pinene, β-pinene, limonene, myrcene, and isoprene, released to the atmosphere, however, remains largely undetermined. Rasmussen and Holdren (1972) have presented evidence that individual monoterpene hydrocarbons are present in the low parts-per-billion range in rural air. Aerosol formation from terpenes, however, has not been thoroughly investigated (National Academy of Sciences, 1977). Only one terpene, α-pinene, has received research attention, and this is very limited.

The formation of condensation nuclei from irradiated mixtures of 0.1 ppm (188 μg m^{-3}) nitrogen dioxide and 0.5 ppm (2780 μg m^{-3}) α-pinene was reported by Ripperton and Lillian (1971). Schwartz (1974) has intensively studied the aerosol products from the reaction of nitrogen oxides and α-pinene. O'Brien et al. (1975) examined the aerosol-forming behavior of a representative group of hydrocarbons under standardized conditions and observed that α-pinene caused a large amount of light scattering (Table 4–11). The National Academy of Sciences, Subcommittee on Ozone and Other Photochemical Oxidants, has concluded that low volatility compounds are probably formed from other olefinic terpenes (for example, β-pinene, limonene), as in the case of α-pinene (Schwartz, observations) and that these materials together probably constitute a major fraction of the blue haze aerosols formed naturally over forested areas (National Academy of Sciences, 1977). Weiss et al. (1977) do not agree, however, and have suggested that the rural haze observed in the Midwest and Southeast results from particulate sulfates rather than organic compounds. These investigators employed humidity-controlled nephelometry during 1975 to obtain information on the dominant optical scattering species in Michigan, Missouri, and Arkansas. They concluded at all sites that submicrometer-sized sulfate particles dominated during all wind and synoptic conditions. They judged that sulfate aerosol was regionally extensive and inferred the importance of determining the fraction of the sulfate due to natural versus anthropogenic sources.

Knights et al. (1975) have presented evidence for the occurrence of α-pinene aerosol products in urban air. It is not clear, however, whether the terpenes were released from urban tree foliage, industrial solvents (for example, turpentine), or were transported into the city from rural forested areas.

3. Other

A variety of studies has indicated that plants release a variety of small particles to the atmosphere. Trace metals, including heavy metals, may be among these particle emissions. Employing radioisotopes and Scotch pine trees, Beauford et al. (1977) provided evidence that cadmium, lead, and zinc can be taken up by the roots and released from the foliage in association with coarse and submicron size particles. The authors suggested two possible mechanisms for particulate loss from leaves. First, epicuticular waxes may function as metal carriers and as wax rodlets are fragmented during leaf expansion, metals may be lost in association with small wax particles. Second, diffusiophoresis associated with water loss during rapid transpiration may result in the release of airborne salt crystals. Based on their studies, Beauford et al. (1977) suggest a potential heavy metal release for 1 km^2 of vegetation of 9 kg zinc and 5 g of lead yr^{-1}.

C. Forest Fires

The burning of forests or forest debris represents a special case where forest systems supply air contaminants to the atmosphere. Fires can be conveniently divided into natural (wild) fires and prescribed (management) fires. Prescribed fires are artificially set and controlled and are intended to fulfill forest management objectives. Included in the latter are one or more of the following: (a) disposal of

Table 4-11. Aerosol Formation from Selected Hydrocarbons.

Hydrocarbon	Maximum light scattering ($\beta_{scat} \times 10^4$ m^{-1})
Glutaraldehyde	0
Ethylbenzene	1
Mesitylene	1
2,6-Octadiene	1
1-Octene	1
trans-4-Octene	1
5-Methyl-1-hexene	1
2,6-Dimethylheptane	1
1-Heptene	1
α-Xylene	8
1,5-Hexadiene	40
Cyclohexene	90
2-Methyl-1,5-hexadiene	110
1,6-Heptadiene	160
1,7-Octadiene	180
α-Pinene	180

Source: O'Brien et al. (1975).

Figure 4-1. Broadcast slash burning in a clear-cut block in a western forest ecosystem.
Source: U.S.D.A. Forest Service.

logging residue (slash), (b) forest fuel reduction to minimize the influence of wildfires, (c) control of unwanted vegetation, (d) reduction of microbial or insect pests, (e) facilitation of crop tree regeneration, and (f) improvement of wildlife habitat.

Hall (1972) presents an overview discussion of the various contaminants produced by forest burning. The visible smoke is very largely water condensed on particulate matter. Darley et al. (1976) and Komarek et al. (1973) have provided scanning electron microscope photomicrographs of these particles which consist largely of carbon. Aerosols, condensation products of terpenoids, phenols, and aldehydes are partially absorbed on the carbon. A variety of hydrocarbons, volatile organic compounds, and carbon oxides are among the most important gaseous components produced. Lande (1987) collected samples from a variety of forest and field burns in Oregon. The primary polynuclear aromatics detected were anthracene, fluoranthene, benzo (b) fluoranthene, phenanthrene, and pyrene. No benzo (a) pyrene (a proven carcinogen and common indicator of polycyclic organic matter) was recorded, although McMahon and Tsoukalas (1978) and White (1985) have associated this chemical with burning forest material. Lande

(1987) recorded additional volatile organics, including methylene chloride, benzene, toluene, xylene, and aldehydes (most importantly, formaldehyde and acetaldehyde), but no pesticides.

1. Prescribed Fires

a. Tree Residue Burning (Slash Fires)

In the northwestern sections of the United States and in western Canada, burning is widely employed to dispose of logging residue. Harvest of old-growth stands of western conifers generates excessive quantities of nonmarketable tree debris. The only economically sound disposal strategy for decades has been burning (Figure 4–1).

Approximately 20 years ago, roughly 81,000 ha (200,000 acres) of logging slash were burned annually west of the Cascade Range in the states of Washington and Oregon (Fritschen et al., 1970; Cramer and Westwood, 1970). This acreage has been decreasing as tree utilization has improved, as harvesting has involved progressively less old-growth timber, and as clean air laws have been implemented. Total forest and range land burned by prescription annually in the West currently approximates one million acres (Cooper, 1976). Individual slash fires average approximately 4–16 ha (10–40 acres) aflame with a fuel loading of 125–495 tons ha^{-1} (50–200 tons $acre^{-1}$). Smoke may be lifted in convective columns 500–1700 m above these fires (Cramer and Graham, 1971; Hedin and Turner, 1977).

Two types of slash burning have been practiced: broadcast and pile. The former leaves the logging debris in place while the latter concentrates it into piles. Broadcast fires burn at a relatively low temperature while pile fires burn at a relatively high temperature (Figure 4-2). Fritschen et al. (1970) analyzed the pollutants produced by broadcast and pile fires set experimentally in Douglas fir slash in Washington in order to explore the hypothesis that pile fires, because they are higher temperature fires, have more complete combustion and produce fewer emissions than broadcast burns. Particulates, at ground level, increased to roughly 10 times background immediately downwind from the broadcast burn. Smoke plume particulates in the vicinity of the fire reduced visibility to 0.5 km, but at a distance of 19 km from the fire, visibility was at the level found over Seattle. High carbon monoxide and carbon dioxide concentrations found at the fire site decreased rapidly to ambient levels in horizontal and vertical directions. Hydrocarbon analyses revealed low concentrations of 25 components. The most important were low molecular weight hydrocarbons and alcohols, including ethylene, ethane, propene, propane, methanol, and ethanol. Several unsaturated components were detected but the amounts were small. Relative to the broadcast fire, the pattern of combustion and emission of the pile fire was more uniform and the carbon monoxide/carbon dioxide ratio lower, but otherwise the differences were not substantial.

Sandberg et al. (1975) sampled the emissions from ponderosa pine logging

Figure 4-2. Windrow burning of pine and hardwood slash in central Georgia. Source: U.S.D.A. Forest Service.

slash. Artificial fuel beds were prepared from material collected from the San Bernardino National Forest in California. The beds were the equivalent of a 125 ton ha^{-1} fuel loading and were similar in size and distribution to actual logging slash. Emission factors are presented in Table 4–12. Of the hydrocarbons, 15–40% were composed of methane and ethylene. Ethane and acetylene were the next most important. Photochemically important compounds represented only a minor portion of the hydrocarbon fraction (Sandberg et al., 1975),

The percentage of Douglas fir and western larch slash that was converted by controlled burns to various contaminants was determined in the field at the Lolo National Forest, Montana, by Malte (1975). His results are presented in Table 4-13.

Table 4-12. Emission Factors for Ponderosa Pine Slash.

Air contaminant	kg ton^{-1} of fuel
Carbon monoxide	66
Hydrocarbons	3.6
Particulates	4

Source: Sandberg et al. (1975).

Table 4-13. Emission Factors for Douglas Fir and Western Larch Slash.

Air contaminant	kg ton^{-1} of fuel
Carbon dioxide	1130
Carbon monoxide	129
Nitrogen dioxide	7.7
Particulates	6.3

Source: Malte (1975).

In an investigation of the contribution of various forms of agricultural burning to California air quality, Darley et al. (1966) included native woody brush in their artificial burnings. Considerable differences in emission were observed depending on whether the brush was "green" or dry (Table 4–14). The yield of olefins in the hydrocarbon fraction of the burned brush was equal to the ethene and saturates plus acetylenes combined.

Adams et al. (1976) reported on field studies of particulate distribution from broadcast slash burns in the Flathead National Forest, Montana. Daily 24-hour high volume suspended particulate concentrations measured were significantly higher at the three downwind sampling sites on prescribed fire days relative to nonfire days. Particulate emission factors were observed by these investigators to vary according to fuel size class (Table 4–15).

b. Management Burning (Nonslash Fires)

In warm and dry areas of the temperate forest, for example, the southern portion of the United States, litter may accumulate and constitute a significant fire hazard in production forests. As a result, prescribed fire is used to reduce the potential damage by wildfire. Managed fire is also employed to protect the commercially important, subclimax pine species from hardwood incursion. Other benefits from artificial fires in the South frequently include site preparation, disease control, improvement of wildlife habitat, and occasionally slash disposal. Over 81 X 10^4 ha (two million acres)(Mobley, 1976) and six million tons of fuel

Table 4-14. Emission Factors for California Woody Brush.

Air contaminant	kg ton^{-1} of fuel	
	Green	Dry
Carbon dioxide	693	1168
Carbon monoxide	61	29
Hydrocarbons	12	2

Source: Darley et al. (1966).

Table 4-15. Particulate Emission Factors for Douglas Fir, Larch, and Spruce of Various Sizes.

Tree material	kg ton^{-1} of fuel
Larch needles	5.8
Larch twigs 0-1 cm	9.3
Larch twigs 1-2.5 cm	3.3
Larch needles and twigs	5.1
Douglas fir needles	4.8
Douglas fir twigs 0.1 cm	3.8
Douglas fir twigs 1-2.5 cm	3.3
Douglas fir needles and twigs	2.8
Duff	7.5
Spruce	3.7

Source: Adams et al. (1976).

(Dietrich, 1971) are prescribed burned annually in the South (Figure 4–3).

Georgia leads all states in the United States in acreage burned for agricultural and forestry purposes, with approximately 41 X 10^4 ha (1 million acres) artificially burned annually. Ward and Elliott (1976) have estimated that these Georgia fires produce 29,000 tons of particulate matter yearly.

Figure 4-3. A low intensity management or prescribed burn in a southeastern loblolly pine stand.
Source: U.S.D.A. Forest Service.

Figure 4-4. Wolf Creek wildfire that burned 41 ha (100 acres) of the Ochoco National Forest.
Source: U.S.D.A. Forest Service.

Typically prescribed fires in the South consume about seven tons of fuel ha^{-1} (three tons acre^{-1}), with a range of roughly 2.5–27 tons (Cooper, 1976; Pharo, 1976). Published emission figures for contaminants produced by these burns are less prevalent than those available for managed fires in the West. Darley et al. (1976) have test burned dry loblolly pine needles sent from Georgia and determined that head fires produced between 24 and 31 kg of particulates ton^{-1} of fuel combusted.

2. Wildfires

Fires of natural origin have historically been extremely common throughout the temperate forests (Figure 4–4). The majority of these were initiated by lightning. Our understanding of the importance of natural fires to the structure and function of forest ecosystems continues to develop. In the United States, prescribed burning and other management strategies have substantially reduced the acreage burned by natural fires.

Wildfires differ substantially from managed fires. In the South a wildfire may consume three times as much fuel as a prescribed burn (22 tons vs. 7 tons ha^{-1}).

Combustion experiments of the U.S.D.A. Forest Service, Southern Forest Fire Laboratory, Macon, Georgia have indicated that emissions from simulated wild-fires are greater than those produced by simulated prescribed fires. Smoke production may be 10 times greater in wildfire situations. Annual particulate production for prescribed fires in the South has been estimated to be 8 kg ton^{-1} of fuel while the wildfire estimate is 26 kg ton^{-1} of fuel. Emission of nitrogen oxides may also be greater in wildfires relative to controlled fires (Cooper, 1976).

Organic soils cover millions of hectares in the United States. The southern United States has approximately 2.8 million ha, with an additional 1 million ha in south Florida. These soils originate from vegetative accumulation in wet areas and soil depths can reach 10 m. During drought periods, or following drainage, these soils dry and can support combustion. Resulting fires can be persistent (months) and may burn down to the water table before going out. McMahon et al. (1980) have presented emissions estimates for organic soil fires (Table 4–16).

3. Forest Fires and Ozone

Air contaminants of forest fire origin that have received the greatest research attention include particulates and the oxides of carbon. Oxidants and nitrogen oxides have received only minor study. In 1972 during the course of air monitoring flights over prescribed burns of Western Australia, Evans et al. (1974) detected elevated concentrations of ozone at the top of some smoke plumes. It is interesting to note that typical controlled burns in Western Australia may involve areas of 4000 ha and may carry a fuel load of 4 tons ha^{-1}. After more thorough analysis, Evans et al. (1977) concluded that high-intensity burns for forest clearing purposes in Australia can produce ozone concentrations in excess of 100 ppb (196 μg m^{-3}) within 1 hour. Specific data from representative fires are presented in Table 4–17. The authors stressed that the ozone content of the plume at the point where the plume reaches the ground would be appreciably less

Table 4-16. Emission factors for organic soils.

Pollutant	μg kg^{-1}	g kg^{-1}
Particulate		
first 24 hrs		30
> 24 hrs		1
Benzo (a) pyrene mean/range	319	
(in particulate matter)	13–1187	
Carbon monoxide		269
Hydrocarbons (total)		23
Nitrogen oxides		1.7

Source: McMahon et al. (1980).

Table 4-17. Highest Ozone Concentrations in the Plume from Three West Australian Prescribed Forest Fires.

Fire no.	Peak ozone ppb	Distance downwind of fire (km)
1	65 (127 μg m^{-3})	29
7	75 (147 μg m^{-3})	15
13	60 (118 μg m^{-3})	27

Source: Evans et al. (1977).

than the upper plume maxima.

4. Commercial and Residential Wood Combustion

Since antiquity, combustion of wood for heat and energy has been useful to humankind. This utility continues in developed nations due to the abundance and renewability of the wood resource, economic competitiveness and supply reliability relative to nonrenewable fossil fuels, and the aesthetic and amenity values associated with residential woodstove and fireplace use. Since emissions from commercial and residential wood combustion may represent significant contributions to tropospheric pollution in certain regions, it is important to introduce this topic.

The need to diversify fuels combusted to produce electricity (Koning and Skog, 1987) and the difficulty of disposing of waste wood in landfills has stimulated the construction of wood-burning electric-generating facilities. During the 1980s a significant number of wood burning power plants have been built in California, Idaho, Maine, Texas, Wyoming, Virginia, and Vermont. A plant proposed for siting in northeastern Connecticut is typical of these recent installations. When built the proposed facility will burn approximately 400,000 tons yr^{-1} of chipped wood fuel generated from land-clearing operations, forest land management, sawmills, and construction/demolition activities. It will market 32 megawatts of electricity to the local utility company. The project will be designed to burn approximately 1200 tons day^{-1} of wood having a heat content approximating 8000–12,000 Btu kg^{-1} (Atkins et al., 1988). Fuel oil, No. 2, with 0.3% sulfur content, will be occasionally used for boiler start-up and flame stabilization during combustion. This facility, using best available control technology, will release in excess of 2000 tons yr^{-1} of criteria pollutants to the atmosphere (Table 4–18).

On a smaller, but very widely distributed scale, wood combustion in residential burners, may generate significant emissions to the atmosphere (Skog and Watterson, 1984). In numerous regions of the United States, woodstove emissions are the dominant single source of particulate pollutants (Burnet et al., 1986). Residential combustion emissions are highly variable due to the wide-ranging differences in woodstoves and fireplace design and operation, burn rates,

Table 4-18. Potential emissions from a 32 megawatt wood-burning power station employing best available control technology.

Pollutant emissions	Tons yr^{-1}
Carbon monoxide	1170
Nitrogen oxides	780
Particulates	78
Sulfur dioxide	46
Sulfuric acid	1
Lead	0

Source: Atkins et al. (1988).

Table 4-19. Emissions from Residential Wood Combustion

Pollutants	Woodstoves	Fireplaces
	g kg^{-1}	
Particulates (total)	8	9
Particulates (respirable)	4	2
Hydrocarbons (gaseous)	2	19
Benzo (a) pyrene	.0025	.00073
Carbon monoxide	160	22
Aldehydes	1	1

Source: Cooper (1980).

fuel type and loading, and fuel moisture. Test data suggest that actual emission for the two primary pollutants, particulates and carbon monoxide, may vary from 3 to 50 g kg^{-1} for the former, and 50–300g kg^{-1} for the latter (Burnet et al., 1986). Other pollutants released during residential wood burning include a variety of organic contaminants (Table 4–19).

The dramatically increasing use of pressure-treated wood in a variety of applications raises a very special issue regarding disposal by combustion. This wood, due to the release of toxic materials during burning, is totally unsuited for commercial or residential use. One of the most common methods of contemporary wood preservation is the use of treating formulations containing copper, chromium, and arsenic salts (chromated copper arsenate or CCA). McMahon et al. (1985) combusted CCA treated wood under a variety of combustion conditions and observed volatilization of 11% of the copper, 15% of the chromium and variable amounts of arsenic (22–77%), depending on the length and temperature of the burn (McMahon et al. 1986). Of the volatilized arsenic recovered, essentially all was condensed onto particulates and consisted of both arsenites and arsenates.

D. Summary

From an air quality perspective, volatile hydrocarbons represent the most important gaseous emission released directly to the atmosphere by forest trees during the course of normal metabolism. An average of 70% of the trees in various United States forests release reactive hydrocarbons. On a global basis, reactive hydrocarbons released by trees exceed the total amount released by anthropogenic sources. The most important compounds released include α-pinene, camphlene, β-pinene, limonene, myrcene, β-phellandrene, and isoprene.

The most important particulate contaminants released by forests, in terms of direct human health significance, are the various pollens. Conifer pollen is distributed by wind for considerable distances but fortunately is generally nonallergenic. Deciduous pollen, which frequently is allergenic, may also be distributed considerable distances by wind and other vectors. While annual pollen production varies greatly due to several factors, an average production of 0.10 kg tree^{-1} may be used for gross estimates of total particulate production. The medical evidence relating to human disease caused by pollen in terms of history, scope, abundance, and confidence of the data exceeds information available for the relationship between any other air contaminant and human health.

Hydrocarbon aerosols are another important particulate contaminant resulting from tree metabolism. Terpenes, particularly, α-pinene, released from tree foliage may react in the atmosphere to form submicrometer particles. The information on the production, transport, and persistence of these particles in natural environments and their role in haze formation of forest regions is not complete but is much improved. Hooper et al. (1987) investigated the regional haze in the Pacific Northwest during summer, 1984 and concluded that organic and elemental carbon associated with vegetative burning was the most important source category, accounting for approximately one-half of the particle-induced atmospheric extinction.

Burning of forests or forest debris by natural or managed fires is a special case where forests may contribute contaminants to the atmosphere. These fires can contribute locally and regionally significant amounts of particulates (Griffin and Goldberg, 1979), carbon oxides, and hydrocarbons to the atmosphere. Limited evidence suggests nitrogen oxides and ozone should be added to this list. Contaminant additions to the atmosphere from wildfires in most regions of temperate forests are less than they were several decades ago due to improved fire management practices. Burning in tropical forests, however, is widespread and may have important global air quality implications. The local and regional importance of contaminants released from controlled forest burning has been reduced again due to improvement in fire management, especially in greater appreciation of meteorological conditions optimal for maximum effluent dispersion.

The U. S. Environmental Protection Agency has estimated that forest wildfires and managed burning in the United States contributed the following percentages of the total estimated national emissions from 1970 to 1976: 4% of the carbon monoxide, 3% of the particulates, 2% of the hydrocarbons, and 0.6% of the oxides of nitrogen (Table 4–20).

On a global basis, carbon monoxide estimates from forest fires range from 11 X 10^6 tons yr^{-1} (Robinson and Robbins, 1968) to 6 X 10^7 tons yr^{-1} (Seller, 1974). Wong estimated the gross input of global carbon dioxide to be 5.7 X 10^9 tons of carbon yr^{-1} (Wong 1978). Fahnestock (1979) argues that the latter figure may be as much as four times too high.

Due to the large emissions of carbon monoxide and nitrogen oxides, the siting of wood-burning electric-generation stations in regions of high carbon monoxide or oxidant loads must be done with extreme caution and clearly only with the application of best available control technology. Also due to the significant contributions of residential wood burners to local particulate loads it is important to bring these sources under national and state, and perhaps local, as necessary, regulatory control. Wood that has been pressure treated with soluble salts, e.g., the CCA process, should not be combusted in standard residential or commercial burners, but should be handled in specialized hazardous waste facilities.

Table 4-20. United States Emission Estimates for Forest Wildfires and Managed Burning, 1970 through 1976.

Year	10^6 tons yr^{-1}			
	Particulates	Nitrogen oxides	Hydrocarbons	Carbon monoxide
1970	0.5	0.1	0.7	3.5
1971	0.7	0.2	1.0	5.1
1972	0.5	0.1	0.7	3.6
1973	0.4	0.1	0.6	2.9
1974	0.5	0.1	0.6	3.9
1975	0.3	0.1	0.4	2.3
1976	0.6	0.2	0.8	4.8

Source: U.S. Environmental Protection Agency (1977).

References

Adams, D.F., R.F. Koppe, and E. Robinson. 1976. Air and surface measurements of constituents of prescribed forest slash smoke. In Air Quality and Smoke from Urban and Forest Fires. National Academy of Sciences, Washington, DC, pp. 105-147.

Anderson, E. F., C.S. Dorsett, and E.O Flemming. 1978. The airborne pollens of Walla Walla, Washington. Ann. Allergy 41: 232-235.

Anderson, S.T. 1967. Tree pollen rain in a mixed deciduous forest in South Jutland (Denmark). Rev. Palaeobot. Palynol. 3: 267-265.

Atkins, R.S., D.T. Grasso, and M.I. Holzman. 1988. Application of the Killingly Energy Limited Partnership to the Connecticut Siting Council for a Certificate of Environmental Compatibility and Public Need for the Construction of a 32.2 MW (met) Wood-Burning Electric Generating Facility. ERL Proj: No. 6322-C11-87. Killingly Energy Limited Partnership, Syracuse, NY.

Beauford, W., J. Barber, and A.R. Barringer. 1977. Release of particles containing metals from vegetation into the atmosphere. Science 195: 571-573.

Brooks, J. 1971. Some chemical and geochemical studies on sporopollenin. In J. Brooks, P.R. Grant, M. Muir, P. van Gijzel, and G. Shaw, eds., Sporopoillenin. Academic Press, New York, pp. 351-407.

Bufalini, J.J. 1980. Impact of Natural Hydrocarbons on Air Quality. U.S. Environmental Protection Agency, Report No. EPA-600/z-80-086, Research Triangle Park, NC, 59 pp.

Burnet, P.G., N.G. Edmisten, P.E. Tiegs, J.E. Houck, and R.A. Yoder. 1986. Particulate, carbon monoxide, and acid emission factors for residential wood burning stoves. J. Air Pollu. Control Assoc. 36: 1012-1018.

Calvert, J.G. 1976. Hydrocarbon involvement in photochemical smog formation in Los Angeles atmosphere. Environ. Sci. Technol. 10: 256-262.

Cooper, J.A. 1980. Environmental impact of residential wood combustion. J. Air Pollu. Control Assoc. 30: 855-861.

Cooper, R.W. 1976. The trade-offs between smoke from wild and prescribed forest fires. In Air Quality and Smoke from Urban and Forest Fires. National Academy of Sciences, Washington, DC, pp. 19-26.

Cramer, O.P. and H.E. Graham. 1971. Cooperative management of smoke from slash fires. J. For. 69: 327-331.

Cramer, O.P. and J.N. Westwood. 1970. Potential impact of air quality restrictions on logging residue burning. U.S.D.A. Forest Service, Pacific Southwest S.W. Forest and Range Exp. Sta., Res. Paper PSW-64, 12 pp.

Dacey, J.W.H. and M.J. Klug. 1979. Methane efflux from lake sediments through water lilies. Science 203: 1253-1254.

Darley, E.F., F.R. Burleson, E.H. Mateer, J.T. Middleton, and V.P. Osteri. 1966. Contribution of burning of agricultural wastes to photochemical air pollution. J. Air Pollu. Control Assoc. 11: 685-690.

Darley, E.F., S. Lerman, G.E. Miller Jr., and J.F. Thompson. 1976. Laboratory testing for gaseous and particulate pollutants from forest and agricultural fuels. In Air Quality and Smoke from Urban and Forest Fires. National Academy of Sciences, Washington, DC, pp. 78-89.

Davies, I.W., R.M. Harrison, R. Perry, D. Ratnayaka, and R.A. Wellings. 1976. Municipal incinerator as source of polynuclear aromatic hydrocarbons in environment. Environ. Sci. Technol. 10: 451-453.

Dieterich, J.H. 1971. Air quality aspects of prescribed burning. In Proc. Prescribed Burning Symposium. U.S.D.A. Forest Service, Southeastern Forest Exp. Sta., Asheville, NC, pp. 139-151.

Erdman, G. 1969. Handbook of Palnology. Hofner, New York, 486 pp.

Evans, L.F., N.K. King, D.R. Packham, and E.T. Stephens. 1974. Ozone measurements in smoke from forest fires. Environ. Sci. Technol. 8: 75-76.

Evans, L.F., I.A. Weeks, A.J. Eccleston, and D.R. Packham. 1977. Photochemical ozone in smoke from prescribed burning of forests. Environ. Sci. Technol. 11: 896-900.

Fahnestock, G.R. 1979. Carbon input to the atmosphere from forest fires. Science 204: 209-210.

Feldstein, M. 1974. A critical review of regulations for the control of hydrocarbon emissions from stationary sources. J. Air Pollu. Control Assoc. 24: 469-478.

Fritschen, L., H. Bovee, K. Buettner, R. Charlson, L. Monteith, S. Pickford, J. Murphy, and E. Darley. 1970. Slash fire atmospheric pollution. U.S.D.A. Forest Service, Pacific Northwest Forest and Range Exp. Sta., Res. Paper No. PNW-97, 42 pp.

Gershenzon, J. 1984. Changes in the levels of plant secondary metabolites under water and nutrient stress. In B.N. Timmermann, C. Steelink and F.A. Loewus, eds., Phytochemical Adaptations to Stress. Plenum Press, NewYork, pp 273-320.

Griffin, J.J. and E.D. Goldberg. 1979. Morphologies and origin of elemental carbon in the environment. Science 206: 563-565.

Hall, J.A. 1972. Forest fuels, prescribed fire, and air quality. U.S.D.A. Forest Service, Pacific Northwest Forest and Range Exp. Sta., Portland, OR, 44 pp.

Hanover, J.W. 1966. Genetics of terpenes. 1. Gene control of monoterpene levels in *Pinus monticola* Dougl. Heredity 21: 73-84.

Hedin, A. and T. Turner. 1977. What is burned in a prescribed fire? Department of Natural Resources, Note No. 16, Olympia, WA, 7 pp.

Hess, D. 1975. Plant Physiology. Springer–Verlag, New York, 333 pp.

Heuss, J.M., G.J. Nebel, and B.A. D'Alleva. 1974. Effects of gasoline aromatic and lead content on exhaust hydrocarbon reactivity. Environ. Sci. Technol. 8: 641-647.

Hooper, M.H., J.E. Core, D.F. Weaver, N.N. Maykut, and J.L. Boylan. 1987. Regional haze in the Pacific Northwest States-panoramas. In Prescribed Forest Burning Emissions — Their Characterization, Control and Effects. Tech. Bull No. 535, National Council Paper Industry Air Stream Improvement, New York, pp. 6-20.

Hyde, H.A. 1951. Pollen output and seed production in forest trees. Q. J. For. 45: 172-175.

Hyde, J.S., N.V. Aroda, C.M. Kumar, and B.S. Moore. 1978. Chronic rhinitis in the pre-school child. Ann. Allergy 41: 216-219.

Knights, R.L., D.R. Cronn, and A.L. Crittenden. 1975. Diurnal patterns of several components of urban particulate air pollution. Paper No. 3, Pittsburgh Conference on Analytical Chemistry and Applied Spectoscopy, Cleveland, OH, March 3, 1975.

Komarek, E.V., B.B. Komarek, and T.C. Carlysle. 1973. The Ecology of Smoke Particulates and Charcoal Residues from Forest and Grassland Fires: A Preliminary Atlas Misc. Publ. No. 3, Tall Timbers Research Sta., Tallahassee, FL, 75 pp.

Koning, J.W. and K.E. Skog. 1987. Use of wood energy in the United States — An opportunity. Biomass 12: 27-36.

Lamb, A.K., H.H. Westberg, T. Quarles, and D.L. Flyckt. 1984. Natural Hydrocarbon Emission Rate Measurements from Selected Forest Sites. U.S. Environmental Protection Agency, Report No. EPA-600/S3-84-001. Research Triangle Park, NC, 6 pp.

Lande, G.E. 1987. A study of air quality sampling during prescribed burning for forest sites treated with desiccants and herbicides. In, Prescribed Forest Burning Emissions — Their Characterization, Control and Effects. Tech. Bull. No. 535, National Council Paper Industry Air and Stream Improvement, New York, pp. 2-5.

Lewis, W.H. and W.E. Imber. 1975. Allergy epidemiology in the St. Louis, Misouri, Area. III. Trees. Ann. Allergy 35-113-119.

Lichti-Federovich, S. and J.C. Ritchie. 1968. Recent pollen assemblages from the western interior of Canada. Rev. Paleobot. Palynol. 7: 297-344.

Maarse, H. and R.E. Kepner. 1970. Changes in composition of volatile terpenes in Douglas fir needles during maturation. J. Agr. Food Chem. 18: 1095-1101.

Mah, R.A., D.M. Ward, L. Baresi, and T.L. Glass. 1977. Biogenesis of methane. Annu. Rev. Microbiol. 31:309-341.

Major, R.T., P. Marchini, and A.J. Boulton. 1963. Observations on the production of a-hexanal by leaves of certain plants. J. Biol. Chem. 238: 1813-1816.

Malte, P.C. 1975. Pollutant production from forest slash burns. Bulletin No. 339, College of Engineering, Washington State Univ. Pullman, WA, 32 pp.

McMahon, C.K., P.B. Bush , and E.A. Woolson. 1985. Relese of copper, chromium, and arsenic from burning wood treated with preservatives. Proc. 78th Annual Meeting Air Pollu. Cont. Assoc., June 16–21, 1985, Detroit, MI.

McMahon, C.K., P.B. Bush, and E.A. Woolson. 1986. How much arsenic is released when CCA treated wood is burned? For. Prod. J. 36: 45-50.

McMahon, C.K., D.D. Wade, and S.N. Tsoukalas. 1980. Combustion characteristics and emissions from burning organic soils. In Proc. 73rd Ann. Meeting Air Pollu. Control Assoc., Montreal, Canada, June 22–27, 1980.

McMahon, C.K. and S.N. Troukalas. 1978. Polynuclear aromatic hydrocarbons in forest fire smoke. In D.W. Jones and R. I. Freudenthal, eds., Vol. 3. Carcinogenesis: A Comprehensive Survey. Raven Press, New York.

Mobley, H.E. 1976. Summary of state regulations as they affect open burning. In Air Quality and Smoke from Urban and Forest Firest. National Academy of Sciences, Washington, DC, pp. 206-212.

Moore, P.D. and J.A. Webb. 1978. An Illustrated Guide to Pollen Analysis. Wiley, New York, 133 pp.

National Academy of Sciences. 1977. Ozone and Other Photochemical Oxidants. National Academy of Sciences, Washington, DC, 719 pp.

Newmark, F.M. 1978. The hay fever plants of Colorado. Ann. Allergy 40: 18-24.

Nilsson, S., J. Praglowski, L. Nilsson, and N.O. Kultur. 1977. Atlas of Airborne Pollen Grains and Spores in Northern Europe. Ljungforetagen, Stockholm, Sweden, 159 pp.

O'Brien, R. J., J.R. Holmes, and A.H. Bockian. 1975. Formation of photochemical aerosol from hydrocarbons. Environ. Sci. Technol. 9: 568-576.

Parsons, J.S. and S. Mitzner. 1975. Gas chromatographic method for concentration and analysis of traces of industrial organic pollutants in environmental air and stacks. Environ. Sci. Technol. 9: 1053-1058.

Pharo, J.A. 1976. Aid for maintaining air quality during prescribed burns in the South. U.S.D.A. Forest Service, Res. Paper No. SE-152, Southeastern Forest Exp. Sta., Asheville, NC, 11 pp.

Radwan, M.A. and W.D. Ellis. 1975. Clonal variation in monoterpene hydrocarbons of vapors of Douglas-fir foliage. For. Sci. 21: 63-67.

Rasmussen, R.A. 1964. Terpenes: Their analysis and fate in the atmosphere. Ph.D. Thesis, Washington Univ., St. Louis, MO.

Rasmussen, R.A. 1970. Isoprene: Identified as a forest-type emission to the atmosphere. Environ. Sci. Technol. 4: 667-671.

Rasmussen, R.A. 1972. What do the hydrocarbons from trees contribute to air pollution? J. Air Pollu. Control Assoc. 22: 537-543.

Rasmussen, R.A. and M.W. Holdren. 1972. Analyses of C_5 to C_{10} hydrocarbons in rural atmosphere. Paper No. 72-19, presented at 65th Annual Meeting of the Air Pollu. Control Assoc., Miami Beach, FL, June 18-22, 1972.

Rasmussen, R.A. and F.W. Went. 1965. Volatile organic material of plant origin in the atmosphere. Proc. Natl. Acad. Sci. U.S.A. 53: 215-220.

Ripperton, L.A. and D.Lilian. 1971. The effect of water vapor on ozone synthesis in the photo-oxidation of alpha-pinene. J. Air Pollu. Control Assoc. 21: 629-635.

Robinson, E. and R.C. Robbins. 1968. Sources Abundance and Fate of Gaseous Atmospheric Pollutants. Final Report SRI Project PR-6755. Stanford Research Institute, Menlo Park, CA.

Rubin, J.M. and N.S. Weiss. 1974. Practical Points in Allergy. Medical Examination, New York, 208 pp.

Sandberg, D.V., S.G. Pickford, and E.F. Darley. 1975. Emssions from slash burning and the influence of flame retardant chemicals. J. Air Pollu. Control Assoc. 25: 278-281.

Schwartz, W. 1974. Chemical Characterization of Model Aerosols. EPA-650-3-74-011. Battelle Memorial Institute, Columbus, OH, 129 pp.

Seiler, W. 1974. The cycle of atmospheric CO. Tellus 26: 116-135.

Shain, L. and W.E. Hillis. 1972. Ethylene production in *Pinus radiata* in response to Sirex-Amylostereum attack. Photopathology 62: 1407-1409.

Silen, R.R. and D.L. Copes. 1972. Douglas-fir seed orchard problems — A progress report. J. For. 70: 145-147.

Smith, R.H. 1982. Xylem Monoterpenes of Some Hard Pines of Western North America: Three Studies. U.S.D.A. Forest Service, Pacific Southwest Forest Range Exp. Sta., Res. Paper No. PSW-160, Berkeley, CA, 7 pp.

Squillace, A.E. 1971. Inheritance of monoterpene composition in cortical oleoresin of slash pine. For. Sci. 17: 381-387.

Stanley, R.G. and H.F. Linskens. 1974. Pollen Biology Biochemistry Management. Springer–Verlag, New York, 307 pp.

Stern, A.C., H.C. Wohlers, R.W. Boubel, and WP. Lowry. 1973. Fundamentals of Air Pollution. Academic Press, New York, 492 pp.

Tauber, H. 1965. Differential pollen dispersion and the interpretation of pollen diagrams, Danm. Geol. Anders. II R. 89: 1-69.

Tauber, H. 1967. Investigations of the mode of pollen transfer in forested areas. Rev. Palaeobot. Palynol. 3: 277-287.

Taylor, R.F. 1984. Bacterial triterpenoids. Microbiol. Rev. 48: 181-198.

Turk, A. and C.J. D'Angio. 1962. Composition of natural fresh air. J. Air Pollu. Control Assoc. 12: 29-33.

U.S.D.A. Forest Service. 1974. Seeds of Woody Plants in the United States. Agr. Handbook No. 450, U.S.D.A., Forest Service, Washington, DC, 833 pp.

U.S. Environmental Protection Agency. 1986. National Air Pollutant Emission Estimates, 1940–1984. Report No. EPA-450/4-85-014. U.S. Environmental Protection Agency, Research Triangle Park, NC.

Ward, D.E. and E.R. Elliott. 1976. Georgia rural air quality: Effect of agricultural and forestry burning. J. Air Pollu. Control Assoc. 26: 216-220.

Weiss, R.E., A.P. Waggoner, R.J. Charlson, and N.C. Ahlquist. 1977. Sulfate aerosol: Its geographical extent in the mid western and southern United States. Science 195: 979-980.

Went, F.W. 1955. Air pollution. Sci. Am. 192: 62-72.

Went, F.W. 1960a. Blue hazes in the atmosphere. Nature 187: 641-643.

Went, F.W. 1960b. Organic matter in the atmosphere and its possible relation to petroleum formation. Proc. Natl. Acad. Sci. U.S.A. 46: 212-221.

Went, F.W. 1964. The nature of Aitken condensation nuclei in the atmosphere. Proc. Natl. Acad. Sci. U.S.A. 51: 1259-1267.

Went, F.W., D.B. Slemmons, and H.N. Mozingo. 1967. The organic nature of atmospheric condensation nuclei. Proc. Natl. Acad. Sci. U.S.A. 58: 69-74.

Westburg, H. and B. Lamb. 1985. Ozone Production and Transport in the Atlanta, Georgia Region. U.S. Environmental Protection Agency, Report No. EPA-600-S3-85-013, Research Triangle Park, NC, 4 pp.

White, J.D. 1985. Validating a simplified determination of benzo [a] pyrene in particulate matter from prescribed forestry burning. Am. Ind. Hyg. Assoc. 46: 299-302.

Wodehouse, R.P. 1945. Hay Fever Plants. Chronica Botanica Co., Waltham, MA, 245 pp.

Wong, C.S. 1978. Atmospheric input of carbon dioxide from burning wood. Science 200: 197-199.

Yoo, T., E. Spitz, and J.L. McGerity. 1975. Conifer pollen allergy: Studies of immunogencity and cross antigenicity of conifer pollens in rabbit and man. Annu. Allergy 34: 87-93.

5
Forests as Sinks for Air Contaminants: Soil Compartment

Air contaminants may be removed from the atmosphere by a variety of mechanisms. The primary processes are precipitation scavenging, chemical reaction, and dry deposition (sedimentation) and absorption (impaction) (Rasmussen et al., 1974). Wet deposition involves transfer to the earth in association with all forms of precipitation. Primary and secondary contaminants are subject to a large number of chemical reactions in the atmosphere that may ultimately transform them into an aerosol or an oxidized or reduced product. Attachment by aerosols and subsequent deposition on the surface of the earth is termed *dry deposition*. Absorption by water bodies, soils, or vegetation at the surface of the earth is an extremely important component of the removal process.

Components of the ecosystems of the earth that remove pollutants from the atmospheric compartment and store, metabolize, or transfer them may be termed *sinks* (Warren, 1973). Forest ecosystems in general, and temperate forest ecosystems in particular, are locally, regionally, and globally important sinks for a wide variety of atmospheric contaminants. Soil and vegetation surfaces represent the major sink for pollutants introduced into terrestrial systems (Little, 1977).

The transfer of contaminants from the atmospheric compartment to the surfaces of soil or vegetation is expressed as a flux (pollutant uptake) rate and is given as the weight of pollutant removed by a given surface area per unit time. Actual determinations of flux rates (sink strengths) are extremely complex and involve an appreciation of atmosphere conditions (wind, turbulence, temperature, humidity), pollutant nature and concentration, sink surface conditions (geometry, presence or absence of moisture), and other parameters (Chapter 2).

The principal repository in terrestrial ecosystems for air contaminants of anthropogenic origin is soil. By virtue of their distribution and physical and chemical characteristics, forest soils may be particularly efficient short- and long-term sinks. For selected trace metals and gases, the evidence supporting the importance of soils as sinks is considerable.

A. Forest Soils as Particulate Sinks

Particles are transferred from the atmosphere to forest soils directly by dry deposition and precipitation scavenging and indirectly via leaf and twig fall. A very large number of human activities generate small particles (.01–5 μm) with high concentrations of trace metals. Depending on weather conditions these particles may remain airborne for days or weeks and be transported 100–1000 km from their source. The evidence that forest soils may be the ultimate or temporary repository for the trace elements associated with these particles is substantial. Soils have a very high affinity for heavy metals, particularly the clay and organic colloidal components (John et al., 1972; Lagerwerff, 1967; Stevenson, 1972; Pertruzzelli et al., 1978; Somers, 1978; Korte et al., 1976; McBride, 1978; Zunino and Martin, 1977a,b).

1. Trace Metals

Galloway et al. (1982) have provided a comprehensive review of trace metal deposition from the atmosphere. These authors examined metal-emission rates (mobilizing factors), ratios of metal concentrations in the atmosphere (enrichment factors), and historical trends in metal deposition to evaluate the relative source strength of natural versus human sources. One or more techniques indicated that human activities caused moderate or large increases in heavy metal deposition from the atmosphere for antimony, cadmium, chromium, copper, lead, molybdenum, nickel, silver, tin, vanadium, and zinc. Small increases caused by human activities were indicated for cobalt, manganese, and mercury.

The processes associated with human release of heavy metals are varied and range from combustion activities to agricultural activities. Particularly important combustion processes include coal burning, solid waste incineration, iron and steel industry burning, metal smelting and motor vehicle combustion. Coal combustion is an especially important source of heavy metal input to the atmosphere. The trace metal content of coal is site and coal bed specific. Antimony, cadmium, lead, mercury, and molybdenum, however, are commonly more abundant in coal than in the average heavy metal composition of the earth's crust. At the beginning of this decade, annual consumption of coal in the United States was about 600 million metric tons. By 1990 this consumption is projected to double (National Research Council, 1980). Coal pretreatment (washing), where major amounts of mineral matter are separated, can reduce the heavy metal content of coal. Current best-available technology for emissions management can collect more than 99% of the solid residues of coal combustion. Tall stacks, emissions management associated with new plants and coal washing have reduced heavy metal deposition to forest, aquatic, and agricultural ecosystems in the local environments of these point sources. Older coal-burning plants, however, will continue to release heavy metals to the atmosphere in large amounts until they are closed or required to install best-available technology for emissions reduction. In addition, however, some fraction of gaseous emissions and submicron size particles will continue to escape emission-control devices of all coal-burning plants and enter the atmosphere. Currently soil disturbance by cultivation is the most significant agricultural heavy metal source; historically it has been pesticide applications.

Parekh and Husain (1981) attempted to characterize the heavy metal deposition in rural New York State during the growing season by human source. They suggested an association of: antimony and zinc deposition with refuse incineration; chromium, iron and manganese deposition with iron and steel industrial activity; and iron and manganese deposition with soil disturbance.

Human activities introduce heavy metals to the atmosphere primarily in particulate form. Only cadmium and mercury are thought to enter the troposphere in the vapor phase. During high-temperature combustion (coal, industrial, or motor vehicle), metal elements and their oxides become volatilized. The elements of high volatility, for example, cadmium, chromium, nickel, lead, thallium, and zinc, show a pronounced concentrating effect as they condense on fine particle surfaces. The concentrations of these elements in fly ash from coal-fired power plants increase greatly with decreasing particle size (Davidson et al., 1974). Due to large surface-to-volume ratios, particles with diameters approximately 1 μm or less contain as much as 80% of their total elemental mass on the surface (Linton et al., 1976).

Preferential association of heavy metals with small particles is not only significant because these small particles may escape emission control, but because these small particles have the longest atmospheric residence times and therefore can be carried long distances. Depending on climate conditions and topography, fine particles may remain airborne for days or weeks, and may be transported 100–1000 km or more from their source. This long-distance transport and subsequent deposition qualify heavy metals as regional-scale air pollutants (Smith, 1985). Ecosystems downwind of major power-generating, industrial, or urban complexes receive atmospheric deposition that has accumulated and integrated heavy metals from fossil-fuel electric power plants, metal smelters, foundaries, steel mills, incinerators, and motor vehicles.

Heavy metal particles are deposited to forests by both wet and dry processes (Swank, 1984). Dry deposition is presumed more effective for large particles and elements such as iron and manganese, while wet deposition is presumed more effective for fine particles and elements such as cadmium and lead (Galloway et al. 1982). Galloway et al. (1982) have provided an excellent inventory of heavy metal deposition rates (kg ha^{-1} yr^{-1}) for urban, rural, and remote environments. Evidence from studies in the eastern United States indicates that wet deposition of certain heavy metals, for example cadmium, lead, manganese, and zinc, is highest during the growing season (warmest months) (Lindberg 1982).

Extensive evidence is available to support the suggestion that heavy metals deposited from the atmosphere to forest systems are accumulated in the upper soil horizons or forest floors (Andresen et al. 1980, Friedland et al. 1984a, Friedland et al. 1984b, Miller and McFee, 1983; Page and Chang, 1979; Siccama et al., 1980; Smith and Siccama, 1981).

2. Lead

Lead is naturally present, in small amounts, in soil, rocks, surface waters, and the atmosphere. Due to its unique properties it has been an element widely useful to humans. This utility has resulted in greatly elevated lead concentrations in certain ecosystems. In locations where lead is being mined, smelted, and refined, where industries are consuming lead, and in urban-suburban complexes the environmental lead level is greatly elevated. It is widely agreed that a primary source in these latter sites is the combustion of gasoline containing lead additives. Other important sources include coal combustion, refuse and sludge incineration, burning or attrition of lead-painted surfaces, and industrial processes. Since the vast majority of atmospheric lead particles are less than 0.5 μm (Smith, 1976), they have been widely distributed to all parts of the earth.

The input of lead, its cycling within forest ecosystems, its transfer in food chains, its rate of loss to downstream aquatic systems, and its residence in soil has received considerable research attention.

Lead deposited from the atmosphere will vary greatly depending on area and regional source strengths. In urban areas, with abundant motor vehicle use, the lead flux rate may exceed 3000 g ha^{-1} yr^{-1} (Chow and Earl, 1970) while areas distant from cities and industrial or power generating sources may have flux rates less than 20 g ha^{-1} yr^{-1} (Chow and Johnstone, 1965).

The lead concentration of the upper soil horizons of unmineralized and uncontaminated areas ("baseline" level) is generally given as approximately 10–20 μg g^{-1} of dry soil. Analyses of forest soils from throughout temperate zone forests, however, frequently show elevated lead amounts associated with the soil compartment. Lead may be added to the soil as components of organic compounds in plant debris. The divalent cationic nature of lead added via precipitation, dry fallout, throughfall, or stemflow may cause the lead to be bound to organic exchange surfaces (Zimdahl and Skogerboe, 1977) abundant in the forest floor. Subsequent reaction of lead with sulfate, phosphate, or carbonate anions may reduce its solubility and impede its downward migration in forest soil profiles. Lead deposited from the atmosphere may have a residence time approximating 5000 yr in the surface organic soil horizons (Benninger, 1975), and long-term concentration increases can be predicted as long as inputs to forest soils exceed outputs.

The author and his colleague, T.G. Siccama, have been involved in an intensive biogeochemical study of several trace metals, including lead, in a northern hardwood forest (Siccama and Smith, 1978; Smith and Siccama, 1980). The study is being conducted on the Hubbard Brook Experimental Forest in central

New Hampshire (elevation 230–1010 m). Hubbard Brook was judged to be an excellent site to do this work because of the potential for quantifying input/output relationship (small watershed technique), the accumulation of related information on other elements, and the long term research plans for the study site (Likens et al., 1977). The Hubbard Brook Experimental Forest is a 3160 ha reserve, located in the White Mountain National Forest of north central New Hampshire, dedicated to the long-term study of forest and stream ecosystems. The forest is very approximately 120 km northwest of Boston and is relatively distant from major sources of heavy metal emission. It has been shown, however, that due to continental air mass movement, storm tracks that have passed over centers of industrial and human activity also frequently pass over northern New England, resulting in heavy metal deposition potential that is higher than might be expected based on the distance of the experimental forest from industrial and urban centers (Schlesinger et al., 1974, Johnson et al., 1972). We presently have 15 years of data on the bulk precipitation input (dry plus wet deposition, winter and summer periods) and streamwater outputs of cadmium, copper, iron, lead, manganese, nickel, and zinc.

The Hubbard Brook forest is typified by an unbroken canopy of second-growth northern hardwoods (sugar maple, American beech, and yellow birch) with patches of red spruce and balsam fir, particularly at higher elevations and along the valley bottoms. The mean annual lead flux (1975–1984) to the forest is 190 g ha^{-1} yr^{-1}. The lead output from the system in streamwater approximates 6 g ha^{-1} yr^{-1} (Smith et al., 1986). The extraordinary disparity in these input-output figures is accounted for by the accumulation of lead in the soil compartment.

The parent material of the Hubbard Brook soil is compact unsorted acidic glacial till derived primarily from local gneissic bedrock. Profiles are strongly developed with typical podsol horizonation. A forest floor of the mor type with well-developed L, F, and H horizons of unincorporated organic matter overlies an A2 of 2–6 cm. The average lead concentration in the forest floor in 1979 was determined to be 89 μg g^{-1}, with total lead of the forest floor averaging 9 kg ha^{-1} (Figure 5-1).

In 1986, some quantitative forest floor samples were exhibiting lead concentrations in excess of 100 μg g^{-1} (Table 5-1). Presumably lead will continue to accumulate in the forest floor at Hubbard Brook, and in other temperate zone forest soils, as long as lead inputs exceed ecosystem outputs. Fortunately, in the United States, lead was made a criteria pollutant and came under Clean Air Act regulation in 1979. Efforts to reduce lead content of gasoline are presumed to be responsible for the decrease in lead input to Hubbard Brook over the last decade. Examination of the input of this element on an annual basis reveals variable concentrations in bulk precipitation but a consistent decrease in input of approximately 70% over the study period (Figure 5-2). This decrease in lead input to the forest correlates extremely well with the decrease in national urban levels (Figure 5-3) and with the decrease in lead consumed in gasoline (Figure 5-4) over the same time period.

Reiners et al. (1975) also have investigated lead retention by forest soils in the White Mountains of New Hampshire in an area they estimated had a lead flux rate of approximately 200 g ha⁻¹ yr⁻¹. Samples were collected from Mt. Moosilauke that represented various vegetative zones from the northern hardwood forest (~700 m) to the alpine tundra (~1400 m). Cores were removed from the organic layers (Oi, Oe, Oa) of the forest floor and analyzed for lead. Table 5-2 presents the soil lead concentrations obtained for the various vegetative types and compares them with various concentrations published in the literature.

These New Hampshire investigations were performed in an area reasonably remote from strong lead sources. The results, however, reveal high lead concentrations in the soil relative to other sites similarly remote from lead sources. The

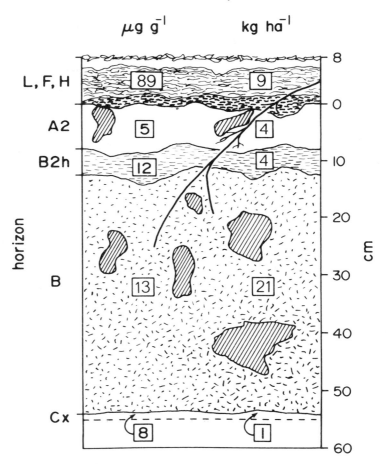

Figure 5-1. Lead distribution in a typical northern hardwood forest soil profile in central New Hampshire. Concentrations and amounts are based on acid extraction of ashed (475°C) samples and do not include lead in mineral crystal complexes not extracted by this procedure.
Source: Smith and Siccama (1980).

Table 5-1. Proton induced x-ray emission analysis of forest floor material (organic layers Oi, Oe, and Oa) collected July 1986 from a mixed stand of red spruce, hemlock, and northern hardwoods at 250 m at the Hubbard Brook Experimental Forest (White Mountain National Forest), West Thornton, NH (N = 8).

	ppm			
Element	Mean	Std. dev.	Min.	Max.
Aluminum	11,853.25	3130.14	7686	16809
Calcium	2364.50	815.09	1603	4241
Chlorine	176.37	34.94	104	213
Chromium	18.85	11.96	8	46
Copper	7.70	1.06	6	9
Gallium	3.29	1.02	2	4
Iron	6618.26	2104.73	4302	10491
Lead	113.14	12.01	95	133
Magnesium	822.69	332.80	438	133
Manganese	269.50	93.28	112	419
Nickel	7.88	1.08	6	9
Phosphorus	994.81	143.97	804	1221
Potassium	3645.75	972.49	2376	5280
Rubidium	27.21	6.41	19	36
Silicon	49,517.38	14,031.57	32,497	70,620
Strontium	29.61	4.05	24	35
Sulfur	1823.81	116.85	1731	2083
Titanium	1075.13	350.86	634	1586
Vanadium	24.15	5.66	18	35
Zinc	63.11	5.00	59	72
Zirconium	54.23	19.38	31	8

authors feel their data is consistent with the hypothesis that the New England mountains are frequently exposed to air masses from industrial and urban areas along with high winds, precipitation, and cloudiness and that the combination of these factors plus horizontal interception at higher elevations contributes to higher rates of aerosol deposition than are likely in other regions.

Van Hook et al. (1977) have examined lead cycling in eastern Tennessee's oak forests in the vicinity (14 km) of coal-fired electric generating facilities. The estimated lead input to the forest was given as 286 g ha^{-1} yr^{-1}. For the four forest types examined, the litter layer was fractioned into two horizons: original leaf form still discernible (O_1) and original form lost (O_2). The lead concentrations and amounts in the litter are presented in Table 5-3. This study calculated the standing pools of lead in the vegetative components of the study forests. From these data it was observed that the litter (O_1 and O_2 horizons), which constituted 13% of the total forest organic matter, contained 71% of the lead contained within the ecosystem. Movement of lead from the O_2 litter to the underlying soil horizons was concluded to be high and estimated at 182 g ha^{-1} yr^{-1}.

As part of the International Biological Program, Heinrichs and Mayer (1977) gathered soil lead data from beech and spruce forests in Germany that are repre-

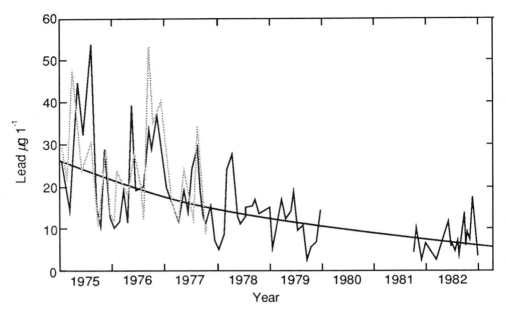

Figure 5-2. Lead concentration in bulk precipitation collected at the Hubbard Brook Experimental Forest, White Mountain National Forest, New Hampshire from 1975 to 1984. The regression line, $Y = ae^{-bx}$, has coefficients for a of 25.53 and b of -0.01589, where $Y = \mu g$ Pb 1^{-1} and x = months. The dashed line is for a collector at 490-m elevation in the south-facing experimental watersheds. The break in the data occurred because of an interruption in funding for the study.

sentative of ecosystems widely distributed in central Europe. The authors suggested that the sampled forests were in a "relatively unpolluted" environment yet the measured lead flux in precipitation below the beech canopy was 365 g ha^{-1} yr^{-1} and below the spruce canopy was 756 g ha^{-1} yr^{-1} (adjacent open field lead input in precipitation was given as 405 g ha^{-1} yr^{1}). Lead concentrations and amounts in the soils of these forests are presented in Table 5-4. All studies summarized to this point were conducted in rural sites in locations considered by their authors to be relatively remote from excessive lead contamination. Despite this fact it is obvious that considerable quantities of lead are accumulating in the soil and that this compartment is serving as an important sink for this trace metal. For locations subject to excessive lead input, for example, urban, industrial, or roadside locations, the accumulation of lead in the soil sink is even more impressive.

Parker et al. (1978) have examined lead distributions in an oak forest in a heavily urbanized and industrialized section of East Chicago, Indiana. Lead flux via precipitation and dry fall out was determined to be 815 g ha^{-1} annually. The lead concentrations for various soil horizons are given in Table 5-5.

Soil samples analyzed from sites 1–2 km from a lead smelter in Kellogg,

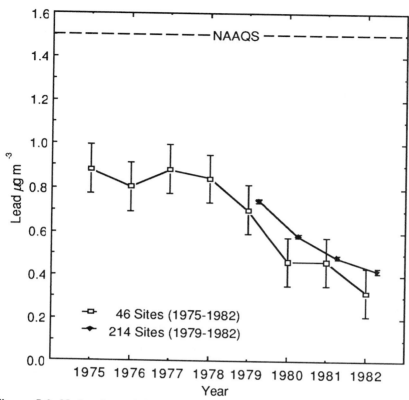

Figure 5-3. National trend in maximum quarterly average lead levels with 95% confidence intervals at 46 urban sites (1975–1982) and 214 urban sites (1979–1982).
Source: U.S.E.P.A. (1984).

Table 5-3. Lead Concentration and Amount Associated with the Litter of Eastern Tennessee Mixed Deciduous Forests.

Litter	Yellow poplar		Chestnut oak		Oak-hickory		Pine	
	Conc. (μg g^{-1})	Amt. (g ha^{-1})	Conc. (μg g^{-1})	Amt. (g ha^{-1})	Conc. (μg g^{-1})	Amt. (g ha^{-1})	Conc. (μg g^{-1})	Amt. (g ha^{-1})
O_1	31	200	27	190	25	220	31	340
O_2	42	320	51	940	35	630	37	580

Source: Van Hook et al. (1973).

Table 5-2. Lead Concentrations of the Organic Horizon of Mt. Moosilauke, New Hampshire, Forest Soils Compared with Various Published Lead Concentrations.

Area	Pb ($\mu g\ g^{-1}$)
Minimum—all sites, layers	11
Maximum—all sites, layers	336
Weighted average, all layers	
Hardwoods	35
Spruce-fir	79
Fir forest	145
Fir krummholz	120
Alpine tundra	38
Mt. Moosilauke bedrock	46
World average (presumably inorganic)	10
Connecticut, U.S.A.	10-15
New Zealand	15
Missouri, U.S.A.	17
West England −	42
Wales	42
Wales	45
Salt marsh, Connecticut	38
Heath, Sweden (above ground litter)	66, 80
Heath, Sweden (below ground litter)	40, 45
Spruce needle litter, Sweden	
Least decomposed	27, 34, 78
Intermediate	49, 61, 102
Most decomposed	66, 105
O_2, Pennsylvania	240-355

Source: Reiners et al. (1975).

Table 5-4. Lead Concentration and Amount for Various Soil Horizons of Central German Beech and Spruce Forests.

Horizon (cm)	Conc. ($\mu g\ g^{-1}$)	Amt. ($kg\ ha^{-1}$)
0-10	61	70.2
10-20	21	25.0
30-40	12	15.6
40-50	12	17.8

Source: Heinrichs and Mayer (1977).

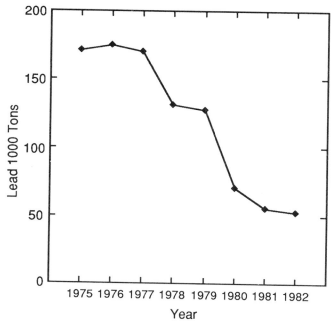

Figure 5-4. Lead consumed in gasoline, 1975–1982. Sales to the military excluded.
Source: U.S.E.P.A. (1984).

Idaho averaged 4640 μg g^{-1} at the surface (0–2 cm) (Ragaini et al., 1977). Soil samples collected from 65 major street intersections in urban locations in southern Ontario revealed a high of 21,000 μg g^{-1} lead in the upper 5 cm of soil (Linzon et al., 1976). These soils were not collected from forest sites but woody vegetation did occur in the general area.

Roadside environments are grossly contaminated with lead as motor vehicles combusting gasoline containing lead alkyls release approximately 80 mg of lead per km driven. The size of the roadside ecosystem approximates 3.04 X 10^7 ha (118,000 square miles) in the United States. Since much of this ecosystem contains woody vegetation, it is important to consider roadside soil lead burden. If 20 μg g^{-1} lead is accepted as a baseline lead concentration for uncontaminated soils, it can be seen that soil samples taken within a few meters of the road surface of a heavily traveled highway may range to more than 30 times baseline. At 10 m distance from the roadway, however, the lead level is typically only 5–15 times baseline. At approximately 20 m distance, several studies suggest that a constant level of soil lead is achieved and the influence of the roadway is lost (Smith, 1976). Invariably, investigations of roadside soil lead concentrations are positively correlated with traffic volume and are negatively correlated with perpendicular distance from the roadway. Significant variations from strict correlation occur, however, due to numerous additional variables. Since lead accumu-

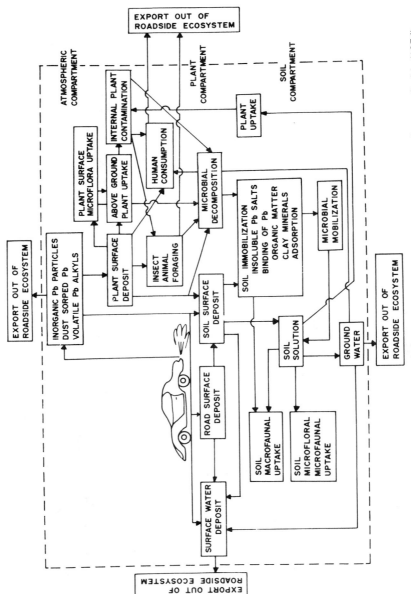

Figure 5-5. Potential lead distribution and transfer routes in the roadside ecosystem. Source: Smith (1976).

Table 5-5. Lead Concentration for Various Soil Horizons from an Urban Oak Forest in East Chicago, Indiana.

Soil horizon	Lead ($\mu g\ g^{-1}$)
Surface litter (O horizon)	400
A1 (0-2.5 cm)	463
A1 (2.5-14 cm)	140
B (14-25 cm)	8

Source: Parker et al. (1978).

lates over time in the soil, road age is important. Soil lead adjacent to an old road with lower traffic volume may exceed soil lead adjacent to a young road with higher traffic volume. If prevailing winds blow normal to the highway, significantly higher soil lead may be found on the lee side of the road. Other important variables causing deviation from generalized correlations include, of course, soil type, vehicle types, topography, and vegetative cover.

Detection of trends in rates of accumulation or loss of trace metals in soils over long periods of time is generally impossible due to the paucity of quantitatively obtained and systematically retained soil samples collected from the same site at different times. Reported changes in lead in mineral soils has generally been limited to measurements of changes in concentration. There have been reports documenting changes in amount as well as concentration in areas not associated with major local sources of metals such as smelters or highways. A search by our laboratory for preserved soil samples from New England forests revealed a unique opportunity to analyze quantitatively sampled and carefully preserved forest floor material collected in 1962 by D.L. Mader, of the University of Massachusetts as part of a study of forest floor characteristics. Lead contents were determined for Oi, Oe, and Oa layers of the forest floor of 10 white pine stands in central Massachusetts collected from the same place in 1962 and 1978. The total lead content was found to have increased significantly. The average lead concentration increased in all layers but the increases were not statistically significant primarily due to the dilution effect of the concurrent increase in the mass of the forest floor. The observed net increase in lead of 30 mg m^{-2} yr^{-1} was approximately 80% of the estimated total annual input of this element via precipitation in this region during the 16 year period (Siccama et al., 1980). The total amount of lead in the forest floor of the pine stands (12 kg ha^{-1}) was greater than the amount in the forest floor of the northern hardwood forest in New Hampshire (9 kg ha^{-1}).

When a trace metal such as lead is added to the soil it may be (a) absorbed on soil particle exchange sites, (b) precipitated as an insoluble compound, (c) leached to lower depths in the soil profile, (d) lost to the atmosphere, (e) metabolized by soil fauna or microbes, or (f) absorbed by plant roots (compare Figure 5-5). In the case of lead and forest soils, it is apparent that mechanisms c through f are generally unimportant, that mechanisms a and b prevail, and as a result the

Table 5-6. Heavy metals are essential and nonessential for plant health and have natural and pollution sources. Capitalized metals are thought, or known, to be especially significant for considerations of forest ecosystem health risk.

Essential	Nonessential	
COPPER (Cu)	Antimony (Sb)	SILVER (Ag)
IRON (Fe)	Bismuth (Bi)	THALLIUM (Te)
MANGANESE (Mn)	CADMIUM (Cd)	Thellurium (Te)
Molybdenum (Mo)	CHROMIUM (Cr)	Thorium (Th)
ZINC (Zn)	Cobalt (Co)	Tin (Sn)
	LEAD (Pb)	Uranium (U)
	MERCURY (Hg)	VANADIUM (V)
	NICKEL (Ni)	

soil compartment, particularly the organic forest floor, is an important sink for atmospheric lead. The flux of lead from the atmosphere to the forest floor is highly variable throughout the temperate zone and may range from approximately 0.01 to 1000 kg ha^{-1} yr^{-1} (See Table 2-18).

The threshold concentration of soil lead for phytotoxicity approximates 600 μg g^{-1} (Linzon et al., 1976). Since most forest ecosystems are well below this threshold, except those in selected urban, industrial, and roadside sites, and since evidence for tree root uptake of lead is meager, this biologically nonessential element is judged not to be responsible for direct impairment of tree health.

3. Other Trace Elements

Forest soils may constitute a sink for a variety of trace elements, in addition to lead. Elements judged to have particular potential for biological significance if accumulated in the terrestrial ecosystems include cadmium, nickel, thallium, copper, fluorine, vanadium, zinc, cobalt, molybdenum, tungsten, mercury, and selenium (Van Hook and Shults, 1977). Under certain conditions manganese, chlorine, chromium, and iron might be appropriate additions to this list (Table 5-6). Numerous of these elements differ substantially from lead in their interaction with plants. Some, for example, iron, chlorine, manganese, zinc, copper, and molybdenum, are required by vegetation in small amounts for normal growth and development. Others, for example, cadmium, nickel, thallium, and tungsten, are very mobile in plants. Cadmium, nickel, fluorine, thallium, vanadium, mercury, and copper have high a potential for phytotoxicity.

Information concerning the ability of forest soils to act as a sink for these elements is very limited. A comparison of Tables 5-7 and 5-8, which contain trace element concentrations for selected temperate forest soils, with Table 5-9, which contains baseline trace element concentrations for a variety of soils in relatively unpolluted areas and from regions lacking major mineral deposits, suggests that certain forest soils may be accumulating elevated levels of zinc, cad-

Table 5-7. Trace Element Concentrations and Amounts for Various Forest Soils.

Element	Location and sample site	Conc. (μg $\bar{g}^1$)	Amt. (g $\bar{ha}^1$)	Reference
Zinc	White Mountains, N.H., soil organic horizons			Reiners et al. (1975)
	Minimum—all sites, layers	19		
	Maximum—all sites, layers	169		
	Weighted average, all layers			
	Hardwoods	100		
	Spruce-fir	53		
	Fir forest	90		
	Fir krummholz	73		
	Alpine tundra	29		
	Eastern Tennessee forests			Van Hook et al. (1977)
	Soil litter layer O_1			
	Yellow poplar	58	380	
	Chestnut oak	42	290	
	Oak-hickory	48	420	
	Pine	56	610	
	Soil litter layer O_2			
	Yellow poplar	130	980	
	Chestnut oak	110	2000	
	Oak-hickory	125	2200	
	Pine	59	930	
	Central German beech and spruce forests, various soil depths			Heinrichs and Mayer (1977)
	0-10 cm	38	44×10^3	
	10-20 cm	50	60×10^3	
	30-40 cm	55	72×10^3	
	40-50 cm	49	73×10^3	
	East Chicago oak forest (urban)			Parker et al. (1978)
	Surface litter (O horizon)	915		
	A1 horizon (0-2.5 cm)	2456		
	A1 horizon (2.5-14 cm)	711		
	B horizon (14-25 cm)	103		
	Central Sweden spruce forest			Tyler (1972)
	Needle litter (O_1)	530		
	Humus (O_2)	2300		
	Palmerton, Pennsylvania, within 1 km of zinc smelter			Buchauer (1973)
	Humus (O_2)	135×10^3	13.5×10^6	
Cadmium	Eastern Tennessee forests			Van Hook et al. (1977)
	Soil litter layer O_1			
	Yellow poplar	1.0	8	
	Chestnut oak	0.8	15	
	Oak-hickory	0.2	4	
	Pine	0.6	9	

Table 5-7. (continued)

Element	Location and sample site	Conc. ($\mu g\ \bar{g}^{1}$)	Amt. ($g\ ha^{-1}$)	Reference
Cadmium	Central German beech and spruce forests, various soil depths			Heinrichs and Mayer (1977)
	0-10 cm	.056	64	
	10-20 cm	.060	71	
	30-40 cm	.074	96	
	40-50 cm	.096	142	
	East Chicago oak forest (urban)			Parker et al. (1978)
	Surface litter (O horizon)	4.7		
	A1 horizon (0-2.5 cm)	10.4		
	A1 horizon (2.5-14 cm)	3.8		
	B horizon (14-25 cm)	0.3		
	Central Sweden spruce forest			Tyler (1972)
	Needle litter (O_1)	24		
	Humus (O_2)	44		
	Palmerton, Pennsylvania, within 1 km of zinc smelter			Buchauer (1973)
	Humus (O_2)	1750	18×10^4	
Copper	Central German beech and spruce forests, various soil depths			Heinrichs and Mayer (1977)
	0-10 cm	24	28×10^3	
	10-20 cm	18	21×10^3	
	30-40 cm	23	30×10^3	
	40-50 cm	17	25×10^3	
	East Chicago oak forest (urban)			
	Surface litter (O horizon)	76		
	A1 horizon (0-2.5 cm)	119		
	A1 horizon (2.5-14 cm)	30		
	B horizon (14-25 cm)	3		
	Central Sweden spruce forest			Tyler (1972)
	Needle litter (O_1)	260		
	Humus (O_2)	660		
	Palmerton, Pennsylvania, within 1 km of zinc smelter			Buchauer (1973)
	Humus (O_2)	2000		
Nickel	Central German beech and spruce forests, various soil depths			Heinrichs and Mayer (1977)
	Needle litter (O_1)	26		
	Humus	36		
	Central Sweden spruce forest			Tyler (1972)
	Needle litter (O_1)	26		
	Humus (O_2)	36		

Table 5-7. (continued)

Element	Location and sample site	Conc. $(\mu g\ g^{-1})$	Amt. $(g\ ha^{-1})$	Reference
Manganese	Eastern Tennessee forests, various soil depths			Van Hook et al. (1973)
	Litter	2000		
	0-40 cm	700		
	40-80 cm	700		
	80-120 cm	75		
	120-160 cm	25		
	> 160 cm	2000		
	Central German beech and spruce forests, various soil depths			Heinrichs and Mayer (1977)
	0-10 cm	330	380×10^3	
	10-20 cm	700	833×10^3	
	30-40 cm	730	949×10^3	
	40-50 cm	740	110×10^4	
Chromium	Eastern Tennessee forests, various soil depths			Van Hook et al. (1973)
	Litter	2		
	0-40 cm	5		
	40-80 cm	17		
	80-120 cm	2		
	120-160 cm	5		
	> 160 cm	50		
	Central German beech and spruce forests, various soil depths			Heinrichs and Mayer (1977)
	0-10 cm	52	60×10^3	
	10-20 cm	59	70×10^3	
	30-40 cm	56	73×10^3	
	40-50 cm	51	76×10^3	
Chlorine	Eastern Tennessee forests, various soil depths			Van Hook et al. (1973)
	Litter	50		
	0-40 cm	55		
	40-80 cm	50		
	80-120 cm	150		
	120-160 cm	150		
	> 160 cm	50		

mium, copper, nickel, manganese, or iron. This is obviously the case for forests within several kilometers of an urban area, for example, the East Chicago situation, or a metal smelter, for example the Palmerton environment. It is most significant, however, that it also appears that forest soils are accumulating elevated

Table 5-8. Trace Element Concentrations and Amounts for Central German Beech and Spruce Forests at Various Soil Depths.

Soil depth	Iron		Cobalt		Vanadium		Mercury		Thallium	
	Conc. ($\times 10^5$ $\mu g\,g^{-1}$)	Amt. (tons ha^{-1})	Conc. ($\mu g\,g^{-1}$)	Amt. (kg ha^{-1})	Conc. ($\mu g\,g^{-1}$)	Amt. (kg ha^{-1})	Conc. ($\mu g\,g^{-1}$)	Amt. (g ha^{-1})	Conc. ($\mu g\,g^{-1}$)	Amt. (g ha^{-1})
0-10	1.6	18.4	7	8.1	55	63	0.12	138	0.5	610
10-20	1.9	22.6	11	13.1	62	74	0.08	100	0.3	381
30-40	1.9	24.7	12	15.6	62	81	0.05	65	0.2	247
40-50	2.0	29.6	13	19.2	63	93	0.02	27	0.2	340

Source: Heinrichs and Mayer (1977).

Table 5-9. Baseline or General Trace Element Concentrations for Soils in Unmineralized and Uncontaminated Areas.

Element	Conc. in soil ($\mu g\ g^{-1}$)	
	Mean	Range
Cadmium	0.3-0.6	0.01-0.7
Chlorine	100	
Chromium	60	5-300
Cobalt	8	1-40
Copper	20	2-100
Fluorine	200	30-300
Iron	38,000	7000-55 $\times$ 10^4
Manganese	550	100-4000
Mercury	0.05	0.001-0.5
Molybdenum	2	0.2-5
Nickel	14	<5-700
Selenium	0.1-2.0	<0.04-1200
Tungsten	1	
Vanadium	80	15-100
Zinc	50	25-65

Sources: Cannon (1974); Bowen (1966); Connor and Shacklette (1975).

concentrations of these trace elements even when located many kilometers from a primary source. The residence time of heavy metals in forest soils is of fundamental importance in judging sink efficiency. Tyler (1978) has examined the leachability of manganese, zinc, cadmium, nickel, vanadium, copper, chromium, and lead from two organic spruce forest soils. One soil was from a site grossly polluted by copper and zinc from a metal smelter and the other from an unpolluted forest location. Both soils were treated in lysimeters with artificial rainwater, acidified to pH 4.2, 3.2, and 2.8. The results of this experiment are presented in Table 5-10. At pH 4.2, relatively commonly measured in the northeastern United States and in northern Europe, the residence times are impressive, particularly in the polluted soil. The high buffer capacity of the latter soil is cited by Tyler as conferring protection against leaching. While extrapolation of lysimeter data to natural soils can only be made with considerable risk, Tyler's experiment is very informative.

It is concluded that, in addition to lead, forest soils, in particular the organic forest floor, function with variable efficiency as a sink for zinc, cadmium, copper, nickel, manganese, vanadium, and chromium associated with particulates input from the atmosphere. Temperate zone fluxes in kg ha^{-1} yr^{-1} for these elements include: cadmium 0.002--0.02 (Table 2-19), copper 0.016–0.24 (Table 2-20), zinc 0.012–0.178 (Table 2-21), manganese 0.08–6.48 (Table 2-22), cobalt 0.005–0.01 (Table 2-23), and nickel 0.008–0.014 (Table 2-23). Judgements concerning the relative efficiency and importance of the retention of these various metals must await greater understanding of their dynamics in a larger range of soil types.

Table 5-10. Residence Times of Heavy Metals in Contaminated and "Clean" Organic Spruce Soils (Sweden) Artificially Leached with Acidified Precipitation Equivalent to an Annual Percolation of 150 liters m^{-2a}.

Metal	pH	Control soil	Polluted soil
Mn	4.2	3	30-40
	3.2	1.5	4
	2.8	0.5	1.5
Zn	4.2	7	9
	3.2	2	3
	2.8	0.8	1.2
Cd	4.2	6	20
	3.2	3	4-5
	2.8	1.3	1.7
Ni	4.2	5	15
	3.2	2	4-5
	2.8	2	2
V	4.2	17	2
	3.2	25-30	6-7
	2.8	9	9
Cu	4.2	13	80-120
	3.2	11	18-20
	2.8	9	6
Cr	4.2	20	100-150
	3.2	18-20	50-70
	2.8	15	50-70
Pb	4.2	70-90	>200
	3.2	40-50	>100
	2.8	20	17

Source: Tyler (1978).

[a] Data are number of years necessary for a 10% decrease of metal contamination.

4. Other Particulates

Dry and wet deposition processes input numerous other materials into forest soil systems. Sulfur (sulfates), nitrogen (nitrates), and hydrogen are among the most important. Total deposition (to vegetation and soils) for numerous forest systems have provided a sulfur input range from 0.7 to 99 kg ha^{-1} yr^{-1} (Table 2-15), a nitrogen input range from 0.5 to 45 kg ha^{-1} yr^{-1} (Table 2-16), and a hydrogen input range from 0.0005 to 2.4 kg ha^{-1} yr^{-1} (Table 2-17). For several forest ecosystems in the United States the annual deposition of sulfate, expressed as sulfate-sulfur (SO_4^{-2}-S), is between 10 and 20 kg ha^{-1} (Likens et al., 1977; Shriner and Henderson, 1978; Swank and Douglas, 1977). Shriner and Henderson (1978) have examined sulfur cycling in various eastern Tennessee forest types in a region they estimated had an annual average deposition of 18.1 kg ha^{-1} sulfate. Eighteen percent of the deposition occurred as dry particulate fallout, while the bulk (82%) was dissolved in rainfall. Of this input, 64% or 11.5 kg ha^{-1} was lost from the forest in streamflow. The retention of 6.6 kg ha^{-1} (net annual accumulation) in this Tennessee forest greatly exceeds the sulfate retention estimated for New Hampshire forests (1.1 kg ha^{-1}; Likens et al., 1977) but approximates the estimate for North Carolina forests (7.6 kg ha^{-1}; Swank and Douglass, 1977).

Shriner and Henderson (1978) estimated that 65% of the net annual sulfur accumulation is located in the soil. Unlike trace metal accumulation, these authors conclude that 92% of the sulfur is located in mineral soil horizons, with only 3% contained in the organic forest floor. The ability of forest soils to serve as a sink for these inputs and the impacts of these inputs on forest health will be discussed in subsequent chapters.

B. Forest Soils as Sinks for Atmospheric Gases

Our understanding of the capacity of various soils to function as sinks for gaseous air contaminants is very limited. This general topic has received only meager research attention over the last 20 years. Obviously soils have considerable capacity to absorb a variety of gases from the atmosphere and to incorporate and transform them in or on the soil through a large number of microbial, other biological, physical, and chemical processes. Specific information is lacking, however, on the relative importance of the source versus sink function, the capacities and rates of various soils for absorption, residence and reaction rate times, the influence of soil physical (mineral and organic matter content, structure, porosity) and chemical (pH, moisture content, exchange capacity) properties and climate on removal rates, and the significance of soil management practices.

Bohn (1972) reviewed the literature on soil sink function and presented some generalizations. He concluded that soils will absorb organic gases more rapidly and in larger quantities with increasing molecular weight and with greater numbers of nitrogen, phosphorus, oxygen, sulfur, and other functional group substitutions in the compound. Absorption of low molecular weight and less substituted organic gases was judged to be dependent on the development of an appropriate microbial population. Soil removal of inorganic gases was concluded to involve primarily chemical and physical processes. The author observed that the literature regarding soil removal of reducing gases (oxides of carbon, sulfur and nitrogen, hydrocarbons, aldehydes) was modest, while the information regarding oxidizing gases (ozone, peroxy compounds, chlorine) was nonexistent.

Unfortunately the literature addressing the ability of forest soils to remove atmospheric gaseous contaminants is especially small. The following discussion will draw heavily on evidence obtained from situations involving agricultural or other nonforest soils.

1. Carbon Monoxide

Carbon monoxide is formed in all combustion processes as a result of the incomplete oxidation of carbon and as a result anthropogenic production is locally and regionally enormous. In excess of 6×10^{14} g of carbon monoxide is annually discharged into the atmosphere (Seiler, 1974). The primary contributors of combustion carbon monoxide are in the United States, Europe, and Japan. As a result most of the anthropogenic emissions are concentrated in the temperate latitudes of the northern hemisphere. As a consequence, a key feature of the global distribution of carbon monoxide is a higher concentration in the northern than in the southern hemisphere.

Carbon monoxide contents are at a maximum in the northern hemisphere during winter and spring, and are at a minimum during the summer. Over the past two decades, winter concentrations have tended to increase while summer levels have remained constant. Despite the geographic and seasonal variations, the available information supports the conclusion that global carbon monoxide concentrations have remained relatively constant in "clean" atmospheres. Some evidence suggests a very slight increase (Chapter 19). Available data generally support the conclusion that during the warm season atmospheric levels of carbon monoxide are regulated by an intense natural sink function (Nozhevnikova and Yurganov, 1978).

A large number of natural sinks have been proposed: (a) absorption by oceans, (b) oxidation to carbon dioxide by OH⁻ in the troposphere, (c) migration to the stratosphere followed by photochemical reaction, (d) reaction with animal hemoprotein, (e) fixation by higher plants, and (f) absorption by soil. In light of available evidence, only the latter two hypotheses can be said to be of general importance.

Vegetative oxidation of carbon monoxide to carbon dioxide (Ducet and Rosenberg, 1962) and fixation as serine (Chappelle and Krall, 1961) have been described. Employing ^{14}CO, Bidwell and Fraser (1972) observed uptake by leaves of bean plants under both light and dark conditions. Numerous other non-tree species were tested for carbon monoxide uptake at low gas concentration in the light. Using their bean plant data, the authors estimated a summer removal capacity of $12–120$ kg km^{-2} day^{-1}, globally $3–30 \times 10^8$ tons yr^{-1} (6-month growing season). With the risks of extrapolation from in vitro work with greenhouse plants to the global scene aside, judgments concerning the significance of vegetation as a carbon monoxide sink remain difficult to make. Inman and Ingersoll (1971) failed to observe any capability of several plants, including seedlings of Monterey and knobcone pine and mimosa, to remove carbon monoxide from the atmosphere. Also the role of plants as producers of carbon monoxide (Nozhevnikova and Urganow, 1978) must be more accurately understood before the role of vegetation as a sink for carbon monoxide can be judged appropriately. While vegetation may play some role in the maintenance of carbon monoxide in the natural atmosphere, soils are concluded to be the most important removal agent.

The first evidence supporting the importance of soil as a sink for carbon monoxide was presented in 1926, and this study along with many that have followed indicate that soil microorganisms are responsible for the removal.

Table 5-11. Carbon Monoxide Uptake from Test Atmospheres [80–130 ppm (9.2–1.5 X 104 µg m⁻³) CO] by Various Soils at 25°C.

Vegetation	Location	mg hr^{-1} m^{-2} of soil
Coast redwoods	CA	16.99
Oak	CA	15.92
Coast redwoods	CA	14.39
Ponderosa pine	CA	13.89
Grass-legume pasture	CA	11.94
Grapefruit	CA	11.48
Grass meadow	CA	10.52
Forest	HI	9.90
Chaparral	CA	6.46
Oak stubble	CA	6.23
Chaparral	CA	4.31
White fir	CA	3.48
Cotton (fallow)	CA	2.82
Almond weeds (fallow)	FL	2.65

Source: Inman et al. (1971).

Inman et al. (1971) conducted preliminary experiments with a greenhouse potting mixture and found that the test soil could deplete carbon monoxide in an experimental atmosphere [containing 120 ppm (13.8 X 10⁴ µm⁻³) CO] to near zero within 3 hr. Treatment of the soil with steam sterilization, antibiotics, salt, and anaerobic conditions all prevented carbon monoxide uptake and indicated the importance of biological processes. A variety of soils from California, Hawaii, and Florida were brought into the laboratory and tested for their ability to remove carbon monoxide (Table 5-11). Inman and coworkers generalized from their results that cultivated soils were less active than natural soils and that higher organic matter and lower pH soils were the most active. These observations clearly support the potential importance of forest soils!

In order to improve the confidence of their observations, Inman and others (Ingersoll et al., 1974) outfitted a mobile laboratory and field tested soils in most major vegetative regions in the United States. Field testing was accomplished by covering a square meter of undisturbed soil and covering vegetation with a gas tight chamber. The carbon monoxide uptake rate showed considerable variation in the field ranging from 7.5–109 mg hr⁻¹ m⁻². As in the previous work, cultivated soils were invariably lower than natural soils. Natural soil uptake rates are presented in Table 5-12. Ingersoll et al. (1974) concluded that the potential rates of carbon monoxide uptake by the soils of the United States and the world were 505 million and 14.3 billion tons per year, respectively (Tables 5-13 and 5-14). Forest soils are indicated to be of particular significance. If Ingersoll's estimates approximate the natural condition, it must be concluded that temperate and tropical forests play extraordinarily important roles as sinks for global carbon monoxide.

Table 5-12. Carbon Monoxide Uptake Rates Determined under Field Conditions (Corrected for Annual Average Temperature at Test Site) at 100 ppm CO for Various Soils under Natural Vegetation in North America.

Vegetation	Location	$mg\ hr^{-1}$ m^{-2} of soil
Flood plain forest	LA	26.2
Deciduous	NY	26.3
Mixed deciduous	OH	21.3
Oak-hickory forest	OH	21.3
Coastal forest	OR	14.1
Grassland	TX	11.9
Oak-hickory	MO	10.2
Boreal forest	Alberta	6.8
Boreal forest	Alberta	4.2
Grassland	KA	3.6
Grassland	Saskatchewan	2.7
Bluestem prairie	KA	2.3

Source: Ingersoll et al. (1974).

Seiler (1974) has determined carbon monoxide uptake rates for several European soils (location and vegetation unspecified) and calculated an average flux rate of 1.5 X 10-11 g cm^{-2} sec^{-1} at 15°C. This rate is very approximately an order of magnitude less than Ingersoll's average. Seiler explained the discrepancy by indicating that Inman's group had employed initial carbon monoxide concentrations of 100 ppm (11.5 X 10^4 μg m^{-3}), while his laboratory had employed 0.20 ppm (230 μg m^{-3}) judged to be the normal ambient concentration in unpolluted areas of Europe. Seiler's estimate for global soil removal was 4.5 X 10^8 tons per year.

The question of an aninadequate account of carbon monoxide evolution by soils has been raised by Smith et al. (1973). This deficiency could contribute to important overestimations of sink function. Perhaps the actual capacity for carbon monoxide removal lies between the estimates of Inman's and Seiler's groups, or perhaps it approximates one or the other. Whatever the case, it can be concluded that forest soils play a role of very significant importance as a repository for atmospheric carbon monoxide. Since the soil removal rate increases with increasing levels of carbon monoxide, forest ecosystems in and around urban and industrial areas may be especially important sinks.

The specific components of the soil microflora involved in the removal of carbon monoxide are not completely understood and a thorough discussion of this topic is beyond the scope of his book. Inman and Ingersoll (1971) reported the identification of 16 fungal isolates belonging to the *Penicillium*, *Aspergillus*, *Mucor*, *Haplosporangium*, and *Mortierella* genera that were capable of removing atmospheric carbon monoxide. As Nozhevnikova and Yurganov (1978) point out, however, all microbes with hemoproteins and cytochrome oxidase or other carbon monoxide reacting enzymes will fix this gas to

Table 5-13. Potential Carbon Monoxide Uptake Rates of the Soils of the Coterminous United States.

Soil-vegetation type	Area (mi^2)	CO uptake $(tons\ yr^{-1}\ mi^{-2})$	Total CO uptake $(tons \times 10^6\ yr^{-1})$
Pasture	616,674	179.3	110.57
Deciduous forest	339,485	254.8	86.50
Appalachian forest	265,519	313.7	84.17
Coastal forest	329,390	200.7	83.93
Cropland	468,000	86.0	40.25
Southern flood plain forest	58,782	595.0	34.95
Montane forest	87,150	242.0	21.10
Sagebrush steppe	264,867	76.7	20.32
Sagebrush	145,047		
Desert scrub	220,723	52.2	19.09
Southern mixed forest	48,644	86.5	4.21
Paved roads	28,100	0	0
Covered area	26,500	0	0
Lakes, rivers	78,267	0	0
Total	2,977,128		505.12

Source: Ingersoll et al. (1974).

Table 5-14. Potential Carbon Monoxide Uptake Rates of the Soils of the World.

Soil-vegetation type	Area $(10^6\ mi^2)$	Average CO uptake $(\times 10\ tons\ yr^{-1}\ mi^{-2})$	Total CO uptake $(\times 10^6\ tons\ yr^{-1})$
Tropical rain forest	6.55	805.6	5,277.5
Tropical grassland	3.45	886.9	3,062.5
Tropical deciduous forest	1.78	1,105.9	1,969.6
Montane forest	4.66	175.7	817.8
Taiga forest	4.71	127.5	600.9
Desert	10.86	52.2	567.0
Mixed and broadleaf forest	1.32	254.8	410.0
Agricultural	4.60	86.0	395.4
Pasture	1.84	179.3	329.7
Temperate grassland	3.10	86.5	268.4
Steppe	3.45	76.7	264.5
Southern pine forest	0.29	505.8	145.2
Tundra	1.15	123.0	141.3
Covered by ice, water, roads, structures	9.20	0	0
Total	56.96		14,250.0

Source: Ingersoll et al. (1974).

some extent. The microbes of importance in this sink function are the organisms for which oxidation of carbon monoxide serves as a source of energy. Bacteria that oxidize carbon monoxide belong to a physiological group termed *carboxydobacteria* and belong to genera that appear to be related to the genus *Pseudomonas*. They are gram-negative, aerobic species that are considered to be widely distributed in nature. Their abundance in urban substrates may reflect a more favorable environment created by elevated carbon monoxide levels (Nozhevnikova and Yurganov, 1978). Their particular abundance in forest soils is unfortunately not clear.

2. Sulfur Dioxide and Hydrogen Sulfide

The sulfur gases currently recognized as important components of tropospheric air include sulfur dioxide (SO_2), hydrogen sulfide (H_2S), carbonyl sulfide (COS), carbon disulfide (CS_2), dimethyl sulfide ($CH_3 SCH_3$), and sulfur hexafluoride (SF_6) (Bremner and Stelle, 1978). Until very recently it was generally assumed that the primary and most important forms of sulfur in the atmosphere were hydrogen sulfide, sulfur dioxide, and sulfates. The information available concerning soil as a sink for atmospheric sulfur has dealt almost exclusively with the latter two.

As previously discussed in this chapter, sulfate will be added to soil largely by precipitation and will become part of the soluble sulfur content held by soil colloids. Soils also have a large capacity to quickly absorb sulfur dioxide from the atmosphere. Factors that tend to increase the soil uptake of sulfur dioxide include fine texture, high soil organic matter content, high pH, presence of free $CaCO_3$, high soil moisture content, and the presence of soil microorganisms (Nyborg, 1978).

Since Alway et al. (1937) presented the first evidence that soils can absorb sulfur dioxide, a large number of studies have followed. Smith et al. (1973) studied the capability of six soils, from Oregon, Iowa, and Saskatchewan with variable chemical and physical properties and found that removal of sulfur dioxide and hydrogen sulfide was much more rapid than the removal of carbon monoxide. It is not clear if any of the soils were from forest ecosystems, but the pH (4.8) and organic carbon percent (93.8%) suggest that the Astoria, Oregon soil may have been (Table 5-15). Clearly soil moisture favors uptake of sulfur dioxide, presumably due to the high solubility of this gas in water. The Astoria soil appears quite average in its ability to remove sulfur dioxide and hydrogen sulfide. The authors cautioned that the uptake rates of Table 5-15 should not be judged to be maxima for natural soils as the conversion to sulfate in vivo would presumably create more absorption sites in natural soils. Bohn (1982) has suggested that the absorption rate of hydrogen sulfide slows with high soil moisture contents, perhaps due to slow diffusion rates in water-filled pores. The fact that hydrogen sulfide absorption capacity increases with higher soil pH may reduce the importance of forest ecosystems in removing this gas from the atmosphere. Despite the relatively large number of papers addressing the soil removal of these

Table 5-15. Sorption Capacity of Several Soils for Sulfur Dioxide and Hydrogen Sulfide Determined under Laboratory Conditions (100 ppm Gas Concentration at 23°C).

Soil	Air-dry		Moist	
	SO_2	H_2S	SO_2	H_2S
Astoria	8.9	62.9	37.1	51.6
Weller	15.3	61.0	31.9	58.1
Okoboji	10.2	52.9	31.4	44.5
Thurman	1.1	15.4	9.3	11.0
Regina	13.2	65.2	50.4	62.5
Harpster	10.2	46.6	66.8	40.6
Mean	9.8	50.7	37.8	44.7

mg g^{-1} of soil

Source: Smith et al. (1973).

two gases (Bremner and Steele, 1978; Moss, 1975), estimates of global and regional removal amounts by soil are relatively few (Rasmussen et al., 1974). Available data suggest that deposition velocities (v) for sulfur dioxide are generally in the range of 0.2–0.7 cm sec^{-1} (Rasmussen et al., 1974), which is less than those for vegetation (Chapter 6). Eriksson (1963) has estimated that the global removal of sulfur dioxide by soil equals 25 X 10^9 kg sulfur dioxide-sulfur annually. Abeles et al. (1971) examined the capacity of soil collected from Waltham, Massachusetts to remove sulfur dioxide under laboratory conditions [100 ppm (26.2 X 10^4 μg m^{-3}) gas] and extrapolated their results to suggest that United States soils may be capable of removing 4 X 10^{13} kg of sulfur dioxide (2 X 10^{13} kg sulfur dioxide-sulfur) per year.

The mechanism of soil removal of sulfur dioxide is not completely understood but is generally concluded to involve both microbial and chemical mechanisms. Since autoclaving only partially attenuated the ability of their Massachusetts soil to removal sulfur dioxide, Abeles et al. (1971) concluded that the major removal mechanism was chemical rather than microbial. Sterilization of test soils has led others to conclude that microbial metabolism of sulfur dioxide may be relatively unimportant (Smith et al., 1973; Ghiorse and Alexander, 1976). Using a soil of unclear origin and $^{35}SO_2$ hr, however, Craker and Manning (1974) studied uptake by soil bacteria and fungi and found that fungi were capable of removing sulfur dioxide. The relative importance of biological and nonbiological removal strategies will be determined as more sensitive and specific analytical techniques for sulfur forms becomes available. Present evidence encourages the conclusion that sorption of sulfur dioxide by soils involves the formation of sulfite and sulfate and that sorption of hydrogen sulfide involves formation of metallic sulfides and elemental sulfur (Bremner and Steele, 1978).

Moss (1975) presented one of a few investigations that have systematically examined sulfur uptake in natural ecosystems adjacent to excessive sulfur release.

He investigated plant and soil sulfate burdens in the vicinity of industrial centers in Sheffield, England, and Welland, Ontario. In the latter situation, the sulfate accumulation was observed to be greater in grassland relative to forest sites at an equivalent distance downwind from the source. Why this was so was unclear and highlights our limited understanding of actual sink capabilities in natural environments.

a. Other Sulfur Gases

The gas chromatographic studies of Bremner and Banwart (1976) showed that air-dry and moist soils had the capacity to sorb dimethyl sulfide, dimethyl disulfide, carbonyl sulfide, and carbon disulfide, but did not sorb sulfur hexafluoride. The first four gases were removed more efficiently by soils. Soil sterilization indicated that soil microorganisms were partially responsible for the removal of these gases. As the rates of removal were substantially less than for sulfur dioxide or hydrogen sulfide, however, the authors conclude that while soils may constitute a sink for low levels of dimethyl sulfide, dimethyl disulfide, carbonyl sulfide, and carbon disulfide, they would not effectively reduce elevated levels of these gases in areas of high anthropogenic emission.

3. Nitrogen Oxides and Other Nitrogen Gases

In the atmosphere, nitric oxide is either oxidized to nitrogen dioxide or photolyzed into nitrogen gas. Nitrogen dioxide reacts photochemically or is removed by precipitation, primarily in the form of nitric acid. Nitric oxide and nitrogen dioxide may also be removed by soils. Nitric oxide is oxidized to nitrogen dioxide in soil, but the former gas does not persist in acid soils as long as it does in basic soils (Bohn, 1972).

Working with Waltham, Massachusetts soil, Abeles et al. (1971) found that the removal rate for nitrogen dioxide was slower than the removal rate for sulfur dioxide. Twenty-four hours were required to reduce nitrogen dioxide concentrations from 100 to 3 ppm, (18.8×10^4 to 56.4×10^2 $\mu g \ m^{-3}$) in test atmospheres. Extrapolation of these laboratory experiments allows these authors to suggest that the soils of the United States may be capable of removing 6×10^{11} kg of nitrogen dioxide per year.

Both Ghiorse and Alexander (1976) and Smith and Mayfield (1978) have documented rapid absorption of nitrogen dioxide by both sterile and nonsterile soil In the latter study, soil from an uncultivated grassland in Ontario absorbed 99% of the nitrogen dioxide introduced into a test vessel at 25° C in 15 min.

Mechanisms of nitrogen dioxide uptake may involve reaction with soil cations to form $NaNO_2$ or KNO_2, reaction with soil water to form HNO_2 and HNO_3, binding with organic matter, or persistence as a gas in interparticle soil spaces (Smith and Mayfield, 1978).

Since ammonia would probably be present in the atmosphere in the form of $(NH_4)_2SO_4$ rather than NH_3 in all but the most unusual environments, direct soil absorption of gaseous NH_3 is probably not important. Where ambient conditions might expose soils to high ammonia levels, however, evidence suggests that acid soils are particularly efficient removal agents (Rasmussen et al., 1974).

The specific capabilities that forest soils may have to remove nitrogenous gases must await further experimentation. Evidence has been presented indicating considerable variation in gas uptake by soils depending on diurnal, seasonal, and vegetative cover variables (Voldner et al, 1986).

4. Hydrocarbons

Hydrocarbons are generally not soluble in water, and as a result, soil uptake where it is important, is concluded to be primarily microbial. The light hydrocarbon from motor vehicles most actively removed by soil is ethylene (Zimmerman and Rasmussen, 1975). Abeles et al. (1971) observed that Maryland soil samples removed ethylene more slowly than other soils removed sulfur and nitrogen dioxides. Soil removal of ethylene, mediated by various microorganisms, was calculated by Abeles' group to approximate 7×10^9 kg of ethylene annually in the United States. Smith et al. (1973) determined that the soil flux rate for acetylene was from 0.24 to 3.12×10^{-1} mole g^{-1} day^{-1}. Their test soil with the lowest pH and highest organic matter (forest soil) was the most active of all soils tested for acetylene removal.

5. Oxidants

There is limited evidence that soils function as a sink for atmospheric ozone (Rasmussen et al., 1974). Aldaz (1969) concluded that the soil and vegetation of the surface of the earth represent a major sink for this gas and estimated the capacity of this sink to be within the range of $1.3–2.1 \times 10^{12}$ kg ozone yr^{-1}.

Most reports of ozone removal have examined plant uptake or plant and soil uptake combined. Turner et al. (1973) have tested the sink capacity of a freshly cultivated sandy loam devoid of vegetation. Their results, which were recorded under field conditions, showed the flux rate of ozone removal varied from 3 to 12 $\times 10^{11}$ mole cm^{-2} sec^{-1}, making bare soil, in the author's judgment, an important sink for ozone.

6. Other Gases

Fang (1978) examined the uptake of mercury vapor by five Montana soils by exposing them to a test atmosphere containing 75.9 μg metallic[203] Hg vapor m^{-3} for 24 hr. The soil with the highest organic matter content had the highest mercury uptake. While mercury vapor is currently only an extremely localized problem, more than 90% of the mercury contained in coal is vaporizing during combustion. Even with relatively low mercury concentrations in coal, widespread or large volume coal combustion may increase the significance of soil retention of this gas.

C. Summary

Forest soils are important sinks for a variety of air contaminants. Retention of particulate lead by organic materials in the forest floor is a most dramatic example. The flux of lead to temperate forest ecosystems downwind of industrial, urban, or roadside sources may approximate 10–1,000,000 g ha^{-1} yr^{-1}, with much of this lead accumulating in the forest floor. Certain forest soils may also serve as a sink for additional trace metals including zinc, cadmium, copper, nickel, manganese, vanadium, and chromium. The efficiency of sink function for the latter metals is generally substantially less than for lead, but may be important, particularly in forest systems close to primary sources. The soil accumulation of sulfate, nitrate and hydrogen ions from the atmosphere may have long-term implications for forest health.

Forest soils remove pollutant gases from the atmosphere via several microbial, chemical, and physical processes. Forest soils function as an especially efficient sink for carbon monoxide and may play a dominant role in regulating the concentration of this gas in the atmosphere. Other gases that may be significantly removed by forest soils include sulfur dioxide, ammonia, some hydrocarbons, and mercury vapor.

References

Abeles, F.B., L.E. Craker, L E. Forrence, and G R. Leather. 1971. Fate of air pollutants: Removal of ethylene, sulfur dioxide and nitrogen dioxide by soil. Science 173: 914-916.

Aldaz, L. 1969. Flux measurements of atmospheric ozone over land and water. J. Geophys. Res. 74: 6943-6946.

Alway, F.J., A.W. Marsh, and W.J. Methley. 1973. Sufficiency of atmospheric sulfur for maximum crop yields. Soil Sci. Soc. Am. Proc. 3: 229-238.

Andresen, A.M., A.H. Johnson, and T.G. Siccama. 1980. Levels of lead, copper and zinc in the forest floor in the northeastern United States. J. Environ. Qual. 9: 293-296.

Benninger, L.K., D.M. Lewis, and K.K. Turekian. 1975. The use of natural Pb-210 as a heavy metal tracer in the river-estuarine system. In T.M. Church, ed., Marine Chemistry and Coastal Environment, American Chemical Society Symposium Series No. 18, pp. 201-210, American Chemical Society Symposium Series No. 18, pp. 201-210, American Chemical Society, Washington, DC.

Bidwell, R.G.S. and D.E. Fraser. 1972. Carbon monoxide uptake and metabolism by leaves. Can. J. Bot. 50: 1435-1439.

Bohn, H.L. 1972. Soil absorption of air pollutants. J. Environ. Qual. 1: 372-377.

Bowen, H.J.M. 1966. Trace Elements in Biochemistry. Academic Press, New York, 241 pp.

Bremner, J.M. and W.L. Banwart. 1976. Sorption of sulfur gases by soils. Soil Biol. Biochem. 8: 79-83.

Bremner, J.M. and C.G. Steele. 1978. Role of microorganisms in the atmospheric sulfur cycle. Adv. Microb. Ecol. 1: 155-201.

Buchauer, M.J. 1973. Contamination of soil and vegetation near a zinc smelter by zinc, cadmium, copper, and lead. Environ. Sci. Technol. 7: 131-135.

Cannon, H.L. 1974. Natural toxicants of geologic origin and their availability to man. In P. L. White and D. Robbins, ed., Environmental Quality and Food Supply. Futura, New York, pp. 143-163.

Chappelle, E.W. and A.R. Krall. 1961. Carbon monoxide fixation by cell- free extracts of green plants. Biochem. Biophys. Acta 49: 578-580.

Chow, T.J. and J.L. Earl. 1970. Lead aerosols in the atmosphere: Increasing concentration. Science 169: 577-580.

Chow, T.J., and M.S. Johnstone. 1965. Lead isotopes in gasoline and aerosols of Los Angeles Basin, California. Science 147: 502-503.

Connor, J.J. and H.T. Shacklette. 1975. Background Geochemistry of Some Rocks, Soils, Plants, and Vegetables in the Conterminous United States. U.S.D.I., Geological Survey Professional Paper 574-F, Washington DC, 168 pp.

Craker, L.E. and W.J. Manning. 1974. SO_2 uptake by soil fungi. Environ. Pollu. 6: 309-311.

Davidson, R.L., D.F.S. Natusch, and J.R. Wallace. 1974. Trace elements in fly ash: Dependence of concentration on particle size. Environ. Sci. Technol. 8: 1107-1113.

Ducet, G. and A.I. Rosenberg. 1962. Leaf respiration. Ann. Rev. Plant Physiol. 13: 171-200.

Eriksson, E. 1963. The yearly circulation of sulfur in nature. J. Geophys. Res. 68: 4001-4008.

Fang, S.C. 1978. Sorption and transformation of mercury vapor by dry soil. Environ. Sci. Technol. 12: 285-288.

Friedland, A.J., A.H. Johnson, and T.G. Siccama. 1984a. Trace metal content of the forest floor in the Green Mountains of Vermont: Spatial and temporal patterns. Water Air Soil Pollu. 21: 161-170.

Friedland, A.J., A.H. Johnson, T.G. Siccama, and D.L. Mader. 1984b. Trace metal profiles in the forest floor of New England. Soil Sci. Soc. Am. J. 48: 422-425.

Galloway, J.N., J.D. Thornton, S.A. Norton, H.L. Volchok, and R.A.N. McLean. 1982. Trace metals in atmospheric deposition: A review and assessment. Atmos. Environ. 16: 1677-1700.

Ghiorse, W.C. and M. Alexander. 1976. Effect of microorganisms on the sorption and fate of sulfur dioxide and nitrogen dioxide in soil. J. Environ. Qual. 5: 227-230.

Heinrichs, H. and R. Mayer. 1977. Distribution and cycling of major and trace elements in two central European forest ecosystems. J. Environ. Qual. 6: 402-407.

Ingersoll, R.B., R.E. Inman, and W.R. Fisher. 1974. Soils potential as a sink for atmospheric carbon monoxide. Tellus 26: 151-158.

Inman, R.E. and R.B. Ingersoll. 1971. Uptake of carbon monoxide by soil fungi. J. Air. Pollu. Control Assoc. 21: 646-657.

Inman, R.E., R.B. Ingersoll, and E.A. Levy. 1971. Soil: A natural sink for carbon monoxide. Science 172: 12291231.

John, M.K., H.H. Chuah, and C.J. Vandaerhoven. 1972. Cadmium and its uptake by oats. Environ. Sci. Technol. 6: 555-557.

Johnson, N.M., R C. Reynolds, and G.E. Likens. 1972. Atmospheric sulfur: Its effect on the chemical weathering of New England. Science 177: 514-516.

Korte, N.E., J. Skopp, W.H. Fuller, E.E. Niebla, and B.A. Alessii. 1976. Trace element movement in soils: Influence of soil physical and chemical properties. Soil Sci. 122: 350-359.

Lagerwerff, J.V. 1967. Heavy metal contamination of soils. In N.C. Brady, ed., Agriculture and the Quality of Our Environment. Am. Assoc. Adv. Sci., Publ. No. 85, Washington, DC, pp. 343-364.

Likens, G.E., F.H. Bormann, R.S. Pierce, J.S. Eaton, and N.M. Johnson. 1977. Biogeochemistry of a Forested Ecosystem. Springer--Verlag, New York, 146 pp.

Lindberg, S.E. 1982. Factors influencing trace metal, sulfate and hydrogen ion concentrations in rain. Atmos. Environ. 16: 1701-1709.

Linton, R.W., A. Loh, D.F. S. Natusch, C.A. Evans Jr., and P. Williams. 1976. Surface predominance of trace elements in airborne particles. Science 191: 852-854.

Linzon, S.N., B.L. Chai, P.J. Temple, R.G. Pearson, and M.L. Smith. 1976. Lead contamination of urban soils and vegetation by emissions from secondary lead industries. J. Air Pollu. Control Assoc. 26: 650-654.

Little, P. 1977. Deposition of 2.75, 5.0 and 8.5 μm particles to plant and soil surfaces. Environ. Pollu. 12: 293-305.

McBride, M.B. 1978. Transition metal bonding in humic acid: An ESR study. Soil Sci. 126: 200-209.

Miller, W.P. and W.W. McFee. 1983. Distribution of cadmium, zinc, copper and lead in soils of industrialized northeastern Indiana. J. Environ. Qual. 12: 29-33.

Moss, M.R. 1975. Spatial patterns of sulfur accumulation by vegetation and soils around industrial centres. J. Biogeography 2: 205-222.

National Research Council. 1980. Trace Element Geochemistry of Coal Resource Development Related to Environmental Quality and Health. National Academy Press, Washington, DC.

Nozhevnikova, A.N. and L.N. Yurganov. 1978. Microbiological aspects of regulating the carbon monoxide content in the earth's atmosphere. Adv. Microbial Ecol. 2: 203-244.

Nyborg, M. 1978. Sulfur pollution and soils. In J.O. Nriagu, ed., Sulfur in the Environment. Part II. Ecological Impacts. Wiley, New York, pp. 359-390.

Page, A.L. and A.C. Chang. 1979. Contamination of soil and vegetation by atmospheric deposition of trace elements. Phytopathology 69: 1007-1011.

Parekh, P.P. and L.Husain. 1981. Trace element concentrations in summer aerosols at rural sites in New York state and their possible sources. Atmos. Environ. 15: 1717-1725.

Parker, G.R., W.W. McFel, and J.M. Kelly. 1978. Metal distribution in forested ecosystems in urban and rural northwestern Indiana. J. Environ. Qual. 7: 337-342.

Petruzelli, G., G. Guidi, and L. Lubrano. 1978. Organic matter as an influencing factor on copper and cadmium absorption by soils. Water Air Soil Pollu. 9: 263-269.

Ragaini, R.C., H.R. Ralston, and N. Roberts. 1977. Environmental trace metal contamination in Kellogg, Idaho, near a lead smelting complex. Environ. Sci. Technol. 11: 773-781.

Rasmussen, K.H., M. Taheri, and R.L. Kabel. 1974. Sources and Natural Removal Processes for Some Atmospheric Pollutants. U.S. Environmental Protection Agency, Publ. No. EPA-60/4-74-032, U.S.E.P.A., Washington, DC, 121 pp.

Reiners, W.A., R.H. Marks, and P.M. Vitousek. 1975. Heavy metals in subalpine and alpine soils of New Hampshire. Oikos 26: 264-275.

Seiler, W. 1974. The cycle of atmospheric CO. Tellus 26: 116-135.

Schlesinger, W.H., W.A. Reiners, and D.S. Knopman. 1974. Heavy metal concentrations and deposition in bulk precipitation in mountain ecosystems of New Hampshire, U.S.A. Environ. Pollu. 6: 39-47.

Shriner, D.S. and G.S. Henderson. 1978. Sulfur distribution and cycling in a deciduous forest watershed. J. Environ. Qual. 7: 392-397.

Siccama, T.G. and W. H. Smith. 1978. Lead accumulation in a northern hardwood forest. Environ. Sci. Technol. 12: 593-594.

Siccama, T.G. W.H. Smith, and D.L. Mader. 1980. Changes in lead, zinc, copper, dry weight and organic matter content of the forest floor of white pine stands in central Massachusetts over 16 years. Environ. Sci. Technol. 14: 54-56.

Smith, E.A. and C.I. Mayfield. 1978. Effects of nitrogen dioxide on selected soil processes. Water Air Soil Pollu. 9: 33-43.

Smith, K.A., J.M. Bremner, and M.A. Tabatabai. 1973. Sorption of gaseous atmospheric pollutants by soils. Soil Sci. 116: 313-319.

Smith, W.H. 1976. Lead contamination of the roadside ecosystem. J. Air Pollu. Control Assoc. 26: 753-766.

Smith, W.H. 1985. Forest quality and air quality. J. For. 83: 82-92.

Smith, W.H. and T.G. Siccama. 1981. The Hubbard Brook Ecosystem Study: Biogeochemistry of lead in the northern hardwood forest. J. Environ. Qual. 10: 323-333.

Smith, W.H., T.G. Siccama, and S.L. Clark. 1986. Atmospheric Deposition of Heavy Metals and Forest Health: An Overview and a Ten-Year Budget for the Input/Output of Seven Heavy Metals to a Northern Hardwood Forest. Publ. No. FWS-8702, Virginia Polytechnic Institute and State University, Blacksburg, VA, 20 pp.

Somers, G.F. 1978. The role of plant residues in the retention of cadmium in ecosystems. Environ. Pollu. 17: 287-295.

Stevenson, F.J. 1972. Role and function of humus in soil with emphasis on adsorption of herbicides and chelation of micronutrients. Bioscience 22: 643-650.

Swank, W.T. 1984. Atmospheric contributions to forest nutrient cycling. Water Res. Bull. 20: 313-321.

Swank, W.T. and J.E. Douglass. 1977. Nutrient budgets for undisturbed and manipulated hardwood forest ecosystems in the mountains of North Carolina. In Watershed Research in Eastern North America. Smithsonian Inst., Edgewater, MD, pp. 343-364.

Turner, N.C., S. Rich, and P.E. Waggoner. 1973. Removal of ozone by soil. J. Environ. Qual. 2: 259-264.

Tyler, G. 1972. Heavy metals pollute nature, may reduce productivity. Ambio 1: 53-59.

Tyler, G. 1978. Leaching rates of heavy metal ions in forest soil. Water Air Soil Pollu. 9: 137-148.

U.S. Environmental Protection Agency. 1984. National Air Quality and Emissions Trends Report, 1982. U.S.E.P.A. Publ. No. EPA-450/4-84-002. Research Triangle Park, NC, 128 pp.

Van Hook, R.I. and W.D. Shults. 1977. Effects of Trace Contaminants from Coal Combustion. Proc. Workshop, Aug. 2-6, 1976, Knoxville, TN, U.S.E.R.D.A. Publ. No. 77-64, U.S. Energy Research and Development Administration, Washington, DC, 79 pp.

Van Hook, R.I., W.F. Harris, and G.S. Henderson. 1977. Cadmium, lead and zinc distributions and cycling in a mixed deciduous forest. Ambio 6: 281-286.

Van Hook, R.I., W.F. Harris, G.S. Henderson, and D.E. Reichle. 1973. Patterns of trace- element distribution in a forested watershed. Proc. 1st Annu. NSF Trace Contaminants Conf., Oak Ridge National Laboratory, Oak Ridge, TN, pp. 640-655.

Voldner, E.C., L.A. Barrie, and A. Sirois. 1986. A literature review of dry deposition of oxides of sulfur and nitrogen with emphasis on long range transport modeling in North America. Atmos. Environ. 20: 2101-2123.

Warren, J.L. 1973. Green Space for Air Pollution Control. Tech. Report No. 50, School of Forest Resources, North Carolina State Univ., Raleigh, NC, 118 pp.

Zimdahl, R.L. and R.K. Skogerboe. 1977. Behavior of lead in soil. Environ. Sci. Technol 9: 1077-1079.

Zimmerman, P. and R. Rasmussen. 1975. Identification of soil denitrification peak as N_2O. Environ. Sci. Technol. 9: 1077-1079.

Zunino, H. and J.P. Martin. 1977a. Metal-binding oranic macromolecules in soil: 1. Hypothesis interpreting the role of soil organic matter in the translocation of metal ions from rocks to biological systems. Soil Sci. 123: 65-76.

Zunino, H. and J.P. Martin. 1977b. Metal-binding organic macromolecules in soil: 2. Characterization of the maximum binding ability of the macromolecules. Soil. Sci. 123: 188-202.

6

Forests as Sinks for Air Contaminants: Vegetative Compartment

In addition to the soil compartment, the vegetative compartment of forest ecosystems functions as a sink for atmospheric contaminants. As in the case of soils, a complex variety of biological, chemical, and physical processes are involved in the transfer of pollutants from the air to the surfaces of vegetation. For certain contaminants, for example, persistent heavy metal particles, the repository functions of vegetation and soils are intimately linked as a portion of the heavy metals input to the soil are derived from vegetative sources contributing litter to the forest floor. Interest in the ability of plants to remove pollutants from the air has grown considerably in recent years as individuals have become increasingly aware of the amenity functions (Heisler, 1975; Smith, 1970a) of woody plants, particularly in urban and suburban areas. The capability of plants to act as a sink for air contaminants has been addressed by a variety of recent reviews, for example, Aubertin and Aubertin (1981), U.S. Environmental Protection Agency (1976a), Smith and Dochinger (1976), Bennett and Hill (1975), Hanson and Thorne (1972), Hill (1971), Environmental Health Science Center (1975), Keller (1978), and Warren (1973). These papers indicate that the surfaces of vegetation provide a major filtration and reaction surface to the atmosphere and importantly function to transfer pollutants from the atmosphere to the biosphere.

A. Forest Vegetation as a Sink for Particulate Contaminants

Much of the understanding of the mechanics of deposition of particles on natural surfaces has been gleaned from studies with particles in the size range 1–50 μm and is reviewed in the excellent papers of Chamberlain (1967, 1970, 1975), Ingold (1971), Gregory (1973), Slinn (1976), and Albritton et al. (1987).

The physics and theory of interception and retention of fine particles by vegetation are well beyond the scope of this book. It is useful for the interpretation of the evidence to follow, however, to have an introduction to some basic observations.

Particles are deposited on plant surfaces by three processes: sedimentation under the influence of gravity, impaction under the influence of eddy currents, and deposition under the influence of precipitation. Sedimentation usually results in the deposition of particles on the upper surfaces of plant parts and is most important with large particles. Sedimentation velocity varies with particle density, shape, and other factors. Impaction occurs when air flows past an obstacle and the airstream divides, but particles in the air tend to continue in a straight path due to their momentum and to strike the obstacle. The efficiency of collection via impaction is the principal means of deposition if (a) particle size is of the order of tens of micrometers or greater, (b) obstacle size is of the order of centimeters or less, (c) approach velocity is of the order of meters per second or more, and (d) the collecting surface is wet, sticky, hairy, or otherwise retentive. Ingold (1971) presented data indicating that leaf petioles are considerably more efficient particulate impactors than either twigs (stems) or leaf lamina. For particles of dimensions 1–5 μm, impaction is not efficient and interception by fine hairs on vegetation is possibly the most efficient retentive mechanism. The efficiency of washout of particles by rain is high for particles approximately 20–30 μm in size. The capturing efficiency of raindrops falls off very sharply for particles of 5 μm or less.

Following deposition, particles may be retained on vegetative surfaces, they may rebound from the surface, or they may be temporarily retained and subsequently removed. If either the particle or the tree surface is wet or sticky, deposited particles are generally retained. Surficial salt accumulation by plants in marine or north-temperate roadside environments (where deicing chemicals are employed) results as vegetation acts to trap salt particles. Fluorine, sulfate, and nitrate molecules associated with moisture droplets (fog) in the atmosphere may be distributed to vegetative surfaces with great efficiency (Chamberlain, 1975).

The transfer of particles from the atmosphere to natural surfaces is commonly expressed via deposition velocity which has been previously discussed (Chapter 2). For small particles, for example, condensation aerosols less than 1 μm, deposition velocities are much less than for large particles, for example, spores and pollen 20–40 μm in diameter.

In addition to spores and pollen, particles in the atmosphere larger than 10 μm are frequently the result of mechanical processes, for example grinding or spraying. Soil particles, process dust, industrial combustion products, and marine salt particles are typically between 1 and 10 μm in diameter. Particles in the 0.1–1 μm range frequently represent gases that have condensed to form nonvolatile products.

The interaction of these variously sized particles with exceedingly diverse vegetative surfaces under conditions of extremely variable microclimate and particle source characteristics suggests an enormously complex relationship. Since this is the case, field evidence to quantify the amounts of natural or anthropogenic particles removed by trees is very sparse. Numerous investigations have studied detached plant parts or seedlings under wind tunnel, growth chamber, or greenhouse conditions. This is an appropriate and necessary initial step and these studies have yielded considerable qualitative perspective on the capacity of plants to filter air. The studies reviewed in the following sections were typically not conceived nor conducted specifically to evaluate the role of plants as repositories for atmospheric contaminants. Nevertheless, the hypothesis that trees are important particulate sinks is supported by evidence obtained from studies dealing with diverse particulates including radioactive, trace element, pollen, spore, salt, precipitation, dust, and other unspecified particles.

1. Radioactive Particles

Because of the considerable interest in the distribution of radioactive material following the use of nuclear weapons or nuclear accidents and because of the ease of counting, several investigations have examined the ability of aboveground plant parts to intercept radioactive aerosols (Chamberlain, 1970; Oak Ridge National Laboratory, 1969).

Witherspoon and Taylor (1969) treated potted, seedling white pine and red oak with 88–175 µm diameter quartz particles tagged with ^{134}Cs under field conditions. Initial particle retention by oak foliage was 35%, while for pine it was only 24%. After 1 hr, however, the oak leaves lost 91% of their initial concentrations while pines lost only 10%. Most pine particles were trapped at the base of needle bundles around branch termini while oak particles were retained in small, hairy recesses along leaf veins. Particle half-lives were calculated at intervals of 0–1 day, 1–7 days, and 7–33 days. For pine the values were 0.25, 5, and 21 days, respectively. For oak they were 0.12, 1, and 25 days, respectively. Particulate loss was primarily attributed to the action of wind and rain.

Contamination of tree foliage with radioactive fallout, despite its obvious disconcerting implications, has provided an especially valuable perspective because it has frequently been examined on large trees in natural environments. Romney et al. (1963) concluded from fallout evidence that Utah juniper foliage principally intercepted particles smaller than 44 µm. Interior canopy elm leaves were shown to be contaminated with elevated levels of radioactivity in selected New England and eastern New York sites following the 1957 atom bomb test series (Bormann et al., 1958). More recently Russell (1974) and Russell and Choquette (1974) have measured concentrations of fission product radionuclides, resulting from megaton range Chinese nuclear explosions in coniferous and deciduous trees in the New England area between 1968 and 1974. Peak contamination was determined to be reached 6–9 months following injection of the stratospheric source the preceding year. It was hypothesized that primary acquirement was due to attachment of rain droplets on leaf surfaces with subsequent diffusion of soluble radionuclides to leaf cuticles, where they were fixed or transported to leaf interiors. Dry deposition was concluded to be relatively unimportant.

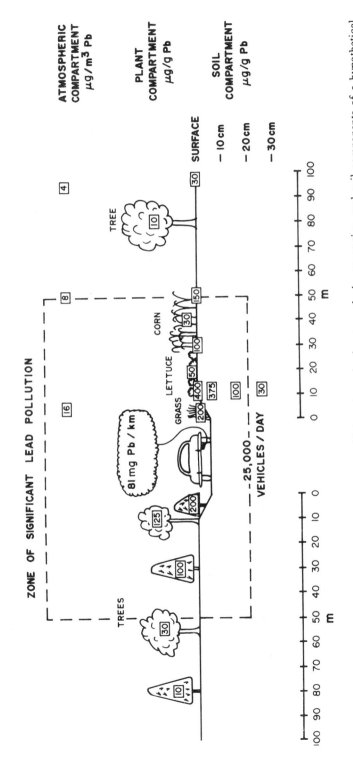

Figure 6-1. Distribution of particulate lead from motor vehicle exhaust in the atmospheric, vegetative, and soil components of a hypothetical roadside ecosystem bisected by a roadway averaging 24,000 vehicles per day. Surface contamination of roadside trees with particulates containing lead may elevate burdens to 200 times baseline levels.

2. Trace Metal Particles

Trace metals, especially heavy metals, are most commonly associated with fine particles in contaminated atmospheres. Trace element investigations conducted in roadside, industrial, and urban environments have dramatically demonstrated the impressive burdens of particulate heavy metals that can accumulate on vegetative surfaces.

In the case of lead in the roadside ecosystem, for example, the increased lead burden of plants, largely due to surface deposition, may be 5–20, 5–200, and 100–200 times baseline (nonroadside environment) lead levels for unwashed agricultural crops, gasses, and trees, respectively (Smith, 1976) (Figure 6-1).

Eastern white pine is widely planted in the roadside environment in New England and its capacity to accumulate fine particles (~7 μm diameter; Heichel and Hankin, 1972) has been shown to be substantial (Smith, 1971). Heichel and Hankin (1976) have investigated the distribution of lead deposited on this species in roadsides and have advanced several important observations. The lead burden of older needles and twigs was consistently greater than that of younger organs and was greater in samples taken adjacent to rather than far from the road. These are consistent with observations we made with the same species (Smith, 1971) and are important, as the former indicates that lead accumulates over time on the trees, while the latter argues against the importance of soil uptake as a mechanism of lead acquisition. Heichel and Hankin further concluded that twigs retained particles more effectively than needles throughout the season. This was judged to be due to the roughness of twigs relative to needles. The authors observed that a 12 m tall white pine growing in a dense planting would have about 15×10^5 cm^2 of foliage surface. Although white pine exposed approximately ten fold more foliage than woody surface, the woody surfaces retained about 20-fold the lead burden of foliage.

Like the roadside environment, urban atmospheres also have elevated amounts of particles containing trace metals. In New Haven, Connecticut, we have examined the surfaces of a variety of city trees and have found substantial accumulation of certain metals, particularly lead, zinc, and iron (Smith, 1973; Smith and Staskawicz, 1977). Observations of the leaves of mature London plane trees in New Haven throughout the growing season indicated nickel and zinc foliar surface amounts remained relatively constant. Aluminum, iron, manganese, and lead, on the other hand, appeared to accumulate through the spring and early summer, and decrease during the late summer and fall on this species (Figure 6-2). This latter decrease and late season decreased particle density indicated by observation with the scanning electron microscope, suggest the particles on foliage are weathered or transported off the leaf. Precipitation, wind, insect activity, and other forces may cause particles to be lost from the leaves and transported by way of the petiole to the twig tissue. Our leaf observations with the scanning electron microscope allowed a variety of additional observations:

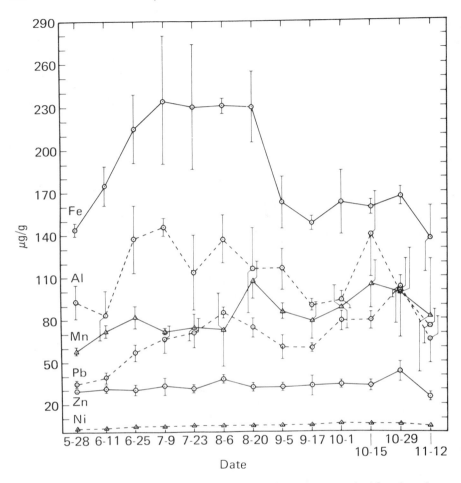

Figure 6-2. Superficial foliar trace metal burden of mature, unwashed London plane leaves collected throughout an entire growing season in New Haven, Connecticut. Concentrations are μg g^{-1} (dry weight basis), vertical bars represent 95% confidence intervals.
Source: Smith and Staskawicz (1977).

1. In regard to spatial distribution throughout the growing season, particles were more prevalent on the adaxial (upper, facing the twig) surface than the abaxial (lower, away from the twig) surface. Peripheral leaf areas were always the cleanest. Midlaminar areas were generally only lightly contaminated. Most particulates were located in the midvein, center portion of the leaves. The greatest particulate burden was located on the adaxial surface at the base of the blade just above the petiole junction. It is probable that precipitation washing plays an important role in this distribution pattern.
2. Late in the growing season, particulate loads could increase to very high levels.

3. Particle morphology was very variable (Figure 6-3). Carbonaceous and aggregate particles were especially common. Visible particles, including pollen grains, fungal spores, and mycelium (Figure 6-4), were particularly prevalent after early July.
4. Particle size was extremely variable. Most particles appeared to fall in the 5–50 μm range. Significant numbers outside this range were observed. Submicron particles could easily be found, particularly on trichomes (leaf hairs). Large aggregates in excess of 100 μm could also be easily found.
5. On the lower leaf surface, complete stomatal blockage by particulates, partial blockage, or contamination could be seen. It was evident, however, that particulate association with stomates was the exception, rather than the rule.

Figure 6-3. Scanning electron microscope micrograph of the adaxial surface of an 8-week-old London plane leaf. Spore, pollen, carbonaceous, angular, and aggregate particles are visible. Scale, 10μm.

6. Trichomes, particularly abundant on the center portion of upper and lower leaf surfaces, are particulate accumulators (Figure 6-5). As the season progresses the trichomes are reduced in size by "weathering" and occasionally are completely broken off.

7. Fungal mycelium, which becomes particularly abundant on leaf surfaces as the growing season progresses, is in intimate association with particulate contaminants. The surficial trace metal burden of London plane leaves appeared to be generally less than the average for several other New Haven deciduous species (Smith, 1973). This disparity may have resulted from a comparison of late season averages from the latter with the more representative "throughout the growing season" average of the former.

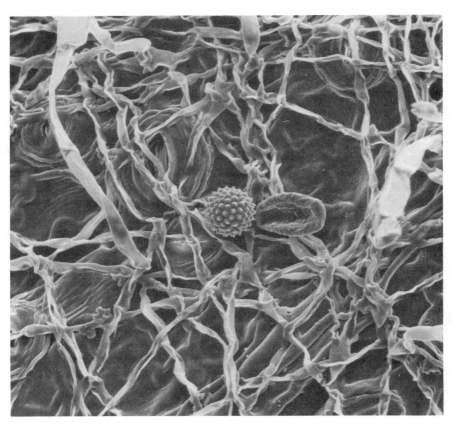

Figure 6-4. Scanning electron microscope micrograph of ragweed pollen (center) and a collapsed fungus spore (right center) and mycelium on the abaxial surface of a 15-week-old London plane leaf. Scale, 10 μm.

The literature is replete with studies demonstrating significant trace metal particle accumulation on trees in roadside, urban, and industrial situations. Industrial regions, particularly those with metal smelters may excessively contaminate surrounding woody vegetation with particles containing trace metals. A representative study is that conducted by Little and Martin (1972) in the Avonmouth industrial complex, Severnside, England. Close to this complex, elm leaves exhibited 8000, 5000, and 50 μg g^{-1} zinc, lead, and cadmium, respectively. Page and Chang (1979) and Helmke et al. (1984) have reviewed the trace element contamination of vegetation in the vicinity of coal-fired power plants. Helmke et al. (1984), in their study of vegetation in the vicinity of a coal burning Wisconsin power plant, observed that a significant portion of the dust deposited on local leaf surfaces was fly ash. Wind and rain removal was minimal and fly ash accumulated during the growing season. No detrimental effects of fly

Figure 6-5. Scanning electron microscope micrograph of a trichome on the adaxial surface of a 17-week-old London plane leaf. The leaf hair has accumulated numerous particles. Scale, 10 μm.

ash on trees were observed.

Data provided, however, rarely permit a quantitative estimate of the sink capability of the woody plants. The required mass or vegetative surface area calculations are not presented, as the purpose of the studies typically did not include sink function assessment. When quantitative estimates of sink capacity are made, the results can be impressive. During the course of interpretation of some of our urban lead data, we combined sugar maple dimension analysis information from New Hampshire with contamination data from New Haven to speculate on the sink capacity of a single sugar maple (Table 6-1).

3. Pollen and Spores

Pollen studies have provided important evidence of vegetative interception of large particles. G.S. Raynor of the Brookhaven National Laboratory, Upton, Long Island, New York, has conducted a series of dispersion experiments employing ragweed pollen emitted from sources at various distances and heights upwind of a forest edge. Pollen loss from the plume occurred in two stages and by two mechanisms, impaction near the forest edge and deposition well within the forest. Pollen loss to the forest was considerably greater than over open terrain (Raynor, 1967; Raynor et al., 1966). Interception of ragweed pollen by a Pennsylvania forest canopy reduced pollen concentration in the forest atmosphere to only 70% of the concentration in a nearby open field (Elder and Hosler, 1954). Neuberger et al. (1967) measured ragweed pollen concentrations in and out of forests and found that 100 m inside a dense coniferous forest over 80% of the pollen had be subtracted from the atmosphere. Data indicated that deciduous species are less effective than conifers in filtration of pollen (Neuberger et al., 1967; Steubing and Klee, 1970). For these large (~20 μm) particles the dominant transfer to vegetative surfaces is via sedimentation and not impaction (Aylor, 1975).

Fungal spore (size range, 1.5–30 μm) interception studies have provided important evidence for understanding particulate capture (Gregory, 1971). Ingold (1971) has concluded that the most efficient plant parts for spore collection are petioles, twigs, and leaf lamina, respectively. Efficiency of spore collection increases with decreasing diameter of the collecting cylinder. Observations of basidiomycete spores in Washington Douglas fir forests have emphasized the extraor

Table 6-1. Calculated Particulate Metal Sink Capacity for the Leaves and Current Twigs of a Single, 30 cm (12-inch) Diameter Urban Sugar Maple During the Course of a Growing Season.

Metal contaminant	Growing season removal (mg tree^{-1})
Lead	5800
Nickel	820
Chromium	140
Cadmium	60

Source: Smith (1974).

dinary importance of microclimate and forest stand structure on the distribution and deposition of these particles in the forest. Wind speed, air temperature, inversions, cloud cover, and forest openings all influenced particle movement (Edmonds and Driver, 1974; Fritschen et al., 1970).

4. Salt Particles

Vegetative interception of saline aerosol (primarily NaCl) and nutrient particles has also contributed to our understanding of plant sink function. Particulate deposition of salt particles occurs in roadside environments where deicing salts are employed, in the vicinity of cooling towers, and in maritime regions (Eaton, 1979; Moser, 1979). Conifers planted close to roads receiving deicing salt applications frequently exhibited needle necrosis due to the accumulation of toxic levels of salt transported from the road to the leaves via the atmosphere (Constantini and Rich, 1973; Hofstra and Hall, 1971; Smith, 1970b). In coastal ecosystems subject to airborne marine salt, accumulation of salt particles by above ground plant parts injures foliage and twigs (Boyce, 1954; Wells and Shunk, 1938; Oosting, 1945; Oosting and Billings, 1942) and may control species success or failure depending on tolerance to salt loading (Martin, 1959). Clayton (1972) described the trapping of particulate salts by *Baccaris* brushlands in coastal California. Woodcock (1953) provided evidence that the shape of plant leaves influences the amount of salt deposited. By employing plates of various shapes, he found that long narrow plates accumulated more salt per unit area than did circular plates. Edwards and Claxton (1964) found over four times the deposition of salt on the windward side of a hedgerow compared to the leeward side.

Where foliar capture of marine particulates is below the threshold of foliar injury, particle accumulation may be an important mechanism for nutrient acquisition (Art, 1971; Art et al., 1974). Numerous investigations, reviewed by White and Turner (1970), have indicated that nonmaritime trees also catch airborne nutrient particles. These authors found that a mixed deciduous forest was capable of annually removing 125 kg ha^{-1} sodium, 6 kg ha^{-1} potassium, 4 kg ha^{-1} calcium, 16 kg ha^{-1} magnesium, and 0.1 kg ha^{-1} phosphorus from the atmosphere. The degree of leaf hairiness was inversely correlated with particle retention. Apparently the small droplets employed had insufficient inertia to penetrate the stable boundary layer created by the hairy leaves. Small diameter branches were more efficient particle collectors than large diameter branches in all species examined.

In their examination of the impact of saline aerosols of cooling tower origin, McCune et al. (1977) emphasized the importance of particle wetness in causing damage to surrounding trees. Dry particles appeared less toxic than hydrated particles. This supports the contention that moist particles are more effectively retained by vegetative surfaces than dry ones.

5. Precipitation, Dust, and Other Particles

Foliar interception of precipitation has been intensively investigated (Zinke, 1967), but the relatively large size of the particles (range, 50–700 μm) makes these data of limited application for considerations of fine particle retention. The enormous importance of rainout in transferring fine particles from the atmosphere to vegetation is recognized (Altshuller, 1984). Numerous precipitation studies support the general observation that conifers intercept more particles than deciduous species, for example, Helvey (1971), who reported canopy interception loss greatest in a spruce-fir-hemlock type, intermediate in pine, and least in broad-leaved deciduous forests.

Numerous additional studies employing dust, synthetic, or unspecified particles have contributed to our understanding of particulate capture by vegetation. Rosinki and Nagamoto (1965) investigated the deposition of 2 μm particles on Rocky Mountain juniper and Douglas fir. At low dosage, particles preferentially accumulated on the windward leaf edge. Eventually a new layer was formed on the previously deposited layer. Thickness increased until an equilibrium was reached. Total deposition was increased when wind exposed different leaf areas for deposition. Langer (1965) concluded that dust deposition on coniferous leaves was not significantly influenced by electrostatic effects. Podgorow (1967) investigated the relative effectiveness of pine, birch, and aspen in filtering dust particulates. Pine proved most effective. Interior crown needles accumulated more and retained more dust than exterior needles. Bach (1972) also presented evidence supporting the superior collecting capacity of pines relative to deciduous species. In an Ohio study, Dochinger (1972) examined dustfall and suspended particulate matter in three areas — treeless, deciduous canopy, and conifer canopy — and concluded that trees have the capacity to reduce particulate pollutants in the ambient atmosphere.

Wedding et al. (1975) found, under controlled wind tunnel conditions, that particulate deposition on rough pubescent sunflower leaves was 10 times greater than on smooth, waxy tulip poplar leaves. In a unique study, Graustein (1978) employed the ratio of strontium isotopes in soil dust to determine strontium input to forested watersheds in New Mexico. Most of the atmospherically transported strontium entered the watershed by impaction of soluble particles on spruce foliage. Aspen, also present in the ecosystem, was judged to trap little, if any, dust. The flux of dust-derived strontium to the forest floor was four times greater than the flux to an unforested area.

In an extremely informative set of experiments, Little (1977) exposed freshly collected leaves of several tree species in a wind tunnel to various sizes of polystyrene aerosols labeled with technetium. Particles sized 2.75, 5.0, and 8.5 μm were tested with leaves from European beech, white poplar, and nettle (*Urtica dioica*). Surface texture was critical in capture efficiency, with the rough and hairy leaves of nettle more effective than the densely tomentose leaves of poplar or the smooth surfaces of beech. For each species there was a strong negative linear correlation between leaf area and deposition velocity, the latter being

Table 6-2. Average Percentage of Total Catch Intercepted by Leaf Laminas, Petioles, and Stems of Freshly Cut European Beech, White Poplar, and Nettle Exposed to Polystyrene Particles in a Wind Tunnel.

Wind speed (cm sec^{-1})	Particle size (μm)	Plant part	Beech	White poplar	Nettle
150	5.0	Leaf laminas	85.01	49.16	90.68
		Petioles	11.18	33.52	4.76
		Stems	3.80	17.32	4.56
250	2.75	Leaf laminas	63.08	73.71	68.30
		Petioles	30.19	11.08	15.56
		Stems	6.73	15.20	16.11
	5.0	Leaf laminas	70.94	57.10	78.59
		Petioles	23.08	26.78	6.66
		Stems	5.99	16.12	7.00
	8.5	Leaf laminas	62.68	45.39	68.30
		Petioles	17.17	29.12	11.29
		Stems	19.61	25.48	20.41
500	2.75	Leaf laminas	63.87	64.81	82.85
		Petioles	28.81	18.81	5.50
		Stems	7.32	16.36	11.06
	5.0	Leaf laminas	73.86	53.39	77.71
		Petioles	14.94	26.12	9.23
		Stems	11.19	20.49	13.06
	8.5	Leaf laminas	90.83	69.27	83.27
		Petioles	3.35	12.58	7.21
		Stems	5.81	18.15	9.82

Source: Little (1977).

smallest for the largest leaves. Deposition was heaviest at the leaf tip and along leaf margins, where a turbulent boundary layer was present. Leaves with complex shapes and the largest circumference to area ratio were the most efficient collectors. Both increased wind speed and particle size were reflected in increased deposition velocities. Deposition velocities to petioles and stems were many times greater than deposition velocities to leaf laminas, even though the majority of the total catch was intercepted by the leaf lamina (Table 6-2). Table 6-2 reveals that nonlaminar catch is significant, however, and Little (1977) suggested that this may cause deposition of atmospheric particles to trees to be relatively high, even during the winter when deciduous species are devoid of leaves.

a. Case Study of Street Tree Particulate Sink Capacity

The U.S. Environmental Protection Agency has developed a demonstration plan to explore the capability of urban vegetation to improve air quality (U.S. Environmental Protection Agency, 1976c). This plan, which utilized pollutant fluxes summarized and extrapolated from the literature and air quality and environmental conditions as they existed in the St. Louis, Missouri area, included an assessment of the particulate removal capacity of selected and hypothetical street trees. It was proposed that trees be planted on both sides of the streets within the city boundaries of St. Louis. The trees would be planted 8.5 m (30 feet) apart. The total street length in St. Louis was determined to be 2316 linear km (6.6 X 10^6 feet), requiring a total of 440,000 trees for complete planting. The three tree species proposed for the street planting included red oak, Norway maple, and linden. An average particulate flux rate of 2.5 X 10^3 μg m^{-2} hr^{-1} guesstimated from the literature was employed (U.S. Environmental Protection Agency, 1976b). Table 6-3 presents the dimensions of the tree species and the estimated quantity of particles that would be removed by the 440,000 street trees. The hypothetical transfer of particles from the atmosphere to the tree surfaces totaled 340 tons annually. The 1980 estimate for total particulate emission in the St. Louis area equaled 126,290 tons. The biological and medical significance of the transfer of the 340 tons from the atmosphere to the vegetation is unclear.

b. Case Study of Tree Particulate Sink Capacity in the Vicinity of a Coal-Fired Power Station

During the conversion of an oil-fired steam generating unit to a coal-fired unit, the Baltimore Gas and Electric Company considered the ability of trees in the vicinity of the power station to remove particulates from the atmosphere (Jashnani, 1988). The station site was on the Patapsco River, Anne Arundel County, Maryland and the proposed boiler was a 135 megawatt unit. Based on a literature survey, particulate removal efficiencies for trees were estimated. Assumptions included: particulate average deposition velocity of 1 cm sec^{-1} for trees and 0.8 cm sec^{-1} for grass and weeds, leaf area index of 5.1 for deciduous trees and 2.3 for conifers, and approximately 2 ha of deciduous tree surface and 1 ha of coniferous tree surface ha^{-1} of land area. Given these assumption particulate removal rates and land area required for removal were estimated (Table 6-4). The annual cost of particulate removal by trees, including land acquisition, tree planting and maintenance, was estimated to be very approximately \$11,000 ton^{-1} of particle pollutants removed (1987-1988 dollars). The comparable cost for conventional control technology (electrostatic precipitation) was \$131 ton^{-1}.

B. Forest Vegetation as a Sink for Gaseous Contaminants

Substantial evidence is available to support the potential that plants in general (Bennett and Hill, 1975; Hill, 1971; Rasmussen et al., 1975) and trees in particular (Smith, 1979; Smith and Dochinger, 1975, 1976; Roberts, 1971; Warren, 1973) have to function as sinks for gaseous pollutants. The latter are transferred from the atmosphere to vegetation by the combined forces of diffusion and flowing air movement (Chapter 2). Once in contact with plants gases may be bound

Table 6-3. Estimation of the Amount of Particulates Absorbed by Hypothetical St. Louis Street Trees.

1. Number of trees planted
   Maple   146,666
   Oak   146,667
   Linden   146,667
       Total   440,000

2. Dimensions of the maple
   Height = 6 m
   Diameter of canopy = 3 m
   Total surface area^a tree^{-1} = 36.8 m^2
   Total surface area of 146,666 trees = 5.40×10^6 m^2

3. Dimensions of the oak
   Height = 6 m
   Diameter of canopy = 3 m
   Total surface area tree^{-1} = 36.1 m^2
   Total surface area of 146,667 trees = 5.30×10^6 m^2

4. Dimension of the linden
   Height = 5 m
   Diameter of canopy = 2.4 m
   Total surface area tree^{-1} (including undergrowth) = 23.0 m^2
   Total surface area for 146,667 trees = 3.40×10^6 m^2

5. Total surface area for the 440,000 trees = 1.4×10^7 m^2

6. Estimated particulate flux to vegetation = 2.5×10^3 μg^{-2} hr^{-1}

7. Calculation to determine the amount of particulates absorbed by street trees
   1.4×10^7 m^2 $\times$ 2.5×10^3 μg m^{-2} hr^{-1} $\times$ gm/10^6 μg $\times$ 1b/453.59 gm $\times$
   T/2000 lbs $\times$ 24 hr/day $\times$ 365 days/yr = 3.40×10^2 tons particulate yr^{-1}

Source: U.S. Environmental Protection Agency (1976a).
^{a} Maple canopy diameter = 3 m
estimated ground area covered by maple canopy = 7.1 m^2
area index for maple = 5.18 (Raunder, 1976)
area index = surface area $\div$ ground area
$\therefore$ 5.18 = x/7.1 m^2
surface area of maple = 36.8 m^2.

Table 6-4. Particulate removal rates and land area required for pollutant removal in the vicinity of a 135 megawatt coal burning power station releasing 186 tons of particulates yr^{-1}.

	deciduous	conifer
	tons ha^{-1} yr^{-1}	
Particulate removal rate	0.38	0.14
	ha	
Land area required	486	1336

Source: Jashnani (1988).

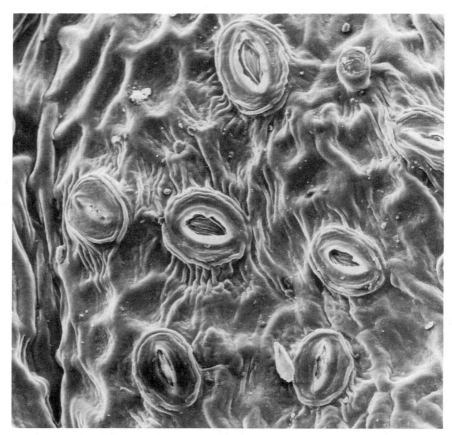

Figure 6-6. Scanning electron microscope micrograph of the abaxial surface of a 3-week-old London plane leaf showing stomates. Scale, 10 μm.

or dissolved on exterior surfaces or be taken up by the plants via stomata. If the surface of the plant is wet and if the gas is water soluble, the former process can be very important. When the plant is dry or in the case of gases with relatively low water solubilities, the latter mechanism is assumed to be the most important.

1. Stomatal Uptake

Stomatal pores are small openings, typically approximately 10 μm in length and 2 –7 μm in width, in the epidermal surface of leaves through which plants naturally exchange carbon dioxide, oxygen, and water vapor with the atmosphere (Figure 6-6). The waxy cuticle of leaf surfaces restricts diffusion so that essentially all gas exchange carried out by leaves is via stomatal openings. Even though these openings make up only approximately 1% of the leaf surface area, their orientation and mechanics prove to be nearly optimal for maximum gas diffusion in and out of the leaf (Salisbury and Ross, 1978). Stomates undergo diurnal opening and closing with the pores of most plants opened within an hour of sunrise and closed by dark. Gas diffusion to and from leaves and the timing and degree of opening of stomatal aperatures is strongly influenced by a number of complex environmental factors.

During daylight periods when plant leaves are releasing water vapor and taking up carbon dioxide, other gases, including trace pollutant gases, in the vicinity of the leaf will also be taken up through the stomates. Once inside the leaf these gases will diffuse into intercellular spaces and will be absorbed on or in the surfaces of palisade or spongy parenchyma cell walls (Figure 6-7).

The rate of pollutant gas transfer from the atmosphere to interior leaf cells is regulated by a series of resistances conveniently thought of as atmospheric, stomatal, and mesophyllic. Factors controlling atmospheric resistance include wind speed, leaf size and geometry, and gas viscosity and diffusivity. Stomatal resistance is regulated by stomatal aperature, which is influenced by water deficit, carbon dioxide concentration, and light intensity. Mesophyllic resistance is regulated by gas solubility in water, gas liquid diffusion, and leaf metabolism (Kabel et al., 1976). Because the rate of pollutant uptake is regulated by numerous forces and conditions, the rate of removal under field conditions is highly variable. If leaf characteristics, wind speed, atmospheric moisture, temperature, and light intensity are quantified, however, the pollutant uptake rate can be estimated (Kabel et al., 1976; Bennett et al., 1973).

2. General Plant Uptake

The fundamental investigations of A. Clyde Hill and Jesse H. Bennett of the University of Utah allow several general conclusions concerning gaseous pollutant uptake (Hill, 1971; Bennett and Hill, 1973, 1975). Their studies have concentrated on alfalfa, oats, barley, and grass. Standard alfalfa canopies removed gaseous pollutants from the atmosphere in rates of the following order: hydrogen fluoride > sulfur dioxide > chlorine > nitrogen dioxide > ozone > peroxyacetylnitrate > nitric oxide > carbon monoxide. In general plant uptake rates increased as the solubility of the pollutant in water increased. Hydrogen fluoride, sulfur dioxide, nitrogen dioxide, and ozone, which are soluble and reactive, were readily absorbed. Nitric oxide and carbon monoxide, which are very insoluble, were absorbed relatively slowly or not at all (Table 6-5). The rate of pollutant removal was found to increase linearly as the concentration of the pollutant was increased over the ranges of concentration that are encountered in ambient air and that were low enough not to cause stomatal closure.

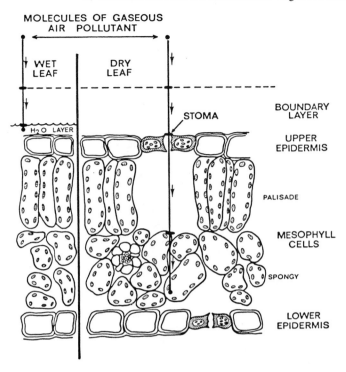

Figure 6-7. Schematic diagram portraying the interaction of molecules of gaseous air pollutants with leaves. When leaf surfaces are wet, the gases may be dissolved in the moisture film on the leaf. When the leaves are dry, the gases are taken up through the stomates and ultimately may react on or in mesophyl cells of the leaf interior. Movement of gas molecules into the leaf involves several "resistances" imposed by the boundary layer (quiescent zone of retarded air flow surrounding the leaf), substomatal cavity (open space below the stoma), and the mesophyll cells themselves. Since these "resistances" can be measured or estimated, the rate of gas uptake can be modeled for a given set of gas, plant, and environmental circumstances.

Under growth chamber conditions, wind velocity, canopy height, and light intensity were shown to affect the rate of pollutant removal by vegetation. As previously stressed, light plays a critical role in determining physiological activities of the leaf and stomatal opening, and as such exerts a great influence on foliar removal of pollutants. Under conditions of adequate soil moisture, however, pollutant uptake by vegetation was judged almost constant throughout the day, as the stomata were fully open. Pollutants were absorbed most efficiently by plant foliage near the canopy surface where light-mediated metabolic and pollutant diffusivity rates were greatest. Sulfur and nitrogen dioxides were taken up by respiring leaves in the dark, but uptake rates were greatly reduced relative to rates in the light.

Table 6-5. Solubility in Water and Uptake Rate of Pollutants by Alfalfa.

Pollutant	Uptake rate by alfalfa at 1 pphm (liters min^{-1} m^{-2})	Equivalent deposition velocity (cm sec^{-1})	Solubility at $20°C$ (cm^3 gas cm^{-3} H_2O)
CO	0.0	0.00	0.02
NO	0.6	0.10	0.05
CO_2	2.0	0.33	0.88
PAN	3.8	0.63	–
O_3	10.0	1.67	0.26
NO_2	11.4	1.90	Decomposes
Cl_2	12.4	2.07	2.30
SO_2	17.0	2.83	39.40
HF	22.6	3.77	446

Source: Hill and Chamberlain (1974).

3. Tree Uptake

a. Sulfur Dioxide

Because of its high solubility in water, large amounts of sulfur dioxide are absorbed to external tree surfaces when they are wet. In the dry condition, sulfur dioxide is readily absorbed by tree leaves and is rapidly oxidized to sulfate in mesophyll cells. At low uptake rates sulfur dioxide is presumed to be oxidized about as rapidly as it is absorbed (Bennett and Hill, 1975).

Roberts (1974) measured sulfur dioxide sorption by single leaves or shoots of several 1 year-old seedlings of numerous woody species. All species examined were capable of reducing high ambient levels within his test chambers (Table 6-6). Because of the large dose employed, 1 ppm (2620 μg m^3) for 1 hr. Roberts reduced the concentration in subsequent trials and examined uptake at concentrations of 0.2 and 0.5 ppm (524 and 1310 μg m^3). At the lower concentration uptake by birch and firethorn was significantly less. It was speculated that higher concentrations of sulfur dioxide may maintain stomatal opening. Under controlled environmental conditions, comparable to those employed by Roberts, Jensen (1975) fumigated hybrid poplar cuttings with sulfur dioxide ranging in concentration from 0.1 to 5 ppm (262–13.1 X 10^3 μg m^3) for periods of 5–80 hr. Uptake was determined by measuring the total sulfur content of the leaves. At low levels of fumigation [0.1 and 0.25 ppm (262 and 655 μg m^{-2}),] leaf sulfur initially increased but then declined to unfumigated levels as fumigation continued. This reduction was judged by the author to be due to one or more of the following: reduction in absorption rate, translocation of sulfur out of the leaves, leaching of sulfate from the roots, or release of hydrogen sulfide by the leaves.

Jensen and Kozlowski (1975) have provided additional perspective on tree seedling uptake of sulfur dioxide in their experiments that exposed 1-year-old

Table 6-6. Foliar Sorption of Sulfur Dioxide by Selected Seedlings Fumigated at 1.0 ppm, (2620 μg m^{-3}) for 1 hr in a Controlled Environment Chamber (27 ± 1°C, 51 ± 7% RH, 1300 ft-c).

Species	SO$_2$ uptake	
	mg SO$_2$ dm^{-2} hr^{-1}	mg SO$_2$ g^{-1} hr^{-1}
Red maple	0.088	0.260
White birch	0.086	0.268
Sweetgum	0.074	0.267
Firethorn	0.072	0.213
Privet	0.068	0.134
Rhododendron	0.056	0.079
White ash	0.046	0.118
Azalea	0.044	0.072

Source: Roberts (1974).

sugar maple, bigtooth aspen, white ash, and yellow birch to 2.75 ppm (7205 μg m^{-3}) for 2 hr. Prefumigation with 0.75 ppm (1965 μg m^{-3}) sulfur dioxide for 20 hr or more reduced the rate of absorption in all species except white ash. The authors speculated that tolerance to sulfur dioxide injury following uptake may be related to the rate at which accumulated sulfur can be moved out of the leaves. Roberts and Krause (1976) monitored sulfur dioxide uptake of intact plants of rhododendron (3-month-old) and firethorn (12-month-old) and suggested that the greater uptake of the latter may have been partially due to abundant trichomes (leaf hairs).

These various controlled-environment studies are important, as they qualify and caution our efforts to extrapolate the sink function to trees in natural environments. Uptake under ambient conditions may be less than under experimental conditions, as the latter frequently employ unnaturally high concentrations of sulfur dioxide. Prefumigation, common in natural situations, may further reduce natural uptake. Uptake rates in the field may decline over time as pollution episodes continue. Even though mentioned by several investigators, the seedling studies have not addressed two very important questions concerning uptake, namely, the relative importance of stomatal uptake versus adsorption to the surface of dry plant parts and the relative uptake of dry plants versus plants with moisture films on their surfaces. Garland and Branson (1977), however, have recently provided evidence supporting the importance of stomates and wet surfaces in uptake. These investigators determined the rates of water vapor and sulfur dioxide conductance in detached Scotch pine needles collected from a 45-year-old stand and in intact trees in a 10-year-old plantation. The similarity of the conductances observed for these two processes led the authors to conclude that they were both controlled by diffusion through the stomata. By analogy with transpiration, the authors further estimated that the deposition velocity of sulfur dioxide to a dry pine canopy would vary from 0.2 to 0.6 cm sec^{-1} during daytime (0.05–0.1 cm sec^{-1} at night) but might be 10 times this rate if the canopy was wet from precipitation or dew. This deposition velocity appears slightly conservative when

Table 6-7. Foliar Sorption of Ozone by Selected Seedlings Fumigated at 0.20 ppm (392 μg m^{-3}) for a few Hours in a Controlled-Environment Chamber (26 $\pm$ 1.5°C, 45 $\pm$ 5% RH, 2100 ft-c).

Species	O_3 uptake	
	mg O_3 dm^{-2} hr^{-1}	mg O_3 g^{-1} hr^{-1}
White oak	0.635	1.318
White birch	0.536	2.347
Coliseum maple	0.502	0.991
Sugar maple	0.371	0.863
Ohio buckeye	0.362	0.927
Redvein maple	0.285	0.911
Sweetgum	0.278	0.854
Red maple	0.272	0.555
White ash	0.239	0.562

Source: Townsend (1974).

compared with the dry deposition rates provided by other investigators (Sheih, 1977; Sheih et al., 1979). Martin and Barber (1971) monitored the sulfur dioxide loss in the immediate vicinity of a large hawthorne hedge (approximately 4 m high X 3m wide) subject to effluent from an electric generating facility. With ambient concentrations generally less than 10 pphm (262 μg m^{-3}), the authors observed a significant loss of sulfur dioxide near (~150 mm) the foliage. The greatest loss was during rain or dew periods when the hedge foliage was wet.

b. Oxidants

Ozone is relatively insoluble in water (0.052 g 100 g^{-1} H$_2$O at 20°C) but readily diffuses into stomatal cavities (Rich and Turner, 1972; Rich et al., 1970; Thorne and Hanson, 1972; Wood and Davis., 1969). The very reactive nature of this gas undoubtedly causes it to rapidly react on the surface of leaf mesophyll cells.

Under controlled environmental conditions, Townsend (1974) has monitored ozone uptake by a variety of seedling tree species (Table 6-7). Ozone sorption exhibited a linear increase up to 0.5 ppm (980 μg m-3) for both white birch and red maple. These two species were also capable of reducing ambient ozone throughout a prolonged 8-hr exposure.

While tree removal of atmospheric peroxyacetylnitrate has not been reported, Garland and Penkett (1976) have suggested that the deposition velocity of this gas to grass was approximately 0.25 cm sec^{-1}, which is lower than the value for ozone (0.8 cm sec^{-1}) or sulfur dioxide (1 cm sec^{-1}).

For the other gases judged to be significantly removed from the atmosphere by vegetation, hydrogen fluoride and nitrogen dioxide, relatively little work has been reported on trees as sinks. Hydrogen fluoride in is very water soluble and very reactive, and can be adsorbed onto plant surfaces and absorbed through stomates. Leaves exposed to hydrogen fluoride may accumulate fluoride to one mil-

Table 6-8. Comparison of Simulated Sulfur Dioxide Uptake for a South Carolina Loblolly Pine Forest with Experimental Uptake by Deciduous Seedlings.

Species	SO$_2$ uptake kg ha^{-1} hr^{-1} ppb^{-1} by volume
Loblolly pine (simulated)	
January	
day 1	1.3×10^{-3}
day 2	2.2×10^{-4}
June	
day 1	2.0×10^{-3}
day 2	2.0×10^{-4}
Seedlings (experimental)	
Red maple	8.8×10^{-5}
White birch	8.6×10^{-6}
Sweetgum	7.4×10^{-5}
White ash	4.6×10^{-5}

Source: Murphy et al. (1977).

lion times the ambient concentration. Nitrogen dioxide dissolved in water yields nitrate and nitrate ions in solution. The latter can be reduced to ammonia in leaf cells (Bennett and Hill, 1975). Rogers et al. (1979) have provided nitrogen dioxide uptake rates for loblolly pine and white oak and have suggested deposition velocities of 0.53 and 0.11 cm sec^{-1}, respectively.

4. Models of Forest Gas Sink Function

Efforts to estimate the sink capability of forest vegetation under natural conditions must consider a complex set of variables, including pollutant concentration and deposition velocity, meteorological parameters, and dimensions (leaf or canopy area, dry weight) and conditions of the trees. The systematic approach of model development is desirable and necessary despite the obvious deficiencies in field data.

Murphy et al. (1977) have modeled the sulfur dioxide uptake of a simulated loblolly pine forest exposed to 50 ppb (131 µg m^3) sulfur dioxide on two clear days in January and June using climate data from a station near Aiken, South Carolina. The simulated uptake compared favorably, but was smaller than the seedling uptake rates reported by Roberts (1974) (Table 6-8). Murphy et al. (1977) also applied their model to regions where forest vegetation was dominant and where actual frequency distributions of sulfur dioxide concentrations were known. At a site on the Savannah River Laboratory with an average sulfur dioxide concentration of 8 ppb (21 µg m^3) during the spring the model predicted an uptake of 11 metric tons day^{-1} over the 778 km^2 area of the southern pine forest site. For Long Island, New York, over an area of 1723 km^2 in June, the model predicted a sulfur dioxide uptake of 103 metric tons day^{-1} (SO$_2$ 32 ppb, 84 µg m^{-3}) for a west wind condition. According to the authors the New York estimate

Table 6-9. Area Ratios for Forest Communities.

Stem (bark) to ground surface
0.5-0.7 for mature, closed forests
0.2-0.4 for small, open forests
0.7-1.0 for dense, young stands
Branch (bark) to ground surface
1.5-1.6 for mature, deciduous forests
Leaf to ground surface
4.0-6.0 for closed, deciduous forests
6.0-7.0 for dense, evergreen forests

Source: Whittaker and Woodwell (1967).

was larger due to the larger land area, higher ambient sulfur dioxide, and greater leaf area employed.

Kabel et al. (1976) have argued, and appropriately so, that the uptake rate of sulfur dioxide on a leaf area basis must be extrapolated to a ground area basis in order to predict large area pollutant removal. In addition deposition velocities, or mass transfer coefficients — as these authors prefer (compare Kabel, 1976), — must be given for uptake of stems and branches as well as leaves. Fortunately, Whittaker and Woodwell (1967) have provided generalized area ratios for temperate forest communities (Table 6-9). Kabel et al. (1976) calculated the following deposition velocities using the appropriate ratios: 0.015 m sec[-1] for dry condition (stomatal plus soil) and 0.21 m sec[-1] for damp canopy condition, yielding a total deposition velocity of 0.23 m sec[-1] for a moist forest canopy. Uptake rates were calculated for a model forest (dry condition) downwind from a sulfur dioxide source. Gas concentration profile and uptake rates are presented in Figure 6-8.

The radioactive measurements of Garland (1977) propose that the deposition velocity for a dry forest canopy varies from 0.001 to 0.006 m sec[-1]. The Kabel et al. (1976) figure is presumably higher than this due to the inclusion of soil uptake in the latter.

Unfortunately models describing sink capabilities of forests for other gaseous contaminants are not readily available. Waggoner (1971, 1975) has made some preliminary observations with ozone, but no formal model has been proposed.

a. Case Studies of Forest Gas Sink Capability

The ability of trees surrounding the coal fired power station described in Section A5b (preceding) to remove sulfur dioxide and nitrogen oxides was also estimated. Using the assumptions previously described, the gaseous removal rates and land area required for removal were calculated (Table 6-10). The annual cost of gaseous pollutant removal by trees, including land acquisition, tree planting and maintenance, was estimated to be very approximately $11,800 ton[-1] and $6000

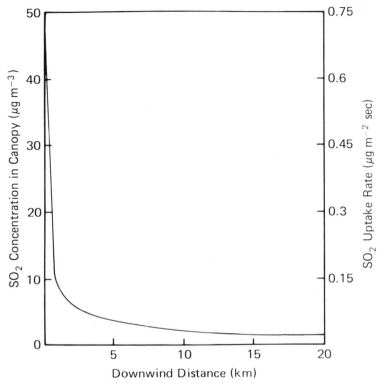

Figure 6-8. Ground level SO_2 concentration and uptake rate over a model forest with a deposition velocity of 0.015 m sec^{-1} and eddy diffusivity of 7 m^2 sec^{-1}.
Source: Kabel et al. (1976).

Table 6-10. Gaseous Removal Rates and Land Area Required for Removal in the Vicinity of a 135 Megawatt Coal Burning Power Station Releasing 4330 tons yr^{-1} Sulfur Dioxide and 2148 tons yr^{-1} Nitrogen Oxides.

	deciduous conifer tons ha^{-1} yr^{-1}	
Pollutant removal rate		
SO_2	0.36	0.12
NO_x	0.69	0.24
Land area required	ha	
SO_2	11,898	33,995
NO_x	2995	8499

Source: Jashnani (1988).

ton^{-1} for sulfur dioxide and nitrogen oxides respectively. The comparable costs for conventional sulfur dioxide and nitrogen oxide management (BACT) was $962 ton^{-1} and $124 ton^{-1}, respectively (Jashnani, 1988).

As part of the U.S. Environmental Protection Agency's demonstration plan to explore the capability of urban trees to improve air quality previously described in this chapter (U.S. Environmental Protection Agency, 1976c), a model forest hectare was developed and employed to estimate gas uptake. The model forest consisted of six species including red oak, Norway maple, linden, poplar, birch, and eastern white pine. The arrangement and spacing of the proposed forest is presented in Figure 6-9. The estimations of total tree surface area (canopy plus woody) at five years after planting were as follows:

Maple	(6 m ht)	36.8 m^2
Oak	(6 m ht)	36.1 m^2
Poplar	(6 m ht)	52.5 m^2
Linden	(5 m ht)	23.0 m^2
Birch	(5 m ht)	27.2 m^2
Pine	(3 m ht)	4.2 m^2

The total number of each species planted in the model forest and total vegetative area was as follows:

69	maple	2.54 X 10^3 m^2
69	oak	2.50 X 10^3 m^2
69	poplar	3.63 X 10^3 m^2
68	linden	1.56 X 10^3 m^2
69	birch	1.88 X 10^3 m^2
700	pine	2.90 X 10^3 m^2
	Total	15.00 X 10^3 m^2

The soil area of the model forest, not covered by tree trunks, was estimated to be 9.98 X 10^3 m^2. Pollutant flux rates guesstimated from the literature and used in the calculations are presented in Table 6-11. Table 6-12 lists the estimated sink capability. The U.S. Environmental Protection Agency concluded that if 122,517 ha (473 mi^2) of the model forest were in place in the St. Louis air quality region studied, this forest could remove 80.5 X 10^6 tons of sulfur dioxide per year and function to maintain the air quality standard for this pollutant.

C. Summary

Our examination of the literature permits us to conclude that abundant field data are available to support the suggestion that tree surfaces accumulate a variety of natural and anthropogenic particles from the atmosphere and that controlled environment and wind tunnel studies allow the following generalizations:

Table 6-11. Guesstimated Gaseous Pollutant Flux Rates for Dry Soil and Vegetative Surfaces.

Pollutant	$\mu g\ m^{-2}\ hr^{-1}$	
	Soil surface	Vegetative surface
Carbon monoxide	1.9×10^4	2.6×10^3
Nitrogen oxides	2.0×10^2	2.3×10^3
Ozone	1.0×10^9	6.2×10^4
Peroxyacetylnitrate	—	1.2×10^3
Sulfur dioxide	7.7×10^6	4.1×10^4

Source: U.S. Environmental Protection Agency (1976b).

Table 6-12. Guesstimated Gaseous Pollutant Removal for Model Forest Hectare[a].

Pollutant	tons yr^{-1}
Ozone	9.6×10^4
Sulfur dioxide	748
Carbon monoxide	2.2
Nitrogen oxides	0.38
Peroxyacetylnitrate	0.17

Source: U.S. Environmental Protection Agency (1976c).
[a] Total includes both soil and tree removal, dry condition.

1. The interception and retention of atmospheric particles by plants is highly variable and is primarily dependent on:
   a. Size, shape, wetness, and surface texture of the particles
   b. Size, shape, wetness, and surface texture of the intercepting plant part
   c. Micro-and ultramicroclimatic conditions surrounding the plant
2. More is known concerning the physical-mechanical aspects of particle deposition under controlled conditions than is known about the relative capture and retention efficiencies of various plants and different species under natural conditions.
3. Generally, greater leaf surface roughness increases particle capture efficiency for particles approximately 5 μm (and less) in diameter. Smooth leaved species (for example, horse chestnut and yellow poplar) are less efficient than rough leaved species (for example, elm and hazel).
4. Surface roughness acts to decrease the stability of the boundary layer (region of retarded air flow) surrounding the leaf and thus acts to increase particle impaction. Leaf hairs and leaf veins are principal contributors to surface roughness.
5. Smaller leaves are generally more efficient particle collectors than larger leaves.
6. Particle deposition (but probably not retention) is heaviest at the leaf tip and along leaf margins where a turbulent boundary layer is present. Leaves

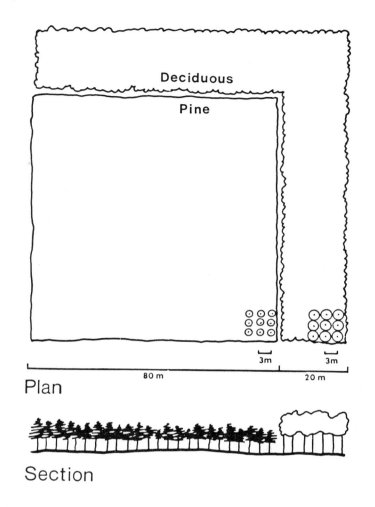

Figure 6-9. Model forest hectare developed by the U.S. Environmental Protection Agency to examine the potential of open space to serve as an air quality management strategy.
Source: U.S. Environmental Protection Agency (1976c).

with complex shapes and large circumference/area ratios collect particles most efficiently.

7. Increased wind speeds and increased particle size typically increase particulate deposition velocities.

8. Deposition velocities to petioles and stems are generally many times greater than deposition velocities to leaf laminas. Collection of atmospheric particles by leafless deciduous species in the winter may remain quite high due to twig and shoot impaction.

9. Conifers are generally more effective particulate sinks than deciduous species.
10. Mechanisms by which particles are resuspended or otherwise removed from tree surfaces must be investigated more thoroughly.

Unfortunately only a modest portion of all experiments conducted with artificially generated particles has been conducted in natural forest environments, and very few models have been developed to estimate the removal capacity of groups of trees. Hosker (1973) applied a standard plume diffusion model to hypothetical sources located at several heights above homogeneous stretches of grassland and forest. He concluded that the amount of effluent physically deposited on the foliage would be significantly larger for the forest than for the field. Slinn (1975) has provided an excellent analysis of the problems associated with developing a model for dry deposition in a plant canopy. He stresses the importance of considering "resuspension" of particles from vegetation in model design. Bache (1979) stresses the importance of considering all trapping mechanisms, for example, sedimentation and impaction, when designing particle trapping models.

It is tempting to conclude that trees may be especially efficient filters of airborne particles because of their large size, high surface to volume ratio of foliage, petioles and twigs, and frequently hairy or rough leaf, twig, or bark surfaces. Because the interior portions of forest stands act to still the air, mean wind speeds are reduced and particle sedimentation will be augmented. We are unable to quantify the filtration capacity of forests at the present time, however, because of deficient field data. Much of the data on particle loading of trees is expressed on a $\mu g\ g^{-1}$ (ppm) dry weight basis. Little (1977) has very appropriately indicated that judgments regarding particle collection efficiencies that are expressed as a function of plant dry weight must be used with caution and differences in area/dry weight ratios should be considered if they are to be meaningful. Further, unless accurate dimension analysis data (leaf, twig dry weight, leaf, and trunk surface area) are available, along with ambient microclimatological information, tree or stand loading can only be speculated. Much of the information on deposition velocities available for particle transfer to natural surfaces was not accumulated from experiments designed to access sink capability and did not employ particles small enough ($<1\ \mu m$) to be particularly relevant to individuals primarily interested in air quality.

We conclude that there is also substantial evidence that trees remove *gaseous* contaminants from the atmosphere. Experiments, again largely performed under controlled environmental circumstances and with seedling or young plants, allow the following generalizations:

1. Plant uptake rates increase as the solubility of the pollutant in water increases. Hydrogen fluoride, sulfur dioxide, nitrogen dioxide, and ozone, which are soluble and reactive, are readily sorbed pollutants. Nitric oxide and carbon monoxide, which are very insoluble, are absorbed relatively slowly or not at all by vegetation.

2. When vegetative surfaces are wet (damp) the pollutant removal rate may increase up to ten fold. Under damp conditions, the entire plant surface — leaves, twigs, branches, and stems — is available for uptake.
3. Light plays a critical role in determining physiological activities of the leaf and stomatal opening and as such exerts a great influence on foliar removal of pollutants. Under conditions of adequate soil moisture, pollutant uptake by vegetation is almost constant throughout the day, as the stomata are fully open. Moisture stress sufficient to limit stomatal opening, and relatively common in various urban environments, would severely restrict gaseous pollutant uptake.
4. Pollutants are absorbed most efficiently by plant foliage near the canopy surface, where light-mediated metabolic and pollutant diffusivity rates are greatest.
5. Sulfur and nitrogen dioxides are taken up by respiring leaves in the dark, but uptake rates are greatly reduced relative to rates in the light.

Models of gas pollutant uptake by forests based on flux rates derived from controlled environment investigations, for example, the U.S. Environmental Protection Agency's model forest hectare, probably overestimate the sink capacity as they employ flux rates determined in environments with unnaturally high pollutant concentrations, with limited or no prefumigation treatments and with moisture, light, and temperature conditions optimal for stomatal function. Models based on pollutant concentrations actually monitored in field situations, for example, those of Murphy et al. (1977) and Kabel et al. (1976), are more informative.

Under certain environmental circumstances, especially when tree surfaces are wet and when leaves are metabolically active, biologically and medically significant reductions in ambient levels of sulfur dioxide, nitrogen dioxide, ozone, and hydrogen fluoride may be realized by stands of trees for extended periods as long as the atmospheric loading of the contaminant gases is not excessive.

The use of forest areas as the exclusive means to reduce ambient pollution associated with point source facilities is not practical because of the large size of wooded hectares required and the associated high costs of establishing or maintaining these areas. The use of greenbelts surrounding industrial or power generating facilities, however, can certainly contribute to improved air quality and their costs can be justified in recognition of the additional, multiple-use benefits realized.

References

Albritton. D., F. Fehsenfeld, B. Hicks, J. Miller, S. Liu, J. Hales, J. Shannon, J. Durham, and A. Patruios. 1987. Atmospheric process. In Atmospheric Processes and Deposition. Interim Assessment. Vol. III. National Acid Precipitation Assessment Program, Washington, DC, pp. 37-59.

Altshuller, A.P. (ed.). 1984. The Acidic Deposition Phenomenon and Its Effects. Vol. I. Atmospheric Sciences. Publ. No. EPA-600-8-83-016AF, U.S.E.P.A., Washington, DC.

Art, H.W. 1971. Atmospheric salts in the functioning of a maritime forest ecosystem. Unpublished Ph.D. Thesis, Yale University, School of Forestry and Environmental Studies, New Haven, Connecticut, 135 pp.

Art, H.W., F.H. Bormann, G K. Voigt, and G.M. Woodwell. 1974. Barrier island forest ecosystem: Role of meteorologic nutrient inputs. Science 184: 60-62.

Aubertin, G.M. and M.P. Aubertin. 1981. Assessment of Non-traditional Controls on Ambient Air Quality. Publ. No. 81-07. Institute of Natural Resources, State of Illinois, Chicago, IL, 284 pp.

Aylor, D.E. 1975. Deposition of particles of ragweed pollen in a plant canopy. J. Appl. Meteorol. 14: 52-57.

Bach, W. 1972. Atmospheric Pollution. McGraw-Hill, New York, 144 pp.

Bache, D.H. 1979. Particle transport within plant canopies — I. A framework for analysis. Atmos. Environ. 13: 1257-1262.

Bennett, J.H. and A.C. Hill. 1973. Absorption of gaseous air pollutants by a standardized plant canopy. J. Air Pollu. Control Assoc. 23: 203-206.

Bennett, J.H. and AC. Hill. 1975. Interactions of air pollutants with canopies of vegetation. In J. B. Mudd and T.T. Kozlowski, eds., Responses of Plants to Air Pollution. Academic Press, New York, pp. 273-306.

Bennett, J.H., A.C. Hill, and D.M. Gates. 1973. A model for gaseous pollutant sorption by leaves. J. Air Pollu. Control Assoc. 23: 957-962.

Bormann, F.H., P.R. Shafer, and D. Mulcahy. 1958. Fallout on the vegetation of New England during the 1957 atom bomb test series. Ecology 39: 376-378.

Boyce, S.G. 1954. The salt spray community. Ecol. Monogr. 24: 29-67.

Chamberlain, A.C. 1967. Deposition of particles to natural surfaces. In P.H. Gregory and J.L. Monteith, eds., Airborne Microbes. 17th Symp. Soc. Gen. Microbiol., Cambridge Univ. Press, London, pp. 138-164.

Chamberlain, A.C. 1970. Interception and retention of radioactive aerosols by vegetation. Atmos. Environ. 4: 57-78.

Chamberlain, A.C. 1975. The movement of particles in plant communities. In J.L. Monteith (Ed.), Vegetation and the Atmosphere, Vol. I. Academic Press, New York, pp. 155-203.

Clayton, J.L. 1972. Salt spray and mineral cycling in two California ecosystems. Ecology 53: 74-81.

Costantini, A.and A.E. Rich. 1973. Comparison of salt injury to four species of coniferous tree seedlings when salt was applied to the potting medium and to the needles with or without an anti-transpirant. Phytopathology 63: 200.

Dochinger, L.S. 1972. Can trees cleanse the air of particulate pollutants? Intl. Shade Tree Conf. Proc. 48: 45-48.

Eaton, T.E. 1979. Natural and artificially altered patterns of salt spray across a forested barrier island. Atmos. Environ. 13: 705-709.

Edmonds, R.L. and C.H. Driver. 1974. Dispersion and Deposition of spores of *Fomes annosus* and fluorescent particles. Phytopathology 64: 1313-1321.

Elder F. and C. Hosler. 1954. Ragweed pollen in the atmosphere. Report, Dept. of Meteorology, Pennsylvania State Univ., University Park, PA.

Environmental Health Science Center. 1975. The Role of Plants in Environmental Purification. Environ. Health Sci. Ctr., Oregon State Univ., Corvallis, OR, 34 pp.

Fritschen, L.J., C.H. Driver, C. Avery, J. Buffo, R. Edmonds, R. Kinerson, and P. Schiess. 1970. Dispersion of air traces into and within a forested area (3). Report No. OSDO1366, College of Forest Resources, Washington Univ., Seattle, WA, 53 pp.

Garland, J.A. 1977. The dry deposition of sulfur dioxide to land and water surfaces. Proc. Royal Soc. London 354: 245-268.

Garland, J.A. and J.R. Branson. 1977. The deposition of sulfur dioxide to pine forest assessed by a radioactive tracer method. Tellus 29: 445-454.

Garland, J.A. and S.A. Penkett. 1976. Absorption of peroxyacetylnitrate and ozone by natural surfaces. Atmos. Environ. 10: 1127-1131.

Graustein, W.D. 1978. Measurement of dust input to a forested watershed using $^{87}Sr/^{86}Sr$ ratios. Geol. Soc. Am. Abst. 10: 411.

Gregory, P.H. 1971. The leaf as a spore trap. In T.F. Preece and C.H. Dickinson, eds., Ecology of Leaf Surface Microorganisms. Academic Press, New York, 640 pp.

Gregory, P.H. 1973. The Microbiology of the Atmosphere. Wiley, New York, 377 pp.

Hanson, G.P. and L. Thorne. 1972. Vegetation to reduce air pollution. Lasca Leaves 20: 60-65.

Heichel, G.H. and L. Hankin. 1972. Particles containing lead, chlorine and bromine detected on trees with an electron microprobe. Environ. Sci. Technol. 6: 1121-1112.

Heichel, G.H. and L. Hankin. 1976. Roadside coniferous windbreaks as sinks for vehicular lead emissions. J. Air. Pollu. Control Assoc. 26: 767-770.

Heisler, G.M. 1975. How trees modify metropolitan climate and noise. In Forestry Issues in Urban America. Proc. 1974 National Convention of Society of American Foresters, New York, pp. 103-112.

Helvey, J.D. 1971. A summary of rainfall interception by certain conifers of North America. In Proc. Biological Effects in the Hydrological Cycle. U.S.D.A. Forest Service, Washington, DC, pp. 103-113.

Helmke, P.A., W.P. Robarge, M.B. Schoenfield, P. Burger, R.D. Koons, and J.E. Thresher. 1984. Impacts of Coal Combustion on Trace Elements in the Environment: Wisconsin Power Plant Impact Study. Publ. No. EPA-600-53-84-070. U.S. Environmental Protection Agency Environmental Research Lab, Duluth, ME.

Hill, A.C. 1971. Vegetation: A sink for atmospheric pollutants. J. Air Pollu. Control Assoc. 21: 341-346.

Hill, A.C. and E.M. Chamberlain Jr. 1974. The removal of water soluble gases from the atmosphere by vegetation. Atmospheric-Surface Exchange of Particulate and Gaseous Pollutants Symp. Richland, WA, Sept. 4-6, 1974, 12 pp.

Hofstra, G. and R. Hall. 1971. Injury on roadside trees: Leaf injury on pine and white cedar in relation to foliar levels of sodium chloride. Can. J. Bot. 49: 613-622.

Hosker, R.P. Jr. 1973. Estimates of dry deposition and plume depletion over forests and grassland. Proc. IAEA Symposium on the Physical Behavior of Radioactive Contaminants in the Atmosphere, Vienna, Austria, Nov. 12-16, 1973.

Ingold, C.T. 1971. Fungal Spores. Clarendon Press, Oxford, 302 pp.

Jashnani, I. 1988. A Study of the Feasibility of Using Trees to Reduce Pollutants Resulting from the Proposed Coal Conversion of Unit No. 2 H.A. Wagner Power Plant, Engineering and Computer Services, 10451 Twin Rivers Road. Columbia, MD, 63 pp.

Jensen, K.F. 1975. Sulfur content of hybrid poplar cuttings fumigated with sulfur dioxide. U.S.D.A. Forest Service, Res. Note No. NE-209, Upper Darby, PA, 4 pp.

Jensen, K.F. and T T. Kozlowski. 1975. Absorption and translocation of sulfur dioxide by seedlings of four forest tree species. J. Environ. Qual. 4: 379-382.

Kabel, R.L. 1976. Natural removal of gaseous pollutants. 3rd Symp. Atmospheric Turbulence, Diffusion and Air Quality. Amer. Meteorological Soc., Oct. 19–22, 1976, Raleigh, NC.

Kabel, R.L., R.A. O'Dell, M. Taheri, and D.D. Davis. 1976. A preliminary model of gaseous pollutant uptake by vegetation. Center for Air Environment Studies, Publ. No. 455-76, Pennsylvania State Univ., University Park, PA, 96 pp.

Keller, T. 1978. How effective are forests in improving air quality? Eighth World Forestry Conference, Jakarta, Indonesia, Oct. 16–28, 1978, 9 pp.

Langer, G. 1965. Particle deposition and re-entrainment from coniferous trees. Part II. Experiments with individual leaves. Kolloid Z.Z. Polym. 204: 119-124.

Little, P. 1977. Deposition of 2.75, 5.0 and 8.5 µm particles to plant and soil surfaces. Environ. Pollu. 12: 293-305.

Little, P. and M.H. Martin. 1972. A survey of zinc, lead and cadmium in soil and natural vegetation around a smelting complex. Environ. Pollu. 3: 241-254.

Martin, A. and F.R. Barber. 1971. Some measurements of loss of atmospheric sulfur dioxide near foliage. Atmos. Environ. 5: 345-352.

Martin, W.E. 1959. The vegetation of Island Beach State Park, New Jersey. Ecol. Monogr. 29: 1-46.

McCure, D.C., D.H. Silberman, R.H. Mandl, L.H. Weinstein, P.C. Freudenthal, and P.A. Giardina. 1977. Studies on the effects of saline aerosols of cooling tower origin on plants. J. Air Pollu. Control Assoc. 27: 319-324.

Moser, B.C. 1979. Airborne salt and spray techniques for experimentation and its effects on vegetation. Phytopathology 69: 1002-1006.

Murphy, C.E. Jr., T.R., Sinclair, and K.R. Knoerr. 1977. An assessment of the use of forests as sinks for the removal of atmospheric sulfur dioxide. J. Environ. Qual. 6: 388-396.

Neuberger, H., C.C. Hosler, and C. Koemond. 1967. Vegetation as an aerosol filter. In S.W. Tromp and W. H. Weihe, eds., Biometeorology 2. Pergamon Press, New York, pp. 693-702.

Oak Ridge National Laboratory. 1969. Progress report in postattack ecology. Interim Progress Report N. ORNL-TM-2466. Oak Ridge, TN, 60 pp.

Oosting, H.J. 1945. Tolerance to salt spray of plants of coastal dunes. Ecology 26: 85-89.

Oosting, H.J. and W.D. Billings. 1942. Factors affecting vegetational zonation on coastal dunes. Ecology 23: 131-142.

Page, A.L. and A.C. Chang. 1979. Contamination of soil and vegetation by atmospheric deposition of trace elements. Phytopathology 69: 1007-1011.

Podgorow, N.W. 1967. Plantings as dust filters. Les. Khoz. 20: 39-40.

Rasmussen, K.H., M. Taheri, and R.L. Kabel. 1975. Global emissions and natural processes for removal of gaseous pollutants. Water Air Soil Pollu. 4: 33-64.

Rauner, J.L. 1976. Deciduous forests. In J.L. Monteith, ed., Vegetation and the Atmosphere. Vol. 2. Academic Press, New York, pp. 241-264.

Raynor, S. 1967. Effects of a forest on particulate dispersion. In C.A. Mawson ed., Proc. USAEC Meteorological Information Meeting, Chalk River Nuclear Laboratories, Chalk River, Ontario, Canada, Sept. 11–14, 1967, pp. 581-586.

Raynor, G.S., M.E. Smith, I.A. Singer, L.A. Cohen, and J.V. Hayes. 1966. The dispersion of ragweed pollen into a forest. Proc. 7th National Conf. Agricultural Meteorology, Aug. 29-Sept. 1, 1966. Rutgers Univ., New Brunswick, NJ.

Rich, S. and N.C. Turner. 1972. Importance of moisture on stomatal behavior of plants subjected to ozone. J. Air. Pollu. Control Assoc. 22: 369-371.

Rich, S., P.E. Waggoner, and H. Tomlinson. 1970. Ozone uptake by bean leaves. Science 169: 79-80.

Roberts, B.R. 1971. Foliar absorption of gaseous air pollutants. Am. Nursery 133: 44-45.

Roberts, B.R. 1974. Foliar sorption of atmospheric sulfur dioxide by woody plants. Environ. Pollu. 7: 133-140.

Roberts, B.R. and C.R. Krause. 1976. Changes in ambient SO_2 by rhododendron and pyracantha. Hort Sci. 11: 111-112.

Rogers, H.H., H.E. Jeffries, and A.M. Witherspoon. 1979. Measuring air pollutant uptake by plants: Nitrogen dioxide. J. Environ. Qual. 8: 551-557.

Romney, E.M., R.G. Lindberg., H.A. Hawthorne, B.G. Bystrom, and K.H. Larson. 1963. Contamination of plant foliage with radioactive fallout. Ecology 44: 343-349.

Rosinki, J. and C.T. Nagamoto. 1965. Particle deposition on and re-entrainment from coniferous trees. Part I. Experiments with trees. Kolloid Z.A. Polym. 204: 111-119.

Russell, I.J. 1974. Some factors affecting beta particle dose to tree populations in the eastern New England area from stratospheric fallout to 1974. Report No.CH-3015-8, Atomic Energy Commission, Chicago, IL, 58 pp.

Russell, I.J. and C.E. Choquette. 1974. Scale factors for foliar contamination by stratospheric sources of fission products in the New England area. Report N. CH-3015-13, Atomic Energy Commission, Chicago, IL, 47 pp.

Salisbury, F.B. and C.W. Ross. 1978. Plant Physiology. Wadsworth, Belmont, CA, 422 pp.

Sheih, C.M. 1977. Application of a statistical trajectory model of the simulation of sulfur pollution over northeastern United States. Atmos. Environ. 11: 173-178.

Sheih, C.M., M.L. Wesely, and B.B. Hicks. 1979. A guide for estimating dry deposition velocities of sulfur over the eastern United States and surrounding regions. Argonne National Laboratory Report No. ANL-RER-79-2, 55 pp.

Slinn, W.G.N. 1975. Dry deposition and resuspension of aerosol particles — A new look at some old problems. Proc. Conf. Atmosphere-Surface Exchange of Particles and Gases, ERDA Conf. Series, No. CONF-740921, Washington, DC, pp. 1-40.

Slinn, W.G.N. 1976. Some approximations for the wet and dry removal of particles and gases from the atmosphere. Atmos. Sciences Dept., Battelle Memorial Institute, Pacific Northwest Laboratory, Richland, WA.

Smith, W.H. 1970a. Technical review: Trees in the city. J. Am. Inst. Planners 6: 429-436.

Smith, W.H. 1970b. Salt contamination of white pine planted adjacent to an interstate highway. Plant Dis. Reptr. 54: 1021-1025.

Smith, W.H. 1971. Lead contamination of roadside white pine. For. Sci. 17: 195-198.

Smith, W.H. 1973. Metal contamination of urban woody plants. Environ. Sci. Technol. 7: 631-636.

Smith, W.H. 1974. Air pollution — Effects on the structure and function of the temperate forest ecosystem. Environ. Pollu. 6: 111-129.

Smith, W.H. 1976. Lead contamination of the roadside ecosystem. J. Air pollu. Control Assoc. 26: 753-766.

Smith, W.H. 1979. Urban vegetation and air quality. In Proc. National Urban Forestry Conference, Washington, DC, Nov. 13-16, 1978, U.S.D.A. Forest Service, Washington, D.C. and State Univ. of New York, Publ. No. 80-003 Syracuse, NY, pp. 284-305.

Smith, W.H. and L.S. Dochinger. 1975. Air Pollution and Metropolitan Woody Vegetation. Pinchot Institute, consortium for Environmental Forestry Research, Publ. No. PIEFR-PA-1. U.S.D.A. Forest Service, Upper Darby, PA, 74 pp.

Smith, W.H. and L.S. Dochinger. 1976. Capability of metropolitan trees to reduce atmospheric contaminants. In H. Gerhold, F. Santamor, and S. Little, eds., Proc. Better Trees for Metropolitan Landscapes, U.S.D.A. Forest Service, Gen. Tech. Report No. NE-22, Upper Darby, PA, pp. 49-59.

Smith, W.H. and B. J. Staskowicz. 1977. Removal of atmospheric particle by leaves and twigs of urban trees: Some preliminary observations and assessment of research needs. Environ. Mamt. 1: 317-328.

Steubing, L. and R. Klee. 1970. Comparative investigations into the dust filtering effects of broad leaved and coniferous woody vegetation. Agnew. Bot. 4: 73-85.

Thorne, L. and G.P. Hansen. 1972. Species differences in rates of vegetal ozone absorption. Environ. Pollu. 3: 303-312.

Townsend, A.M. 1974. Sorption of ozone by nine shade tree species. J. Am. Soc. Hort. Sci. 99: 206-208.

U. S. Environmental Protection Agency. 1976a. Open Space as an Air Resource Management Measure. Vol. I. Sink Factors. U.S.E.P.A. Publ. No. EPA-450/3/76-028a, Research Triangle Park, NC.

U.S. Environmental Protection Agency. 1976b. Open Space as an Air Resource Management Measure. Vol. II. Design Criteria. U.S.E.P.A. Publ. No. EPA-450/3/76-028b, Research Triangle Park, NC.

U.S. Environmental Protection Agency. 1976c. Open Space as an Air Resource Management Measure. Vol. III. Demonstration Plan (St. Louis, MO). U.S.E.P.A. Publ. No. EPA-450/3-76/028c, Research Triangle Park, NC.

Waggoner, P.E. 1971, Plants and polluted air. Bioscience 21: 455-459.

Waggoner, P.E. 1975. Micrometeorological models. In J. L. Monteith, ed., Vegetation and the Atmosphere, Vol. I. Academic Press, New York, pp. 205-228.

Warren, J.L. 1973. Green space for air pollution control. School of Forest Resources, Tech. Rep. No. 50, North Carolina State Univ., Raleigh, NC, 118 pp.

Wedding, J.B., R.W. Carlson, J.H. Stukel, and F.A. Bazzaz. 1975. Aerosol deposition on plant leaves. Environ. Sci. Tech. 9: 151-153.

Wells, B.W. and I.V. Shunk. 1938. Salt spray: An important factor in coastal ecology. Torr. Bot. Club Bull. 65: 485-492.

White, E.J. and F. Turner. 1970. Method of estimating income of nutrients in catch of airborne particles by a woodland canopy. J. Appl. Ecol. 7: 441-461.

Whittaker, R.H. and G.M. Woodwell. 1967. Surface area relations of woody plants and forest communities. Am. J. Bot. 8: 931-939.

Witherspoon, J.P. and F.G. Taylor, Jr. 1969. Retention of a fallout simulant containing ^{134}Cs by pine and oak trees. Health Phys. 17: 825-829.

Wood, F.A. and D.D. Davis. 1969. Sensitivity to ozone determined for trees. Pennsylvania State Univ., Sci. Agr. 17: 4-5.

Woodcock, A.H. 1953. Salt nuclei in marine air as a function of altitude and wind force. J. Meteorol. 10: 362-371.

Zinke, P.J. 1967. Forest interception studies in the United States. In Forest Hydrology, Pergamon Press, Oxford, England, pp. 137-160.

7

Class I Summary: Relative Importance of Forest Source and Sink Strength and Some Potential Consequences of These Functions

The Class I relationship between forest ecosystems and air pollution is of primary importance when the atmospheric load of air contaminants from anthropogenic sources is relatively low. This situation exists locally and regionally when the sources of air pollutants produced by the activities of human beings are not operating or are operating at a low level, or when meteorological conditions are not conducive to atmospheric accumulation. On a global scale, the Class I relationship may be extensive throughout those regions relatively remote from the activities of people. The specific concentration of air contaminants under "low" conditions is variable depending on the pollutant, but in general is meant to approximate "background," clean-air concentration as, for example, presented by Rasmussen et al. (1975) for the major trace gases in $\mu g \ m^{-3}$: sulfur dioxide (1–4), hydrogen sulfide (0.3), dinitrogen oxide (460–490), nitric oxide (0.3–2.5), nitrogen dioxide (2–2.5), ammonia (4), carbon monoxide (100), ozone (20–60), and reactive hydrocarbons (< 1). Since the majority of air contaminants of greatest significance to vegetative and human health (Table 2-1) originate from, and are removed by, both anthropogenic and natural agents, it is essential to evaluate the importance of forest ecosystems in the latter group.

The evidence supporting the hypothesis that forest systems are important sources and sinks of air contaminants (Table 7-1) has been represented in Chapters 3 through 6.

Table 7-1. Interaction of Air Pollution and Temperate Forest Ecosystems Under Conditions of Low Air Contaminant Load, Designated a Class I Interaction.

Forest soil and vegetation: Activity and response	Ecosystem consequence and impact
1. Forest soils and vegetation release particulate and gaseous contaminants to the atmosphere	1. Atmospheric burden of contaminants from anthropogenic sources supplemented by forest additions—scale may be local, regional, or global
2. Forest soils and vegetation remove particulate and gaseous contaminants from the atmosphere	2. Air contaminants transferred from the atmosphere to the biosphere, forest ecosystems supplement natural removal mechanisms
3. No or minimal alteration of structure or metabolism of forest soils or vegetation	3. No adverse ecosystem change discernible, slight fertilization possible

A. Forests as Sources of Pollutants

Efforts to assess the global importance of forest ecosystems as direct and indirect sources of air pollutants are frustrated by the very great imprecision of global estimates. On a global scale, air monitoring is absent and calculation based on approximation and intuition is abundant. Table 7-2 contains, at best, approximate order of magnitude guesstimations of air pollution materials released from anthropogenic relative to forest sources. For the purpose of these estimates the size of the global forest as estimated by Woodwell (1978) (5700 X 10^6 ha), Persson (1974) (4030 X 10^6 ha), and Ingersoll et al. (1974) (5010 X 10^6 ha) were averaged to yield 5000 X 10^6 ha global forest or very approximately one third of global terrestrial ecosystems. While little absolute faith should be placed in any of the figures of Table 7-2, the table does suggest that the hypothesis that global forests are important sources of air contaminants is supported. In the case of carbon dioxide, trace gases containing sulfur and nitrogen, and reactive hydrocarbons, the quantities released directly or indirectly by forests may exceed those produced by people on a global scale. Only in the cases of carbon monoxide and particulates do anthropogenic sources potentially exceed forest sources.

A comparison of air pollutants produced from the activities of people in the United States relative to production by American forests (Table 7-3) also suggests that nationwide, forests are important sources. The size of this forest is estimated to approximate 300 X 10^6 ha (U.S. Department of Agriculture Forest Service, 1978, 1979), which is roughly one third of United States terrestrial ecosystems. Total forest generated carbon dioxide, sulfur and nitrogen containing gases, and reactive hydrocarbons may exceed amounts resulting from human activity.

On a global basis, the forest-related production of carbon dioxide and hydrocarbons may be of particular significance. The latter may importantly influence regional atmospheric oxidant patterns. The consequences to forest ecosystems of air contaminants of the global atmosphere are generally not considered adverse, as the concentrations are generally well below the thresholds of subtle (Class II) or acute (Class III) effects. The primary exception, however, may be the continually increasing radiatively active trace gas concentrations of the global atmosphere (Chapter 19).

1. Elevated Carbon Dioxide, Global Warming, and Forest Health

It is presumed that a primary result of an elevated global atmospheric concentration of carbon dioxide will be warming (Schneider, 1975). While incoming solar radiation is uninfluenced by atmospheric carbon dioxide, portions of outgoing infrared radiation returned from earth to space are adsorbed by carbon dioxide. Over time, the net influence of reduced loss of infrared radiation from the earth should act to warm the planet (Baes et al., 1977). While the forces controlling global temperature change are varied and complex, the increase of 0.5°C since the mid 1800s is generally agreed to be at least partially caused by increased carbon dioxide. By 2000 it may increase an additional 0.5°C (McLean, 1978). Numerous models have been advanced to estimate the amount of the average global increase in temperature per doubling of carbon dioxide and these projections are in the range of 0.7–9.6°C (Schneider, 1975). Despite considerable uncertainty, it is reasonable to conclude that in the early part of the next century the average temperature of the earth will be beyond the limits experienced during the last 1000 years (Broecker, 1975).

The consequences of a warmer global climate, with even a very modest temperature increase, on the development of forest ecosystems could be profound. A complete discussion of this important topic is provided in Chapter 18. Trace gases, other than carbon dioxide, have the potential to contribute to global warming (Chapter 19). Elevation of temperature, particularly with increased carbon dioxide available in the atmosphere, might enhance forest growth. This suggestion, however, is largely based on observations in greenhouses and other "controlled" growth facilities, where numerous factors influencing plant development are regulated. In nature, especially with long-lived plants, the interaction of vegetation with climate is very complex and sensitive to subtle alteration over several years. It has been suggested that the numerous die-back and decline diseases that periodically influence a large number of forest species throughout the temperate zone are at least partially a reflection of species "out-of-phase" with optimal climatic conditions (Chapter 18).

Physiological processes of plants, perhaps most importantly photosynthesis, transpiration, respiration, and reproduction, are extremely sensitive to temperature change. With general warming, respiration and decomposition may increase faster than photosynthetic production. Transpiration and evaporation in-

Table 7-2. Guesstimated Annual Emission Strength for Global Forest Ecosystems and Anthropogenic Sources[a].

Pollutant	Forest ecosystems		Anthropogenic sources	
	Mechanism	kg	kg	Mechanism
Carbon dioxide	Forest destruction, humus oxidation forest burning	78×10^{12} [b]	3.5×10^{12} [l]	Fossil fuel combustion, cement production
Carbon monoxide	Total of a and b below	36×10^{10}	60×10^{10} [m]	Motor vehicle exhaust, other combustion
	a. Metabolism and chlorophyll loss	30×10^{10} [c]		
	b. Forest burning	6×10^{10} [d]		
Sulfur (inorganic and organic volatiles)	Forest soil release	36×10^{12} [e]	0.08×10^{12} [n]	Coal combustion, petroleum refining, smelting, other industrial
Nitrogen (in NO_x and NH_3)	Forest soil release	17×10^{10} [f]	2.3×10^{10} [o]	Coal and other combustion, fertilizer, waste treatment
Hydrocarbons (reactive)	Tree foliage release	175×10^{9} [g]	27×10^{9} [p]	Motor vehicle exhaust, oil combustion
Hydrocarbons (nonreactive)	Forest soil CH_4 release	?	70×10^{9}	Motor vehicle exhaust, oil combustion
Particulates	Total of a–d below	191×10^{9}	296×10^{9} [k]	Combustion, industrial operations, gas-particle conversion, photochemical
	a. Pollen shed	31×10^{9} [h]		
	b. Gas-particle conversion	91×10^{9} [i]		
	c. Photochemical (terpenes)	66×10^{9} [j]		
	d. Forest burning	3×10^{9} [k]		

[a] Calculations based on estimates provided in references reviewed in Chapters 2 and 3.
[b] [(Woodwell, 1978; and Wong, 1978) + (Bolin, 1977)] ÷ 2 (for mean of gross release).
[c] (Nozhevnikova and Yurganov, 1978) x 0.33 (forest contribution).
[d] Seiler, 1974.
[e] (Adams et al., 1979) x 5000 x 10^6 (forest area ha).
[f] [(Kim, 1973) + (Focht and Verstraete, 1977)] ÷ 2 (for mean) x 5000 x 10^6 (forest area ha).
[g] Rasmussen, 1972.
[h] (0.1 kg $tree^{-1}$) x (62 trees ha^{-1}) x 5000 x 10^6 (forest area ha).
[i] (Robinson and Robbins, 1971) x 0.1 (forest contribution).
[j] (Robinson and Robbins, 1971) x 0.33 (forest contribution).
[k] Robinson and Robbins, 1971.
[l] Bolin, 1977.
[m] Seiler, 1974.
[n] Bremner and Steele, 1978.
[o] Rasmussen et al., 1975 (NH_3, NO_x converted to N, NO_x = ½ NO_2, ½ NO).
[p] Rasmussen et al., 1975.

creases may impose enhanced stress on drier sites. Reproductive strategies may be altered due to changes in population dynamics of pollinating insect species or seedling survival. The geographic or host ranges of significant exotic microbial pathogens or insect pests may be caused to expand. Previously innocuous endemic microbes or insects may be elevated to important pest status following climatic warming. The significance of forest ecosystems as sources of nitrous oxide (Chapter 3) and methane (Chapter 4) are unresolved but also potentially significant in global climate considerations.

2. Regional Importance of Forest Emissions: Forests as Sources of Oxidant Precursors

On a regional scale, primary forest emissions may be converted to secondary pollutants. The hypothesis has been advanced that hydrocarbons released from vegetation may play a role in ozone synthesis in the atmosphere (Middleton, 1967). While the chemistry of oxidant synthesis is exceedingly complex in detail, it is straightforward in summary outline (Figure 7-1).

In the presence of sunlight nitrogen dioxide is dissociated and forms equal numbers of nitric oxide molecules and oxygen atoms. The oxygen atoms rapidly combine with molecular oxygen to form ozone. This ozone then reacts with the nitric oxide, on a one-to-one basis, to reform nitrogen dioxide. The steady-state concentrations, of ozone that is produced by this cycle is very small. When hydrocarbons, aldehydes, or other reactive atmospheric constituents are present, however, they can form peroxy radicals that oxidize the nitric oxide back to nitrogen dioxide. With reduced nitric oxide available to react with the ozone, the latter may accumulate to relatively high concentrations (Maugh, 1975; Spedding, 1974; National Academy of Sciences, 1977).

Table 7-3. Guesstimated Annual Emission Strength for United States Forest Ecosystems and Anthropogenic Sources[a].

Pollutant	Forest ecosystems		Anthropogenic sources	
	Mechanism	kg	kg	Mechanism
Carbon dioxide	Forest destruction humus oxidation, forest burning	5.2×10^{12} b	1.75×10^{12} n	Fossil fuel combustion
Carbon monoxide	Total of a and b below	2.2×10^{10}	8.7×10^{10} o	Motor vehicle exhaust, other combustion
	a. Metabolism and chlorophyll loss	1.8×10^{10} c		
	b. Forest burning	0.4×10^{10} d		
Sulfur (in inorganic and organic volatiles)	Forest soil release	2.2×10^{12} e	$.014 \times 10^{12}$ p	Coal combustion, petroleum refining, smelting, other industrial
Nitrogen (in NO_x and NH_3)	Forest soil release	9.9×10^9 f	8.9×10^9 q	Coal and other combustion, motor vehicle exhaust
Hydrocarbons (reactive)	Total of a and b below	10.6×10^9	7.8×10^9 r	Motor vehicle exhaust, refining, industrial solvents
	a. Tree foliage	10.5×10^9 g		
	b. Forest burning	0.09×10^9 h		
Hydrocarbons (nonreactive)	Total of a and b below	?	20×19^9 s	Motor vehicle exhaust, refining, industrial solvents
	a. Forest soil CH_4	?		
	b. Forest burning	0.21×10^9 i		
Particulates	Total of a-d below	13×10^9	12.8×10^9 t	Motor vehicle exhaust, refining, industrial solvents
	a. Pollen shed	3×10^9 j		
	b. Gas-particle conversion	5.5×10^9 k		
	c. Photochemical (terpenes)	4×10^9 l		
	d. Forest burning	0.45×10^9 m		

[a] Calculations based on estimates provided in references reviewed in Chapters 2 and 3.

[b] (Woodwell, 1978) (net global temperate forest) x 0.25 (for net U.S. temperate forest) x 4 (net to gross conversion).

[c] Table 6-1 (global forest CO metabolism + chlorophyll loss) x 0.06 (for U.S. forest contribution).

[d] 71 kg CO ton^{-1} fuel burned x 56 x 10^6 tons (U.S. total forest fuel burned).

[e] (Adams, 1979) x 300 x 10^6 (forest area ha).

[f] [(Kim, 1973) + (Focht and Verstraete, 1977)] ÷ 2 (for mean) x 1300 x 10^6 (forest area ha).

[g] Table 6-1 (global forest reactive HC tree foliage) x 0.06 (for U.S. forest contribution).

[h] 5.9 kg HC ton^{-1} fuel burned x 56 x 10^6 tons (U.S. total forest fuel burned) x 0.3 (for portion reactive).

[i] 5.9 kg HC ton^{-1} fuel burned x 56 x 10^6 tons (U.S. total forest fuel burned) x 0.7 (for portion nonreactive).

[j] (0.1 kg tree^{-1}) x (100 trees ha^{-1}) x 300 x 10^6 (forest area ha).

[k] (Robinson and Robbins, 1971) x 0.1 (global forest contribution) x 0.06 (U.S. forest contribution).

[l] (Robinson and Robbins, 1971) x 0.33 (global forest contribution) x 0.06 (U.S. forest contribution).

[m] [5.15 kg particulates ton^{-1} fuel burned x 56 x 10^6 tons (U.S. total forest fuel burned) + 0.6 (U.S. Environmental Protection Agency, 1977)] ÷ 2 (for mean).

[n] Table 6-1 (global CO$_2$ production) x 0.5 (U.S. contribution).

[o] U.S. Environmental Protection Agency (1977).

[p] 26.9 x 10^9 kg SO$_2$ (U.S. Environmental Protection Agency, 1977) x 0.5 (for S).

[q] 23.0 x 10^9 kg NO$_x$ (U.S. Environmental Protection Agency, 1977) x 0.385 (for N).

[r] 27.9 x 10^9 kg HC (U.S. Environmental Protection Agency, 1977) x 0.28 (for reactive HC).

[s] 27.9 x 10^9 kg HC (U.S. Environmental Protection Agency, 1977) x 0.72 (for unreactive HC).

[t] 13.4 x 10^9 kg particulates (U.S. Environmental Protection Agency, 1977) − 0.6 x 10^9 kg forest fire particulates (U.S. Environmental Protection Agency, 1977).

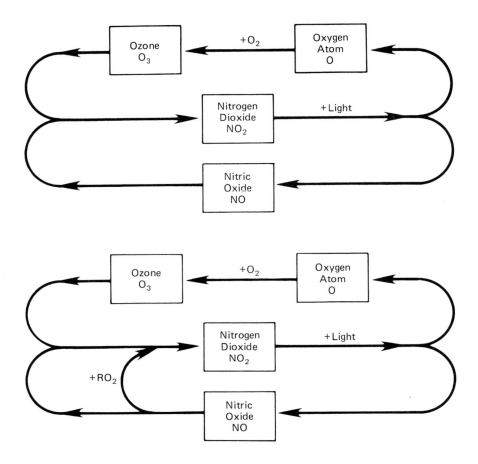

Figure 7-1. The nitrogen oxide-ozone cycles. The upper cycle portrays the atmosphere containing only nitrogen oxides and oxygen. The lower cycle portrays the situation in the presence of volatile organic carbon.
Source: NAS (1977).

a. Laboratory Evidence

In 1962, Stephens and Scott reported that terpenes could react photochemically with oxides of nitrogen to form ozone and peroxyacetylnitrates. They demonstrated that the monoterpenes pinene and α-phellandrene both showed the high reactivity predicted by their olefinic structure. Isoprene oxidation of nitric oxide to nitrogen dioxide was found by Glasson and Tuesday (1970) to be at a rate intermediate between the rates for ethylene and *trans*-2-butene. Ripperton and Lillian (1971) reported the formation of oxidants, ozone, condensation nuclei, and nitric oxide, following irradiation of systems containing 0.1 ppm (188 μg

Table 7-4. Photooxidation Rates of Monoterpene Hydrocarbons.

Hydrocarbon	Photooxidation rate at 7 ppb NO_x and 10 ppb HC ($sec^{-1} \times 10^4$)
p-Methane	0.11
p-Cymene	0.25
Isobutene	0.84
β-Pinene	1.1
Isoprene	1.3
α-Pinene	1.3
3-Carene	1.4
β-Phellandrene	1.9
Carvomethene	2.1
Limonene	2.4
Dihydromyrcene	3.0
Mycene	3.8
Oximene	5.3
Terpinolene	11
α-Phellandrene	12
α-Terpinene	55-110

Source: Grimsrud et al. (1975).

m^{-3}) nitrogen dioxide and 0.5 ppm (2780 μg m^{-3}) α-pinene. Japar et al. (1974) have provided rate constants for α-pinene and terpinolene.

Some of the initial experiments on terpene reactivity were conducted with gas concentrations quite dissimilar from ambient forest conditions. Westberg and Rasmussen (1972) and Grimsrud et al. (1975) employed realistic concentrations of 10 ppb (56 μg m^{-3}) hydrocarbons and approximately 7 ppb (12 μg m^{-3}) nitrogen oxides in their determinations of monoterpene reactivity. The data from these studies suggest that olefinic terpenes react with ozone at rates similar to those of the most reactive alkenes (Bufalini et al., 1976) and that structure plays a minor role in reactivity (Table 7-4).

This brief review of laboratory evidence clearly suggests that photochemical reactions of terpenes and nitric oxide can occur and therefore supports the hypothesis that forests may be involved in ozone synthesis in nature.

b. Field Evidence

If forest hydrocarbons do participate in the generation of ozone under ambient conditions, it should be possible to find elevated ozone concentrations in forest (non urban) regions. High rural ozone has been detected in numerous situations: Maryland and West Virginia (Research Triangle Institute, 1973), West Virginia (Richter, 1970), California (Miller et al., 1972), Minnesota (Pratt et al., 1983), and New York (Stasiuk and Coffey, 1974). The latter study simultaneously monitored ozone concentrations at eight rural and urban sites throughout New York State. Ozone concentrations at rural sites were found to be comparable to or in excess of ozone amounts in urban areas. The authors considered three potential sources for the high rural ozone they observed: (a) transport from the stratosphere, (b) generation from anthropogenic precursors, and (c) generation from hydrocarbons released from vegetation. The latter explanation appeared most plausible to Stasiuk and Coffey.

Sandberg et al. (1978) examined correlations between ozone levels in the San Francisco air basin and winter precipitation. These authors concluded that the good correlation between precipitation and ozone they found supported the hypothesis that vegetative precursors played a major role in the high oxidant levels of San Francisco.

Analyses of daily ozone concentrations in New Jersey and New York throughout the week have revealed high ozone levels on summer weekends relative to weekdays, despite considerably attenuated motor vehicle use during the former periods (Bruntz et al., 1974; Cleveland et al., 1974). Graedel et al. (1977) have employed detailed chemical kinetic computations representing workdays and Sundays in Hudson County, New Jersey to evaluate this so-called "weekend or Sunday effect." These authors conclude that weekday ozone is subject to greater scavenging by oxides of nitrogen and thereby infer that while greater ozone may be formed during the week relative to weekends, it does not accumulate . They also, however, suggest that "weekend ozone" may have an enhanced component of oxidant advected from less urban areas. Cleveland and McRae (1978) have expanded the observations of daily ozone levels beyond northern New Jersey and New York City and have included Connecticut and Massachusetts. By examining a larger region it was apparent that while a reduction of primary emissions (weekend situation) may not lead to a reduction in ozone concentrations in the metropolitan region itself, it may result in reduced ozone concentrations downwind from the primary source area. The authors showed that sites downwind of the New York City region with respect to summer prevailing winds (Connecticut and Massachusetts) generally show a reduction of daily maximum ozone concentrations of approximately 10–25% on weekends. As both Connecticut and Massachusetts are very approximately 75% forested, it is possible that widespread production of oxidant precursors occurs in these woodlands.

One of the greatest difficulties in defining the specific origin of rural and urban ozone is the realization that oxidants or their precursors may be transported considerable distances from the site of origin (Table 7-5). Ancora, a small nonindustrial, low-traffic community in southern New Jersey frequently had daily maximum ozone concentrations in excess of the 0.08 ppm (157 $\mu g\ m^{-3}$) federal standard during summer 1973. Cleveland and Kleiner (1975) investigated this enigma and concluded that Ancora's ozone was transported 37 km from the Camden-Philadelphia urban complex. With appropriate wind direction, ozone from this urban complex was moved to sites as far as 49 km away. During this same year, Coffey and Stasiuk (1975) were commonly recording concentrations of ozone in excess of 0.08 ppm in rural areas of New York State. These authors argued that this high rural ozone was transported *into* urban areas via advection

Table 7-5. Maximum Downwind Movement of Urban Generated Oxidants.

Area	Downwind Distance (km)	Reference
Camden-Philadelphia	49	Cleveland and Kleiner (1975)
St. Louis	160	White et al. (1976)
Los Angeles	161	Altshuller (1975)
New York	300	Cleveland et al. (1976)
General mesoscale transport	320	U.S.E.P.A (1986)

and vertical mixing and contributed to the total oxidant level in the urban areas of New York, Syracuse, and Buffalo.

Chameides and Stedman (1976) presented theoretical model evidence supporting the hypothesis that it was the interaction of anthropogenic nitrogen oxides generated over urban areas with photochemical methane oxidation chain products generated over natural areas that caused elevated rural ozone levels. Isaksen et al. (1978), on the other hand, argued that anthropogenic or natural hydrocarbons, other than methane, must be emitted to raise ambient ozone above 0.08 ppm in rural areas. When these nonmethane hydrocarbons are present, and during anticyclonic weather situations, Isaken et al. concluded that ozone concentrations in excess of 0.1 ppm (196 μg m^{-3}) may form in rural regions.

The data of Wolff et al. (1977) support the obvious conclusion that long-range ozone transport can be significant. Their observations of ozone in the urban corridor from Washington, D.C. to Boston, Massachusetts, suggest that an additive effect of sequential urban emissions on ambient ozone concentrations may in fact occur in the Northeast. Cleveland and Graedel (1979) have pointed out that areas as far as 45–60 km downwind of major sources of primary emissions,which themselves have lesser emissions, can have the highest regional ozone concentrations.

Salop et al. (1983) investigated the potential contribution that hydrocarbons from natural ecosystems made to ozone concentrations in southeastern Virginia. These investigations combined literature values of natural hydrocarbon emission rates with Landsat biomass data. They concluded that average daily oxidant levels could be explained by the anthropogenic hydrocarbon inventory alone. Certain summer peak episodic ozone levels, on the other hand, appeared to require supplemental hydrocarbon emissions comparable to those estimated to be released from vegetated systems.

Relative to urban areas, rural areas generally have low levels of hydrocarbons, perhaps 0.006–0.150 ppm carbon (U.S.E.P.A. 1986). Concentrations of individual species rarely exceed 0.010 ppm carbon; monoterpene concentrations are typically $\leq$ 0.020 ppm carbon, while summer isoprene concentrations are generally in the range 0.030–0.040 U.S.E.P.A. 1986). Isoprene, relative to monoterpene, is more efficient in producing ozone via photooxidation in the presence of nitrogen oxides. Since rural areas typically have very low levels of nitrogen oxides, Bufalini (1980) suggested that natural hydrocarbons may be relatively unimportant in ozone synthesis and that vegetative emissions may in fact act as a sink for ozone. Atshuller (1986) has concluded that photochemically generated ozone should equal or exceed ozone injected from the stratosphere in rural locations (low elevation) and further that photochemically generated ozone from human emissions probably constitutes most of the ozone measured at more polluted remote locations during the growing season.

Clearly, resolution of the origin of photochemical oxidant precursors in forested or urban regions within several hundred kilometers of each other must await the development of even greater sophistication in our understanding of atmospheric chemical and transport processes. An additional effort that would greatly facilitate our estimate of the importance of plant communities in ozone generation would be the direct monitoring of plant hydrocarbon release under field conditions. While little information of this kind exists for tree species, Lonneman et al. (1978) have monitored natural hydrocarbon emissions from citrus trees in Florida. Isoprene, which has been detected in other rural samples and is generally associated with deciduous vegetation, was the only hydrocarbon of natural origin detected in the vicinity of the leaves. The authors concluded that citrus production of hydrocarbons was low and that they could not possibly contribute to the production of significant levels of ambient ozone.

B. Forests as Sinks for Air Pollutants

The sink capability of United States ecosystems is guesstimated in Table 7-6. The figures of this table are every bit as approximate and imprecise as those of Tables 7-2 and 7-3, but nevertheless do encourage some reasoned speculation. The role of forests in the carbon dioxide global cycle must be refined. While the efficiency of the photosynthetic sink for carbon dioxide removal may balance the liberation of this gas via humus oxidation and burning in temperate forest zones, the extensive destruction and burning of tropical forests may be contributing excess carbon dioxide to the atmosphere. The Woodwell hypothesis must be pursued, and additional evidence supporting or rejecting this critically important possibility must be developed.

The most striking feature of the forest ecosystem sink review is the apparent importance of the soil compartment of forest ecosystems as a sink for air contaminants. The quality and quantity of the evidence provided are strongest in the case of particles containing lead, and a few additional trace metal elements, and carbon monoxide. It is suggested by at least some evidence that the sink capabilities of forest soils may be very substantial, and well in excess of the capability of forest vegetation, in the case of sulfur dioxide, nitrogen oxides, and ozone.

Forest ecosystems on a global basis appear to have the capacity to transfer air contaminants from the atmosphere in amounts approximately equal or in excess of current anthropogenic production levels, in the instance of carbon monoxide, sulfur dioxide, nitrogen oxides, and hydrocarbons.

Balance sheets of global source and sink strength are academically interesting but practically of limited value as air contaminants are not uniformly produced nor distributed around the globe. Review of the United States forest sink and anthropogenic production are more interesting. United States forests may have the capability to remove roughly 25% of the annual carbon monoxide production and most of the sulfur and nitrogen oxide production. Our uncertainty concerning particulate and hydrocarbon removal is very great as the data are grossly incomplete. Again because air contaminant distribution over forest ecosystems is not uniform throughout the country, generalizations concerning efficiency may be misleading. Excessive local generation of air contaminants in urban and industrial regions results in atmospheric accumulation as natural sinks do not exist, are saturated, or are inefficient.

Table 7-6. Guesstimated Annual Sink Strength for Global and United States Forest Ecosystems[a].

Pollutant	Mechanism	Global (kg)		United States (kg)	
Carbon dioxide	Photosynthesis	70	$\times 10^{12}$ [b]	4.2	$\times 10^{12}$ [c]
Carbon monoxide	Total of a and b below	55	$\times 10^{10}$	2.3	$\times 10^{10}$
	a. Vegetation	5	$\times 10^{10}$ [d]	0.3	$\times 10^{10}$ [e]
	b. Soils	50	$\times 10^{10}$ [f]	2	$\times 10^{10}$ [f]
Sulfur dioxide	Total of a and b below	201	$\times 10^{12}$	13	$\times 10^{12}$
	a. Vegetation	0.675	$\times 10^{12}$ [g]	0.041	$\times 10^{12}$ [h]
	b. Soils	200	$\times 10^{12}$ [i]	13	$\times 10^{12}$ [j]
Nitrogen oxides	Total of a and b below	36	$\times 10^{11}$	2.1	$\times 10^{11}$
	a. Vegetation	1	$\times 10^{11}$ [k]	0.06	$\times 10^{11}$ [l]
	b. Soils	35	$\times 10^{11}$ [m]	2	$\times 10^{11}$ [n]
Hydrocarbons (ethylene only)	Soil	40	$\times 10^{9}$ [o]	2.3	$\times 10^{9}$ [p]
Ozone	Total of a and b below	45	$\times 10^{13}$	2.7	$\times 10^{13}$
	a. Vegetation	.003	$\times 10^{13}$ [q]	.0002	$\times 10^{13}$ [r]
	b. Soils	45	$\times 10^{13}$ [s]	2.7	$\times 10^{13}$ [t]
Particulates (lead only)	Soil	100	$\times 10^{6}$ [u]	6	$\times 10^{6}$ [v]

[a] Calculations based on estimates provided in references reviewed in Chapters 4 and 5.
[b] 50×10^{15} g C terrestrial plant fixation (Woodwell, 1978) x 0.38 (forest contribution) ÷ 0.27 (conversion C to CO_2).
[c] Footnote b x 0.006 (U.S. forest contribution).
[d] Bidwell and Fraser (1972) mean CO vegetation uptake x 0.1 (controlled environment to nature extrapolation) x 0.38 (forest contribution).
[e] Footnote d x 0.06 (U.S. forest contribution).

[f]Ingersoll et al. (1974).

[g]Murphy et al. (1977) mean of Savanna River and N.Y. predictions extrapolated to 1 yr = 0.135 tons SO_2 ha^{-1} x 5000 x 10^6 ha.

[h]Murphy et al. (1977) mean of Savanna River and N.Y. predictions extrapolated to 1 yr = 0.135 tons SO_2 ha^{-1} x 300 x 10^6 ha.

[i]Abeles et al. (1971) (footnote j) ÷ 300 x 10^6 ha x 5000 x 10^6 ha (conversion to global forest).

[j]Abeles et al. (1971) x 0.33 (forest contribution).

[k]NO_x vegetation uptake average (U.S. Environmental Protection Agency, 1976) extrapolated to 1 year and 1 ha x 5000 x 10^6 ha (global forest) x 0.1 (controlled environment to nature extrapolation).

[l]NO_x vegetation uptake average (U.S. Environmental Protection Agency, 1976) extrapolated to 1 year and 1 ha x 300 x 10^6 ha (U.S. forest) x 0.1 (controlled environment to nature extrapolation).

[m]Abeles et al. (1971) (footnote n) ÷ 300 x 10^6 ha (conversion to global forest).

[n]Abeles et al. (1971) x 0.33 (forest contribution).

[o]Abeles et al. (1971) (footnote q) ÷ 300 x 10^6 ha x 5000 x 10^6 ha (conversion to global forest).

[p]Abeles et al. (1971) x 0.33 (forest contribution).

[q]O_3 vegetation uptake average (U.S. Environmental Protection Agency, 1976) extrapolated to 1 year and 1 ha x 5000 x 10^6 ha (global forest) x 0.001 (controlled environment to nature extrapolation).

[r]O_3 vegetation uptake average (U.S. Environmental Protection Agency, 1976) extrapolated to 1 year and 1 ha x 300 x 10^6 ha (U..S. forest) x 0.001 (controlled environment to nature extrapolation).

[s]O_3 soil uptake average (U.S. Environmental Protection Agency, 1976) extrapolated to 1 year and 1 ha x 5000 x 10^6 ha (global forest) x 0.001 (controlled environment to nature extrapolation).

[t]O_3 soil uptake average (U.S. Environmental Protection Agency, 1976) extrapolated to 1 year and 1 ha x 300 x 10^6 ha (U.S. forest) x 0.001 (controlled environment to nature extrapolation).

[u]0.2 kg ha^{-1} yr^{-1} (Chap. 4) x 5000 x 10^6 ha (global forest) x 0.1 (eastern U.S. forest to global forest conversion).

[v]0.2 kg ha^{-1} yr^{-1} (Chap. 4) x 300 x 10^6 ha (U.S. forest) x 0.1 (eastern U.S. forest to U.S. forest conversion).

1. Forest Fertilization as a Consequence of Sink Function

Forest trees require 16 essential elements for growth. Ten of these elements, capitalized below may be added to forest ecosystems as air pollutants in the particulate or gaseous form and have an impact on tree growth.

Macroelements required	Microelements required
CARBON	BORON
hydrogen	CHLORINE
oxygen	COPPER
NITROGEN	IRON
SULFUR	MANGANESE
phosphorus	MOLYBDENUM
potassium	ZINC
magnesium	
calcium	

The indicated elements are potentially introduced to forest ecosystems in the following states: carbon as gaseous carbon monoxide and dioxide; nitrogen as gaseous oxides or ammonia and particulate nitrates; sulfur as sulfur dioxide, hydrogen sulfide, or organic gases or particulate sulfates; and the microelements generally in association with particles. It has been hypothesized that the addition of essential elements to forests via air pollution, at levels below those causing Class II relationships, may act to stimulate the growth of forests by fertilization. What is the evidence?

a. Carbon

A great amount of data shows that plants grow better in atmospheres enriched in carbon dioxide (Dahlman et al., 1985; Kramer, 1981). Most of this information has been gleaned from experiments with various herbaceous agricultural and ornamental plants. Wright (1974), however, examined the photosynthetic response of several California woody plants, under controlled environmental conditions, to a range of carbon dioxide concentrations spanning past and likely future ambient concentrations. Small increases in atmospheric carbon dioxide generally produced significant increases in the rate of net photosynthesis. Wright observed considerable difference in response to carbon dioxide enrichment for his various San Bernadino Mountain tree species according to experimental run and conditioning temperatures. He concluded that rising carbon dioxide would tend to favor angiosperms over gymnoserms. Wright felt that the carbon dioxide "fertilization" he documented in the laboratory would be equally great under field conditions and did, in fact, show that enhancement of carbon dioxide up to about 100 ppm (18 X 104 μg m^{-3}) above ambient levels produced a sharp linear rise in photosynthesis of leafy twigs in a Long Island, New York oak-pine forest (Wright and Woodwell, 1970). Tinus (1972) grew ponderosa pine and blue spruce in the greenhouse at carbon dioxide concentrations four times ambient and observed enhanced growth.

Botkin (1977) very correctly pointed out the considerable risks associated with extrapolating carbon dioxide enrichment data from controlled environment to natural situations. In the former, environmental factors, most importantly temperature, moisture, and light, are not limiting, and the rate of photosynthesis is controlled by the rate at which carbon dioxide can diffuse into leaves. In natural situations, moisture and temperature may impose restrictions on photosynthetic rates as well as carbon dioxide availability. Also, natural forest systems are frequently comprised of a diverse number of species, which may exhibit a differential response to carbon dioxide enrichment. Botkin applied his computer model of northern hardwood forest growth to an assessment of carbon dioxide fertilization and presented several interesting conclusions. Over time, shade tolerant species, such as sugar maple and red spruce, were favored over shade intolerant species, such as white birch. Further, unless the effect of the carbon dioxide increase was large enough to stimulate the annual increment of each tree approximately 50% above normal, changes were not observed in the productivity or biomass of the forest.

Recent evidence indicates a strong correlation between enhanced carbon dioxide supply and numerous "use efficiencies" in trees, including: water use efficiency (Norby et al., 1986a), nutrient efficiency (Luxmoore et al., 1986; Norby et al., 1986a,b), and mycorrhizal efficiency (O'Neill et al., 1987). If these increased efficiencies are comparable under field conditions, some of the reservations raised by Botkin may be diminished.

In summary, the modest information currently available suggests that the fertilizing effect of increased atmospheric carbon dioxide may be less important in natural forest ecosystems in increasing biomass than in altering the capability of tree species to compete with one another. The demonstration that soil and meteorological factors are not limiting tree photosynthesis and that tree water and nutrient use efficiencies are greater in natural forests would be necessary to establish that increased atmospheric carbon dioxide was acting as a "fertilizer."

b. Nitrogen

Soils very infrequently contain enough nitrogen for maximum plant growth. Coniferous forest soils in the north temperate zone are commonly deficient in tree-available nitrogen. Forest trees respond more favorably to nitrogen than to any other fertilizer element. Can the atmospheric input of ammonia, nitrogen dioxide, nitric acid, potassium nitrate, sodium nitrate, or calcium nitrate result in significant forest fertilization?

Commercial rates of forest fertilization with nitrogen approximate 220 kg N ha^{-1} (Cochran, 1978). Approximately 99% of the combined nitrogen in soil is contained in organic matter (Thompson and Troeh,1978). The latter supplies exchangeable nitrogen to plants but does not increase the total nitrogen content of the ecosystem. Biological fixation and atmospheric deposition are the sole external sources of nitrogen to forest ecosystems (Noggle et al., 1978) (Figure 7-2).

The anticipated experiment dealing with nitrogen dioxide uptake by an agricultural species under controlled environmental conditions can be found. Troiano and Leone (1977) and Wellburn et al. (1980) have demonstrated increased nitrogen in tomato plant tissue following exposure to "realistic" concentrations of atmospheric nitrogen dioxide. Rogers et al. (1979) have shown that nitrogen dioxide taken up by snap beans is rapidly metabolized and incorporated into organic nitrogen compounds.

Roberts et al. (1986) exposed black locust, black cherry, Siberian elm, and yellow poplar to 100 ppb (188 μg m^{-3}) NO_2 7 hr daily for 8 wks and did not detect an increase in seedling nitrogen content. This exposure is in substantial excess of ambient forest NO_2 amounts (see Table 2-12). Applications of nitric acid to the planting soil of longleaf pine have been shown to stimulate height growth (Kais et al., 1984).

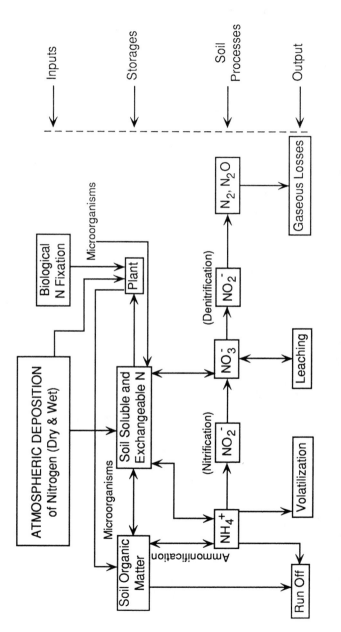

Figure 7-2. The forest nitrogen cycle.   Source: Modified from Naggle et al. (1978).

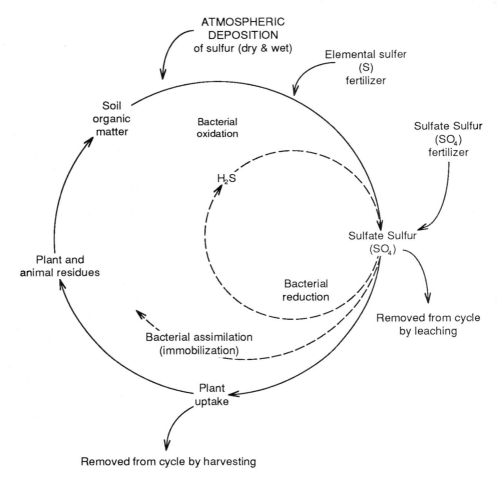

Figure 7-3. The forest sulfur cycle. Source: Modified from Large (1980).

c. Sulfur

Sulfur is similar to nitrogen in many respects; it is held in the soil in association with organic compounds, is most available to plants in its most oxidized
state, and is readily lost via leaching in the soil solution (Thompson and Troeh,
1978) (Figure 7-3). It differs from nitrogen, however, in that it is required by
trees in substantially lesser amounts and its presence in the soils is usually adequate to supply the needs of tree growth. Commercial rates of forest fertilization
with sulfur approximate 30 kg S ha^{-1} (Cochran, 1978), but the annual S requirement of many forest ecosystems may only approximate 10 kg ha^{-1} (Maugh,
1979). Sandy soils, which are highly leached and are low in organic matter, are
the most likely to be sulfur deficient. Regionally, forest soils can be deficient in
sulfur. Many soils of the Pacific Northwest (Will and Youngberg, 1978) and in

the Southeast (U.S. Department of Agriculture, 1964), for example, are deficient in this element. Sulfur deficiencies in agricultural soils may become more prevalent in the future as sulfur percentages in fertilizers have declined steadily in recent years (Thompson and Troeh, 1978).

Plants absorb sulfur from their leaves or from the soil as the sulfate ion or directly from the atmosphere as sulfur dioxide. A variety of controlled environment studies has demonstrated the efficiency of forest tree seedling sulfur dioxide uptake (for example, Jensen, 1975; Jensen and Kozlowski, 1975). Wainwright (1978) made field collections of sycamore leaves and litter in polluted and unpolluted sites in England. Leaves collected from the vicinity of a chemical coking plant had 118 μg g^{-1} elemental sulfur, 213 μg g^{-1} tetrathionate, and 3.7 mg g^{-1} sulfate while leaves from an unpolluted location exhibited no elemental sulfur nor tetrathionate and only 0.2 mg g^{-1} sulfate. Fertilization trials with pole-size ponderosa pine on the Deschutes National Forest in Oregon failed to demonstrate a stimulatory effect from artificial applications of sulfur containing fertilizer (Cochran, 1978).

The Tennessee Valley Authority has conducted studies to measure the amount of sulfur that is contributed by the atmosphere to Tennessee Valley agricultural ecosystems (Noggle and Jones, 1979; Maugh, 1979). Cotton grown 4 km and 3 km from coal-fired power plants accumulated 125 mg and 245 mg sulfur 100 g^{-1}, while fescue accumulated 65 mg and 58 mg 100 g^{-1} at the same locations. As a result of sulfur fertilization, cotton produced more biomass near power plants than that grown at locations remote from sulfur dioxide sources.

d. Micronutrients

All micronutrients are required by plants in relatively small amounts. As a result, deficiencies are the exception rather than the rule. Since all the micronutrients are potential components of anthropogenic effluents, it is possible that atmospheric input of these elements might function to supply a portion of the microelement requirement in areas deficient in these elements and subject to urban or industrial contamination.

Boron deficiencies are widespread in humid regions. In the United States they are prevalent in the uplands of the Atlantic coastal plain, the Great Lakes region, and the coastal region of the Pacific Northwest. Copper deficiencies are most common in organic soils and have been identified in the United States in Washington, California, Florida, South Carolina, and in the Great Lakes region. Iron is one of the most commonly deficient micronutrients because it is frequently unavailable to plants even though it is present in the soil. Manganese deficiencies generally occur in humid areas and in acid sandy soils. In the United States deficiencies have been documented in California, along the Atlantic coastal plain, and in the muck soils of the Great Lakes region. Molybdenum plant requirements are so small that deficiencies are probably infrequent, but do occur, especially in acid soils. Zinc deficiencies may also occur in acid soils and occur in small, widely scattered areas throughout the United States. Chlorine is so abundant that deficiencies are thought to be extremely rare (Thompson and Troeh, 1978). While fluoride is not considered a required plant nutrient, some evidence has been presented indicating that very small amounts of fluoride deposited from the atmosphere may stimulate tree growth (Bunce, 1985).

The importance of acid soils in the incidence of the deficiencies cited above suggests that forest vegetation might be especially benefited by the input of micronutrients from the atmosphere. There is, however, no direct evidence to support this suggestion.

e. Watershed Studies

Approximately two decades ago several ecosystem-scale, forest watershed studies were initiated in the United States. The accumulated record of these investigations provides the best evidence available regarding the importance of atmospheric input of elements to forest nutrient cycles (Likens et al., 1977a, b; Swank and Henderson, 1976; Swank, 1984).

Bormann and Likens (1979) have attempted to determine the relative importance of biological fixation versus atmospheric deposition in their study of the northern hardwood forest in central New Hampshire. They concluded that 70% of the nitrogen added to the ecosystem was added via biological fixation and that 30% was added via precipitation. The range of nitrogen input via bulk deposition (wetfall plus dryfall) for numerous watershed studies is 1–13 kg ha^{-1} yr^{-1} (Swank, 1984) (Table 7-7). This nitrogen is equivalent to at least 70% of the nitrogen incorporated annually in above-ground woody trunks of some temperate deciduous forests (Swank, 1984). Atmospheric sources of calcium and potassium may supply between 20% and 40% of these nutrients combined in woody increments (Swank, 1984) (Table 7-7). Phosphorus input from the atmosphere is generally small.

The northern hardwood forest studies of Likens et al. (1977) provide us with one of the first quantitative field estimates of atmospheric input of sulfur to a forested ecosystem. These investigators suggested that 6.1 kg ha^{-1} yr^{-1} of sulfate are provided to the northern hardwood forest via particulate deposition and direct absorption of gaseous sulfur. Throughfall and stemflow data suggested that sulfur aerosols may be impacted on tree surfaces in large amounts during the summer! The net gaseous sulfur input was judged to be about double that from aerosol deposition on an annual basis.

Lindberg and Harriss (1981) have indicated that the wet and dry deposition of sulfur to the Walker Branch Watershed (TN) may approximate 19 kg ha^{-1} yr^{-1} (Table 7-8). Quantification of manganese and zinc input to the forests of eastern Tennessee are also provided in Table 7-8.

In further studies of the Walker Branch hardwood forests, Lindberg et al. (1986) have estimated that atmospheric deposition supplies nitrogen and calcium at a rate equal to 40% of the annual woody increment of the forest and supplies sulfur at a rate exceeding 100% of the increment. By quantifying the relative size of nutrient inputs via wet and dry deposition (Table 7-9), these investigators have stressed the significance of dry deposition processes for nutrient inputs to forests from the atmosphere. The significance of dry deposition was more important for nitrate than sulfate. Most atmospheric nitrate at the Walker Branch site exists as nitric acid (HNO_2) vapor. Nitric acid is a major sink for atmospheric NO_2 and is characterized by very efficient deposition to forest canopies. When NO and NO_2 inputs are included, the dry deposition component of nitrogen input to the Walker Branch forest may exceed 80% (Lindberg et al., 1986). The contribution of dry deposition to the total flux of potassium and calcium was 60–70%. Wet deposition dominated the input of ammonium. The relative importance of wet versus dry deposition was judged to be seasonably dependent. Dry deposition of ions is most important during the growing season, while wet deposition is more important during the dormant season when most particulates in the air are at a minimum, and sulfur dioxide and nitric acid gas are only moderately higher than summer levels.

It is clear, especially in view of the watershed evidence, that atmospheric deposition of numerous macro- and micronutrients is significant for certain forest ecosystems. Swank (1984) cautions that these inputs have considerable temporal and spatial variability. We also cannot partition the specific contribution of natural versus anthropogenic components in wet and dry deposition. Despite these cautions, it appears reasonable to conclude that some forest ecosystems in the temperate zone are being fertilized by atmospheric deposition. In the case of nitrogen and sulfur deposition, the amounts deposited from the atmosphere — Table 2-16 and Table 2-15 respectively — approach or exceed the commercial rates of forest fertilization. In view of the contribution elements deposited from the atmosphere make to the total annual woody increment as revealed by watershed studies, and assuring no short- or long-term adverse consequences to the ecosystems, fertilization is recognized as an important Class I response.

References

Abeles, F.B., L.E. Craker, L.E. Forrence, and G.R. Leather. 1971. Fate of air pollutants: Removal of ethylene, sulfur dioxide, and nitrogen dioxide by soil. Science 175: 914-916.

Adams, D.F., S.O. Farwell, M.R. Pack, and W.L. Banesberger. 1979. Preliminary measurements of biogenic sulfur-containing gas emissions from soils. J. Air. Pollu. Control Assoc. 29: 380-383.

Altshuller, A.P. 1975. Evaluation of oxidant results at CAMP sites in the United States. J. Air Pollu. Control Assoc. 25: 19-24.

Altshuller, A.P. 1986. Review paper: The role of nitrogen oxides in nonurban ozone formation in the planetary boundary layer over N. America, W. Europe and adjacent areas of ocean. Atmos. Environ. 5: 39-64.

Baes, C. F. Jr., H.E. Goeller, J.S. Olson, and R.M. Rotty. 1977. Carbon dioxide and climate: The uncontrolled experiment. Am. Scien. 65: 310-320.

Bidwell, R.G.S. and D.E. Fraser. 1972. Carbon monoxide uptake and metabolism by leaves. Can. J. Bot. 50: 1435-1439.

Bolin, B. 1977. Changes of land biota and their importance for the carbon cycle. Science 196: 613-615.

Bormann, F.H. and G.E. Likens. 1979. Pattern and Processes in a Forested Ecosystem. Springer–Verlag, New York, 253 pp.

Botkin, D.B. 1977. Forests, lakes, and the anthropogenic production of carbon dioxide. Bio. Sci. 27: 325-331.

Bremner, J.M. and C.G. Steele. 1978. Role of microorganisms in the atmospheric sulfur cycle. Adv. Microb. Ecol. 2: 155-201.

Broecker, W.S. 1975. Climate change: Are we on the brink of pronounced global warming? Science 189: 460-463.

Bruce, H.W.F. 1985. Apparent stimulation of tree growth by low ambient levels of fluoride in the atmosphere. J. Air Pollu. Control Assoc. 35: 46-48.

Bruntz, S.M., W.S. Cleveland, T.E. Graedel, B. Kleiner, and J.L. Warner. 1974. Ozone concentrations in New Jersey and New York: Statistical association with related variables. Science 186: 257-258.

Bufalini, J.J. 1980. Impact of National Hydrocarbons on Air Quality. Publica. No. EPA-600-2-80-086. U.S. Environmental Protection Agency, Research Triangle Park, NC.

Bufalini, J.J., T.A. Walter, and M.M. Bufalini. 1976. Ozone formation potential of organic compounds. Environ. Sci. Technol. 10: 908-912.

Chameides, W.L. and D.H. Stedman. 1976. Ozone formation from NO_x in "clean air." Environ. Sci. Technol. 10: 150-153.

Cleveland, W.S. and T.E. Graedel. 1979. Photochemical air pollution in the northeast United States. Science 204: 1273-1278.

Clevelend, W.S. and B. Kleiner. 1975. Transport of photochemical air pollution from Camden-Philadelphia urban complex. Environ. Sci. Technol. 9:869-872.

Cleveland, S.S. and J.E. McRae. 1978. Weekday-weekend ozone concentrations in the northeast United States. Environ. Sci. Technol. 12: 558-563.

Cleveland, W.S., T.E. Graedel, B. Kleiner, and J.L. Warner. 1974. Sunday and workday variations in photochemical air pollutants in New Jersey and New York. Science 186: 1037-1038.

Cleveland, W.S.,B. Kleiner, J.E. McRae, and J.L. Warner. 1976. Photochemical air pollution: Transport from the New York City area into Connecticut and Massachusetts. Science 191: 179-181.

Cochran, P.H. 1978. Response of a pole-size ponderosa pine stand to nitrogen, phosphorus and sulfur. U.S.D.A. Forest Service, Research Note No. PNW-319, Portland, OR, 8 pp.

Coffey, P.E. and W.N. Stasiuk. 1975. Evidence of atmospheric transport of ozone into urban areas. Environ. Sci. Technol. 9: 59-62.

Dahlman, R.C., B.R. Strain, and H. Rogers. 1985. Research on the response of vegetation to elevated carbon dioxide. J. Environ. Qual. 14: 1-8.

Focht, D.D. and W. Verstraete. 1977. Biochemical ecology of nitrification and denitrification. Adv. Microb. Ecol. 1: 135-214.

Glasson, W.A. and C.S. Tuesday. 1970. Hydrocarbon reactivities in the atmospheric photooxidation of nitric acid. Environ. Sci. Technol. 4: 916-924.

Graedel, T.E., L.A. Farrow, and T.A. Weber. 1977. Photochemistry of the "Sunday Effect." Environ. Sci. Technol. 7: 690-694.

Grimsrud, E.P., H.H. Westberg, and R.A. Rasmussen. 1975. Atmospheric reactivity of monoterpene hydrocarbons. NO_x photooxidation and ozonolysis. Int. J. Chem. Kinet. Symp. 1: 183-195.

Ingersoll, R.B., R.E. Inman, and W.R. Fisher. 1974. Soils potential as a sink for atmospheric carbon monoxide. Tellus 26: 151-158.

Isaksen, I.S., Ø. Hov, and E. Hessvedt. 1978. Ozone generation over rural areas. Environ. Sci. Technol. 12: 1279-1284.

Japar, S.M., C.H. Wu, and H. Niki. 1974. Rate constants for the gaseous reaction of ozone with a-pinene and terpinolene. Environ. Lett. 7: 245-249.

Jensen, K.F. 1975. Sulfur content of hybrid poplar cuttings fumigated with sulfur dioxide. U.S.D.A. Forest Service Research Note No. NE-209, Broomall, PA, 4 pp.

Jensen, K.F. and T.T. Kozlowski. 1975. Absorption and translocation of sulfur dioxide by seedlings of four forest tree species. J. Environ. Qual. 4: 379-382.

Kais, A.G., R.C. Hall, and J.P. Barnett. 1984. Nitric acid and benomyl stimulate rapid height growth of longleaf pine. U.S.D.A. Forest Service, Publ. No. SO-307, Southern Forest Exp. Sta. New Orleans, LA, 4 pp.

Kim, C.M. 1973. Influence of vegetation types on the intensity of ammonia and nitrogen dioxide liberation from soil. Soil Biol. Biochem. 5: 163-166.

Kramer, P.J. 1981. Carbon dioxide concentration, photosynthesis, and dry matter production. Bioscience 31: 29-33.

Likens, G.E., F.H. Bormann, R.S. Pierce, J.S. Eaton, and N.M. Johnson. 1977a. Biochemistry of a Forested Ecosystem. Springer–Verlag, New York. 146 pp.

Likens, G.E., J.S. Eaton, and J.N. Galloway. 1977b. Precipitation as a source of nutrients for terrestrial and aquatic ecosystems. In Precipitation Scavenging. ERDA Symposium Series CONF-741003. Tech. Inform Am., Springfield, VA, pp. 552-570.

Lindberg, S.E. and R.C. Harriss. 1981. The role of atmospheric deposition in an eastern United States deciduous forest. Water Air Soil Pollu. 16: 13-31.

Lindberg, S.E., G.M. Lovett, D.D. Richter, and D.W. Johnson. 1986. Atmospheric deposition and canopy interactions of major ions in a forest. Science 231: 141-145.

Lonneman, W.A, R.L. Seila, and J.J. Bufalini. 1978. Ambient air hydrocarbon concentrations in Florida. Environ. Sci. Technol. 12: 459-463.

Luxmoore, R.J., E.G. O'Neill, J.M. Ells, and H.H. Rogers. 1986. Nutrient uptake and growth response of Virginia pine to elevated atmospheric carbon dioxide. J. Environ. Qual. 15: 244-251.

Maugh, T.H. 1975. Air pollution: Where do hydrocarbons come from? Science 189: 277-278.

Maugh, T.H. 1979. SO_2 pollution may be good for plants. Science 205: 383.

McLean, D.M. 1978. A terminal Mesozoic "greenhouse": Lessons from the past. Science 201: 401-406.

Middleton, J.T. 1967. Air — An essential resource for agriculture. In N.C. Brady, ed., Agriculture and the Quality of Our Environment. Am. Assoc. Adv. Sci., Publ. No. 85, AAAS, Washington, DC, pp. 3-9.

Miller, P.R., M.H. McCutchan, and H.D. Milligan. 1972. Oxidant air pollution in the Central Valley, Sierra Nevada Foothills and Mineral King Valley of California. Atmos. Environ. 6: 623.

Murphy, C.E. Jr., T.R. Sinclair, and K.R. Knoerr. 1977. An assessment of the use of forests as sinks for the removal of atmospheric sulfur dioxide. J. Environ. Qual. 6: 388-396.

National Academy of Sciences. 1977. Ozone and Other Photochemical Oxidants. National Academy of Sciences, Washington, DC, 717 pp.

Noggle, J.C. and H.C. Jones. 1979. Accumulation of Atmospheric Sulfur by Plants and Sulfur-Supplying Capacity of Soils. U.S. Environmental Protection Agency, Publ. No. 600/7-79-109, Washington, DC, 37 pp.

Noggle, J.C., H.C. Jones, and J.M. Kelly. 1978. Effects of accumulated nitrates on plants. Paper presented at the 71st Annual Meeting, Air Pollution Control Association, June 25–30, 1978. Houston, Texas, 16 pp.

Norby, R.J., E.G. O'Neill, and R.J. Luxmoore. 1986a. Effects of atmospheric CO_2 enrichment on the growth and mineral nutrition of Quercus alba seedlings in nutrient-poor soil. Plant Physiol. 82: 83-89.

Norby, R.J., J. Pastor, and J. M. Melillo. 1986b. Carbon-nitrogen interactions in CO_2-enriched white oaks: Physiological and long-term perspectives. Tree Physiol. 2: 233-241.

Nozhevnikova, A.N. and L.N. Yurganov. 1978. Microbiological aspects of regulating the carbon monoxide content in the earth's atmosphere. Adv. Microb. Ecol. 2: 203-244.

O'Neill, E.G., R.J. Luxmoore, and R.J. Norby. 1987. Increases in mycorrhizal colonization and seedling growth in *Pinus echinata* and *Quercus alba* in an enriched CO_2 atmosphere. Can. J. For. Res. 17: 878-883.

Perrson, R. 1974. World Forest Resources. Review of the World's Forest Resources in the Early 1970s. Research Note No. 17, Royal College of Forestry, Stockholm, Sweden.

Pratt, G.C., R.C. Hendrickson, B.I. Chevone, D.A. Christopherson, M.V. O'Brien, and S.V. Krupa. Ozone and oxides of nitrogen in the rural upper-midwestern U.S.A. Atmos. Environ. 17: 2013-2023.

Rasmussen, R.A. 1972. What do the hydrocarbons from trees contribute to air pollution? J. Air Pollu. Control Assoc. 22: 537-543.

Rasmussen, K.H., M. Taheri, and R.L. Kabel. 1975. Global emissions and natural processes form removal of gaseous pollutants. Water Air Soil Pollu. 4: 33-64.

Research Triangle Institute. 1973. Investigation of high ozone concentration in the vicinity of Garrett County, Maryland and Preston County, West Virginia. RTI Publ. NTIS No. PB-218540, Research Triangle Park, NC.

Richter, H.G. 1970. Special ozone and oxidant measurements in vicinity of Mount Storm, West Virginia Research Triangle Institute, Research Triangle Park, NC.

Ripperton, L.A. and D. Lillian. 1971. The effect of water vapor on ozone synthesis in the photo-oxidation of alpha-pinene. J. Air. Pollu. Control Assoc. 21: 629-635.

Roberts, B.R., L.S. Dochinger, and A.M. Townsend. 1986. Effects of atmospheric deposition on sulfur and nitrogen content of four urban tree species. J. Arbor. 12: 209-212.

Robinson, E. and R.C. Robbins. 1971. Emission Concentration and Fate of Particulate Atmospheric Pollutants. Final Report, SRI Project SCC-8507. Stanford Research Institute, Menlo Park, CA.

Rogers, H.H., J.C. Campbell, and R.J. Volk. 1979. Nitrogen-15 dioxide uptake and incorporation by *Phaseolus vulgaris* (L.). Science 206: 333-335.

Salop, J., N.T. Wakelyn, G.F. Levy, E.M. Middleton, and J.C. Gerwin. 1983. The application of forest classification from Landsat data as basis for natural hydrocarbon emission estimation and photochemical oxidant model simulations in southeastern Virginia J. Air Pollu. Control Assoc. 33: 17-607.

Sandberg, J.S., M.J. Basso, and B.A. Okin. 1978. Winter rain and summer ozone: A predictive relationship. Science 200: 1051-1054.

Schneider, S.H. 1975. On the carbon dioxide-climate confusion. J. Atmos. Sci. 32: 2060-2066.

Seiler, W. 1974. The cycle of atmospheric CO. Tellus 26: 116-135.

Spedding, D.J. 1974. Air Pollution. Oxford Chemistry Series. Clarendon Press, Oxford, 76 pp.

Stasiuk, W.N. Jr. and P.E. Coffey. 1974. Rural and urban ozone relationships in New York State. J. Air Pollu. Control Assoc. 24: 564-568.

Stephens, E.R. and W.E. Scott. 1962. Relative reactivity of various hydrocarbons in polluted atmospheres. Am. Petroleum Institute Proc. 42: 655-670.

Swank, W.T. 1984. Atmospheric contributions to forest nutrient cycling. Water Resour. Bull. 20: 313-321.

Swank, W.T. and G.S. Henderson. 1976. Atmospheric input of some cations and anions to forest ecosystems in North Carolina and Tennessee. Water Resour. Res. 12: 541-546.

Thompson, L.M. and F.R. Troeh. 1978. Soils and Soil Fertility. McGraw-Hill, New York, 516 pp.

Tinus, R.W. 1972. CO_2 enriched atmosphere speeds growth of ponderosa pine and blue spruce seedlings. Tree Planters' Notes (1972): 12-15.

Troiano, J.J. and I.A. Leone. 1977. Changes in growth rate and nitrogen content of tomato plants after exposure to NO_2 hr. Phytopathology 67: 1130-1133.

U.S. Department of Agriculture. 1964. Sulfur as Plant Nutrient in the Southern United States. Tech. Bull. No. 1297, Washington, DC, 45 pp.

U.S. Environmental Protection Agency. 1986. Air Quality Criteria for Ozone and Other Photochemical Oxidants. Vol. I. Publ. No. EPA-600-8-84-020aF. U.S.E. P.A., Research Triangle Park, NC.

Wainwright, M. 1978. Distribution of sulfur oxidation products in soils and on *Acer pseudoplatanus* L. growing close to sources of atmospheric pollution. Environ. Pollu. 17: 153-160.

Westberg, H.H. and R.A. Rasmussen. 1972. Atmospheric photochemical reactivity of monoterpene hydrocarbons. Chemosphere 4: 163-168.

Wellburn, A.R., J. Wilson, and P.H. Aldridge. 1980. Biochemical responses of plants to nitric oxide polluted atmospheres. Environ. Pollu. 22: 219-228.

White, W.H., J.A. Anderson, D.L. Blumenthal, R.B. Husar, N.V. Gillani, J.D. Huser, and W.E. Wilson Jr. 1976. Formation and transport of secondary air pollutants: Ozone and aerosols in the St. Louis urban plume. Science 194: 187-189.

Will, G.M. and C.T. Youngberg. 1978. Sulfur status of some central OR pumice soils. Soil Sci. Soc. Am. 42: 132-134.

Wolff, G.T., P.J. Lioy, R.E. Meyers, R.T. Cederwell, G.D. Wight, R.E. Pasceri, and R. S. Taylor. 1977. Anatomy of two ozone transport episodes in the Washington, DC to Boston, Mass. corridor. Environ. Sci. Technol. 11: 506-510.

Wong, C.S. 1978. Atmospheric input of carbon dioxide from burning wood. Science 200: 197-200.

Woodwell, G.M. 1978. The carbon dioxide question. Sci. Am. 238: 34-43.

Woodwell, G.M., R.H. Whittaker, W.A. Reiners, G.E. Likens, C.C. Delwiche, and D.B. Botkin. 1978. The biota and the world carbon budget, Science 199: 141-146.

Wright, R.D. 1974. Rising atmospheric CO2 and photosynthesis of San Bernadino Mountain Plants. Am. Mid. Natural 91: 360-370.

Wright, R.D. and G.M. Woodwell. 1970. Effect of increased CO2 on carbon fixation by a forest. Brookhaven National Laboratory, Publ. No. 1444, Upton, NY, 7 pp.

U.S. Department of Agriculture, Forest Service. 1978. Forest Statistics of the U.S. 1977. U.S.D.A., Washington, DC, 133 pp.

U.S. Department of Agriculture, Forest Service. 1979. A Report to Congress on the Nation's Renewable Resources. RPA Assessment and Alternative Program Directions. U.S.D.A., Washington, DC, 209 pp.

U.S. Environmental Protection Agency. 1976. Open Space as an Air Resource Management Measure. Vol. II. Design Criteria. Publ. No. EPA-450/3-76-028b. U.S.E.P.A., Research Triangle Park, NC.

U.S. Environmental Protection Agency. 1977. National Air Quality and Emissions Trend Report, 1976. U.S.E.P.A. Publ. No. EPA-450/1-77-002, Research Triangle Park, NC.

SECTION II

FORESTS ARE INFLUENCED BY AIR CONTAMINANTS IN A SUBTLE MANNER — CLASS II INTERACTIONS

8
Forest Tree Reproduction: Influence of Air Pollutants

Sexual reproduction of forest trees is critically important for maintenance of genetic flexibility and the persistence of most species in natural forest communities. Reproductive strategies, however, are typically beset by a variety of "weak points," and reproductive growth of many forest trees is, at best, irregular and quite unpredictable. Generally there is a very good correlation between tree vigor and the capacity for flowering and fruiting (Kozlowski, 1971). A variety of environmental constraints may reduce tree vigor and, in view of the fact that numerous potential points of interaction have been identified between air pollutants and reproductive elements (Figure 8–1), it has been hypothesized that air contaminants may impact forest ecosystems by influencing reproductive processes.

A. Pollen Production and Function

Copious quantities of pollen must be produced by trees in order to accomplish the infrequent occurrence of successful gamete transfer. Pollen grains must be transferred in a viable form from stamens to pistils in angiosperms and from staminate to ovulate strobili in gymnosperms. Adequate pollen germination and tube growth must occur in order to realize fertilization of ovaries.

Unfortunately we have little information on the influence of air pollution on the quantity of pollen produced and on pollen distribution under either artificial or natural conditions. As summarized in Chapter 4, pollen production by trees is naturally quite irregular and is subject to a variety of environmental influences. This results in considerable variation in the specific time of pollen release from year to year. Once initiated, however, the process of pollen shed is relatively rapid, ranging from a few hours to a few days (Kozowsky, 1971), and

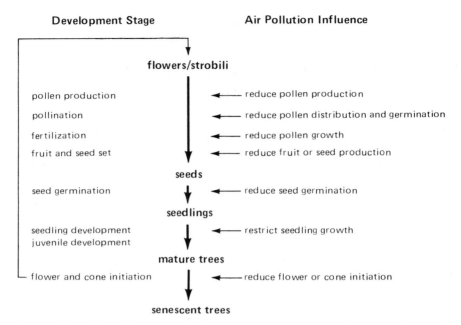

Figure 8-1. Potential points of interaction between air pollutants and sexual reproduction of forest trees.

it should be possible to design experiments that would attempt to correlate air quality with pollen yield. Experimental design, however, would have to carefully account for variations in pollen production due to environmental factors other than air pollution.

We are equally uncertain about the influence of air quality on pollen distribution as well as production. In the instance of wind pollinated species — for example, gymnosperms, *Populus, Quercus, Fraxinus, Ulmus, Carya,* and *Platanus* species — it is presumed that air contaminants are unimportant in pollen movement. In the instance of tree species dependent on animal pollen transfer, however, it has been speculated that air pollutants may impact pollination if they exert an adverse influence on pollinating insects, birds, or other animal vectors. This consideration would apply to the important forest species of the *Acer, Salix,* and *Tilia* genera and, of course, fruit species such as *Malus, Pyrus, Ficus,* and others. Since forest ecosystems with a preponderance of animal-pollinated species are our most species diverse ecosystems (Ostler and Harper, 1978), an adverse impact on animal pollinators might potentially result in competitive weakening and a reduction in species diversity.

Bees, by virtue of their ecology and anatomy, constitute the most important group of insect pollinators. Flies and midges follow bees in importance and are in turn followed by a variety of other insect orders including butterflies, moths, thrips, and beetles. Elevated fluoride concentrations in pollinators including bumblebees, honeybees, sphinx moths, and wood nymph butterflies, have been reported in the vicinity of an aluminum reduction plant (Carlson and Dewey, 1971; Dewey, 1973). We are deficient, however, in our understanding of the relationships between air pollutants and these various pollinating insects. Some evidence for toxicity of arsenic, fluoride, sulfur dioxide, and ozone to bees will be presented (Chapter 13). We are also deficient in our appreciation of which specific insect species are most important for forest tree fertilization. Future research must clarify these relationships as well as explore the impact of various pollutants on pollinating insect physiology and ecology (Bromenshenk and Carlson, 1975).

1. Pollen Germination and Tube Elongation

The greatest research effort regarding air contaminants and pollen physiology has dealt with in vitro and some in vivo efforts to assess pollutant influence on germination percentage and pollen tube elongation. Once deposited on the pistil or ovulate cone, the pollen grain must germinate and produce a tube several millimeters in length in order to reach the ovule. These complex processes are adequately reviewed in Heslop–Harrison (1973) and Street and Öpik (1976). Wolters and Martens (1987) have provided a comprehensive review of air contaminant effects on pollen. These authors indicate that pollutants may influence pollen by altering gas exchange, chemistry, mitosis (DNA synthesis), morphology (weight), number per stigma, tube growth, and germination.

Table 8-1 presents a set of representative pollen studies. Sulzbach and Pack (1972) recorded reduced pollen grains retained on stigma, pollen germination, and pollen tubes reaching ovules when tomato plants were grown in growth chambers and then exposed to hydrogen fluoride doses of 4.2 μg m^{-3} for several weeks. This is, of course, an excessive dose and restricts extrapolation to natural environments. Working with corn in a greenhouse environment, Mumford et al. (1972) employed realistic doses of ozone: 3, 6, and 12 pphm (95, 118, and 235 μg m^{-3}) for 5.5 hr day^{-1} for 60 days to examine various biochemical changes induced in pollen. They concluded that ozone exposure caused autolysis of structural glycoproteins and stimulated amino acid synthesis. Free amino acids of the corn pollen increased 50% following exposure to ozone at 3 pphm (59 μg m^{-3}). The higher doses further enhanced amino acid and peptide accumulation and inhibited germination 40–90%. Ultrastructural examination of petunia pollen, exhibiting reduced germination percentage following ozone exposure to 50 pphm (980 μg m^{-3}) for 3 hr, demonstrated a movement of organelles "away from" the plasma membrane in pollen tubes (Harrison and Feder, 1974). The authors felt this change may have been associated with impaired germination and pollen tube growth. Masaru et al. (1976) collected pollen from greenhouse lilies and exposed the grains to various doses of several air contaminants singly and in combination. Synergistic interaction between various combinations of sulfur dioxide, nitrogen dioxide, ozone, and aldehydes was evident in inhibition of tube elongation. Doses employed were variable but generally were related to typical

Table 8-1. Response of Pollen to Air Pollution under Laboratory and Field Conditions.

Plant	Pollutant	Pollen Parameter Suppressed	Reference
Petunia	Ozone	Germination, tube elongation	Feder (1975)
Tobacco	Ozone	Germination, tube elongation	Feder (1975)
Corn	Ozone	Germination, tube elongation	Feder (1975)
Tomato	Hydrogen fluoride	Germination tube elongation	Sulzbach & Pack(1972)
Lily	Sulfur dioxide, acrolein nitrogen dioxide, formaldehyde	Tube elongation	Masaru et al. (1976)
Tobacco	Motor vehicle exhaust (ambient)	Germination, tube elongation	Flückiger et al. (1978)
Cottonwood, red pine, austrian pine, blue spruce	Sulfur dioxide	Germination, tube elongation	Karnosky & Stairs (1974)
Red pine, white pine	Sulfur dioxide (ambient)	Germination tube elongation (red pine only)	Houston & Dochinger (1977)
Scotch pine	Sulfur dioxide	Grain size	Mamajev & Shkarlet (1972)
White fir	Sulfur dioxide	Germination	Keller (1976)
Mugo pine, black pine, Scotch pine	Sulfur dioxide	Germination	Keller & Beda (1984)
Red pine	Heavy Metals	Germination	Chaney & Strickland (1984)
Camellia	Acidity	Tube elongation	Masuru et al. (1980)
13 forest species	Acidity	Germination	Cox (1983)
4 forest species	Acidity	Germination, tube elongation	Van Ryn et al. (1986)
Red maple	Acidity	Germination, tube number	Van Ryn et al. (1988)
Fraser fir	Acidity	Production, germination	Feret et al. (1988)

urban ambient levels. Flückiger et al. (1978) monitored tobacco pollen germination and tube growth under ambient atmospheric conditions at a road edge, and 30 and 200 m from a highway. After 8 hr exposure, germination was inhibited 98% and tube growth 89% at the road edge. At 200 m tube elongation was still reduced by 2%. Keller and Beda (1984) determined that 24 hr exposure to 225 ppb (590 μg m^{-3}) sulfur dioxide consistently reduced germination of mugo, black, and Scotch pine pollen in vitro. At 75 ppb (197 μg m^{-3}) exposure (typical of ambient), black and Scotch pine pollen were more sensitive than mugo.

Karnosky and Stairs (1974) collected pollen from several forest tree species and observed germination and tube elongation on agar following exposure to sulfur dioxide. Moist quaking aspen pollen germination was reduced at sulfur dioxide concentrations of 0.75 ppm (1965 μg m^{-3}) and above for 4 hr. Highly significant decreases in tube length occurred at 0.30 ppm (786 μg m^{-3}) for 4 hr. A 4-hr exposure to 1.4 ppm (3668 μg m) sulfur dioxide severely restricted moist pollen germination and tube elongation of red pine, Austrian pine, and blue spruce. The authors cautioned, however, that much of the inhibition of germination and pollen tube elongation may have been due to absorption of sulfur dioxide by the agar media and resulting acidification from pH 7 to 5. The authors speculate that if similar acidification of stigmatic or micropylar tissue occurs in nature, low levels of sulfur dioxide may effectively limit seed production.

A series of recent studies have directly examined the impact of acidity on pollen. Masaru et al. (1980) examined the influence of acid rain simulant on camellia (*Camellia japonica*) pollen. Hydrogen ion concentration was observed to suppress pollen tube elongation only when the medium pH was less than 3.2. Cox (1983) investigated the in vitro sensitivity of pollen of 13 forest tree species to simulated acid rain. He observed that deciduous pollen was more sensitive to pH than conifer pollen and that understory species tested were of intermediate sensitivity. Two-thirds of the species tested by Cox had pollen significantly inhibited in germination by pH 3.0, but little impact on pollen metabolism was recorded at pH 4.6. In a similar study, Van Ryn et al. (1986) examined the influence of acidity on the *in vitro* germination and tube elongation of four northeastern United States forest tree species. Treatments included media acidification ranging from 5.0 to 2.6. At pH 3.0 and 2.6, pollen germination of all species was inhibited. Between pH 4.2 and 3.4, germination was reduced by more than 50% of that at pH 5.0. Mean tube length was reduced at the lowest pH at which germination was recorded. The pollen impact threshold was pH 4.2. In a companion experiment, the influence of mist acidity on in vitro pollen germination and tube elongation of red maple was studied (Van Ryn et al. 1988). Branches, with female flowers, were subjected to mist simulants at pH values 5.6, 4.6, 3.6, and 2.6. Following exposure, stigmas weere pollinated with untreated pollen. Both the number of grains germinating and the number of tubes reaching the base of the style decreased as acidity increased. The authors observed that acidity effects on pollen *in vivo* were similar to, but not as severe as, their results obtained in vitro.

Unfortunately very few studies of pollen health have been conducted in the field. Houston and Dochinger (1977) collected pollen from stands of red and white pines growing in areas of high and low air pollution incidence in central Ohio. The germination percentage of white pine pollen was higher in the material collected from the "cleaner" site. In the case of red pine, both germination percentage and average pollen tube length was greater in pollen gathered from the low pollution incidence region. Feret et al. (1988) collected male and female strobili from 10 mature Fraser fir trees over 2 years from Mt. Mitchell, NC (forest exhibiting stress symptoms) and from Mt. Rogers, VA (forest without stress symptoms). Pollen production and germination did not vary significantly by location for either year, but did vary within locations for both years.

B. Flower, Cone, and Seed Production

Working with various nonwoody species, including duckweed, carnation, geranium, and petunia, W.A. Feder and coworkers from the University of Massachusetts, Waltham, Massachusetts (Feder and Campbell, 1968; Feder and Sullivan 1969; Feder, 1970) have shown that ozone at low doses, 10 pphm (196 μg m^{-3}) for 5–7 hr for 1–3 months, can reduce flower production.

One of the earliest observations of decreased seed production by forest tree species in response to air contamination was provided by G.G. Hedgcock in 1912. He recorded a few or no seed borne on conifers close to the Washoe smelter at Anaconda, Montana (Hedgcock, 1912). Observations of reduced cone production by ponderosa pine presumably caused by field exposure to ozone (Miller, 1973) and sulfur dioxide (Scheffer and Hedgcock, 1955) have been reported. The latter observations were made near ore smelters at Trail, British Columbia and Anaconda, Montana. Forest Service data collected in the late 1920s and 1930s from the Colville National Forest close to smelter effluent revealed sparse cone production by western larch, lodgepole pine, and Douglas fir.

Mamajev and Shkarlet (1972) have reviewed the European literature on the impact of air pollution on cone production and presented their own data concerning Scotch pine response to sulfur dioxide from smelting operations in the Urals. In an area of high, approximately 1–3 μg m^{-3}, sulfur dioxide concentration for decades the following reductions in Scotch pine cone dimensions were measured: 16–19% in mature female cone length and 37–50% in weight, 12–15% in seed weight, and 23–32% in staminate cone weights (Table 8-2). Houston and Dochinger (1977) made similar measurements of cone parameters in their investigation of white and red pine in Ohio (Table 8-3).

Additional evidence for reductions in fruit production by a wide variety of agricultural crops and fruit trees in response to controlled environmental and ambient exposure to hydrogen fluoride and oxidants has been presented (Pack, 1972; Pack and Sulzback, 1976; Thompson and Taylor, 1969; Thompson and Kats, 1975). Thompson and his colleagues of the Air Pollution Research Center in Riverside, California have demonstrated substantial citrus fruit yield reductions occasioned by photchemical oxidants but reported little effect from ambient

Table 8-2. Reduction in Scotch Pine Cone Dimensions and Seed Weight in the Vicinity of Russian Copper Smelters.

Parameter	High SO_2 incidence	Low SO_2 incidence
Ripe average cone length (mm)	32.1-35.6	39.5-42.4
Ripe average cone weight (g)	2.39-3.03	3.81-5.91
1000 seed average weight (g)	4.85-5.72	5.47-6.67
Staminate average cone length (mm)	17-21	22-31

Source: Mamajev and Shkarlet (1972).

Table 8-3. Reduction of White and Red Pine Cone Dimensions and Seed Number, Weight, and Germination in Ohio Areas of High and Low Ambient Air Pollution.

Parameter	High pollution incidence		Low pollution incidence	
	White pine	Red pine	White pine	Red pine
Aver. cone length (mm)	122	44	124	47[a]
Aver. cone width (mm)	20	23	21	23
Aver. no. seeds cone^{-1}	55		67[a]	
Aver. 100 seed wt (g)	1532	0.666	1.850[b]	0.805[a]
% filled seed	85	50	84	68[a]
% seed germination	70	50	70	66[a]

Source: Houston and Dochinger (1977).
[a] significant 0.01 level.
[b] significant 0.05 level.

fluorine compounds at concentrations varying from 0–1.2 μg m^{-3}. Brewer at al. (1960, 1967), on the other hand, obtained reduced yields of orange fruit from trees exposed continuously to 1–5 ppb (0.8–4 μg m^{-3}) hydrogen fluoride gas for a 2-year period and from trees sprayed periodically with hydrogen fluoride or sodium fluoride solutions to achieve foliar concentrations of 75–150 ppm fluoride. The most common response of fruiting of several agricultural species to hydrogen fluoride was the development of fewer seeds (Pack and Sulzback, 1976). Pepper and corn plants exhibited reduced flower development in this latter study. Bonte and Garrec (1984) have suggested that fluoride accumulation

Table 8-4. Fecundity of Conifers at Five sites Downwind of a Phosphorus Plant in Newfoundland, Canada.

	n	Control 18.7	1.4	4.6	5.8	8.0	10.3	Correlation with mean F level in air
Balsam fir								
Percent fertile trees	100	80	7	12	48	56	60	$p < 0.05$
Cones per fertile tree	20	47	18	30	21	34	20	NS
		(22)	(12)	(9)	(7)	(35)	(15)	
Seeds per cone	25	247	207^a	217^a	212^a	251	192^a	NS
		(17)	(19)	(17)	(13)	(20)	(13)	
Black spruce								
Percent fertile trees	100	82	11	27	40	44	60	$p < 0.05$
Cones per fertile tree	20	106	55	64	117	84	30^a	NS
		(61)	(25)	(34)	(44)	(31)	(16)	
Seeds per cone	25	47	18^a	4^a	45	39^a	38^a	NS
		(11)	(11)	(3)	(11)	(10)	(8)	
Larch								
Percent fertile trees	100	28	$-^b$	4	3	4	3	NS
Cones per fertile tree	20	416	–	28^a	241^a	212^a	345^a	$p < 0.05$
		(32)	–	(18)	(37)	(20)	(42)	
Seeds per cone	25	4		2	4	7^a	5	NS
		(2)		(2)	(4)	(5)	(5)	

Note: Values are means with standard deviations in parentheses. NS, not significant.
[a]Mean value different from that of the control site ($p < 0.05$).
[b]No mature trees.
Source: Sidhu et al. (1986).

on stigmata may be responsible for fruiting failure in strawberry plants. In their study of Scotch pine reproduction, Roques et al. (1980) emphasized that reproductive stress may be greatest when sulfur dioxide and fluoride contamination occur simultaneously.

Sidhu and Staniforth (1986) studied conifer mortality, and subsequent replacement with deciduous species, in the vicinity of a phosphorus plant on the Avalon Peninsula, Newfoundland, Canada. These authors concluded that reproductive failure, especially stress associated with seed dynamics, was due to fluoride contamination and that this was, at least in part, the reason for the change in forest composition (Table 8-4).

Direct effects of acid deposition on forest tree seed are probably not great. At certain stages of plant development, acidity exposure may influence the vigor and viability of agricultural seed (Troiano et al., 1982). Germination of forest tree seed, however, may be stimulated by acidity (Hare, 1981). Raynal et al. (1982) examined the influence of substrate acidity on the germination of five important tree species from the Adirondack region of New York and observed a wide range of sensitivities. Inhibition of germination was recorded at pH 4.0 and 3.0 for red maple and at pH 3.0 for yellow birch. No inhibition was recorded at these pH levels for sugar maple and eastern hemlock, and there was stimulation of eastern

Table 8-5. Maximum Percent Seed Germination for Five Adirondack Forest Tree Species Subjected to Different Substrate Acidity Levels.

Species	Treatment pH			
	5.6	4.0	3.0	2.4
Sugar maple	99.3a	98.7a	99.0a	–
Red maple	5.5a	2.2b	1.4b	–
Yellow birch	67.1a	62.5a	50.7b	–
Hemlock	37.7a	38.0a	35.7a	4.3b
White pine	9.1a	14.1a	35.9b	41.9b

Dashes indicate no treatment. Means followed by the same letter within each row are not significantly different ($p > 0.05$) by Tukey's test from other treatment levels. Source: Raynal et al. (1982).

white pine germination as substrate pH was decreased (Table 8-5). In high-elevation conifer stands exhibiting stress symptoms in the eastern United States, some observations of seed characteristics have been made. Feret et al. (1988) collected female strobili from mature Fraser fir trees from Mt. Mitchell, North Carolina (forest exhibiting stress symptoms) and from Mt. Rogers, Virginia (forest without stress symptoms). Strobili collected in 1986 did not differ with regard to the number of seed produced or total strobilus weight, but seeds from Mt. Mitchell were generally larger. Mt. Mitchell strobili further contained more full seed, less insect-damaged seed, and had seed with a higher germination rate than Mt. Rogers seed. In examining seed of red spruce collected from a variety of high-elevation eastern United States sites, Agmata and Bonner (1988) observed that filled seed per cone values (1.2–4.3) were very low, but perhaps within the normal range for the elevations sampled. No evidence of decreased red spruce seed quality or of air pollution influence was uncovered.

Even if seeds are produced and distributed, their chances for successful germination and initial development are extremely limited by a wide variety of natural constraints. Paramount among these are water and oxygen supply, temperature, salt, microbial infection, and insect infestation or consumption by rodents, birds, or other animals. Maguire (1972) suggested that air pollution should also be added to this list. Specific evidence relating seed germination and air quality is largely nonexistent. A notable exception is the impact of trace elements, especially heavy metals, on seed germination. Jordan (1975) has thoroughly studied the Lehigh Gap area of Pennsylvania in the vicinity of a smelter complex at Palmerton, Pennsylvania, where zinc ores have been smelted since 1898. Within 2 km of the primary smelter, up to 8% zinc, 1500 ppm cadmium, 1200 ppm copper, and 1100 ppm lead were found at the surface of the A1 soil horizon. As expected, very few tree seedlings were found near the smelters, and the author hypothesized that inhibition of seed germination or seedling growth by high levels of soil zinc was responsible. Solution concentrations of up to 100 ppm zinc and 10 ppm cadmium did not affect seed germination of red oak, gray birch, and quaking aspen, but these concentrations did preclude radicle elongation. Radicle elongation occurred, but was significantly reduced at ≤1 ppm zinc or ≥ 5 ppm cadmium in solution culture. Jordan concluded that fire and zinc have interacted to decimate the forest in the smelter vicinity, as high soil zinc levels have inhibited sexual reproduction and have prevented normal succession of vegetative cover following burning.

C. Seedling Development

Forest trees may be especially vulnerable to air pollution stress in the seedling stage. During this period growth is rapid and gaseous air contaminants may be rapidly absorbed. Above ground organs lack complete protective coatings and are fragile. Epigeous germinating species (most gymnosperms and *Acer, Fagus, Cornus,* and *Robinia*) that push their cotyledon above ground by elongation of the hypocotyl may be more severely injured than hypogeous germinating species (*Quercus, Juglans,* and *Aesculus*), whose cotyledons remain underground while the epicotyl grows upward and develops leaves. All seedlings remain vulnerable, however, over the first 100 days as foliar expansion and development remain rapid. Leaf production by ponderosa pine seedling during this period may average one to two leaves per day (Berlyn, 1972).

While we do not have abundant data concerning young seedling response to air pollution, we do have several studies that have examined young trees several months or years old. Townsend and Dochinger (1974) studied the relationship between red maple seed source and susceptibility to ozone damage and recorded significant foliar damage during early development stages (Table 8-6). Overall leaf injury was greatest in the youngest seedling stages. Fumigation of 2-week-old red pine seedlings in the cotyledon stage at four sulfur dioxide concentrations, 0.5, 1, 3, or 4 ppm (1310, 2620, 7860, or 10,480 μg m^{-3}) and four exposure times (15, 30, 60, or 120 min) adversely impacted seedling development (Constantinidou et al., 1976). Sulfur dioxide decreased chlorophyll content and dry weight of both cotyledons and primary needles. The authors observed that the pollutant concentrations employed were high, but indicated that the exposure times, compared with field situations, were short and that they found increasing

Table 8-6. Average Foliar Damage of Red Maple Seedlings from Four Seed Sources Following Ozone Fumigation with 75 pphm (1470 μg m^{-3}) for 7 hr day^{-1} for 3 days.

Seed source	% of leaf damaged	Seedling height (cm)
1	6	4
2	34	11
3	18	24
4	22	32

Source: Townsend and Dochinger (1974).

the time of sulfur dioxide exposure was much more harmful than increasing the gas concentration. Because of their results the authors judged that even at low dosages, continuous exposure to sulfur dioxide in the field may impact seedling development and may restrict regeneration of pine communities. Davis et al. (1977) exposed 2- to 3-year-old black cherry seedlings biweekly to either 0.9 ppm (1764 μg m^{-3}) or 0.10 ppm (196 μg m^{-3}) ozone for 2, 4, 6, and 8 hr. Greatest ozone sensitivity occurred when the foliage was between 4 and 8 weeks old.

Lee and Weber (1979) exposed young seedlings of 11 woody species to 2.3 cm wk^{-1} of simulated acid rain at pH levels of 3.0, 3.5, 4.0, or 5.6. The seedling response was very variable. Emergence was stimulated by at least one acid treatment for four species and was inhibited for one species. Top growth was stimulated by at least one acid treatment for four species (Table 8-7), while root growth was inhibited for one species. Raynal et al. (1982) investigated sugar maple seedling response to acid deposition in controlled facilities. Radicle elongation was reduced when exposed to simulated throughfall at pH 3.0 or less in a laboratory treatment apparatus. In addition, the susceptibility of emerging radicles to bacterial infection increased with exposure to increasing acidity of throughfall simulants. Seedling growth in soil was not influenced by acidic throughfall until after 4 weeks of weekly exposure to pH 3.0 treatment. After this time both promotive and inhibitory effects were observed simultaneously. Percy (1986) examined the response of 11 commercially important North American tree species to precipitation simulants in growth chambers. Species included: white, red, and black spruce; white, red, Scots, and jack pine; balsam fir; paper and yellow birch; and red maple. Germinative capacity was only

Table. 8-7. Effect of Simulated Acid Deposition on the Top Growth of Various Tree Species.

Species	Average top dry weight per treatment of				Species average	Significance level[a]
	pH 3.0	pH 3.5	pH 4.0	pH 5.7		
			—mg —		mg	
Douglas fir	56	46	48	41	48	0.01
Eastern white pine	50	49	49	49	49	NS
Eastern red cedar	23	20	22	20	21	NS
Flowering dogwood	69	66	65	65	66	NS
Staghorn sumac[b]	371	237	–	385	329	NS
Red alder	47	119	109	92	93	NS
Shagbark hickory	335	379	328	260	312	0.10
Yellow birch	58	69	89	197	82	NS
Sugar maple	92	121	120	92	104	0.05
American beech	115	125	111	114	116	NS
Tulip poplar	48	54	102	54	68	0.01

[a]As determined from F test.
[b]Weight of single plant harvested at pH 4.0 not used in analysis.
Source: Lee and Weber (1979).

weakly responsive to rain pH. Seedling survival was more sensitive. No visible seedling symptoms were obvious at rain pH greater than 2.6. Treatment with rain simulants at pH ≤ 4.6 were sufficient to cause growth reductions and morphological changes in conifer seedlings. Deciduous species generally were substantially more resistant than conifer seedlings.

Field studies with tree seedlings have yielded more ambiguous results than controlled environment studies. One-year old seedlings of white birch (sulfur dioxide sensitive) and pin oak (sulfur dioxide resistant) were potted and placed in a field location in Akron, Ohio (average annual ambient sulfur dioxide high, 71 $\mu g \ m^{-3}$) and in Delaware, Ohio (average annual ambient sulfur dioxide low, <10 $\mu g \ m^{-3}$) (Roberts, 1975). After 3 months the overall growth of white birch was greater in the higher, but subphytotoxic environment, and the growth of pin oak was greater at the low sulfur dioxide site. The author suggested that oak stomatal closure in the high sulfur dioxide location may have restricted growth. Two-year old white pine seedlings were field grown for 5 months in either high or low ambient sulfur dioxide environments in Cleveland and Delaware, Ohio, respectively (Roberts, 1976). The seedlings consisted of susceptible (chlorotic dwarf syndrome) and tolerant clones. Foliar injury and reduced leaf growth were observed on susceptible clones growing in the high (annual average, 92 $\mu g \ m^{-3}$) sulfur dioxide environment.

D. Summary

Reproduction of forest trees may be adversely impacted by a variety of air-contaminants at numerous points in the reproductive cycle. We have substantial information suggesting inimical changes in the biochemistry and morphology of pollen that may result in reduced pollen germination or reduced pollen tube elongation. Experimental work suggest in vivo pollen responses to air pollution are not as severe as in vitro responses. Stigmatic surfaces may buffer or other wise protect pollen in nature (Wolters and Martens, 1987). On the other hand, pollen germination and growth in vivo are influenced by pollutant concentrations and acidity levels that are characteristic of some polluted regions in the temperate zone.

We have reviewed several papers suggesting reduced cone and fruit production under field conditions. These data must be interpreted with caution, however, as most are observational in nature and in several instances did not appear to account for non-air pollution phenomena that may have also reduced fruit production. In any case, few of these studies evaluated the specific cause of cone reduction, that is, whether it was a direct or indirect consequence of air pollution exposure.

Solid evidence has been presented to indicate that selected heavy metals in the forest floor may reduce seed germination. It is probable, however, that this only occurs in those limited environments subject to excessive input from smelters or other major sources. Controlled environment fumigations with ambient and above doses of numerous gaseous contaminants strongly indicate a potential for field damage of very young tree seedlings.

If one or more of these various reproductive stress mechanisms is operative in natural forest ecosystems, it is possible that changes in species composition may ultimately occur. Brandt and Rhoades (1972) investigated the effects of accumulated limestone dust on forest community structure in southwestern Virginia. The species composition and structure of control and dusty sites were intensively studied. The control site had an abundance of shrubs, good reproductive efficiency in all strata, and was assumed to be undergoing normal succession. The dusty site, however, evidenced disruptions in structure and composition resulting from dust accumulation. The authors judged that reproductive efficiency of some species had been decreased and the course of succession altered. Dominance in the tree stratum of the control site by chestnut oak, red oak, and red maple was being altered in the dusty site to ultimately include yellow poplar, sugar maple, and possibly chinkapin oak. In their study of ozone impact on the understory vegetation of an aspen ecosystem, Harward and Treshow (1975) concluded that only 1 or 2 years of ozone exposure might be sufficient to cause shifts in community composition because of seed production responses to ozone exposure. Largely due to the reproductive sensitivity of conifers to fluoride pollution, Sidhu and Staniforth (1986) predict a conversion of forest type from sensitive larch, black spruce and balsam fir to more tolerant older, white birch, dogwood, and bunchberry in the vicinity of a phosphorus plant near Long Harbour, Newfoundland.

In controlled environment studies, it is clear that acid rain simulants can have significant direct and indirect impacts on seed germination, radicle elongation, and seedling vitality. While conifer seedlings generally appear most susceptible, responses vary greatly by species. The threshold of acid-induced seedling effects is generally within the range of ambient precipitation pH values.

References

Agmata, A. and F.T. Bonner. 1988. Seed quality of red spruce and Fraser fir in high-elevation stands. G.D. Hertel, ed., Effects of Atmospheric Pollutants on the Spruce-fir Forests of the Eastern United States and Federal Republic of Germany. U.S.D.A. Forest Service, Gen. Tech. Rep., No. NE- 225, Broomall, PA.

Berlyn, G.P. 1972. Seed germination and morphogenesis. In T.T. Kozlowski, ed., Seed Biology. Academic Press, New York, pp. 223-312.

Bonte, J. and J.P. Garsec. 1984. Atmospheric pollution by fluoreinated compounds and fruiting of *Fragaria* L.: Demonstration by the use of an electronic microprobe of a big accumulation of fluoride on stigmata. Sciences Naturelles (Series D) 290: 815-818.

Brandt, C.J. and R.W. Rhoades. 1972. Effects of limestone dust accumulation on composition of a forest community. Environ. Pollu. 3: 217-225.

Brewer, R.F., F.H. Sutherland, F.B. Guillemet, and R.K. Creveling. 1960. Some effects of hydrogen fluoride gas on bearing navel orange trees. Proc. Am. Soc. Hort. Sci. 76: 208-214.

Brewer, R.F., M.J. Garber, F.B. Guillemet, and F.H. Sutherland. 1967. The effects of accumulated fluoride on yields and fruit quality of "Washington" navel oranges. Proc. Am. Soc. Hort. Sci. 91: 150-156.

Bromenshenk, J.J. and C.E. Carlson. 1975. Reduced reproduction-impact on insect pollinators. In W. H. Smith and L.S. Dochinger, eds., Air pollution and Metropolitan Woody Vegetation, U.S.D.A. Forest Service, Publ. No. PIEFR-PA-1, Upper Darby, PA, pp. 26-28.

Carlson, C.E. and J.E. Dewey. 1971. Environmental pollution by fluorides in Flathead National Forest and Glacier National Park. U.S.D.A. Forest Service, Div. State and Private Forestry, Missoula, MT, 57 pp.

Chaney, W.R. and R.C. Strickland. 1984. Relative toxicity of heavy metals to red pine pollen germination and germ tube elongation. Environ. Qual. 13: 391-396.

Constantinidou, H., T.T. Kozlowski, and K. Jensen. 1976. Effects of sulfur dioxide on *Pinus resinosa* seedlings in the cotyledon stage. J. Environ. Qual. 5: 141-144.

Cox, R.M. 1983. Sensitivity of forest plant reproduction to long range transported air pollutants: *In vitro* sensitivity of pollen to simulated acid rain. New Phytol. 95: 269-276.

Davis, D.D., C.A. Miller, and J.B. Coppolino. 1977. Foliar response of eleven woody species to ozone (O_3) with emphasis on black cherry. Proc. Am. Phytopath. Soc. 4: 185.

Dewey, J.E. 1973. Accumulation of fluorides by insects near an emission source in western Montana. Environ. Entomol. 2: 179-182,

Feder, W.A. 1970. Plant response to chronic exposure of low levels of oxidant type air pollution. Environ. Pollut. 1: 73-79.

Feder, W.A. 1975. Abnormal pollen, flower or seed development. In W.H. Smith and L.S. Dochinger, eds., Air Pollution and Metropolitan Woody Vegetation, Publ. No. PIEFR-PA-1, U.S.D.A. Forest Service, Upper Darby, PA, pp. 28-30.

Feder, W.A. and F.J. Campbell. 1968. Influence of low levels of ozone on flowering of carnations. Phytopathology 58: 1038-1039.

Feder, W.A. and F. Sullivan. 1969. Ozone: Depression of frond multiplication and floral production in duckweed. Science 165: 1373-1374.

Feret, P.P., T.L. Sharik, B.I. Chevane, D.J. Paganelli, and P. Moldenhawer. 1988. A comparison of reproductive fitness between two Fraser fir populations in the southern Appalachians. G.D. Hertel, ed., Effects of Atmospheric Pollutants on the Spruce-Fir Forests of the Eastern United States and Federal Republic of Germany. U.S.D.A. Forest Service, Gen. Tech. Rep. No. NE-255, Broomall, PA.

Flückiger, W., S. Braun, and J.J. Oertli. 1978. Effect of air pollution caused by traffic on germination and tube growth of pollen by *Nicotiana sylvestris*. Environ. Pollu. 16: 73-80.

Hare, R.C. 1981. Nitric acid promotes pine seed germination. U.S.D.A. Forest Service. Res. Note No. SO-281. Southern Forest Exp. Sta., New Orleans, LA, 2 pp.

Harrison, B.H. and W.A. Feder. 1974. Ultrastructural changes in pollen exposed to ozone. Phytopathology 64: 257-258.

Harward, M. and M. Treshow. 1975. Impact of ozone on the growth and reproduction of understory plants in the aspen zone of western USA. Environ. Conserva. 2: 17-23.

Hedgcock, G.G. 1912. Winter-killing and smelter-injury in the forests of Montana. Torrya 12: 25-30.

Heslop–Harrison, J., ed. 1973. Pollen: Development and Physiology. Butterworth, London, 338 pp.

Houston, D.B. and L.S. Dochinger. 1977. Effects of ambient air pollution on cone, seed, and pollen characteristics in eastern white and red pines. Environ. Pollu. 12: 1-5.

Jordan, M.J. 1975. Effects of zinc smelter emissions and fire on a chestnut-oak woodland. Ecology 56: 78-91.

Karnosky, D.F. and G.R. Stairs. 1974. The effects of SO_2 on *in vitro* forest tree pollen germination and tube elongation. J. Environ. Qual. 3: 406-409.

Keller, T. 1976. Personal communication and preprint. Intl. Union For. Res. Organ. World Congress, Oslo, Norway, 12 pp.

Keller, T. and H. Beda. 1984. Effects of SO2 on the germination of conifer pollen. Environ. Pollu. 3: 237-243.

Kozlowski, T.T. 1971. Growth and Development of Trees. Vol. II. Cambial Growth, Root Growth, and Reproductive Growth. Academic Press, New York, 514 pp.

Lee, J.J. and D.E. Weber. 1979. The effect of simulated acid rain on seedling emergence and growth of eleven woody species. For. Sci. 25: 393-398.

Maguire, J.D. 1972. Physiological disorders in germinating seeds induced by the environment. In W. Heydecker, ed., Seed Ecology. Pennsylvania State Univ. Press, University Park, PA, pp. 289-310.

Mamajev, S. A. and O.D. Shkarlet. 1972. Effects of air and soil pollution by industrial waste on the fructification of Scotch pine in the Urals. Mitteil. Forst. Bundes-Versuch. Wien 97: 443-450.

Masaru, N., F. Syozo, and K. Soburo. 1976. Effects of exposure to various injurious gases on germination of lily pollen. Environ. Pollu. 11: 181-187.

Masuru, N., F. Katsuhisa, T. Sankichi, and W. Yutaka. 1980. Effects of inorganic components in acid rain on tube elongation of *Camellia* pollen. Environ. Pollu. 21: 51-57.

Miller, P.L. 1973. Oxidant-induced community change in a mixed conifer forest. In J.A. Naegele, ed., Air Pollution Damage to Vegetation. Adv. Chem. Series No. 122, Am. Chem. Soc., Washington, DC, pp. 101-117.

Mumford, R.A., H. Lipke, D.A. Laufer, and W.A. Feder. 1972. Ozone-induced changes in corn pollen. Environ. Sci. Technol. 6: 427-430.

Ostler, W.K. and K.T. Harper. 1978. Floral ecology in relation to plant species diversity in the Wasatch Mountains of Utah and Idaho. Ecology 59: 848-861.

Pack, M.R. 1972. Response of strawberry fruiting to hydrogen fluoride fumigation. J. Air Pollu. Control Assoc. 22: 714-717.

Pack, M.R. and C.W. Sulzbach. 1976. Response of plant fruiting to hydrogen fluoride fumigation. Atmos. Environ. 10:73-81.

Percy, K. 1986. The effects of simulated acid rain on germinative capacity, growth and morphology of forest tree seedlings. New Phytol. 104: 473-484.

Raynal, D.J., J.R. Roman, and W.M. Eichenlaub. 1982a. Response of tree seedlings to acid precipitation. I. Effect of substrate acidity on seed germination. Environ. Exper. Bot. 22: 377-383.

Raynal, D.J., J.R. Roman, and W.M. Eichenlaub. 1982b. Response of tree seedlings to acid precipitation. II. Effect of simulated acidified canopy throughfall on sugar maple seedling growth. Environ. Exper. Bot. 22: 385-392.

Roberts, B.R. 1975. The influence of sulfur dioxide concentration on growth of potted white birch and pin oak seedlings in the field. J. Am. Soc. Hort. Sci. 100: 640-642.

Roberts, B.R. 1976. The response of field-grown white seedlings to different sulphur dioxide environments. Environ. Pollu. 11: 175-180.

Roques, A., M. Kerjean, and D. Auclair. 1980. Effets de la pollution atmospherique par le fluor et le dioxyde de soufre sur l'appareil reproducteur femelle de *Pinus silvestris* en foret de Roumare (Seine-Maritine, France). Environ. Pollu. 21: 191-201.

Scheffer, T.C. and G.G. Hedgcock. 1955. Injury to Northwestern Forest Trees by Sulfur Dioxide from Smelters. U.S.D.A. Forest Service, Tech. Bull. No. 1117, Washington, DC, 49 pp.

Street, H.E. and H. Öpik. 1976. The Physiology of Flowering Plants: Their Growth and Development, Elsevier, New York, 280 pp.

Sulzbach, C.W. and M.R. Pack. 1972. Effects of fluoride on pollen germination, pollen tube growth and fruit development in tomato and cucumber. Phytopathology 62: 1247-1253.

Thompson, C.R. and G. Kats. 1975. Effects of ambient concentrations of peroxyacetylnitrate on navel orange trees. Environ. Sci. Technol. 9: 35-38.

Thompson, C.R. and O.C. Taylor. 1969. Effects of air pollutants on growth, leaf drop, fruit drop, and yield on citrus trees. Environ. Sci. Technol. 3: 934-940.

Townsend, A.M. and L.S. Dochinger. 1974. Relationship of seed source and developmental stage to the ozone tolerance of *Acer rubrum* seedings. Atmos. Environ. 8: 957-964.

Troiano, J., L. Colavito, L. Heller, and D.C. McCune. 1982. Viability, vigor, and maturity of seed harvested from two soybean cultivars exposed to simulated acidic rain and photochemical oxidants. Agri. Environ. 7: 275-283.

Van Ryn, D.M., J.S. Jacobson, and J.P. Lassoie. 1986. Effects of acidity on *in vitro* pollen germination and tube elongation in four hardwood species. Can. J. For. Res. 16: 397-400.

Van Ryn, D.M., J.P. Lassoie, and J.S. Jacobson. 1988. Effects of acid mist on *in vivo* pollen tube growth in red mapole. Can. J. For. Res. (in press).

Wolters, J.H.B. and M.J.M. Marteus. 1987. Effects of air pollutants on pollen. Bot. Rev. 53: 372-414.

9
Forest Nutrient Cycling: Toxic Ions

Nutrients must move into, within, and out of forest ecosystems in appropriate amounts, at appropriate rates, and along established pathways for normal forest growth to occur. The two major sources of nutrients for temperate forest ecosystems are (a) meteorologic input of dissolved, particulate, and gaseous chemicals from outside the ecosystem; and (b) release by weathering of nutrients from primary and secondary minerals stored within the ecosystem (Bormann and Likens, 1979). Healthy forest ecosystems conserve these nutrients and continually recycle them through the system via an elaborate litterfall-decomposition-uptake intrasystem cycle.

The high productivity of forest ecosystems is achieved and maintained through efficient nutrient recycling. For most forest ecosystems essential elements required to maintain productivity cannot be sustained by annual increments from deposition and mineral substrates alone. Decomposition, mineralization, and reuptake are a must. Root uptake requirements will be met by release from detritus in those temperate forest ecosystems completing litter decay in 1 year. In cooler temperate regions, litter input will exceed decay and humus will accumulate. In warmer temperate forests, litter decay may exceed detrital input (Witkamp and Ausmus, 1976).

In the nutrient cycle, primary produced organic matter in the form of litter-fall, leachates, root exudates, and sloughage is remineralized by a variety of soil macro- and microorganisms through a series of fractionation and solubilization decomposition steps (Bond et al., 1976; Mason, 1977). Once nutrients are transferred from organic matter to the available nutrient compartment, they may again be taken up by forest vegetation.

A large number of hypotheses have been proposed for air pollution interference with forest nutrient cycles. These hypotheses are concerned with direct and indirect impacts of pollutants on tree root health and on one or more soil processes or functions. As previously indicated, the deposition of nutrient elements to forest ecosystems may have a beneficial fertilizing influence (Chapter 7). On the other hand, extended term exposure to atmospheric contamination may cause adverse impacts (Class II responses) on forest nutrient cycling by decreasing decomposition rates; decreasing nutrient uptake rates; increasing toxic ion availability; and increasing leaching losses, and/or detrimental influences on symbiont associations (Figure 9-1). Symbiont interactions will be reviewed in Chapter 11, leaching phenomena in Chapter 10, and the potential role of toxic ions will be presented in this chapter.

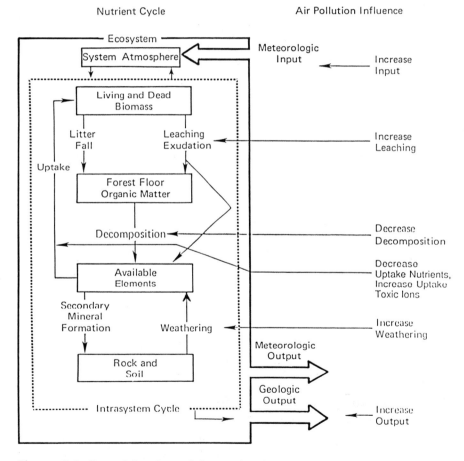

Figure 9-1. Potential points of interaction between nutrient cycling in forest ecosystems and air pollution.
Source: Modified from Bormann and Likens (1979).

A. Heavy Metal Ion Toxicity

Forest floors, with persistent organic matter accumulation, represent important sinks for heavy metals from any source including atmospheric deposition (Chapter 5). Since this accumulation is commonly in the soil horizons with maximum root activity and maximum activity of the soil micro- and macro-biota, it is appropriate to consider the potential for heavy metal toxicity to tree roots and other biotic components of the soil ecosystem. Many studies provide detail on the biochemical and physiological toxicity of heavy metals to higher plants (e.g., Antonovics et al., 1971; Bowen, 1966; Energy Research and Development Administration, 1975; Foy et al., 1978; Foy, 1983) and to microorganisms (e.g., Jernelov and Martin, 1975; Somers, 1961; Summers and Silver, 1978). It is not necessary to review this information, but it is appropriate to summarize selected aspects of our understanding of direct tree toxicity and toxicity to the soil biota caused by heavy metal loading of the upper forest soil profile.

1. Forest Soil Organic Matter Decomposition

The release of inorganic nutrients during decomposition of forest soil organic matter is of fundamental and profound importance in the maintenance of intrasystem nutrient cycling. It has been estimated that 80–90% of net production in terrestrial ecosystems is eventually converted by decomposer organisms (Odum, 1971; Witkamp, 1971; Whittaker, 1975).

The primary input to the forest-soil organic-matter compartment is foliar litterfall, which is concentrated in the autumn in temperate regions. Cromack and Monk (1975) investigated litter production in a mixed deciduous forest in North Carolina. The total annual litter production in the hardwood forest was 4369 kg ha^{-1} yr^{-1} of which 2773 was leaf litter. Total annual litter production in a young white pine forest was 3525 kg ha^{-1} yr^{-1}, 98% of which was needle litter.
A most informative comparison of nutrient elements stored in forest floor organic matter of eastern United States forests and respective turnover times has been presented by Lang and Forman (1978). Likens et al. (1977) have provided detailed information on the chemistry and amounts of various nutrient inputs to the forest floor for the northern hardwood forest. Mader et al. (1977) have characterized the forest floor for a variety of forests throughout the northeastern United States.

Decreasing turnover rates for element release are correlated with increasing latitude and can be related to general climatic change (Table 9-1). In cold temperate environments the activities of decomposing agents are slow and, as a result, forest floor organic matter accumulation is relatively large. In the warm, humid climates of the lower latitudes, decomposing agent activity is high, decomposition and nutrient element release are fast, and organic matter accumulation is relatively small. Temperate forest soils in regions of moderate rainfall and with well-drained, but nutrient-deficient parent material, undergo podsolization and

Table 9-1. Annual Production of Total Litter in Relation to Latitude.

Latitude (N or S)	tons ha^{-1} yr^{-1}	
(degrees)	Mean	Max
0-10	9-11	13.5
30-40	3.5-8	
40-50	2-4	6
50-60	1.5-3.5	4.5
60-65	0.5-1	

Source: Kühnelt and Walker (1976).

produce *mor*-type organic matter. Deciduous forests developing on more nutrient-rich parent material develop brown forest soils with *mull*-type organic matter (Etherington, 1975). Bormann and Likens (1979) have judged that the mor-type forest floor of the New Hampshire northern hardwood forest exerts a major regulating influence on forest development through its capacity for net storage of nutrients and energy. It is not clear if the forest floor of mull-type forest soils exerts a similar regulatory influence of the same magnitude.

It is clear, however, that the decomposition and mineralization rate is a critical component of all intrasystem forest nutrient cycles and it has been hypothesized that any suppression or interruption of the decomposition cycle might lead to a decrease in the supply of available plant nutrients that will ultimately limit tree growth rates (Bond et al., 1976).

a. Organisms Involved in the Decomposition Process

Decomposition of forest organic matter proceeds in stages and involves a complex variety of soil organisms. Microfloral components with primary roles include bacteria, actinomycetes, and fungi. Animal components with important roles include the following micro- and macrofauna: protozoa, nematodes, earthworms, Enchytraeidae, mollusks, Acari, Collembola, Dipteria, and other arthropods.

Bacteria are the most abundant soil organisms with 10^6–10^9 viable cells cm^{-3} but due to their small size (approximately 1 μm) are generally not the major components of soil biomass (Lynch, 1979). They have primary roles in the soil transformations of carbon, nitrogen, phosphorus, iron, and sulfur. Goodfellow and Dawson (1978) have investigated the bacteria colonizing the litter of a mature Sitka spruce stand in Hamsterley Forest, England. Most bacteria were found in the H layer and the fewest in the F layer. Over 90% of the 525 isolates identified belonged to the *Arthrobacter, Bacillus, Micrococcus,* and *Streptomyces* genera. Other studies have suggested the importance of the *Achromobacter, Flavobacterium,* and *Pseudomonas* genera.

Fungi are at least as important, perhaps more important in acid forest soils in the decomposition process as bacteria. A very large number of species are involved with members of the *Aspergillus, Chaetomium, Fusarium, Gliomastix,*

Memnoniella, Penicillium, Stachybotrys, and *Trichoderma* genera, especially abundant in a variety of ecosystems (Garrett, 1963; Griffin, 1972). The microbial biomass in the forest floor of a black spruce ecosystem averaged 5.7 g m^{-2} in the L and F layers and was composed of 85% fungi and 15% bacteria (weight basis; Flanagan and Van Cleve, 1977).

A high proportion of the protozoan soil population is usually encysted, especially if the soil is dry. Our understanding of the ecology of soil protozoa is extremely incomplete, but a prime importance of these organisms is presumed to be as predators of bacteria.

The micro- and macrofauna are important in the decomposition of organic matter as they reduce the size of the litter, improve soil structure, and graze the microflora (Reichle, 1977). Soil animals fragment litter from large to small pieces. The animals also create pore spaces of various dimensions in the soil that allow circulation of air and water. The walls of these same pore spaces provide habitats for bacteria and fungi (Kühnelt and Walker, 1976).

Nematodes are most abundant in well drained and well aerated forest soils. Enchytraeid worms reach maximum numbers in wet, cool, acid soils of high organic matter content. Lumbricid worms (earthworms), because of their large size, are able to move all but the largest soil particles. Mor-type forest floors have extreme horizon differentiation due to minimal mixing by soil animals, particularly earthworms. Mull-type forest floors, on the other hand, generally undergo continual horizon mixing due to the activity of a profile earthworm fauna.

Termites, springtails (*Collembola*), and oribatid mites (*Acarina*) are commonly abundant in numerous forest soils. Ausmus (1977) has pointed out that one of the most important functions of invertebrates in temperate forest ecosystems is in the regulation of wood mineralization.

Carbon is the energy substrate for decomposition. Heterotrophic metabolism is a very important component of ecosystem metabolism and may range in value from 34–57% of the total respiratory carbon flux in the system. Heterotrophic community and microbial respiration may account for approximately 90% (excluding roots) of the total respiration from the soil (Reichle, 1977). Reichle has presented a comprehensive calculation of annual carbon dioxide release by soil decomposers for a yellow poplar (mesic, deciduous) forest (Table 9-2).

b. Measurement of Organic Matter Decomposition Rate and Microbial Biomass

Assessment of impacts on this diverse soil biota commonly involve measurement of metabolic activity (for example, respiration rate), microbial biomass, or both. Unfortunately a prediction of one from the other is difficult as the correlation between the two may not be good under field conditions (Nannipieri et al., 1978). Generally respiration is a better measure of soil metabolic activity than is biomass alone (Reichle, 1977).

The most widely used measure of soil activity is soil respiration, either as oxygen "uptake" or carbon dioxide "evolution" (Nannipieri et al., 1978). Other

Table 9-2. Calculation of Annual Carbon Dioxide Respiration by Soil and Litter Invertebrate Decomposers for a Mesic Deciduous Forest (Yellow Poplar).

Decomposer taxon	Mean body weight (mg dw ind.$^{-1}$)	Mean annual biomass (mg dw m^{-2})	O$_2$ uptake rate at 15 C° (cm^3 O$_2$ g^{-1} dw day^{-1})	Annual CO$_2$ efflux population (g CO$_2$ m^{-2} yr^{-1})	Totals (g CO$_2$ m^{-2} yr^{-1})
MICROFLORA		124,000		2291	2291
NEMATODA	0.0003	950	90.7	49.43	49.43
PULMONATA	37.98	222.46	9.55	1.22	1.22
ARTHROPODA					139.76
Phalangida	0.30	5.79	24.1	0.080	
Pseudoscorpionida	0.413	9.3	22.6	0.120	
Chilopoda	0.69	32.32	20.5	0.380	
Diplopoda	8.757	249.6	12.6	1.80	
Araneae	0.199	115.5	26.1	1.73	
Acarina					
Gamasina	0.280	2,836.3	24.4	39.7	
Uropodina	0.182	237.7	26.5	3.61	
Oribatei	0.099	3,678.2	29.8	62.9	
Prostigmata	0.004	59.3	55.2	1.87	
Pauropoda	0.006	10.3	51.0	0.30	
Symphyla	0.089	104.1	30.4	1.82	
Protura	0.003	8.3	58.3	0.28	
Diplura	0.081	58.3	31.0	1.03	
Insecta					
Collembola					
Onychiuridae	0.005	37.6	52.9	1.14	
Poduridae	0.008	12.4	48.3	0.34	
Isotomidae	0.017	150.4	41.8	3.60	
Entomobryidae	0.045	295.6	34.7	5.88	

Sminthuridae	0.008	6.3	48.3	0.17
Orthoptera	7.69	26.91	12.9	0.20
Psocoptera	0.068	2.3	23.0	0.04
Coleoptera (larvae)	0.209	134.4	25.8	1.99
Hymenoptera	0.110	16.92	29.2	0.28
Lepidoptera (larvae)	0.103	5.73	29.6	0.10
Diptera (larvae)	0.564	849.6	21.3	10.4
ANNELIDA				55.6
Enchytraeidae	0.080	500	31.0	8.90
Lumbricidae	118	10,640	7.65	46.7

Source: Reichle (1977).

determinations of soil activity may be made by monitoring mineralization rates of common biopolymers, carbohydrates, or other organic materials; sulfur oxidation; phosphorus solubilization; and activities of certain soil enzymes (Atlas et al., 1978; Lewis et al., 1978).

Newer strategies for estimating soil biomass include microscopic observation and counting techniques, soil ATP content, agar-film technique, hexosamine (chitin) assay, and enzyme activity (Frankland and Lindley, 1978; Nannipieri et al., 1978; Todd et al., 1973). All these recent studies support the conclusion of Ausmus (1973) that it is too simplistic to attempt to use only one or two indices as a general means of estimating biomass or metabolism in the soil system.

Under natural conditions, and in the absence of anthropogenic pollutants, the two primary controls of litter decomposition rate are presumed to be the prevailing climatic environment and susceptibility of the substrate to attack by specialized decomposers, that is, substrate quality (Meentemeyer, 1978). Meentemeyer concluded, after examining litter decay rates from locations ranging in climate from subpolar to warm temperate, that actual evapotranspiration and lignin content are useful predictors of litter decomposition rates in unpolluted environments.

Considerable evidence that trace metal pollution of forest floors may reduce litter decomposition rates has been presented.

c. Trace Metal Influence on Litter Decomposition

The strongest evidence available to support the importance of air pollution to forest nutrient cycling comes from studies that have been concerned with the impact of trace metals on components of the soil biota and the processes they perform. Investigators active in this research area have enjoyed the considerable benefit of an enormous literature concerning the relationship between trace metals and soil organisms, especially microbes. This literature has developed because of the importance certain trace metals have in the nutrition of microorganisms (Perlman, 1949; Weinberg, 1970; Zajic, 1969), the importance of trace metals as pesticidal components for the control of microorganisms (Horsfall, 1956; Somers, 1961), and the relatively recently recognized significance of microorganisms in the environmental metabolism (transformation) of metals (Jernelöv and Martin, 1975; Saxena and Howard, 1977; Summers, 1978).

The extraordinary accumulation of trace metals, particularly lead, cadmium, zinc, and copper, in the organic horizon of the forest floor (Jackson et al., 1978a), as reviewed in Chapter 5, has led to the hypothesis that heavy metals depress decomposition rates. Tyler (1972) has proposed that decomposition of forest litter and remobilization of nutrients will be slower or less complete as heavy-metal ions bind with colloidal organic matter and increase resistance to decomposition or exert a toxic effect directly on decomposing microbes or the enzymes they produce.

d. Microbes

Many studies have been conducted that have examined the influence of lead on soil microbes. Doelman (1978) has provided a recent review. He concluded that lead impacts appear greater in sandy soils relative to clay or peat soils, that lead influences are restricted to short periods of time, and that soil functions resist alteration and tend to return to a steady state. He cautioned, however, that return to the steady-state condition may be accomplished with a different microbiota in a changed microhabitat in response to lead amendment. Doelman reviewed several studies that artificially applied lead salts to soils and then monitored the impact on fungal or bacterial species. Generally these studies employed lead concentrations ranging from 500 to 5000 ppm. The results were very variable and unfortunately somewhat ambiguous. In polluted soils, bacteria were less sensitive to lead than bacteria from non-lead-polluted soils. Gram-negative bacteria appeared to be less sensitive to lead than Coryneform bacteria. While Doelman did report that the growth rate of some strains of *Arthrobacter globiformis* might be decreased by lead concentrations as low as 1–5 ppm, the more general bacterial threshold of growth impact appeared to be in the range of 2000–5000 ppm lead. Almost no information is available on the in vivo or in vitro influence of lead on actinomycetes. Limited research on fungal reaction to soil lead show a variable response but generally support the suggestion that fungi are more resistant to lead influence than bacteria. Jensen (1976) reported that amendment of soil with 5000 ppm of lead nitrate reduced bacterial counts but increased fungal counts.

Cadmium, because it is one of the most toxic contaminants introduced into soil, has also received considerable attention by soil microbiologists and others. Babich and Stotzky (1978) have provided a most comprehensive review of the cadmium impact on microbiota. Their summary presents several pertinent generalizations. Cadmium toxicity to microbes appears potentiated at elevated soil pH, which may suggest a reduced significance in forest soils. The toxicity of cadmium to yeasts appears greater in aerobic than anaerobic conditions. Evidence has been presented to show that cadmium can decrease and prolong the logarithmic rate of microbes, reduce microbial respiration, inhibit formation of fungal spores, induce abnormal microbial morphologies, inhibit bacterial transformation, and reduce fungal spore germination. Bond et al. (1976) have employed microcosms to examine the influence of cadmium on the biota of forest litter. In experiments with Oregon Douglas fir litter amended with up to 10 ppm cadmium and observed for 4 weeks, they detected no change in bacterial and fungal densities.

In addition to lead and cadmium, zinc and copper have been implicated in altered decomposition rates. Despite the fact that these elements are required nutrients for normal fungal metabolism (Devi, 1962), their extraordinarily high concentrations in soils adjacent to metal smelters has proved to be inimical to microbial mineralization.

Jackson and Watson (1977) have studied the influence of lead, cadmium, zinc, and copper added to hardwood forest soil in the Clark National Forest from a lead smelter in southeastern Missouri. Annual deposition rates within 0.4 km from the smelter for lead, cadmium, zinc, and copper were 103, 0.72, 6.4, and 2.1 g m^{-2}, respectively. Soil samples were taken at 0.4, 0.8, 1.2, and 2.0 km intervals in line with the prevailing wind direction. Effects on accumulation of litter and the soil biota were not apparent at the 1.2- and 2.0-km sampling stations. At 0.4 and 0.8 km, however, there was an indication of depleted soil and litter nutrient pools, and evidence of depressed decomposer communities and nutrient translocation (Table 9-3). Significant accumulation of 02 litter (01 litter has original conformation of plant material, while 02 litter is fragmented and unrecognizable to species; after Lutz and Chandler, 1946) was measured at all sites along the transect relative to the control site (Figure 9-2).

Jordan and Lechevalier (1975) also recorded a significantly greater 02 horizon weight nearby, relative to some distance from, a zinc smelter in Palmerton, Pennsylvania. Within 2 km of the smelter, 13.5% zinc was measured in the 02 horizon and 8% zinc, 1500 ppm cadmium, 1200 ppm copper, and 1100 ppm lead

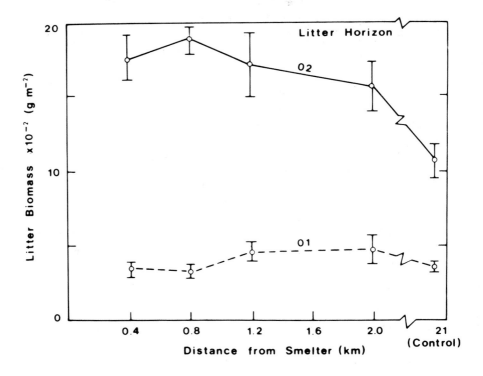

Figure 9-2. Biomass of forest floor litter of the Clark National Forest at Various distances from a lead smelter.
Source: Jackson and Watson (1977).

Table 9-3. Concentration of Trace Metals, Fungal Biomass, and Standing Oak Tree Foliar Calcium Concentration in the Clark National Forest at Various Distances from a Lead Smelter.

Distance from smelter (km)	Forest floor litter (μg g^{-1})				Fungal biomass (mg m^{-2})	Oak leaf calcium concentration (mg g^{-1})
	Pb	Cd	Zn	Cu		
0.4	88,349[a]	129[a]	2,189[a]	1,315[a]	0.8[b]	7.6[a]
0.8	30,420[a]	59[a]	917[a]	448[a]	2.4[b]	9.5[a]
1.2	11,872[a]	36[a]	522[a]	183[a]	5.8	11.6
2.0	6,856[a]	21[a]	351[a]	113[a]	16.1	10.4
21.0 (control)	398	2	111	26	22.1	13.0

Source: Jackson and Watson (1977).
[a] Significantly different from control ($P \leqslant 0.05$).
[b] Significantly different from control ($P \leqslant 0.1$).

were found in the A1 soil horizon. The total number of bacteria, actinomycetes, and fungi were counted by the dilution plate technique and were found to be greatly reduced in the most severely zinc contaminated soils when compared with control soils. The authors judged that the reduction of microbial populations may be a partial cause of the decreased rate of litter decomposition observed in the vicinity of the smelter. The thresholds of zinc toxicity for various microbial groups isolated from smelter influenced and-noninfluenced soils as determined by Jordan and Lechevalier are presented in Table 9-4.

Williams et al. (1977) compared the litter and soil microbiota from the site of an abandoned Wales metal mine and a control site 500 m distant. Retarded litter decomposition was observed at the mine site. The dilution plate count revealed depressed microbial populations in the soil, but not in the litter (Table 9-5).

Table 9-4. Minimum Zinc Concentration Toxic to Various Microorganisms as Determined on Pablum Extract Agar.

Microbe	Isolates from non-Zn-contaminated soil (ppm)	Isolates from Zn-contaminated soil (ppm)
Bacillus spp.	200	100-200
Non-spore-forming bacteria	100-200	600-2000
Actinomycetes	100	600-2000
Fungi	100-1000	100-2000

Source: Jordan and Lechavalier (1975).

Table 9-5. Lead and Zinc Concentrations and Dilution Plate Counts of Litter and Soil from the Site of an Abandoned Mine and a Control Pasture in Wales.

	Soil		Litter	
	Mine waste	Control	Mine waste	Control
	μg g^{-1}			
Lead	21,320	274[a]	14,207	172[a]
Zinc	1,273	79[a]	406	84[a]
	number g^{-1} x 10^5			
Bacteria	1.64	169[a]	504	247
Actinomycetes	0.42	70[a]	466	55
Fungi	0.10	69[a]	46	117

Source: Williams et al. (1977).
[a] Means of two sites significantly different ($P \leqslant 0.05$).

e. Soil Animals

Micro- and macrosoil animals are presumed not to play primary roles in the chemical decomposition of soil organic matter. As summarized earlier, however, they perform critically important changes in physical properties in the sequence of organic matter destruction. Fortunately, several investigators have included some of these organisms in their field assessments of trace metal impact.

Arthropod biomass in the 02 litter of the Clark National Forest was significantly reduced in those areas subject to excessive trace metal contamination by a lead smelter (Watson et al., 1976; Jackson and Watson, 1977). Significant reduction of arthropod predators, detritivores, and fungivores were recorded within 0.8 km of the smelter (Table 9-6). Williams et al. (1977) observed that the commonest animals in the litter of an abandoned mine site grossly contaminated with trace metals and a nearby control site were mites (*Acarina*) and springtails (*Collembola*) (Table 9-7). They observed fewer animals in the litter on the mine waste, primarily due to a considerable reduction in the mite population.

Table 9-6. Average 02 Litter Biomass of Arthropod Predators, Detritivores, and Fungivores in the Clark National Forest at Various Distances from a Lead Smelter.

Distance from smelter (km)	Biomass ($mg\ m^{-2}$)		
	Predator	Detritivore	Fungivore
0.4	2.1[a]	2.3[a]	0.8[a]
0.8	6.8[a]	16.6[a]	2.4[a]
1.2	14.0	12.6[a]	5.8
2.0	87.6	92.6	16.1
21.0 (control)	17.3	61.1	22.1

Source: Watson et al. (1976).

[a] Significantly different from control ($P \leqslant 0.1$).

Table 9-7. Microfauna Extracted from Litter from an Abandoned Mine Contaminated with Trace Metals and a Control Pasture in Wales.

Soil animal	Mine waste	Control
Acarina	30	239
Collembola	120	49
Coleoptera	3	7
Total	153	295

Source: Williams et al. (1977).

Collembola numbers were greater on the mine waste litter. Joosse and Buker (1979) also presented evidence supporting the notion that collembola may be tolerant of trace metal pollution. These investigators fed *Orchesella cincta* green algae contaminated with more than 10,000 ppm lead and analyzed collembola collected from a roadside environment and concluded that high levels of lead in the food of collembola had no apparent harmful short-term effect. While they did not specify the contaminants involved, Singh and Tripathi (1978) have recorded fewer soil microarthropods (28,107 m^{-2}) close to, as opposed to away from, (39,217 m^{-2}) a chemical plant in Varanasi, India. Acarina were the most prevalent group in both the control and polluted sites, and comprised approximately one-half the total fauna in both cases.

2. Soil Processes

Considerable evidence to support the hypothesis that trace metal contamination retards decomposition and nutrient recycling has been provided from studies that have measured soil processes in the laboratory or field.

a. Respiration

The level of aerobic respiration is the most widely used measurement for the general biological activity of soils. Changes in oxygen consumption or carbon dioxide evolution have been extensively employed to assess heavy metal impact on the soil biota. Rühling and Tyler (1973) made comparisons of the carbon dioxide evolution rates of different fractions of spruce needle litter from numerous sites around two metal processing industries in central and southeastern Sweden emitting copper, zinc, cadmium, nickel, vanadium, and lead. Under controlled laboratory conditions, highly significant negative correlations were observed between carbon dioxide release and high concentrations of lead, nickel, cadmium, and vanadium. The authors concluded that the decomposition of acid forest litter would be limited by elevated concentrations of those heavy metals during those periods of the year when soil moisture and temperature were not limiting factors. Ebregt and Boldewijm (1977) have also reported a negative correlation between soil respiration and lead concentration in spruce forest soil. Bhuiya and Cornfield (1972) recorded carbon dioxide release from soils amended with straw and 1000 ppm of various trace metals. Zinc appeared to have no influence on soil respiration, but carbon dioxide evolution was decreased slightly by lead and to a considerable extent by copper and nickel.

The influence of cadmium on forest soil respiration has been investigated by Bond et al. (1976) using Douglas fir litter from the Oregon Coast Mountain Range in microcosms. At various moisture contents, 10 ppm cadmium reduced oxygen and carbon dioxide respiration by 40%. At 0.01 ppm amendment, oxygen consumption was stimulated! Spalding (1979) has also examined carbon dioxide evolution from Douglas fir needle litter. In the laboratory, mercury, cadmium, lead, nickel, zinc, and copper chlorides were added to the litter at rates of

10, 100, and 1000 $\mu g \ g^{-1}$. At 100 ppm only mercury exhibited an inhibitory effect. At 1000 ppm, however, all metals except lead inhibited respiration.

Several studies have presented contrary evidence and have indicated that even relatively high concentrations of trace metals do not appear to reduce respiration in certain soils. Doelman (1978) amended two sandy soils, a clay soil, and a peat soil with lead chloride under controlled environmental conditions. While 2000 ppm lead chloride reduced oxygen consumption in the sandy soils by 50%, the clay soil decrease was only 15%, and the respiration of the peat soil was not inhibited at all. An increase of lead chloride to 10,000 ppm did not influence the respiration of the peat soil. A concentration of 5000 ppm lead did not reduce the carbon dioxide production of a clay soil enriched with glucose and ammonium nitrate in experiments reported by Mikkelsen (1974). This is consistent with the data of Fujihara et al. (1973), who could not detect any inhibition in carbon dioxide release from a silt loam soil enriched with starch and ammonium nitrate and amended with 100 ppm lead.

b. Nitrification

Nitrification is the process of ammonium oxidation to nitrate in soils. It is generally assumed that the microbes primarily responsible are chemoautotrophic bacteria belonging to the *Nitrosomonas* and *Nitrobacter* genera. Because most varieties of these bacteria have pH optima at 7 or above, their importance in acid forest soils has been questioned and fungi may be important nitrifying agents in the latter situation (McLaren and Peterson, 1967; Wilde, 1958). Belser (1979) has recently reviewed the wide array of factors that may influence the population size of nitrifying bacteria. There is some evidence to suggest that trace metal contamination of soil should be added to Belser's list. Morissey et al. (1974) amended a fine sandy loam with various concentrations of lead and incubated it at 30°C for 6 weeks; at 500 ppm nitrification was initially lowered but recovery occurred in 2 weeks, at 1000 ppm the inhibitory effect disappeared after 6 weeks; and it took 10,000 ppm lead to permanently suppress nitrification. Bhuiya and Cornfield (1972) found the autotrophic nitrification process to be more sensitive than the heterotrophic mineralization process. If this is true, forest soils may be more resistant than higher pH soils to altered nitrification occasioned by heavy metals. Tyler et al. (1974) amended a Swedish meadow soil with various moderate concentrations of lead, cadmium, and sodium salts and observed significantly increased nitrification after laboratory incubation for 2–8 weeks. Liang and Tabatabai (1977), however, examined the laboratory response of the nitrification rate of four different soils and observed that amendment with any one of 19 trace elements did not stimulate nitrogen mineralization. Silver, mercury, and cadmium, in fact, inhibited nitrification in all four of the soils tested. In the most acid soil examined, nickel, chromium, iron, aluminum, boron, and tin also inhibited nitrification. Wilson (1977) has presented evidence that nitrification can be significantly inhibited in soils amended with sludge containing high concentrations of lead, cadmium, and zinc. At a concentration of 5 $\mu moles \ g^{-1}$ of soil,

Table 9-8. Average Percentage Inhibition of Nitrification in Three Soils by Various Trace Elements at 5 μmoles g^{-1} following 10 Days Incubation under Laboratory Conditions.

Element	% inhibition	Element	% inhibition
Mercury	96	Cobalt	45
Silver	93	Copper	45
Selenium	91	Tin	41
Arsenic (III)[a]	83	Iron (II)	40
Chromium	81	Zinc	40
Boron	80	Vanadium	37
Cadmium	77	Arsenic (V)	37
Nickel	64	Tungsten	33
Aluminum	60	Manganese	24
Molybdenum	54	Lead	14
Iron (III)	49		

Source: Liang and Tabatabai (1978).
[a] Oxidation state.

trace metals inhibited nitrification in three soils an average of 14–96% (Table 9-8) following laboratory incubation for 10 days (Liang and Tabatabai, 1978).

A few recent investigations have employed litter bags and microcosms to examine trace metal influence on soil decomposition processes. These studies supplement laboratory research and strengthen attempts to extrapolate to natural environments. Inman and Parker (1978) examined decomposition rates of black oak, quaking aspen, and starry false Solomon's seal litter in urban and rural ecosystems in northwestern Indiana. Surface soils of the urban site averaged 10, 2456, 463, and 119 ppm of cadmium, zinc, lead, and copper, respectively. These concentrations are appreciably less than those of the smelter studies discussed previously. Black oak litter at the rural site lost twice as much weight as that of the urban site after 11 months. Solomon's seal and quaking aspen litter at the rural site also lost more weight than at the urban site after 6 and 7 months, respectively. While the authors felt that trace metal contamination, especially by copper and lead, may have been involved in the observed depression of decomposition rate, multiple regression analysis did not support this suggestion.

Jackson et al. (1978b) employed a microcosm to evaluate the impact of heavy metals on a forest soil. Baghouse dust from a Missouri smelter was added to a hardwood forest soil to approximate one annual deposition of metals at a distance of 0.4 km from the smelter. At this high level of contamination, export of essential elements in soil leachate was increased as a result of heavy metal impaction. Extractable nutrient pools in soil were also lowered at the end of 20 months. Using soil cores as microcosm analogs, Jackson et al. (1977) observed nutrient efflux from forest cores treated with 100 μg arsenic cm^{-2}. While they detected no disturbance to micropopulations of the soil, they did record a significant increase in the loss of calcium and nitrate oxygen.

3. Soil Enzymes

At the turn of the century, the first reports of extracellular soil enzyme activity (catalase) appeared in the literature. Since that time a very large number of enzymes have been shown to participate in important extracellular soil processes. These extracellular activities include not only free extracellular enzymes and enzymes bound to inert soil components, but also active enzymes within dead cells and others associated with nonliving cell fragments. While the sources of these enzymes are believed to include animal, plant, and microbial organisms, the latter are generally recognized as the most important source. Additional general information on soil enzymes may be found in the fine review edited by Burns (1978).

In recognition of the interference heavy metals may exert on enzyme activity, the relationship between selected soil enzymes and trace metals has been examined. Metal ions may inhibit reactions by complexing the substrate, by combining with the active group of the enzymes, or by reacting with the enzyme-substrate complex.

The activity of at least 10 forest soil enzymes have been evaluated with respect to their susceptibility to influence by trace metal contaminants in soil (Table 9-9).

a. Dehydrogenase

Dehydrogenases are unspecific oxidative enzymes that are produced by a variety of soil organisms and catalyze the transfer of hydrogen from organic substances to molecular oxygen. Rühling and Tyler (1973) examined dehydrogenase activity in spruce needle litter they collected from various sites around metal

Table 9-9. Forest Soil Enzymes That Have Been Evaluated for Influence by Trace Metal Soil Contaminants.

Enzyme	Reference
Amylase	Spalding (1979), Ebregt and Boldewijm (1977)
Arylsulfatase	Al-Khafaji and Tabatabai (1979)
β-Glucosidase	Tyler (1974), Spalding (1979)
Cellulase	Spalding (1979)
Dehydrogenase	Rühling and Tyler (1973)
Invertase	Spalding (1979)
Phosphatase	Tyler (1974), Tyler (1976), Juma and
Acid phosphatase	and Tabatabai (1977)
Alkaline Phosphatase	
Polyphenoloxidase	Spalding (1979)
Urease	Tyler (1974), Bondietti (1976), Williams et al. (1977)
Xylanase	Spalding (1979)

smelters in central and southeastern Sweden. Under conditions of laboratory analysis, a strong negative correlation was measured between dehydrogenase activity and high concentrations of nickel, lead, cadmium, and vanadium. The utility of dehydrogenase measurements as an indicator of general soil biological activity is unclear. Some investigators have found a positive correlation between dehydrogenase activity and either the rate of carbon dioxide production or oxygen uptake, while others have not (Burns, 1978).

b. Phosphatase

Phosphatases are important soil enzymes that catalyze the decomposition of organic phosphorus compounds in soil. These latter compounds may constitute 30–70% of the total soil phosphate. Phosphatases apparently arise from widely diverse sources and have different pH optima (Burns, 1978).

Acid phosphatase activity of spruce litter was markedly inhibited by high concentrations of copper and zinc (Tyler, 1974, 1976). A highly significant inverse linear relationship between the log of the sum of copper and zinc litter

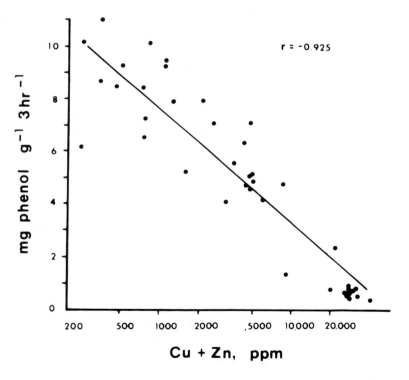

Figure 9-3. Relationship between copper + zinc concentration and phosphatase activity of the mor horizon of a conifer forest surrounding a brass mill in Sweden. Source: Tyler (1976).

concentration and phosphatase activity was presented (Figure 9-3). Tyler (1976) judged that at equal concentrations, copper was more toxic than zinc. Concentrations of vanadium added to spruce litter at 30, 50, 100, and 1000 mg kg^{-1} inhibited phosphatase activity by 20, 40, 47, and 68%, respectively.

Juma and Tabatabai (1977) studied the influence of 20 trace elements on the activity of acid and alkaline phosphatases from three soils of various physical and chemical properties. The relative effectiveness of trace element inhibition was found to be dependent on the soil and enzyme type. At 25 μmoles g^{-1} soil, the most effective average inhibition of acid phosphatase in the three soils was by mercury, arsenic, tungsten, and molybdenum. The most effective inhibitors of alkaline phosphatase activity in soils were silver, cadmium, vanadium, and arsenic.

As in the case of dehydrogenase, attempts to relate soil phosphatase activity to microbial numbers have yielded contradictory results (Burns, 1978).

c. b-Glucosidase

b-Glucosidase is an important member of the soil carbohydrases. In his investigation of spruce litter enzymes, Tyler (1974) recorded that b-glucosidase activity was not measurably reduced at concentrations of copper and zinc up to 40 mg g^{-1}. Spalding (1979) measured extractable enzyme activity of Douglas fir needle litter subject to mercury, cadmium, lead, nickel, zinc, and copper chlorides at rates of 10, 100, and 1000 μg g^{-1} at 1-day, 2-week, and 4-week intervals following treatment with 1000 μg g^{-1} mercury and after 4 wks by the 100 μg g^{-1} mercury amendment. Initially 1000 μg g^{-1} cadmium stimulated b-glucosidase activity, but this effect was not maintained at 2 to 4 weeks.

d. Urease

Because of the agricultural importance of urease as a decomposing agent for urea, which is widely employed as a fertilizer, urease has been the most widely studied soil enzyme. It occurs widely in higher plants and microbes and catalyzes the hydrolysis of urea to carbon dioxide and ammonia. The primary source of soil urea is presumed to be microbial (Burns, 1978).

As in the case of acid phosphatase, Tyler (1974) found a highly significant negative correlation between urease activity in spruce litter and high concentrations of copper and zinc. In their comparison of mine waste soils, contaminated with excessive zinc and lead and uncontaminated pasture soil, Williams et al. (1977) evaluated urease activity. They observed no significant difference in the enzyme activity of the two soils. If urea was added to the two soils, however, the ammonium nitrogen released was significantly greater from the pasture relative to the mine waste soil. Even though the Williams study did not include a forest soil, it is pertinent because of the increasing use of urea as a forest fertilizer.

Several studies employing water-soluble metallic salts to determine the influence on urease activity generally rank, in decreasing order of impact on the enzyme, as follows (50 ppm soil basis): silver > mercury > gold > copper > cobalt > lead > arsenic > chromium > nickel (Bremner and Mulvaney, 1978).

e. Amylase

Alpha- and beta-amylases accumulate in soil and hydrolyze starch. Spalding (1977) has observed significant correlations between carbon dioxide evolution rates from coniferous litter samples and amylase activity. Ebregt and Boldewijm (1977) assessed heavy metal contamination from a brass foundry at Gusam, Sweden on starch decomposition in spruce litter. A linear, negative correlation was determined between amylase activity and the sum of copper, zinc, lead, and cadmium concentrations of the soil. Although amylase activity exhibited the best correlation with respiration of coniferous leaf litter (Spalding, 1977), it was influenced only by cadmium or lead at 1000 μg g^{-1} after 4 weeks in subsequent studies by this author (Spalding, 1979).

f. Cellulase

Cellulolytic activity appears to be inducible in soils and soils may not have measurable amounts of extracellular cellulases. In Spalding's (1979) study, cellulase activity in Douglas fir litter decreased in samples treated with 1000 μg g^{-1} mercury. Reduced activity was again observed after 4 weeks at 1000 μg g^{-1} of both mercury and cadmium.

g. Xylanase

Xylanase activity, which is strongly correlated with cellulase activity (Spalding, 1977), was depressed by mercury at a concentration of 1000 μg g^{-1} in Douglas fir litter after 4 weeks (Spalding, 1979).

h. Invertase

Invertase mediates the hydrolysis of glucose to fructose. Invertase activities in soil are not consistently related to microorganisms present in soils (Burns, 1978). Spalding (1977) found that extractable invertase activity was unrelated to respiration rates of coniferous litter. In his trace metal tests with Douglas fir litter, Spalding (1979) found invertase activity was severely decreased by mercury at both 100 and 1000 μg g^{-1}, and was sustained for 2 and 4 weeks in the latter treatment. Invertase activity was initially stimulated by 10 or 1000 μg g^{-1} zinc and cadmium, but this effect was not sustained at 2 and 4 weeks.

i. Arylsulfatase

Arylsulfatase catalyzes the hydrolysis of the arylsulfate anion. Its presence in soils is believed responsible for sulfur cycling. Al-Khafaji and Tabatabai (1979) studied the influence of 21 trace elements on arylsulfatase activity in four soils. When the trace elements were compared by employing 25 μmoles g^{-1} of soil, the average inhibition was highest with silver, mercury, boron, vanadium, and molybdenum.

It should be clearly kept in mind that the specific abundance and activity of extracellular enzymes in soil are extremely variable and are dependent on vegetation, season, depth of sampling, fertilization, and pesticide use (Burns, 1978). Doelman (1978) has further observed that enzymes in soil are known to be more resistant to inactivation by various inhibitory agents than enzymes tested in vitro.

4. Direct Tree Toxicity

The direct influence of heavy metal ions on tree metabolism, especially root physiology, has the potential to reduce nutrient uptake.

All of the heavy metals, biologically essential or nonessential, can be toxic to forest trees at some threshold level of dose. Only a few metals, however, have been documented to cause direct phytotoxicity in actual field situations. Copper, nickel, and zinc toxicities have occurred frequently. Cadmium, cobalt, and lead toxicities have occurred less frequently and under more unusual conditions. Chromium, silver and tin — in solution culture under experimental conditions — have not been demonstrated to be phytotoxic in field situations even at high doses (Foy et al., 1978).

Direct heavy metal phytotoxicity will result only if the metal can move from the soil to the root or from the plant surface to the plant interior. Movement via diffusion or mass flow of heavy metals from soils to roots, from roots to shoots, or from plant exteriors (leaf surfaces) to plant interiors remains poorly characterized under natural conditions. Heavy metals that do enter plants and cause abnormal physiology do not cause unique or specific symptoms that are useful in field diagnoses.

General metal toxicity symptoms include hypotrophy (growth suppression — stunting) and chlorosis (yellowing). Metal-induced hypotrophy may result from toxicity to one or more specific metabolic pathways, uptake antagonism with one or more required nutrients, and/or inhibition of root growth and development. Initial symptoms of heavy metal stress are frequently associated with root tips. Lateral root development may be acutely restricted. Root length restriction may cause the uptake of phosphorus, potassium, and iron to drop below levels necessary to sustain normal metabolism. Chlorosis caused by heavy metal toxicity appears symptomatically similar to iron-deficiency chlorosis. Evidence is available to link chlorosis due to excess cadmium, copper, nickel, and zinc with interference with foliar iron metabolism (Foy, 1983; Foy et al., 1978).

Direct and acute heavy metal toxicities have been described for forest trees only in the immediate vicinity of point sources. Lower heavy metal exposures associated with regional-scale deposition are linked primarily with demonstrated and potential interactions with nutrient cycling processes, as reviewed by Smith (1984, 1985) and as summarized in the following section.

5. Heavy Metal Availability

Direct toxicity to trees and direct and indirect toxicity to other biotic components of the forest soil by heavy metals are exposure-dependent. Exposure is a function of both deposition from the atmosphere and chemical availability. Heavy metals not available for ready exchange from binding sites or in solution are not available for root or microbial uptake. More than 90% of certain heavy metals deposited from the atmosphere may be biologically not available.

Heavy metals may be adsorbed or chelated by organic matter (humic, fulvic acids), clays, and/or hydrous oxides of aluminum, iron, or manganese. Heavy metals also may be complexed with soluble low-molecular weight compounds. Soluble cadmium, copper, and zinc may be chelated in excess of 99%. Adsorbed heavy metals remain in equilibrium with chelated metals (Foy, 1983). Heavy metals may also be precipitated in organic compounds of low solubility such as oxides, phosphates, or sulfates. Miller and McFee (1983), for example, have suggested that lead may be present in the soil profile in the following forms: bound to organic matter, 43%; bound to ferro-manganese hydrous oxides, 39%; as insoluble precipitates, 10%; and biologically available (exchangeable), 8%.

Adsorption, chelation, and precipitation are strongly regulated by soil pH. As pH decreases and soils become more acid, heavy metals generally become more available for biological uptake. Natural forest soils generally become more acid as they mature. Acidification in excess of natural processes is possible, especially in soils with a pH greater than 5. Under this circumstance, soil acidification associated with acid deposition may result in increased biological availability of heavy metals in the forest floor.

Rhizosphere processes also may transform heavy metals from an unavailable pool to an available pool (Smith, 1987). The single most important zone of the soil ecosystem for forest tree health and growth is the rhizosphere. This zone, while only extending approximately 2 mm from root surfaces, is a unique environment and has extraordinary significance for nutrient dynamics and root metabolism. Microorganisms are stimulated in the rhizosphere because of the ready availability of organic nutrients from plant mucilages, mucigel, lysates, secretions, and root exudates. While bacteria and fungi exhibit the greatest stimulation, a large number of additional organisms are also enhanced. Chemical and physical properties of rhizosphere soil also differ from these properties in soil devoid of roots. Typically, rhizosphere soil has lower pH, lower water potential, lower osmotic potential, lower redox potential and higher bulk density than soil away from roots (Table 9-10). These characteristics enable the rhizosphere to exert dominant regulation over element uptake by forest tree roots, root disease re-

Table 9-10. Biological, Chemical, and Physical Differences of Bulk Soil (soil devoid of roots) and Rhizosphere Soil (soil in the immediate vicinity of roots).

	Rhizosphere Soil	Bulk Soil
Biological		
Microorganisms	diverse and abundant	limited and less
Respiration	high	low
Symbiont habitat	excellent	poor
Mycorrhizas	abundant	non existent, dormant
Root pathogens	active	dormant
Chemical		
Low molecular wt organic compounds	diverse and abundant	restricted
Organic acids	nonhumified	humified
pH	lower	higher
Water potential	lower (more negative)	higher (less negative)
Osmotic potential	lower	higher
O_2/CO_2 ratio	lower	higher
Redox potential	lower	higher
Physical		
Pore space	less	more
Bulk density	higher	lower

sulting from biotic agents, and microbial root symbiont and saprophyte ecology. Atmospheric deposition may alter rhizosphere regulation of heavy metal uptake. It is generally assumed that heavy metals accumulated from the atmosphere in forest floors are retained in bulk soil in largely insoluble form as hydroxide, sulfide, or other low solubility precipitates or are strongly chelated by humic acids. In this form, the heavy metals are not available for root or microbial uptake and the potential for toxicity is presumed low. The rhizosphere processes of pH regulation, solubilization, reduction, and complexation, however, may transform unavailable metals to available metals. Rhizosphere processes that may be particularly important in lead uptake include pH, hydrogen ion availability, organic acid availability, and phosphate availability. Organic acids capable of complexing copper and zinc in the rhizosphere may be important in root uptake. Manganese reduction to the divalent form may allow excess uptake of this heavy metal. Complexing agents and reduced rhizosphere pH may also facilitate cadmium uptake. Again, any alteration of rhizosphere processes caused by chemical changes induced by air pollution could easily alter the availability and uptake of these potential toxic elements.

B. Hydrogen Ion Acidity and Toxicity

Comprehensive reviews of soil acidity and hydrogen ion dynamics in soil solutions are available (McFee et al., 1984; Rosenquist et al., 1980; Krug and Frink, 1983; Ulrich, 1986; Van Breeman et al., 1984; Legge and Crowther,

Table 9-11. Soil Acidity Designations by pH.

Descriptor	pH
Extremely acid	below 4.5
Very strongly acid	4.5–5.0
Strongly acid	5.1–5.5
Medium acid	5.6–6.0
Slightly acid	6.1–6.5
Neutral	6.6–7.3
Mildly alkaline	7.4–7.8
Moderately alkaline	7.9–8.4
Strongly alkaline	8.5–9.0
Very strongly alkaline	9.1 and above

1987; Reuss and Johnson, 1986). Total soil acidity is the sum of undissociated forms of acidity in soil plus free hydrogen ions. The degree of soil acidity is typically represented by pH values that are related to the activity of free hydrogen ions in solution. Values for pH of soil horizons are extremely important for soil classification, identification of soils in the field, and assessment of risks from hydrogen ion toxicity. Generally soil horizons vary in pH from a little below pH 3.5 to a little above pH 9.5 (Table 9-11). Temperate zone forest soils are generally classified as acid soils and have a pH of 6.5 or less.

1. Forest Soil Acidity and Acidification

Natural processes make forest soils acid. These processes act by controlling the chemistry of the soil cation-exchange complex. This complex consists of negative charges located on clay minerals in the soil or on soil organic matter. On clay minerals, negative charges generally result from substitution, within the mineral lattice, of a cation of lower positive charge for one of higher charge. On organic matter, negative charges result from the ionization of hydrogen ions from carboxyl, phenol, and enol groups (Reuss and Johnson, 1986). In acid mineral soils the cation-exchange complex is typically dominated by aluminum species, for example, Al^{3+}, $Al(OH)^{2+}$, and $Al(OH)_2^+$, formed by the dissolution of soil minerals. In acid organic soils, the hydrogen ion may dominate the cation-exchange complex (Reuss and Johnson, 1986). Processes that acidify forest soils include those that act to increase the number of negative charges, such as organic matter accumulation or clay formation, or those that remove basic cations, such as leaching of bases in association with an acid anion. In forest soils weathering by carbonic acid, organic acids (podzolization), humification, and cation uptake by roots all act to increase the negative charge (Krug and Frink, 1983).

Reuss and Johnson (1986) have emphasized that "soil acidification" is a complex of processes and cannot be quantitatively described by any single index. They emphasize the utility of using "capacity" and "intensity" factors. *Capacity* refers to the storage of hydrogen ions, aluminum ions, or base cations on the soil exchange complex or in weatherable minerals. *Intensity* refers to the soil solution concentration, at any point in time, in the case of the hydrogen ion, it refers to the soil solution pH. In forest regions receiving acid deposition from the atmosphere, Reuss and Johnson (1986) suggest, in terms of capacity factors, that the most likely effect will be an increase in the exchange acidity and a reduction of exchangeable bases (Figure 9-4). The former is increased directly by hydrogen ion input or, more likely, by increasing exchangeable aluminum through the reaction of hydrogen ions with soil minerals. The latter, reduction of exchangeable bases, results via replacement of base cations on the exchange complex by aluminum species. The base cations are subsequently leached from the soil horizon in association with strong acid anions (Chapter 10).

The evidence for acidification of soils at the present rates of acid deposition is not great. McFee et al. (1984) conclude that over the next few decades the risk of significant acidification is restricted to those limited soil types characterized by no renewal by fresh soil deposits, low cation exchange capacity, low clay and organic matter content; low sulfate adsorption capacity, high input of acidic de-

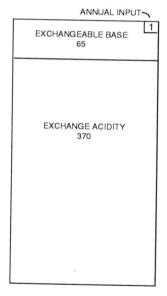

Figure 9-4. Forest soil pool sizes of exchange acidity (hydrogen + aluminum species) and exchangeable bases relative to annual hydrogen ion input via acid deposition. Assumptions include an input of 1000 μm pH 4.2 rainfall plus an equal amount of acidity in dry deposition to 300 mm soil with bulk density 1.2, cation exchange capacity (CEC) of 0.15 eq kg^{-1}, and 15% base saturation. Source: Reuss and Johnson (1986).

position without significant basic cation deposition, high present pH (5.5–6.5), and deficiency of easily weatherable materials to 1 m depth.

2. Soil Organisms

Soil acidification affects numerous components of the soil ecosystem, but it is extremely difficult to distinguish hydrogen ion toxicity effects from associated effects resulting from acid-induced changes in soil environments and effects associated with anionic components of deposition, nitrate and sulfate.

The general sensitivity of blue-green algae to acidity suggests they may be susceptible to stress imposed by acidic deposition (McFee et al., 1984). Green algae appear to be less sensitive (Legge and Crowther, 1987). The presumed dominance of fungi over bacteria in acid forest soils may partially result from a greater sensitivity of heterotrophic bacteria to hydrogen ion stress (Alexander, 1980). Fungi exhibit little sensitivity to pH change unless it is extremely high or low. Soil bacteria, except for sulfur-oxidizing bacteria, and actinomycetes are inhibited at pH levels below 5.0. Protozoans appear to have little sensitivity to pH except under extreme conditions similar to fungi (Legge and Crowther, 1987).

Myrold (1987) has provided a comprehensive review of hydrogen ion impacts on soil populations. He emphasizes that most forests currently receive less than 1 kg H^+ ha^{-1} yr^{-1} (Chapter 2). At this rate, impacts on soil organisms are judged to be generally insignificant, although some shifts in species composition have be recorded. At moderate (1–10 kg H^+ ha^{-1} yr^{-1}) rates of input, effects on soil populations are more common. With regard to mesofauna, most studies have concentrated on Enchytraeidae (potworms), Collembola (springtails), and Acarina (mites). Studies with artificial precipitation generally suggest decreases in potworm numbers at moderate to high (>10 kg H^+ ha^{-1} yr^{-1}) input, but increases in springtails and mites (Table 9-12) (Myrold, 1987). With regard to microorganisms, the Myrold review is consistent with other observations. Bacterial populations are subject to impact especially at moderate to high deposition rates and especially in the forest floor horizon. Populations may be reduced, unchanged in numbers, or altered in species composition. Low to moderate deposition rates have increased starch hydrolyzers and spore formers and decreased nitrite-oxidizers, denitrifiers, and gram negative bacteria. Fungal populations, on the other hand, are generally not influenced by moderate to high acidic inputs, even in the upper soil horizons.

3. Soil Processes

All microbially mediated processes, excepting sulfur oxidation, operate optimally at neutral or nearly neutral pH levels. Specialized bacteria conducting nitrogen fixation and nitrification appear to be sensitive to acidity while fungal processes such as ammonification and decomposition are not (Legge and

Table 9-12. Impact of Acidic Input on Soil Animal Populations.

Organism	H$^+$ load (kg ha^{-1} yr^{-1})[a]		
	< 1	1–10	> 10
Enchytraeidae			
Cognettia sphagnetorum	o	--	--
Total	o	--	--
Collembola			
Anurida pygmaea	+	o	o
Anurophorus binoculatus	o	-	
Anurophorus septentrionalis	o	o	
Folsomia litsteri	o	-	
Isotoma notablilis	+	-	--
Isotomiella minor	-	-	--
Mesaphorura yosii	+	+	++
Neanura muscorum	o	o	+
Neelus minimus	-	o	--
Onychiurus absoloni	+	o	
Onychiurus armatus	o	o	--
Tullgergia krausbaueri	+	++	
Willemia anophthalma	-	-	-
Xenylla borneri	-	-	
Total	+	+	+/-
Acarina			
Astigmata	o	+/-	+
Cryptostigmata (Oribatei)			
Brachychthoniidae spp.	-	+/-	+/-
Caranbodes spp.	o	-	
Oppia nova	o	+/-	+/-
Oppia obseleta	+/-	+/-	++
Oppia ornata	o	o	o
Oppioidea spp.	o	+	
Steganacarus spp.	o	+	++
Suctobelba spp.	o	-	-
Tectocepheus velatus	+	++	++
Total	+	++	++
Mesostigmata	o	o	o
Protostigmata	o	+/-	+/-
Total	+	+	+
Protura			
Total	o	o	o

[a] ++ significant increase   - decrease
   + increase   -- significant decreas
   o no difference   +/- mixed results in different studies
Source: Myrold (1987).

Table 9-13. Threshold Soil pH Levels that Impact Various Microbial Groups and Microbially Mediated Soil Processes

Microbial Group or Soil Process	pH
Nitrogen fixation	6.0
Nitrification	6.0
Ectomycorrhizal	uncertain
Vesicular-abuscular mycorrhizal	6.0
Organic decomposition	2–4
Ammonification	3.0
Soil respiration	3.0
Carbon mineralization	3.0
Community structure	3–4
Soil enzymes	~2.0 (uncertain)

Source: Visser et al. (1987).

Crowther, 1987). Nitrification in forest soils may not be reduced by acidity because of the acid tolerance of heterotrophic nitrifiers (McFee et al., 1984). Sulfur oxidation appears insensitive to soil pH since it can be performed by acid-tolerant bacteria and fungi. Visser et al. (1987) have comprehensively reviewed acid deposition effects on soil microbes and microbial processes and have established thresholds of soil pH impacts on organism types and soil process (Table 9-13). The thresholds of Table 9-13 were generated in laboratory studies and may not reflect thresholds characteristic of natural soils.

In his comprehensive review of acidity impacts on soil processes, Myrold (1987) has summarized studies concerned with soil respiration, nitrogen dynamics, and soil enzymes. Laboratory studies have generally concluded that low to moderate acid inputs do not affect or tend to enhance litter decomposition. Most field studies found no impact on decomposition, even at high deposition rates. Net nitrogen mineralization was generally unimpacted by acid deposition. Autotrophic nitrification, on the other hand, was typically inhibited by acid inputs. Dehydrogenase and cellulase exhibited stimulation at low deposition rates, mixed effects at moderate rates, and typically depressed activities at high rates. Urease and protease, and arylsulfatase activity have been shown to decrease with high input but to be unimpacted at low to moderate deposition rates. Phosphatase was generally inhibited at high deposition rates and was influenced by low and moderate rates in a variable manner (Myrold, 1987).

4. Direct Tree Toxicity

Hydrogen ions are not presumed to be directly toxic to tree roots. Toxic element release, for example aluminum or manganese, associated with hydrogen ion increase, however, may be significant. If soil pH is low enough (< pH 5.0–5.5) in mineral soils to cause the dissolution of aluminum-and manganese containing soil minerals, hydrogen ion input will increase the release of aluminum and manganese to the soil solution.

C. Aluminum Toxicity

Aluminum toxicity may influence forest tree root health where acid deposition plus natural acidifying processes increase soil acidity enough to increase the amount of aluminum available for root uptake.

1. Aluminum in Forest Soils

Aluminum is the third most abundant element in the crust of the earthwhere it occurs primary in aluminosilicate minerals, most commonly as feldspars in metamorphic and igneous rocks, and as clay minerals in weathered soils. In northern temperate zones, podzolic forest soils naturally mobilize aluminum with organic acids resulting from the decomposition of organic matter and leach the aluminum from upper to lower soil horizons where it precipitates (McFee et al., 1984). Despite high concentrations of total aluminum in forest soils, aluminum toxicity is generally presumed unimportant, as the element is present in tightly bound or insoluble form and is not available for root uptake.

2. Aluminum Mobilization by Acid Deposition

Atmospheric additions of strong mineral acids to forest soils may mobilize aluminum by dissolution of aluminum containing minerals or may remobilize aluminum previously precipitated within the soil during podzolization or held on soil exchange sites (Cronan and Schofield, 1979; Ulrich et al., 1980). The chemistry of soil aluminum is complicated and is comprehensively reviewed in David and Driscoll (1984), Lee (1985), Ubet (1987), Ulrich (1987), and Driscoll (1984). The forms (speciation) and dynamics of aluminum equilibria in forest soils are not fully clear. Nilsson (1981) has indicated that either amorphous or crystalline basaluminite, $Al_4 (OH)_{10} SO_4$, seems most likely to be the mineral controlling aluminum solubility in the B horizon of certain Swedish podzols and acid brown forest soils. Other aluminum oxyhydroxides that have been suggested to regulate Al^{3+} activity in soil solution include nordstandite and pseudoboehmite. Dissolution of the solid phase gibbsite [$Al (HO)_3$] is commonly used in calculations of soil pH buffering and Al^{3+} activity in soil solutions (Wolt, 1987).

It is clear that only dissolved forms of aluminum will be readily moved in the soil system and will be accessible for root uptake. In fresh water, dissolved aluminum is known to exist mainly as free ions (Al^{3+}) and as complexes with hydroxide, fluoride, sulfate and dissolved organic matter. In aquatic systems, the formation of mononuclear (e.g., $Al OH^{2+}$, $Al (OH)_2{}^+$, $Al (OH)_4{}^-$) and polynuclear (e.g., $Al_2 (OH)_2^{4+}$, $Al_6 (OH)_{15}^{3+}$, $Al_{10} (OH)_{22}^{8+}$) hydroxide complexes have been intensively studied. Monomeric species are concluded to be especially important and, as a result, the following equilibria are assumed to be especially significant (Lee, 1985):

$$Al\,OH^{2+} + H^+ \rightleftharpoons H_2O + Al^{3+}$$
$$Al\,(OH)_2^+ + 2H^+ \rightleftharpoons 2H_2O + Al^{3+}$$
$$Al\,(OH)_4^- + 4H^+ \rightleftharpoons 4H_2O + Al^{3+}$$
$$Al\,F^{2+} \rightleftharpoons F^- + Al^{3+}$$
$$Al\,SO_4^+ \rightleftharpoons SO_4^{2-} + Al^{3+}$$
$$HF \rightleftharpoons H^+ + F^-$$
$$Al\text{-organic}^{(n-3)} \rightleftharpoons org^{\,n-} + Al^{3+}$$

Soil aluminum speciation is regulated by ligand abundance and pH. As forest soil pH falls below 5, the rate of aluminum release from silicate lattices, especially clay minerals, becomes greater. The liberated aluminum is transformed primarily into polynuclear hydroxide cations and Al^{3+} ions. The polynuclear hydroxide cations are soluble and become attached to negative charge sites but are nonexchangeable. Their residence on exchange sites results in a reduction of cation exchange capacity (CEC). Aluminum ions (Al^{3+}) also become attached to negative charges sites, but are exchangeable. Due to the very selective binding of Al^{3+} on exchangeable sites, Al^{3+} saturation can reach high levels at relatively low Al^{3+} concentrations in the soil solution. At pH < 4.2, the solubility of Al-hydroxide compounds increases to an extent that Al^{3+} may become the dominating cation in the soil solution (Ulrich, 1986, 1987).

It is important to emphasize that bulk soil (soil devoid of roots) chemistry may be substantially different from rhizosphere (soil in the immediate 2 mm vicinity of roots) soil and that cation availability may be higher in the latter relative to the former. Rhizosphere soil is dramatically different from non rhizosphere soil in a variety of biological, chemical, and physical characteristics (Smith, 1987). Microorganisms are stimulated in the rhizosphere because it is a very eutrophic environment. Unique organic compounds are supplied to the rhizosphere by plant mucilages, mucigel, lyates, secretions, and root exudates. The latter are particularly important as they supply an abundant and diversified supply of low-molecular weight, soluble carbon compounds (Smith, 1976). The organic fraction of forest tree root exudates are dominated by organic acids. High molecular weight organic acids (fulvic and humic acids) are commonly associated with complexed aluminum in soil solution. Low molecular weight organic acids, however, also complex aluminum. The approximate order of the stability of the complexes is: citric > oxalic > malic > tannic > aspartic > p-hydrobenzoic > acetic (Wolt, 1987). Hue et al. (1985) have determined that 76–93% of total soil solution aluminum, in two acid subsoils, was complexed with low molecular weight organic acids. In addition, other chemical and physical characteristics distinguish rhizosphere soil from bulk soil. Some of the most important include: lower pH, lower water potential, lower osmotic potential, lower redox potential, and higher bulk density in the former relative to the latter (Table 9-10). As a result, rhizosphere depletion and root uptake of aluminum could result from increased mobilization due to these factors or to the interaction of these factors (see Table 9-10).

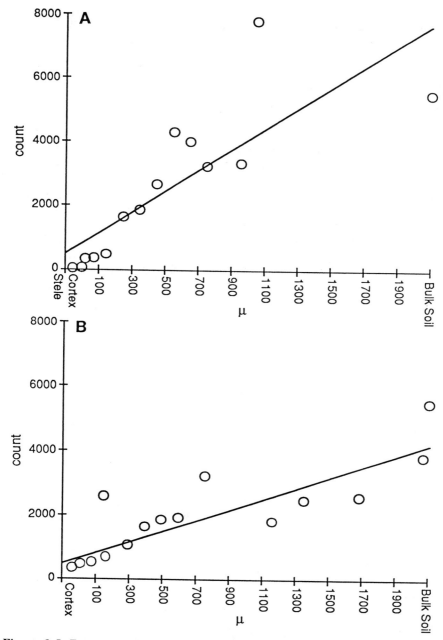

Figure 9-5. Energy dispersive x-ray spectrometer counts of aluminum in the root, rhizosphere soil, and bulk soil, of mature red spruce trees. A represents unamended soil, while B represents forest floor root substrate amended with aluminum chloride sufficient to increase the aluminum concentration 500 ppm over the in vitro forest floor concentration.

In a recent investigation of mature red spruce roots, we found that aluminum concentrations exhibited a strongly descending gradient from bulk soil through the rhizosphere to the root (Smith, 1988) (Figure 9-5). Estimated concentrations ranged from 1000 ppm within 20 μ of the root to 10 times this amount in soil beyond 2000 μ. This gradient is consistent with active aluminum mobilization and root uptake. The reduced pH of red spruce rhizospheres and the abundance of low molecular weight organic ligands may strongly interact to make aluminum available for root uptake and may thereby deplete the aluminum in the near-root zone. We observed rhizosphere soil to be approximately 0.5 pH unit lower than bulk soil. As previously stressed, the solubility of bound aluminum increases dramatically as pH decreases, especially below pH 4.0 (Cronan et al., 1978; Krug and Frink, 1983). Acidification of red spruce rhizospheres may result primarily from the form of nitrogen uptake. Preferential uptake of NH_4^+-N may result in excess cation over anion uptake and may stimulate a compensatory net proton release, resulting in acidification of the inner rhizosphere (Keltjens and Ulden, 1987; Marschener and Romheld, 1983; Schaller and Fisher, 1985). As emphasized, aluminum, as well as heavy metals, in soils are commonly complexed with silicate clay minerals, hydroxides, or large molecular weight humified organic acids. While organic ligands can greatly enhance the solubility of elements may be tightly bound and may not be readily exchangeable (Bloom et al., 1979; Hargrove and Thomas, 1984; Joslin et al., 1987). Organic and amino acids of the rhizosphere, however, may function to enhance both solubility and exchangeability (Stevenson, 1982). Inskeep and Comfort (1986) combined root exudate data (amino and organic acids) from the literature with root morphological data to estimate concentrations of exudates in hypothetical rhizosphere solutions. The authors concluded that amino and organic acids released as root exudates could make a significant contribution to the total soluble amounts of iron, copper and zinc. Presumably a similar case could be argued for aluminum.

3. Aluminum Toxicity

Substantial evidence has been presented indicating that aluminum, once it enters forest tree roots, accumulates in root tissue (Schaedle et al., 1986; Thornton et al., 1987; Vogt et al., 1987a,b). Root retention of aluminum, however, may be focused in vascular tissues (Shortle and Stienen, 1987; Vogt et al., 1987).

A large volume of evidence has been accumulated concerning mechanisms of aluminum toxicity to plants. Direct root effects include: reduced cell division associated with aluminum binding of DNA, reduced root growth caused by inhibition of cell elongation, and destruction of epidermal and cortical cells (Matsumoto et al., 1976; Hecht–Buchholz and Foy, 1981; Schier, 1985; Wallace and Anderson, 1984). Indirect effects of aluminum stress, especially in forest

Table 9-14. Abnormal Physiology of Forest Tree Seedlings Potentially Associated with Aluminum Toxicity as Determined by Laboratory Studies.

Cellular responses
  (1) Inhibition of cell division
  (2) Decreased water permeability in cell membranes
Shoot responses
  (1) Delay in budbreak and leaf expansion
  (2) Decreased stem and leaf production
  (3) Decreased shoot biomass
  (4) Decreased tissue Ca and Mg concentrations
  (5) Increased tissue Al concentration
  (6) Altered tissue P concentration
Root responses
  (1) Increased tissue Al concentration
  (2) Reduced plant water uptake
  (3) Decreased root biomass
  (4) Decreased fine root branching
  (5) Decreased terminal root elongation
  (6) Decreased tissue Ca and Mg concentrations
  (7) Altered tissue P concentration

Source: Cronan (1985, 1989).

trees, may involve interference with the uptake, translocation, and/or utilization of required nutrients such as calcium, magnesium, phosphorus, potassium, or other essential elements (Edwards et al., 1976; Foy et al., 1978) (Table 9-14).

Ulrich et al. (1980) have stressed the relationship between acid deposition, aluminum mobilization, and forest health. In their investigations of forest stress in the Solling region of the Federal Republic of Germany, they proposed that nitrification, during warm-dry periods, produces excess protons that reduce the soil buffering capacity and leave soils at risk to acidification from hydrogen ion pulses from atmospheric deposition. Aluminum mobilized at such times would result in fine root toxicity. Field studies with experimental acidification of forest soils have not always provided evidence in support of this hypothesis (Abrahamsen, 1980; Tviete, 1980). Tree species, however, may vary widely in their tolerance to available soil aluminum (Foy, 1984; Foy et al., 1978). Symptom expression of various plant species occur at Al^{3+} concentrations varying over three orders of magnitude (Table 9-15; Figure 9-6).

Reductions in calcium uptake by roots have been associated with increases in aluminum uptake (Clarkson and Sanderson, 1971). As a result, an important toxic effect of aluminum on forest tree roots could be calcium deficiency (Mayz de Manzi and Cartwright, 1984; Siegel and Haug, 1983; Schier, 1985; Edwards et al., 1976; Hutterman, 1983). Mature forest trees have a high calcium requirement relative to agricultural crops (Rennie, 1955). In view of this hypothesis, investigators have become aware that aluminum/calcium ratios in tree tissues may help quantify aluminum stress. Seedling evidence accumulated under laboratory conditions has indicated an adverse influence on fine roots when alu-

Table 9-15. Threshold of Al^{3+} Toxicity for Root Elongation.

Plant	(Al^{3+}) at Phytotoxic Threshold mmol l^{-1}	Solution
Cotton	0.0015^a	nutrient and soil
Coffee	0.004^a	nutrient and soil
Gramineae spp.	$< 0.009^a$	nutrient and soil
Norway spruce	3.3^b	nutrient
Red spruce, balsam fir	1.0^c	nutrient
Hybrid poplar	0.3^c	nutrient
Autumn olive	1.3^c	nutrient
Pine, oak, birch	3.7^c	nutrient

[a] (Al^{3+}) computed by authors
[b] Total Al concentration
[c] (Al^{3+}) estimated from solution composition
Source: Walt (1987).

minum/calcium ratios in nutrient solutions approximate one or greater. Higher soil solution molar ratios of aluminum/calcium have been observed in New England sites with symptomatic red spruce relative to sites with healthy red spruce (Shortle and Stienen, 1987). Elemental ratios of aluminum:calcium, however, did not reveal a consistent trend in forest floor samples collected from a variety of United States sites (Johnson et al., 1988). Recent evidence, however, continues to associate forest tree fine root morbidity and mortality, and interference with major cation ion uptake, with higher aluminum:calcium ratios (Matzner and Ulrich, 1985; Matzner et al., 1986; Ulrich, 1987.)

Shortle and Smith (1988) have proposed an integrated hypothesis concerning aluminum interference with calcium uptake in red spruce in the northeastern United States. These investigators have stressed that calcium is incorporated at a constant rate per unit volume of wood produced and is not recovered from sapwood as it matures into heartwood. As a result, when aluminum and calcium are present in approximately equimolar concentrations within the soil solution, or when the aluminum/calcium ratio exceeds one, aluminum will reduce calcium uptake by competition for binding sites in the cortical apoplant of fine roots. Reduced calcium uptake will suppress cambial growth (annual ring widths) and predispose trees to disease and injury by biotic stress agents when the functional sapwood becomes less than 25% of the cross-sectional area of the stem. Ulrich (personal communication) has a comparable integrated hypothesis suggesting that elevated available soil aluminum leads to reduced magnesium uptake, resulting in a smaller and more shallow root system. This reduction in root biomass leads to increased stress from winter dessication. Aluminum interference with nutrient cation uptake is not a form of aluminum toxicity. It is, however, a form of nutrient exclusion that may have profound significance for tree health in very acid soils.

In our own investigation of aluminum dynamics in red spruce rhizospheres (Smith, 1988), we have found that aluminum concentrations exhibit a strongly

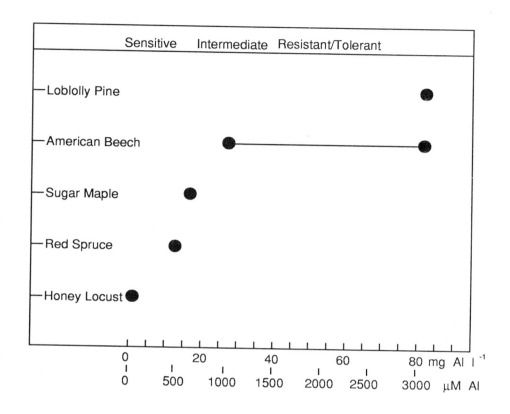

Figure 9-6. Aluminum toxicity thresholds (visible symptoms), determined in solution culture for five tree species. Beech response was variable and loblolly pine exhibited little response.
Source: Cronan (1985).

descending gradient from bulk soil through the rhizosphere to root. Calcium distribution, without artificial aluminum amendment, was relatively constant through the rhizosphere, but with aluminum amendment, calcium exhibited a sharply decreasing gradient near the root.

Schier (1985) has studied the response of red spruce and balsam fir seedlings to aluminum toxicity in solution culture. Seedlings were exposed to aluminum at 0.25, 50, 100, and 200 mg l^{-1} at pH 3.8. Red spruce were less tolerant of high aluminum concentrations than balsam fir. Although inhibition of root elongation by aluminum was similar in both species, extensive root deterioration occurred only in spruce. In the case of red spruce, aluminum stress may be manifest indirectly, i.e., cation competition, and directly via root toxicity.

D. Manganese Toxicity

In soils having pH levels less than 5.5, manganese reduction may cause this element to also reach toxic level. Soils vary significantly, however, in the manganese concentrations of parent materials (McFee et al., 1984). The significance of increased manganese availability in forest soils subject to acid deposition is unresolved.

E. Summary

The high productivity of forest ecosystems relative to other terrestrial ecosystems is largely the result of efficient recycling of the elements essential for growth. Forest ecosystems are nutrient element accumulating systems in which elements are continuously recombined in a variety of organic and inorganic compounds. The release of nutrients from organic compounds is accomplished by a large number of soil microbes and animals via a complex series of decomposition and mineralization processes. The rate of litter decomposition appears to control the rate of nutrient release. As nutrients are probably always limiting, the rate of decomposition also controls the rate of primary production in forests (Witkamp and Ausmus, 1976). Any significant reduction in the rate of decomposition, therefore, has the potential to importantly impact forest growth.

Over the last decade a large amount of evidence has been provided to address the hypothesis that trace metal contaminants of the soil reduce forest litter decomposition and mineralization rates. We judge that this hypothesis has been supported by the evidence provided, but only for *excessively* contaminated environments in the immediate vicinity of metal processing industries or other extreme sources of metal contaminants such as some urban roadside environments. Interference with decomposition process is certainly a Class II relationship in that its influence would be very subtle at moderate dose levels. It turns out, however, that current evidence can only support its importance in Class III, or high air-pollution exposure, situations.

The thresholds of toxicity for numerous bacterial, fungal, insect, and other components of the soil biota are in the range of 1000–10,000 ppm metal cation on a soil dry weight basis. These concentrations are two and three orders of magnitude greater than the concentrations of heavy metals throughout most of temperate forest ecosystems. In addition, fungi probably play a larger role in nutrient cycling in forests than other microbial groups, and they appear more resistant to trace metal influence than other microbes. In addition, there is considerable evidence for microbial adaptation to high trace metal exposure. While components of the forest floor soil biota may be impacted by trace metals, decomposition rates may remain unchanged due to adaptation or shifts in species composition unreflected in gross soil process activities.

Many of the investigations reviewed report data from studies that have amended soils or laboratory media with soluble salts containing the trace cations of interest. In natural soils, trace metals may not be contained in soluble compounds and natural availability, and hence dissolved concentrations, may be appreciably less than in experimental designs. The experimental protocols were also seen to frequently employ optimal temperature and moisture conditions for decomposition. In nature, soil climate may impose more severe restrictions on decomposition than metal pollutants.

Future research efforts should concentrate on field observations that introduce as little artificiality as possible. Investigators should examine several parameters, for example, soil biomass, soil process, and soil enzyme concentration or activity, concurrently; they should also consider the influence of gaseous contaminants on decomposition and mineralization processes. Grant et al. (1979) have recently observed that exposure of a forest soil to 1.0 ppm (2620 μg m^{-3}) sulfur dioxide reduced the rate of glucose decomposition and that nitrate (at 5 mg N g^{-1} soil) inhibited oxygen consumption and carbon dioxide evolution. Continuous fumigation of an acid soil with excessive sulfur dioxide (26,000 μg m^{-3}, 10 ppm) reduced nitrification in the experiments of Labeda and Alexander (1978). These investigators also found that sustained fumigation with 9400 μg m^{-3} (5 ppm) nitrogen dioxide inhibited the rate of ammonium disappearance, led to greater rates of nitrate formation, and resulted in nitrate accumulation. Can ambient concentrations of sulfur and nitrogen dioxides adversely impact forest nutrient cycling?

With regard to direct toxicity of heavy metal ions to plant roots, copper, nickel, and zinc toxicities have occurred frequently. Cadmium, cobalt and lead toxicities have occurred less frequently. In forest ecosystems, in excess of 90% of the heavy metals deposited from the atmosphere may not be available for root uptake. Rhizosphere processes may, however, transform heavy metals from an unavailable to an available pool. Acute, direct toxicity caused by heavy metal deposition and/or mobilization by acidification probably does not occur in forest trees located outside of urban, roadside, or selected point source industrial and electric-generating environments.

The role of increased hydrogen ion input to forest soils is difficult to assess. Most forest soils are acid and are naturally dominated by hydrogen ions. Only certain forest soils are judged to be at risk to acidification from hydrogen ions from the atmosphere. These include those soil types with low cation exchange capacity, low clay or organic matter content, low sulfate adsorption capacity, and currently pH in the 5.5–6.5 range. While numerous soil organisms and soil processes are influenced by hydrogen ion concentrations, it is extremely difficult to partition hydrogen ion effects from other soil chemistry effects in natural profiles. Based on laboratory and some field evidence, the influence of hydrogen ion deposition on soil organisms (animals and microbes), soil processes, and soil enzymes is detrimental only under conditions of moderate deposition (1–10 kg H$^+$ ha^{-1} yr^{-1}) or high deposition (>10 kg H$^+$ ha$^-$ yr^{-1}). Most forest systems currently receive low hydrogen ion input (< 1 kg H$^+$ ha$^-$ yr^{-1}). Hydrogen ions are presumed not to be directly toxic to tree roots. The most significant influence of increased hydrogen ion input to the soil profile may be the enhancement of the availability of toxic ions such as manganese and aluminum.

Aluminum stress imposed on forest tree roots is concluded to be a significant result of hydrogen ion deposition from the atmosphere. This stress may manifest as both direct toxicity resulting from Al^{3+} uptake and as reduced major nutrient cation uptake resulting from Al^{3+} competition with calcium, magnesium or potassium. Substantial evidence indicates that aluminum antagonism of calcium uptake may result in forest growth reduction and in predisposition to both abiotic and biotic stress factors.

References

Abrahamsen, G., J. Hovland, and S. Hagvar. 1980. Effects of artificial acid rain and burning on soil organisms and the decomposition of organic matter. In T.C. Hutchenson and M. Havas, eds., Effects of Acid Precipitation on Terrestrial Ecosystems. Plenum Pres, New York, pp. 341-362.

Alexander, M. 1980. Effects of acidity on microorganisms and microbial processes in soil. In T.C. Hutchenson and M. Havas, eds., Effects of Acid Precipitation on Terrestrial Ecosystems. Plenum Press, New York, pp. 363-374.

Al-Khafaji, A.A.,and M.A. Tabatabai. 1979. Effects of trace elements on arylsulfatase activity in soils. Soil Sci. 127: 129-133.

Antonovics, J., A.D. Bradshaw, and R.G. Turner. 1971. Heavy metal tolerance in plants. Adv. Ecol. Res. 7: 1-85.

Atlas, R.M., D. Pramer, and R. Bartha. 1978. Assessment of pesticide effects on non-target soil microorganisms. Soil Biol. Biochem. 10: 231-239.

Ausmus, B.S. 1973. The use of ATP assay in terrestrial decomposition studies. Bull. Ecol. Res. Commun. (Stockholm) 17: 223-234.

Ausmus, B.S. 1977. Regulation of wood decomposition rates by arthropod and annelid populations. In U. Lohm and T. Persson, eds., Soil Organisms as Components of Ecosystems. Ecolog. Bull. (Stockholm) 25: 180-192.

Babich, H. and G. Stotzky. 1978. Effects of cadmium on the biota: Influence of environmental factors. Adv. Appl. Microbiol. 23: 55-117.

Belser, L.W. 1979. Population ecology of nitrifying bacteria. Annu. Rev. Microbiol. 33: 309-333.

Bhuiya, M.R.H. and A.H. Cornfield. 1972. Effects of addition of 1000 ppm Cu, Ni, Pb, and Zn on carbon dioxide release during incubation of soil alone and after treatment with straw. Environ. Pollu. 3: 173-177.

Bloom, P.R., M.B. McBride, and R.M. Weaver. 1979. Aluminum organic matter in acid soils: Salt-extractable aluminum. Soil Sci. Soc. Am. J. 43: 813-815.

Bond, H., B. Lighthart, R. Shimabuku, and L. Russell. 1976. Some effects of cadmium on coniferous forest soil and litter microcosms. Soil Sci. 121: 278-287.

Bondietti, E.A. 1976. Percent amino sugars and urease enzyme activity in litter as a function of distance from the smelter stack on Crooked Creek Watershed. In R. I. Van Hook and W.D. Shults, eds., Ecology and Analysis of Trace Contaminants. Progress Report. Oct. 1974 – Dec. 1975. ORNL/NSF/EATC-22 Oak Ridge Natl. Lab., Oak Ridge, TN, p. 102.

Bormann, F.H. and G.E. Likens. 1979. Pattern and Process in a Forested Ecosystem. Springer–Verlag, New York, 253 pp.

Bowen, H.J.M. 1966. Trace Elements in Biochemistry. Academic Press, NewYork, 241 pp.

Bremner, J.M. and R.L. Mulvaney. 1978. Urease activity in soils. In R.G. Burns, ed., Soil Enzymes, Academic Press, New York, pp. 149-196.

Burns, R.G. 1978. Soil Enzymes. Academic Press, New York, 380 pp.

Clarkson, D.T. and J. Sanderson. 1971. Inhibition of the uptake and long-distance transport of calcium by aluminum and other polyvalent cations. J. Exp. Bot. 22: 837-851.

Cromack, K. Jr. and C.D. Monk. 1975. Litter production, decomposition, and nutrient cycling in a mixed hardwood watershed and a white pine watershed. In F.G. Howell, J.B. Gentry, and M.H. Smith, eds., Mineral Cycling in Southeastern Ecosystems. ERDA Symposium Series No. CONF-740513, pp. 609-624.

Cronan, C.S. 1985. A Comparative Inter-Regional Analysis of Aluminum Biogeochemistry in Forested Watersheds Exposed to Acidic Deposition. ALBIOS Annual Report for 1985. Elective Power Research Institute Proj. No. RP 2365-01. Orono, ME, 37 pp.

Cronan, C.S. et al. 1989. Aluminum toxicity in forests exposed to acidic deposition: The Albios results. Water Air Soil Pollu. (in press).

Cronan, C.S. and C.L. Schofield. 1979. Aluminum leaching response to acid precipitation: Effects on high-elevation watershed in the northeast. Science 204: 304-306.

Cronan, C.S., W.A. Reiners, R.C. Reynolds Jr., and G.E. Lang. 1978. Forest floor leaching: Contributions from mineral, organic, and carbonic acids in New Hampshire subalpine forests. Science 200: 309-311.

David, M.B. and C.T. Driscoll. 1984. Aluminum speciation and equilibria in soil solutions of a haplorthod in the Adirondack Mountains (New York, U.S.A.) Geoderma 33: 297-318.

Devi, L.S. 1962. Nutritional Requirements of Fungi. Univesity of Madras, Madras, India, 29 pp.

Doelman, P. 1978. Lead and terrestrial microbiota. In J.O. Nriagu, ed., The Biogeochemistry of Lead in the Environment. Part B. Biological Effects. Elsevier-North-Holland Biomedical Press, New York, pp. 343-353.

Driscoll, C.T. 1984. Aluminum chemistry in dilute acid waters. In Chapter 4. The Acidic Deposition Phenomenon and Its Effects. Vol. II. Effects Sciences. U.S. Environmental Protection Agency, Washington, DC.

Ebregt, A. and J.M. A.M. Boldewijm. 1977. Influence of heavy metals in spruce forest soil on amylase activity, CO_2 evolution from starch and soil respiration. Plant Soil 47: 137-148.

Edwards, J.H., B.H. Horton, and H.C. Kirkpatrick. 1976. Aluminum toxicity symptoms in peach seedlings. J. Am. Soc. Hort. Sci. 101: 139-142.

Energy Research and Development Administration. 1975. Biological Implications of Metals in the Environment. ERDA Symposium Series No. 42. Washington, DC, 682 pp.

Etherington, J.R. 1975. Environment and Plant Ecology. Wiley, New York, 347 pp.

Flanagan, P.W. and K. Van Cleve. 1977. Microbial biomass, respiration and nutrient cycling in a black spruce taiga ecosystem. In V. Lohnn and T. Persson, eds., Soil Organisms as Components of Ecosystems. Ecolog. Bull. (Stockholm) 25: 261-273.

Foy, C.D. 1983. The physiology of plant adaptation to mineral stress. Iowa State J. Res. 57: 355-391.

Foy, C.D. 1984. Physiological effects of hydrogen, aluminum, and manganese toxicities in acid soil. In Soil Acidity and Liming, Agronomy Monograph No. 12, Soil Sci. Soc. Am. Madison, WI, pp. 57-97.

Foy, C.D., R.L. Chaney, and M.C. White. 1978. The physiology of metal toxicity in plants. Annu. Rev. Plant Physiol. 29: 511-566.

Frankland, J.C. and D.K. Lindley. 1978. A comparison of two methods for the estimation of mycelial biomass in leaf litter. Soil Biol. Biochem. 10: 323-333.

Fujihara, M.P., T.R. Gorland, R.E. Wildung, and H. Drucker. 1973. Response of microbiota to the presence of heavy metals in soils. Proc. 1973 Annu. Meet. Am. Soc. Microbiol.

Garrett, S.D. 1963. Soil Fungi and Soil Fertility. Pergamon Press, New York, 165 pp.

Goodfellow, M. and D. Dawson. 1978. Qualitative and quantitative studies of bacteria colonizing *Picea sitchensis* litter. Soil Biol. Biochem. 10: 303-307.

Grant, I.F., K. Bancroft, and M. Alexander. 1979. SO_2 and NO_2 effects on microbial activity in an acid forest soil. Microbiol. Ecol. 5: 85-89.

Griffin, D.M. 1972. Ecology of Soil Fungi. Syracuse Univ. Press, Syracuse, New York, 193 pp.

Hargrove, W.L. and G.W. Thomas. 1984. Extraction of aluminum from aluminum — organic matter in relation to titratable acidity. Soil Sci. Soc. Am. J. 48: 1458-1460.

Hecht-Buchholz, C. and C.D. Foy. 1981. Effect of aluminum toxicity on root morphology of barley. Plant Soil 63: 93-95.

Horsfall, J.G. 1956. Principles of Fungicidal Action. Chronicá Botanic Co., Waltham, MA, 279 pp.

Hue, N.V., G.R. Craddock, and F. Adams. 1985. Effect of organic acids on aluminum toxicity of subsoils. Agron. Abstr. 77: 8.

Hutterman, A. 1983. The effects of acid deposition on the physiology of the soil and root system in forest ecosystems. Allgem. Forstz. Ztg 27: 663-664.

Inman, J.C. and G.R. Parker. 1978. Decomposition and heavy metal dynamics of forest litter in northwestern Indiana. Environ. Pollu. 17: 39-51.

Inskeep, W.P. and S.D. Comfort. 1986. Thermodynamic predictions for the effects of root exudates on metal speciation in the rhizosphere. J. Pl. Nut. 9: 567-586.

Jackson, D.R. and A.P. Watson. 1977. Disruption of nutrient pools and transport of heavy metals in a forested watershed near a lead smelter. J. Environ. Qual. 6: 331-338.

Jackson, D.R., C.D. Washburne, and B.S. Ausmus. 1977. Loss of Ca and NO_3-N from terrestrial microcosms as an indicator of soil pollution. Water Air Soil Pollu. 8: 279-284.

Jackson, D.R., W.J. Selvidge, and B.S. Ausmus. 1978a. Behavior of heavy metals in forest microcosms: I. Transport and distribution among components. Water Air Soil Pollu. 10: 3-11.

Jackson, D.R., W.J. Selvidge, and B.S. Ausmus. 1978b. Behavior of heavy metals in forest microcosms: II. Effects on nutrient cycling processes. Water Air Soil Pollu. 10: 13-18.

Jensen, V. 1976. Effects of lead on biodegradation of hydrocarbons in soil. Oikos 28: 220-224.

Jernelöv, A. and A.L. Martin. 1975. Ecological implications of metal metabolism by microorganisms. Annu. Rev. Microbiol. 29: 61-77.

Johnson, D.W., A. J. Friedland, H. Van Miegroet, R.B. Harrison, E. Miller, S.E. Lindberg, D. W. Cole, D.A. Schaefer, and D.E. Todd. 1988. Nutrient status of some contrasting high elevation forests in the eastern and western United States. In G.D. Hertel, ed., Proc. Effects of Atmospheric Pollutants on the Spruce-Fir Forests of the Eastern United States and Federal Republic of Germany. Gen. Tech. Rep. No. 255, U.S.D.A. Forest Service, Northeastern Forest Exp. Sta., Broomall, PA, 250 pp.

Joose, E.N.G. and J.B. Buker. 1979. Uptake and excretion of lead by litter-dwelling collembola. Environ. Pollu. 18: 235-240.

Jordan, M.J. and M.P. Lechevalier. 1975. Effects of zinc-smelter emissions on forest soil microflora. Can. J. Microbiol. 21: 1855-1865.

Joslin, J.D., P.A. Mays, M.H. Wolfe, J.M. Kelly, R.W. Garber, and P.F. Brewer. 1987. Chemistry of tension lysimeter water and lateral flow in spruce and hardwood stands. J. Environ. Qual. 16: 152-160.

Juma, N.G. and M.A. Tabatabai. 1977. Effects of trace elements on phosphatase activity in soils. Soil Sci. Soc. Am. J. 41: 343-346.

Keltjens, W.G. and P.S.R. van Ulden. 1987. Effects of Al on nitrogen ($NH_4^=$and NO_3^-) uptake, nitrate reductase activity and proton release in two sorghum cultivars differing in Al tolerance. Plant Soil 104: 227-234.

Krug, E.C. and C.R. Frink. 1983. Acid rain on acid soil: A new perspective. Science 221: 520-525.

Kühnelt, W. and N. Walker. 1976. Soil Biology. Michigan State Univ. Press, East Lansing, MI, 483 pp.

Labeda, D.P. and M. Alexander. 1978. Effects of SO_2 and NO_2 on nitrification in soil. J. Environ. Qual. 7: 523-526.

Lang, G.E. and R.T.T. Forman. 1978. Detrital dynamics in a mature oak forest: Hutcheson Memorial Forest, New Jersey. Ecology 59: 580-595.

Lee, Y.H. 1985. Aluminum speciation in different water types. Ecol. Bull. (Stockholm) 37: 109-119.

Legge, A.H. and R.A. Crowther. 1987. Acidic Deposition and the Environment. A Literature Overview. Centre for Environmental Research, University of Calgary, Calgary, Alberta, Canada, 235 pp.

Lewis, J.A., G.C. Papavizas, and T.S. Hora. 1978. Effect of some herbicides on microbial activity in soil. Soil Biol. Biochem. 10: 137-141.

Liang, C.N. and M.A. Tabatabai. 1977. Effects of trace elements on nitrogen mineralization in soils. Environ. Pollu. 12: 141-147.

Liang, C.N. and M.A. Tabatabai. 1978. Effects of trace elements on nitrification in soils. J. Environ. Qual. 7: 291-293.

Likens, G.E., F.H. Bormann, R.S. Pierce, J.S. Eaton, and N.M. Johnson. 1977. Biogeochemistry of a Forested Ecosystem. Springer–Verlag, New York, 146 pp.

Lutz, J.H. and R.F. Chandler Jr. 1946. Forest Soils. Wiley, New York, 514 pp.

Lynch, J.M. 1979. Micro-organisms in their natural environments. The terrestrial environment. In J.M. Lynch and N.J. Poole, eds., Microbial Ecology: A Conceptual Approach. Wiley, New York, pp. 67-91.

Mader, D.L., H.W. Lull, and E.I. Swenson. 1977. Humus Accumulation in Hardwood Stands in the Northeast. Mass. Exp. Sta. Res. Bull. No. 648, Univ. of MA, Amherst, MA, 37 pp.

Marshner, H. and V. Romheld. 1983. In vitro measurement of root-induced pH changes at the soil-root interface: Effect of plant species and nitrogen sources. Z. Pflanzenphysiol. 111: 241-251.

Mason, C.F. 1977. Decomposition. Institute of Biology's, Studies in Biology No. 74, Edward Arnold, London, 58 pp.

Mater, E., D. Murach, and H. Fortmann. 1986. Soil acidity and its relationship to root growth in declining forest stands in Germany. Water Air Soil Pollu. 31: 273-282.

Matsumoto, H., F. Hrasawa, E. Torikai, and E. Takahashi. 1976. Location and absorption of aluminum in pea roots and its binding to nucleic acids. Plant Cell Physiol. 17: 127-137.

Matzner, E. and B. Ulrich. 1985. Implications of the chemical soil conditions for forest decline. Experientia 41: 578-584.

Mayz de Manzi. J. and R.M. Cartwright. 1984. The effects of pH and aluminum toxicity on the growth and symbiotic development of cowpea (Vigna unguiculata L. Walp). Plant Soil 80: 423-430.

McFee, W.W., F. Adams, C.S. Cronan, M.K. Firestone, C.D. Foy, R.D. Harter, and D.W. Johnson. 1984. Effects on soil systems. In The Acidic Deposition

Phenomenon and its Effects. Vol. II. Effects Sciences. U.S. Environmental Protection Agency, Washington, DC.

McLaren, A.D. and G.H. Peterson. 1967. Soil Biochemistry. Dekker, New York, 509 pp.

Meentemeyer, V. 1978. Macroclimate and lignin control of litter decomposition rates. Ecology 59: 465-472.

Miller, W.P. and W.W. McFee. 1983. Distribution of cadmium, zinc, copper, and lead in soils of industrialized norwestern Indiana. J. Environ. Qual. 12: 29-33.

Mikkelsen, J.P. 1974. Effects of lead on the microbiological activity in soil. Tidster Plant 78: 509-516.

Morissey, R.F., E.P. Dugan, and J.S. Koths. 1974. Inhibition of nitrification by incorporation of select heavy metals in soil. Proc. Annu. Meet. Am. Soc. Microbiol. 74: 2.

Myrold, D.D. 1987. Effects of acidic deposition on soil organisms. In Acidic Deposition and Forest Soil Biology. National Council of the Paper Industry for Air and Stream Improvement, Tech. Bull. No. 527, New York, pp. 1-29.

Nannipieri, P., R.L. Johnson, and E.A. Paul. 1978. Criteria for measurement of microbial growth and activity in soil. Soil Biol. Biochem. 10: 223-229.

Nilsson, T. 1981. Groundwater-level discharge relationship in the area of Gårdsjön, Bohuslän, Sweden. Section of Hydrology, Department of Physical Geography, University of Uppsala, Uppsala, Sweden, 38 pp.

Odum, E.P. 1971. Fundamentals of Ecology. Saunders, Philadelphia, 574 pp.

Perlman, D. 1949. Effects of minor elements on the physiology of fungi. Bot. Rev. 15: 195-220.

Rennie, P.J. 1955. The uptake of nutrients by mature forest growth. Plant and Soil 7: 49-95.

Reuss, J.O. and D.W. Johnson. 1986. Acid Deposition and the Acidification of Soils and Waters. Vol. 59. Ecological Studies. Springer-Verlag, New York, 119 pp.

Riechle, D.E. 1977. The role of soil invertebrates in nutrient cycling. In U. Lohn and T. Persson, eds., Soil Organisms as Components of Ecosystems. Ecol. Bull. (Stockholm) 25: 145-156.

Rosenquist, I.T., P. Jorgensen, and H. Rueslatten. 1980. The importance of natural H+ production for acidity in soil and water. In D. Drablos and A. Tollen, eds., Proc. Ecological Impact of Acid Precipitation. SNSF Project, Sandefjord, Norway.

Rühling, Å. and G. Tyler. 1973. Heavy metal pollution and decomposition of spruce needle litter. Oikos 24: 402-416.

Saxena, J. and P.H. Howard. 1977. Environmental transformation of alkylated and inorganic forms of certain metals. Adv. Appl. Microbio. 21: 185-226.

Schaedle, M., F.C. Thornton, and D.J. Raynal. 1986. Non-metabolic binding of aluminum to roots of loblolly pine and honey locust. J. Plant Nut. 9: 1227-1238.

Schaller, G. and W.R. Fischer. 1985. pH — Anderungen in der Rhizosphare von Mais—und Erdnusswurzeln. Z. Pflanzenernahr. Bodenk 148: 306-320.

Schier, G.A. 1985. Response of red spruce and balsam fir seedlings to aluminum toxicity in nutrient solutions. Can. J. For. Res. 15: 29-33.

Shortle, W.C. and K.T. Smith. 1988. Aluminum-induced calcium deficiency syndrome in declining red spruce. Science 240: 1017-1018.

Shortle, W.C. and H. Stienen. 1987. Role of ions in the etiology of spruce decline. In G.D. Hertel, ed., Effects of Atmospheric Pollutants on the Spruce-fir Forests in the Eastern United States and Federal Republic of Germany, Gen. Tech. Rep. No.255, U.S.D.A. Forest Service, Northeastern Forest Exp. Sta., Broomall, PA.

Siegel, N. and A. Haug. 1983. Calmodulium-dependent formation of membrane potential in barley root plasma membrane vesicles: A biochemical model of aluminum toxicity in plants. Physiol. Plant 59: 285-291.

Singh, U.R. and B.D. Tripathi. 1978. Effects of industrial effluents on the population density of soil microarthropods. Environ. Conserv. 5: 229-231.

Smith, W.H. 1984. Ecosystem pathology: A new perspective for phytopathology. For. Ecol. Mamt. 9: 193-219.

Smith, W.H. 1987. The Atmosphere and the Rhizosphere: Linkages with Potential Significance for Forest Tree Health. National Council of the Paper Industry for Air and Stream Improvement, Tech. Bull. No. 527, New York, pp. 30-94.

Smith, W.H. 1988. Red spruce rhizosphere dynamics: Spatial distribution of aluminum and zinc in the near-root soil zone. For. Sci. (in press).

Smith, W.H., T.G. Siccama, and S.L. Clark. 1986. Atmospheric deposition of heavy metals and forest health: An overview and a ten-year budget for the input/output of seven heavy metals to a northern hardwood forest. Publ. No. FWS-87-02. Virginia Polytechnic Institute and State University, Blacksburg, VA, 27 pp.

Somers, E. 1961. The fungitoxicity of metal ions. Annu. Appl. Bio. 49: 246-253.

Spalding, B.P. 1977. Enzymatic activities related to the decomposition of coniferous leaf litter. Soil Sci. Soc. Am. J. 41: 622-627.

Spalding, B.P. 1979. Effects of divalent metal chlorides on respiration and extractable enzymatic activities of Douglas-fir needle litter. J. Environ. Qual. 8: 105-109.

Stevenson, F.J. 1982. Humus Chemistry. John Wiley and Sons, New York, 443 pp.

Summers, A.O. 1978. Microbial transformations of metals. Annu. Rev. Microbiol. 32: 637-672.

Summers, A.O. and S. Silver. 1978. Microbial transformations of metals. Annu. Rev. Microbiol. 32: 637-672.

Thornton, F.C., M. Schaedle, and D.J. Raynal. 1987. Effects of aluminum on red spruce seedlings in soil culture. Environ. Exper. Bot. 27: 489-498.

Todd, R.L., K. Cromack, and J.C. Stormer. 1973. Chemical exploration of the microhabitat by electron probe microanalysis of decomposer organisms. Nature 243: 544-546.

Tveite, B. 1980. Effects of acid precipitation on soil and forest. 8. Foliar nutrient concentrations in field experiments. In D. Drablos and A. Tollan, eds., Ecological Impact of Acid Precipitation. Proc. Internat. Conf., Sandefjord, Norway, SNSF Project, Oslo, pp. 204-205.

Tyler, G. 1972. Heavy metals pollute nature, may reduce productivity. Ambio 1: 52-59.

Tyler, G. 1974. Heavy metal pollution and soil enzymatic activity. Plant Soil 41: 303-311.

Tyler, G. 1976. Heavy metal pollution, phosphatase activity, and mineralization of organic phosphorus in forest soils. Soil Biol. Biochem. 8: 327-332.

Tyler, G., B. Mörnsjö and B. Nilsson. 1974. Effects of cadmium, lead, and sodium salts on nitrification in a mull soil. Plant Soil 40: 237-242.

Ulrich, B. 1986. Natural and anthropogenic components of soil acidification. Z. Pflanzenernaebr. Bodenk. 149: 702-717.

Ulrich, B. 1987. Stability, elasticity, and resilience of terrestrial ecosystems with respect to matter balance. Ecol. Studies 61: 11-49.

Ulrich, B., R. Mayer, and P.K. Khanna. 1980. Chemical changes due to acid precipitation in a loess-derived soil in central Europe. Soil Sci. 130: 193-199.

van Breemen, N., C.T. Driscoll, and J. Mulder. 1984. Acidic deposition and internal proton sources in acidification of soils and waters. Nature 307: 599-604.

Visser, S., R.S. Danielson, and J.F. Parr. 1987. Effects of acid-forming emissions on soil microorganisms and microbially-mediated processes. Publica. No. ADRP-B-02-87. Centre for Environmental Research, University of Calgary, Calgary, Alberta, Canada, 86 pp.

Vogt, K.A., R. Dahlgren, F. Ugolini, D. Zabowski, E.E. Moore, and R. Zasoski. 1987a. Aluminum, Fe, Ca, Mg, K, Mn, Cu, Zn, and P in- and belowground biomass. I. *Abies amabilis* and *Tsuga metensiana*. Biogeochemistry 4: 277-294.

Vogt, K.A., R. Dahlgren, F. Ugolini, D. Zabowski, E.E. Moore, and R. Zasoski. 1987b. Aluminum, Fe, Ca, Mg, K, Mn, Cu, Zn, and P in- and belowground biomass. II. *Abies amabilis* stand. Biochemistry 4: 295-311.

Wallace, S.V. and I.C. Anderson. 1984. Aluminum toxicity and DNA synthesis in wheat roots. Agronomy J. 76: 5-8.

Watson, A.P., R.I. Van Hook, and D.E. Reichle. 1976. Impact of lead mining smelting complex on the forest-floor litter arthropod fauna in the new lead belt region of southeast, Missouri. Environ. Sci. Div. Publ. No. 881, Oak Ridge National Laboratory, Oak Ridge, TN, 163 pp.

Weinberg, E.D. 1970. Biosynthesis of secondary metabolites: Roles of trace metals. Adv. Microbiol. Physiol. 4: 1-44.

Whittaker, R.H. 1975. Communities and Ecosystems. Macmillan, London, 385 pp.

Wilde, S.A. 1958. Forest Soils. Ronald Press, New York, 537 pp.

Williams, S.T., T. McNeilly, and E.M.H. Wellington. 1977. The decomposition of vegetation growing on metal mine waste. Soil Biol. Biochem. 9: 271-275.

Wilson, D.O. 1977. Nitrification in soil treated with domestic and industrial sewage sludge. Environ. Pollu. 12: 73-82.

Witkamp, M. 1971. Soil as components of ecosystems. Annu. Rev. Ecol. Syst. 2: 85-110.

Witkamp, M. and B.S. Ausmus. 1976. Processes in decomposition and nutrient transfer in forest systems. In J.M. Anderson and A. Macfayden eds., The Role of Terrestrial and Aquatic Organisms in Decomposition Processes. 17th Symp. Br. Ecol. Soc. Blackwell, London, pp. 375-376.

Wolt, J. 1987. Effects of acidic deposition on the chemical form and bioavailability of soil aluminum and manganese. Tech. Bull. No. 518, National Council of the Paper Industry for Air and Stream Improvement. New York, 46 pp.

Zajic, J.E. 1969. Microbial Biogeochemistry. Academic Press, New York, 345 pp.

10

Forest Nutrient Cycling: Leaching and Weathering

As suggested in Figure 9-1, a large number of processes important in forest nutrient cycling are potentially at risk of alternation from air pollutant deposition. It has been hypothesized that acid deposition may accelerate leaching of nutrients from forest foliage and forest soils, and may alter weathering rates of forest soil minerals.

A. Vegetative Leaching

Vegetative leaching refers to the removal of substances from plants by the action of aqueous solutions such as rain, dew, mist, and fog. Precipitation washout of chemicals from trees has been appreciated for some time (for example, Ovington, 1962). The review of Tukey (1970) has presented numerous, pertinent generalizations. Inorganic chemicals leached from plants include all the essential macro- and microelements. Potassium, calcium, magnesium, and manganese are typically leached in the greatest quantities. A variety of organic compounds, including sugars, amino acids, organic acids, hormones, vitamins, pectic and phenolic substances, and others, are also leached from vegetation. As the maturity of leaves increases, susceptibility to nutrient loss via leaching also increases and peaks at senescence. Leaves from healthy plants are more resistant to leaching than leaves that are injured, infected with microbes, infested with insects, or otherwise under stress.

Deciduous trees lose more nutrients from foliage than do coniferous species during the growing season. Conifers, however, continue to lose nutrients throughout the dormant season. The stems and branches of all woody plants lose nutrients during both the growing and dormant seasons. Tamm (1951) has

estimated that 2–3 kg ha^{-1} each of potassium, sodium, and calcium were transferred by rain from spruce and pine foliage to the forest floor during 30 days in
the autumn. In his studies of the coniferous forests of southern, coastal British
Columbia, Feller (1977) has judged that the extent to which precipitation leached
nutrients from standing trees decreased in the order: potassium > calcium >
sodium > magnesium.

The mechanism of leaching is presumed to be primarily a passive process.
Cations are lost from "free space" areas within the plant. Under uncontaminated
natural environmental conditions, little if any cations are thought to be lost from
within cells or cell walls. On the leaf surface it has been demonstrated that
leaching of cations involves exchange reactions in which cations on exchange
sites of the cuticle are exchanged by hydrogen from leaching solutions. Cations
may move directly from the translocation stream within the leaf into the
leaching solution by diffusion and mass flow through areas devoid of cuticle
(Tukey, 1970).

1. Acid Deposition and Foliar Leaching

Since acid precipitation may increase hydrogen ion activity by one to two orders
of magnitude (pH 5.6 to pH 3.6) due to increasing concentrations of sulfuric and
nitric acids in precipitation (Galloway and Cowling, 1978), and since damage to
cuticles and epidermal cells may also result from exposure to acid precipitation
(Tamm and Cowling, 1976), the potential for accelerated leaching under this
stress is obvious.

Grennfelt et al. (1978) have very appropriately pointed out that the phrase
acid precipitation infers that acid in rain and snow is the sole acidifying agent input to ecosystems. They correctly point out that any substance capable of increasing the hydrogen-ion activity in the ecosystem is important and that all
compounds capable of this should be termed *acidifying substances*. For coniferous forests of southern Sweden, a comprehensive inventory of all acidifying substances would include all materials listed in Table 10–1. What is the evidence
that these acidifying substances enhance forest tree leaching?

a. Seedling Studies

Colleagues at the School of Forestry and Environmental Studies, Yale
University have employed simulated acid rain applied to seedling forest species
in order to evaluate the potential for foliar leaching. Field collected first-year
sugar maple seedlings were exposed to simulated rain ranging in acidity from pH
5.0 to pH 2.3. Foliar leachate was collected and analyzed for sodium, potassium, magnesium, and calcium cations. As the acidity of the artificial mist increased, the loss of cations increased. Significant increases in leaching were
measured at pH 3.3 and pH 4.0, where no foliar injury symptoms appeared. At
pH 3.0 and below, the seedlings were visibly damaged and tissue destruction may
have contributed to foliar leaching (Figure 10–1) (Wood and Bormann, 1975).

Table 10-1. Estimation of Acidifying Substances Deposited in a Coniferous Forest Ecosystem in a Rural Area of Southern Sweden.

Substance	Deposition $(mmole\ m^{-2}\ yr^{-1})$	S deposition $(kg\ ha^{-1}\ yr^{-1})$	N deposition $(kg\ ha^{-1}\ yr^{-1})$
Gases			
SO_2	32	10.2	
H_2S	<0.5	<0.2	
NO	<3		<0.4
NO_2	24		3.4
HNO_2	<0.5		<0.1
HNO_3	8		1.1
NH_3	<2		<0.3
Particles			
SO_4^{2-}	9.5	3.0	
NH_4^+	16		2.2
NO_3^-	3.2		0.4
Mist and fog			
SO_4^{2-}	2.5	0.8	
NH_4^+	1.7		0.2
NO_3^-	1.7		0.2
Precipitation			
SO_4^{2-}	30	9.6	
NH_4^+	20		2.8
NO_3^-	20		2.8
Total		23.6-23.8	13.1-13.9

Source: Grennfelt et al. (1978).

Additional studies have established threshold doses of acid precipitation necessary to cause foliar tissue damage for forest tree seedlings. It is presumed that this damage would increase the loss of cations via leaching (Table 10–2).

b. Field Studies

Field investigations of foliar leaching have typically compared throughfall and stemflow chemistry to direct deposition chemistry in order to evaluate leaching

Table 10-2. Threshold of Acute Damage of Acid Rain to Forest Tree Seedling Foliage.

Species	Simulant pH	Exposure	Period	Reference
Yellow birch	3.0	Intermittent	11 weeks	Wood and Bormann (1974)
White pine	2.3	Intermittent	20 weeks	Wood and Bormann (1976, 1977)
Hybrid poplar	3.4	—	5 days	Evans et al. (1978)

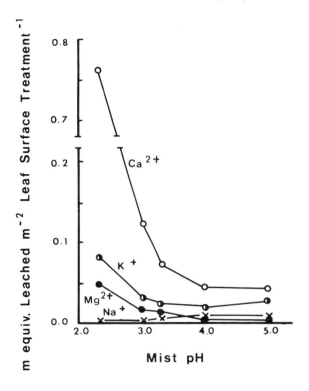

Figure 10-1. Effect of artificial acid mist on the leaching of Na^+, K^+, Mg^{2+}, and Ca^{2+} from first-year sugar maple seedlings
Source: Wood and Bormann (1975).

(Figure 10–2). If cation enrichment is detected in stemflow or throughfall, relative to precipitation collected in the open, leaching is presumed. Parker (1983, 1987) and Miller (1984) have provided comprehensive reviews of canopy leaching and hydrology.

One of the initial efforts that attempted to quantify increases in tree leaching by acidifying substances was conducted in Norway by Abrahamsen et al. (1976a). These investigators compared throughfall collected in forest ecosystems in southern Norway (higher pollution loads) with systems in northern Norway (lower pollution loads). Throughfall enrichment of sulfate, calcium, and potassium was greater in southern Norway than in northern Norway. The authors judged, however, that conclusive interpretation of their data was impossible and that it was probable that a larger part of the throughfall enrichment in chloride, sulfate, calcium, and sodium was derived from dry deposition than from leached metabolites.

There is evidence of acidification of throughfall, stemflow, and tree surfaces in natural forests. Baker et al. (1976) have examined forest ecosystems in

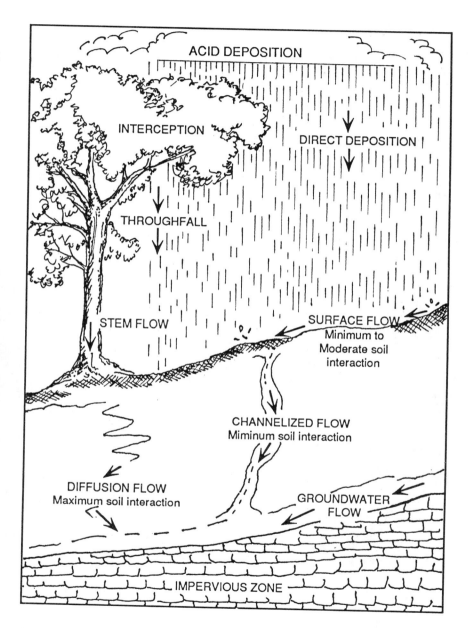

Figure 10-2. Movement of precipitation through a forest system.
Source: U.S.E.P.A. (1985).

Alberta, Canada adjacent to large industrial sources of sulfur dioxide. The latter gas appeared to have an acidifying effect on open rainfall, throughfall, and stemflow near the sources relative to sampling sites more distant from the sources. Grodzinska (1976) was able to find a positive correlation between tree bark pH and regional sulfur dioxide concentration in Poland, and Staxäng (1969) found a correlation between tree bark pH and regional pollution in Sweden. Grether (1976), on the other hand, was unable to find significant increases in bark acidity of trees in the vicinity of an electric generating facility in Minnesota.

Cronan and Reiners (1983) proposed that neutralization of acid precipitation in forest canopies occurs by two primary processes: ion exchange removal of free hydrogen ions (H+) by foliage; and Brønsted base leaching from the tree canopy. Based on their studies of comparative throughfall chemistry in New Hampshire, where precipitation pH averaged 4.1, they observed that northern hardwood forest canopies produced a throughfall chemistry that is less acid and higher in basic cations than either direct precipitation or throughfall collected from nearby conifer (balsam fir) forests. In a comparable study of throughfall chemistry in paired hardwood and conifer forests in New York (average precipitation pH 4.06), Mollitor and Raynal (1983) also recorded that the hardwood canopy produced a less acid throughfall than the conifer canopy, but that greater amounts of potassium, calcium, magnesium, and sodium were contained in conifer throughfall. Evidence from the Acadia Forest Experiment Station, central New Brunswick, Canada — mean pH 4.75 — also supported the enhanced ability of hardwood canopies relative to conifer canopies to neutralize precipitation deposited hydrogen ions (Mahendrappa 1983) (Table 10-3).

Subalpine conifer forests in the eastern United States receive significant canopy input via cloud droplet deposition as well as incident precipitation. Recent studies have investigated the throughfall chemistry of these systems. Reiners et al. (1985) have indicated that balsam fir throughfall and stemflow solutions are typically highly enriched in calcium, magnesium, potassium and sul-

Table 10-3. Percent precipitation hydrogen ion neutralized by various tree canopies of the Acadia Forest Experiment Station, New Brunswick, Canada, during field studies conducted 1977–1981.

Species	Percent H+ neutralized
Red spruce	46
White spruce	67
Red pine	43
White pine	47
Balsam fir	53
Larch	21
Maple	69
Birch	40
Aspen	80

Source: Mahendrappa (1983).

fate; slightly enriched in hydrogen, sodium and occasionally nitrate; and depleted in ammonium and occasionally nitrate (Figure 10-3). Similar enrichment of cations was indicated in quantification of red spruce and Fraser fir throughfall and stemflow on Mt. Mitchell, NC (Robarge et al. 1987). Joslin and Brewer (1986) indicated that the more acidic the cloud event, the more calcium concentration increased in red spruce throughfall on Whitetop Mountain, VA.

Foliar loss of nutrients via leaching has been implicated in forest stress symptomology. Mies and Zöttl (1985), in their studies of spruce in the Black Forest, FRG, indicated that older needle yellowing was due to the mobilization and loss of magnesium and zinc.

There are, however, major limitations on our understanding of atmospheric deposition and canopy leaching. Parker (1987) details these limitations. First, throughfall and stemflow input to forest soil is a normal and common feature of forest nutrient cycles and is not solely due to atmospheric stress. Second, and more important, field studies of throughfall and stemflow generally do not dis-

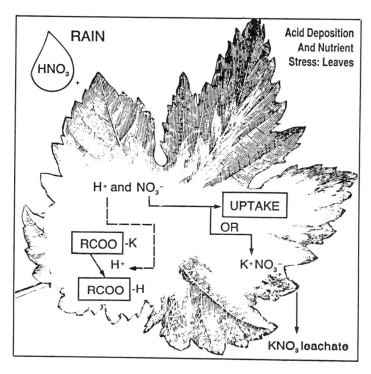

Figure 10-3. Foliar leaching can result from the deposition of nitric acid. Hydrogen ions from nitric acid may displace nutrient cations held on leaf cell-wall exchange sites. Nitrate ions from the acid, if not taken up by the leaf or leaf microorganisms, may be available to combine with potassium and to remove this nutrient from the leaf. If potassium ions are not resupplied by the roots, forest trees may suffer from nutrient stress.

tinguish ions actually leached from foliar surfaces from those dry deposited on canopy surfaces and subsequently washed off the foliage during precipitation events. Third, reductions of hydrogen ion input by the canopy may not be reflected in reduced hydrogen ion loading of the soil as rhizosphere acidification may result as replacement cations are taken up from the soil solution by impacted trees. Fourth, the long-term significance of foliar cation loss is regulated by the ability of forest soils to provide replacement nutrients to the available nutrient pool.

Lindberg et al. (1986) have investigated nutrient cycling in a mature oak-hickory forest at Walker Branch, Tennessee, and propose that dry deposition of atmospheric inputs represents a major input of elements to the forest. Dry deposits of sulfur, nitrogen, and free acidity were primarily in the gas form, and calcium and potassium were primarily in the particulate form. During the growing season, the oak-hickory canopy was estimated to remove over 70% percent of the total wet plus dry deposition of free hydrogen ion via ion-exchange and weak-base buffering reactions. The authors estimated that 40–60% of the nutrients leached from the forest canopy were due to exchange for deposited free acidity. What component of these nutrients had been contained within the leaf, or simply dry deposited to the leaf surface, cannot be determined. The accelerated movement of mobile nutrients from the canopy may lead to rhizosphere acidification as protons are released in association with increased ion uptake by roots (Leonardi and Flückiger, 1987; Matzner, 1985) (Chapter 9).

2. Pollutant Synergism and Foliar Leaching

Pollutant exposure may predispose foliage to leaching loss by cuticular erosion, membrane dysfunction, or metabolic abnormality.

Epicuticular wax, the outermost layer of plant leaves, consists of a complex and variable mixture of long-chain alkanes, alkenes, aromatic hydrocarbons, fatty acids, ketones, aldehydes, alcohols and esters (Martin and Juniper, 1970). The composition and integrity of this layer is strongly controlled by climate, foliar age, and air contaminants. Several reports, based on field observations in polluted environments, have emphasized the eroded appearance of surface wax (Percy and Riding, 1978; Karhu and Huttunen, 1986; Grill et al., 1987).

Trimble et al. (1982) exposed ozone tolerant and sensitive clones of 3 yearr old eastern white pine to 300 ppb (588 μg m^{-3}) ozone for 6 hr day^{-1} for 7 days. Alkane concentration (hentricontane-C$_{31}$ was the dominant alkane) was significantly greater for the tolerant clone regardless of ozone exposure. Karhu and Huttenen (1986) exposed Norway spruce foliage to ozone at lower exposures and found needle surface injury similar to field observations. Keller (1986) also treated Norway spruce foliage with water, sulfur dioxide, and ozone stress, and found that water stress and sulfur dioxide, but not ozone, increased leaching loss. Long-term exposure of Norway spruce clones to ozone, sulfur dioxide, or both gases together suggested sulfur dioxide exposure was most effective in enhancing leaching of calcium, magnesium and potassium (Guderian et al., 1987). Vogels et al. (1986), on the other hand, reported that spruce foliage exposed to ozone had higher magnesium and calcium leaching rates than those exposed to sulfur dioxide. Glatzel et al. (1987) have suggested that the leaching losses of magnesium, potassium, manganese, and zinc are especially high when precipitation water has high ammonium concentration and low pH. Zoettle and Huettl (1986) have stressed the synergism between oxidants and acid precipitation, especially fog, in leaching losses from forest trees.

In a related hypothesis, Makela and Huttunen (1987) have proposed that sulfur dioxide exposure causes cuticular erosion, which leads to excessive foliar water loss with the associated risk of winter moisture stress.

B. Soil Leaching

The chemistry of soil solutions is very variable and depends on a complex series of equilibrium, adsorption, displacement, immobilization, weathering, and decomposition reactions well beyond the scope of this book. It is critically important, however, to consider soil leaching, as changes in soil solution chemistry have been judged to depend mainly on transfers by leaching (Feller, 1977), and leaching rates have been hypothesized to be influenced by acid precipitation.

Most metallic nutrient elements taken up by trees are absorbed as cations and exist in three forms in the soil: (a) as slightly soluble components of mineral or organic material, (b) adsorbed into the cation exchange complex of clay and organic matter, and (c) in small quantities in the soil solution (Etherington, 1975). As water migrates through the soil profile, the movement of cations from any of the above three compartments is known as *cation desaturation* or *leaching*. An excellent overview of the leaching process and its role in nutrient cycling has been provided by Trudgill (1977).

The relative mobilites of cations leached from decomposing litter are typically sodium > potassium > calcium > magnesium. The leaching rate is largely dependent on the generation of a supply of mobile anions. In forest soils the anions are produced along with hydrogen cations and occur as acids. The hydrogen cations have a very powerful substituting capability and can readily replace other cations adsorbed to the soil. The mobile anions function to move released cations through the soil (Feller, 1977). In soils of low atmospheric deposition and minimal human disturbance, carbonic and organic acids dominate the leaching process (Cronan et al., 1978). In their review of cation leaching of forest soils, Johnson et al. (1983) emphasized that carbonic acid leaching dominates natural leaching processes in tropical and temperate coniferous sites; nitric acid, resulting from nitrification, dominates in nitrogen fixing temperate deciduous sites; and organic acids dominate surface soil leaching in subalpine sites and contribute to leaching in numerous additional sites. In regions with sufficient moisture, cations will be transported from the forest floor in proportion to the availability of HCO_3^- or organic acid anions or both (Cronan et al., 1978; Feller, 1977; Graustein et al., 1977).

In forest soils that are subject to the deposition of air pollutants, it has been proposed that sulfuric and nitric acids may provide the primary, or a significant, source of H^+ for cation displacement and mobile anions for cation transport. Extensive reviews are available on this topic (Johnson et al. 1983, Johnson et al. 1982, U.S.E.P.A. 1978, 1984; Voigt, 1980, Frink and Voigt, 1976). What is the evidence?

1. Sulfate Dynamics

As mentioned, soil cation leaching requires the mobility of the anion associated with the leaching acid. This results from the requirement for charge balance in soil solutions, a requirement that disallows cation leaching without associated mobile anions (Johnson et al., 1982). Nitric and sulfuric acid deposited from the atmosphere supply mobile anions in the form of nitrate (NO_3^-) and sulfate (SO_4^{2-}) respectively. In most soils, rapid biological uptake may immobilize the NO_3^- anion. As a result, risk of cation loss associated with this anion may be restricted to ecosystems rich in nitrogen where biological immobilization of NO_3^- is minimal (Johnson et al., 1982). In a similar manner, SO_4^{2-} may be immobilized in weathered soils by adsorption to free iron and aluminum oxides. In other soils, however, especially those low in free iron or aluminum, or high in organic matter, which appears to block SO_4^{2-} absorption sites, SO_4^{2-} may readily combine with nutrient cations and leach the elements beyond the rooting zone (Figure 10–4).

Adsorbed SO_4^{2-} is readily available to plants and in numerous soils it provides the major supply of plant sulfur. Sulfur requirements of most forests are modest (< 5 kg ha^{-1} yr^{-1} for net vegetative increment), however, and in polluted regions where sulfur deposition is high (10–80 kg ha^{-1} yr^{-1}, Table 2-15), the ecosystem requirement for sulfur is readily saturated (Johnson, 1984). Where biological demand is saturated and retentive capacity is low, leaching losses may occur. Eastern North American podzolic zones may be at special risk (Foster et al., 1983).

2. Laboratory and Microcosm Studies

When cation leaching has been enhanced by acid inputs in laboratory studies, Ca^{2+} and Mg^{2+} appear most affected, while K^+ is least affected (Johnson et al., 1982).

Pots containing eastern white pine seedlings were subjected to one weekly 6-hr application of artificial acid rain adjusted to pH 5.6, 4.0, 3.3, 3.0, and 2.3 for a 20-week period (Wood and Bormann, 1977). Leaching of magnesium and calcium steadily increased with "rain" acidity through the pH range 5.6–2.3. Potassium losses did not increase until the pH was lowered to 3.0 or 2.3 (Figure 10-5). Declines in exchangeable potassium, magnesium, and calcium were observed at pH levels 3.0 and below.

Cronon (1980b) prepared microcosms using forest floor material collected from the subalpine zone (1462 m) of Mt. Moosilauke, New Hampshire. When

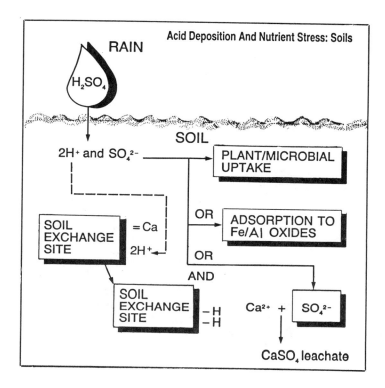

Figure 10-4. Leaching of forest soil can result from deposition of sulfuric acid. Hydrogen ions from the acid may displace nutrient cations held on organic or clay exchange sites. Sulfate ions from sulfuric acid, if not taken up by the soil biota or adsorbed to iron or aluminum oxides, may be available to combine with calcium and to remove this nutrient from the rooting zone. If a sufficient amount of calcium ions is not resupplied by decomposition or soil weathering, forest trees may suffer from nutrient stress.

acidity of throughfall inputs to the microcosms was increased, the forest floors exhibited increased leaching losses of Ca^{2+}, Mg^{2+}, K^+ and NH_4^+. Based on his results, Cronon estimated that with a throughfall pH of 3.5, and in the absence of net plant uptake and recycling, approximately 0.8% of the total forest floor calcium pool could be lost annually. Similar calculations for magnesium, potassium, and nitrogen suggested leaching losses approximating 3%, 85%, and 6–16% respectively.

A variety of microcosm studies, especially when employing treatment solutions of low pH (~3.5), emphasize the leaching potential of sulfuric acid (Fernandez, 1987; Fernandez and Kosian, 1986; Kelly and Strickland 1986, Kelly and Strickland, 1986; Stroo and Alexander, 1986). McClenahen (1987) emphasized that the SO_4^{2-} component of simulant throughfall had a greater impact than pH in leaching from deciduous forest soils.

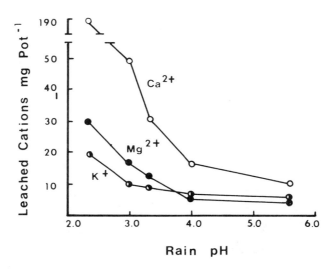

Figure 10-5. Estimated total leaching losses of potassium, magnesium, and calcium from pots containing white pine seedlings following 6-hr weekly treatments of artificial rain of various pHs for 20 weeks.
Source: Wood and Bormann (1977).

3. Lysimeter and Forest Studies

Overrein (1972) subjected 40-cm deep forest soil profiles with differing physical and chemical properties, collected from southern Norway, to precipitation with pH levels adjusted from 2.0 to 5.0. When the pH of simulated precipitation was less than 4, he observed a sharp increase in the rate of calcium leaching relative to distilled water controls. At pH 2, soil analyses made over the 40-day experimental period revealed a gradual acidification of the soil starting at the profile top and gradually progressing through the entire column. Mayer and Ulrich (1976) have monitored for several years lysimeters subject to natural precipitation in spruce and beech forests in central Germany. The hydrogen ion input to the mineral soil of the beech forest was determined to be 2 kiloequivalents ha^{-1} yr^{-1}, of which 60% was from precipitation input and 40% from mineralization of nitrogen and sulfur compounds via litter decomposition. Aluminum, manganese, sodium, potassium, calcium, and magnesium were lost from the forest. Johnson and Cole (1976) have leached lysimeters established in second-growth Douglas fir in Washington with aliquots of sulfuric acid ranging in concentration from 10 to 1000 times higher (10^{-3}, 10^{-2}, and 10^{-1} N H_2SO_4) than ambient precipitation at present. Cation and sulfate concentrations were monitored for 3 months after treatment. Approximately two thirds of the estimated supply of sodium, calcium, and potassium were removed from the forest floor and A horizon, but almost none of the cations were moved beyond 50 cm in the B horizon. This restriction may have been due to the limited mobility of sulfate in this horizon, and the authors cautioned that failure to consider sulfate mobility could lead to overestimations of cation transfer due to acid precipitation. Abrahamsen et al. (1976b) have recorded decreased cation saturation, mainly due to leaching of calcium and magnesium in lysimeters containing Norwegian forest soil and subject to 50 mm month^{-1} of simulated rain water of pH 3. Tamm et al. (1976) applied sulfuric acid to the soil of a young pine forest north of Stockholm at a rate equivalent to 100 kg of acid ha^{-1}. The soil appeared relatively resistant to leaching, and much of the sulfate ions and cations were retained in the lysimeter soil. The addition of ammonium nitrate to fertilized lysimeters led to an increased leaching of hydrogen ions and cations.

Cronan et al. (1978) and Cronan (1980a) have analyzed soil water and groundwater samples in subalpine balsam fir forests in New Hampshire for HCO_3^-, SO_4^{2-}, and organic acid anions in an effort to determine the relative importance of these compounds as sources of H^+ and mobile anions. Their results indicated that sulfate anions supplied 76% of the electrical charge balance in the leaching solutions, suggesting that atmospheric input of sulfuric acid provided the primary mechanism of cation replacement and transport in this region. Preliminary analyses of additional field and laboratory data led these authors to suggest that cation leaching in the subalpine New England fir forest increases as the rainfall pH drops below 4–4.5. Baker et al. (1976) have studied the impact of sulfur dioxide released from natural gas treatment plants in Alberta, Canada on the soils of the surrounding forest ecosystem. Soils collected near (within 18 km or less) and far (32–78 km) from the point source were compared for exchangeable cations (Table 10-4). While the differences in cation characteristics between the sites were not extreme, the decreased calcium and increased aluminum close to the source are significant. Aluminum ions rapidly react with water to form insoluble aluminum hydroxide and hydrogen ions (Etherington, 1975). This highlights the difficulty of assigning relative importance to the var-

Table 10-4. pH and Exchangeable Cations in 1 N KCl Extracts of Forest Soils Sampled Close to and Far from a Sulfur Dioxide Point Source in Alberta, Canada.

| Location | Horizon | pH | Cations (meq 100 g^{-1}) | | | | | |
			Al	Ca	Mg	Fe	Mn	Total bases
Distant sites	L-F-H	6.0	1.1	47.0	7.5	0.3	0.1	54.9
(mean)	0-5 cm	5.2	0.5	10.6	1.9	nil	nil	12.5
	5-15 cm	5.1	0.7	7.7	1.4	nil	nil	9.1
	15-30 cm	5.0	0.5	9.4	2.0	nil	nil	11.4
Near sites	L-F-H	4.5	1.5	29.4	5.2	0.6	1.6	36.8
(mean)	0-5 cm	4.2	3.2	8.5	1.8	0.1	0.1	10.5
	5-15 cm	4.4	2.2	6.4	1.5	0.1	nil	8.0
	15-30 cm	4.4	1.7	8.6	2.2	nil	nil	10.8

Source: Baker et al. (1976).

ious sources of hydrogen ions in acid forest soils subject to air pollution stress. Exchangeable calcium and magnesium concentrations were relatively low in soils of lowland forests in southeastern Ohio most affected by acid mine drainage, but were significantly higher in soils less affected by this drainage (Cribbin and Scacchetti, 1976). The authors judged that altered nutrient status, along with high acidity and aluminum concentrations, of the soil exerted strong selection pressures in certain Ohio river-bottom forest communities.

Cronan and Schofield (1979) have pointed out an additional aspect of soil aluminum chemistry in fir zone soils in New England that may be important in other noncalcareous forest ecosystems subject to acid precipitation. In most podzolized soils, acidification of the soil solution and organometallic complexation mediated by organic acids results in movement of aluminum and iron into lower horizons. In New Hampshire balsam fir forests, however, the soils exhibited limited increases in soil solution pH with depth, and amorphous and exchangeable aluminum were mobilized and tended to be transported in solution out of the profile and ultimately into streams.

Mollitor and Raynal (1982) studied cation leaching in hardwood and conifer stands in Huntington Forest, Newcomb, New York. Cation leaching from the deciduous site appeared equally influenced by SO_4^{2-} and organic anions. Sulfate and organic anion concentrations were greater in the conifer site, but organic anion leaching was judged to dominate in these stands.

Studies that add vegetation to soil lysimeters and that explore the potential interactive nature of multiple pollutants (e.g., Seufert et al., 1988) are sorely needed.

Despite the efficacy of SO_4^{2-} leaching in selected soils, it must be realized that forest soils vary greatly in their susceptibility to SO_4^{2-} leaching (Table 10-5), that mineral weathering and deep rooting constantly replenish base cations for tree growth, and that, even on sensitive soils, leaching losses affect cation reserves only on extended time scales (Richter et al., 1983).

Reinhard Hüttl, Institute of Soil Science and Forest Nutrition, University of Freiburg, FRG has studied Norway spruce health in southwestern Germany. Hüttl and his colleagues have developed an integrated hypothesis for Norway spruce stress that importantly involves both foliar leaching and soil leaching with associated nutrient stress symptoms. Multiple pollutants, including acid deposition and oxidants, are presumed to be involved. Stress symptoms in Norway spruce could be ameliorated by fertilization, especially with manganese (Hüettl 1986, Hüttl and Wisniewski, 1987; Huettl et al., 1987).

C. Soil Weathering

Chemical weathering involves hydrolysis, hydration, oxidation-reduction reactions, carbonation, and solution of compounds and elements from parent soil material (Frink and Voigt, 1976). Numerous reactants participate in weathering processes. Commonly the most significant reactant is the hydrogen ion. The latter in solution reacts with a cation in a mineral and replaces it in the crystal

Table 10-5. General sensitivity of soils to cation leaching.

| | Sensitivity | | |
	Low	Moderate	High
1. Soil Chemistry			
a. Exchangeable bases	>15 mEq 100 g^{-1}	6–15 mEq 100g^{-1}	< 6 mEq 100g^{-1}
b. pH	clayey, pH > 5.0 loamy, pH > 5.5 all calcareous soils	clayey, pH 4.5–5.0 loamy, pH 5.0–5.5 sandy, pH 5.5	clayey, pH < 4.5 loamy, pH < 5.0 sandy, pH < 5.5
c. Texture	clay, silty clay, sandy clay (> 35% clay)	silty clay loam, clay loam, sandy clay loam, silt loam, loam 10–35% clay)	silty, sandy loam loamy sand loamy sand (< 10% clay)
d. Cation exchange capacity (CEC)	> 25 mEq 100g^{-1}	10–25 mEq 100g^{-1}	< 10 mEq 100g^{-1}
e. SO$_4^{2-}$ adsorption capacity as determined by:	low organic matter and high Al$_2$O$_3$ and/or Fe$_2$O$_3$ + Fe$_3$O$_4$ (high SO$_4^{2-}$ adsorption)		high organic matter and/or low Al$_2$O$_3$ and/or Fe$_2$O$_3$ + Fe$_3$O$_4$ (low SO$_4^{2-}$ adsorption)
2. Soil Depth	> 25 cm	> 25 cm	< 25 cm
3. Underlying Material			
a. Parent Material	carbonate bearing	non-carbonate bearing	non-carbonate bearing
b. Bedrock Material	limestone, dolomite and metamorphic equivalents, calcarcous clastic rocks, carbonate rocks interbedded with noncarbonate rocks	volcanic rocks, shales, greywacke, sandstones, ultramafic rocks, gabbro, mudston, meta-equivalents	granite, granite gneiss, orthoquartzite, syenite

Source: Olson et al. (1981).

lattice of the mineral. Soil acids are the primary source of hydrogen ions for weathering (Trudgill, 1977). In soils, mineral weathering is the initial source of required nutrients other than nitrogen. In the absence of fertilization, nutrient supply by weathering assumes a special importance in forest ecosystems. The hypothesis that mineral acids resulting from air pollution can importantly supplement weathering reactions principally due to carbonic acid and organic acids is an important topic. White et al. (1988) have provided a comprehensive review of this issue.

In their pot trials of seedling white pine subjected to artificial precipitation acidified to various pH levels, Wood and Bormann (1977) observed that the treatment appeared to accelerate the weathering rate of their sandy loam greenhouse soil. Estimated weathering inputs of potassium and magnesium increased linearly with logarithmic increases in the acidity of the simulated rain. Three- to fourfold increases coincided with 1500-fold increases in acidity from pH 5.6 to 2.3. The weathering of calcium responded more logarithmically with an approximate 22-fold increase over the same pH range.

In the forest ecosystems in the United States known to have soil leaching processes dominated by mineral acids, the New England region, the role of sulfuric and nitric acids in chemical weathering has been judged to be relatively small (Johnson et al., 1972; Johnson, 1979). By including biomass accumulation in the estimation of cationic denudation rates for New Hampshire forests, however, the values of Johnson et al. (1972) are doubled and approximate 2.0×10^3 equivalents ha^{-1} yr^{-1} (Likens et al., 1977). The latter authors have estimated that this cationic denudation is balanced by H$^+$ ions produced in equal amounts by internal generation (chemical reactions of soil nitrogen, carbon, and sulfur) and external generation (meteorologic input of mineral acids). This infers considerable importance to weathering associated with acid precipitation.

Cronan (1985) estimated weathering rates for three eastern North American forest soils and found a fivefold variation ranging from 0.5 Keq ha^{-1} yr^{-1} for the sandy Adams spodosol, to 1.2 Keq ha^{-1} yr.$^{-1}$ for the sandy loam Becket spodosol, to 2.7 Keq ha^{-1} yr^{-1} for the silty Unadilla inceptisol. Major differences in weathering rates were found to occur between soil horizons with A horizon rates exceeding B rates. Cronan also explored the influence of selected environmental variables on the weathering process. Soil column cation denudation increased in the spodosol and inceptisol as treatment inputs of acid deposition increased from 0.04 Keq ha^{-1} yr^{-1}. Unique characteristics of biogeochemical cycling associated with upland New England forest systems and the unusually high input of acid precipitation in this region preclude extrapolation of the circumstance of this location to the more general temperate forest situation.

White et al. (1988) stress, however, that despite large differences in experimental approaches and assumptions, weathering rates reported in the literature are reasonably consistent (Table 10-6).

D. Summary

Anthropogenic release of sulfur and nitrogen oxides has increased the acidity of precipitation to 10–100 times preindustrial levels. In certain sections of the world, most notably the northeastern United States, eastern Canada, northern Europe, and Scandinavia, large areas of forest are currently subjected to precipitation of dramatically reduced pH.

The data reviewed to support the hypothesis that acid precipitation accelerates leaching loss of nutrients from forest foliage are not convincing. The limited analyses of throughfall in natural environments have not adequately differentiated

Table 10-6. Nutrient release (weathering) data from mass balance studies of forest watersheds.

Location	Parent materials	Na	K	Ca	M
			-----------------kg ha^{-1} yr^{-1}-----------------		
White Mts., CA	dolomite	2	4	86	52
	qtz monzonite	1	8	17	2
Cascades, OR	tuffs/breccias	28	1.6	47	11.6
Luxembourg	metashale	9.1	0.2	8.7	15.7
Piedmont, MD	schist	2.6	2.3	1.3	1.7
	serpentinite	tr	tr	tr	34.1
Hubbard Brook, NH	moraine/gneiss	5.8	7.1	21.1	3.5
Brookhaven, NY	glacial outwash	6.7	11.1	24.2	8.4
Central Wisconsin	glacial outwash	–	6.9	22	5.0
SW Idaho	qtz monzonite	13.5	4.3	20.1	2.4
Piedmont, VA	granite	24.8	1.3	7.9	2.6
Boheimian Massif	biotite-muscovite gneiss	5.0	9.4	3.7	2.9
Czechoslovakia	biotite gneiss	13.0	23.0	19.0	14.0
	quartzitic gneiss	6.0	13.0	8.5	6.3
Scotland	till	10.1	2.37	17.2	5.25
	gabbro	2.26	1.18	20.2	5.94
Range low		tr	tr	tr	1.7
high		28	23	86	52
Average		8.65	5.98	20.24	10.84
Standard deviation		8.31	6.05	20.89	13.67

Source: White et al. (1988).

material actually leached from the plants from material deposited on, and then subsequently washed from, the foliage. Studies that have employed seedling trees subjected to artificially acidified rain have revealed a potential for foliar nutrient loss, but only at pH levels of 4.0 or less. In the presence of foliar damage to the cuticle occasioned by acid rain, nutrient loss from leaves could be substantial. The threshold for this damage, however, appears to be approximately pH 3 or less for numerous forest trees, and this intensity of precipitation acidification is not widespread in natural environments at the present time. Ozone exposure may increase leakage of foliar metabolites by increasing membrane permeability. This may enhance foliar leaching associated with acid deposition.

The movement of nutrient cations via leaching in forest soil profiles is an extremely important component of forest nutrient cycling. The evidence that has been provided by numerous experiments subjecting soil lysimeters to natural or artificially acidified precipitation indicates a potential for a meaningful acid precipitation influence on the soil leaching process. The threshold for significant increases in the rate of movement of calcium, potassium, and magnesium appears to require precipitation in the pH range of 3–4 for most systems examined. For certain forests, for example, subalpine balsam fir in New England, the threshold of increased leaching may be higher and in the range of pH 4.0–4.5.

Lysimeter data reflects an integration of many complex soil processes. As a result, this evidence does not provide answers to numerous important questions concerning acid deposition impact on leaching. What are the relative efficiencies of cation ion transfer between the forest floor and the A soil horizon and between the A and B soil horizons? What are the specific sources of H^+ ions in the soil profile? It is clear that sulfate soil dynamics are of fundamental importance in regulating strong mineral acid leaching of forest soils. Immobilization of sulfate can effectively prevent cation leaching. Soils at risk to nutrient loss via leaching by sulfuric acid, therefore, are restricted to soils low in free iron and aluminum oxides or high in inorganic matter. In addition, nutrient depletion via leaching by acid deposition would be an extended term, decades to centuries, process.

Mineral weathering represents a significant soil resistance mechanism to the impacts of acid deposition. Weathering reactions function as a sink for protons and as sources of plant nutrients. If weathering rates increase with increased deposition, these reactions may act to partially compensate for adverse soil influences.

Forest soils have been judged more vulnerable to influence by acid precipitation than agricultural soils. Forest soils supporting early regeneration following harvest or severe natural events may be especially vulnerable to an adverse impact on nutrient cycling by acid rain as the system "controls" on nutrient conservation are weakest at this time (Tamm, 1976; Likens et al., 1977). The mitigation of forest stress symptomology by fertilization (Zöettl 1987a,b, Zöettl and Huettl, 1986, Huettl and Wisniewski, 1987) is consistent with the hypothesis that regional scale air pollutants may be adversely influencing forest nutrient cycles. The bulk of the current evidence, however, remains consistent with the conclusion of Frink and Voigt (1976) that unless the acidity of precipitation increases substantially or the buffering capacity of forest soils declines significantly, acid rain influence will not quickly nor dramatically alter the productivity of most temperate forest soils.

References

Abrahamsen, G., R. Horntveldt, and B. Tveite.1976a. Impacts of acid precipitation on coniferous forest ecosystems. In L.S. Dochinger and T.A. Seliga, eds., Proc. 1st Intl. Symp. Acid Precipitation and the Forest Ecosystem. U.S.D.A. Forest Service, Gen. Tech. Rep. No. NE-23, Upper Darby, PA, pp. 991-1009.

Abrahamsen, G., K. Bjor, R. Horntveldt, and B. Tveite. 1976b. Effects of acid precipitation on coniferous forests. In F.H. Braeke, ed., Impact of Acid Precipitation on Forest and Freshwater Ecosystems in Norway. Research Report No. 6 SNF Project, Oslo, Norway, pp. 37-63.

Baker, J., D. Hocking, and M. Nyborg. 1976. Acidity of open and intercepted precipitation in forests and effects on forest soils in Alberta, Canada. In L.S. Dochinger and T.A. Seliga, eds., Proc. 1st Intl. Symp. Acid Precipitation and the Forest

Ecosystem. U.S.D.A. Forest Service, Gen. Tech. Rep. No. NE-23, Upper Darby, PA, pp. 779-790.

Cogbill, C.V. 1975. The history and character of acid precipitation in eastern North America. In L.S. Dochinger and T.A. Seliga, eds., Proc. 1st Intl. Symp. Acid Precipitation and the Forest Ecosystem. U.S.D.A. Forest Service, Gen. Tech. Rep. No. NE-23, Upper Darby, PA, pp. 363-370.

Cribbin, L.D. and D.D. Scacchetti. 1976. Diversity in tree species in southeastern Ohio Betula nigra L. communities. In L.S. Dochinger and T.A. Seliga, eds., 1st Intl. Symp. Acid Precipitation and the Forest Ecosystem, U.S.D.A. Forest Service, Gen. Tech. Rep. No. NE-23, Upper Darby, PA, pp. 779-790.

Cronan, C.S. 1980a. Solution chemistry of a New Hampshire subalpine ecosystem: A biogeochemical analysis. Oikos 34: 272-281.

Cronan, C.S. 1980b. Controls on leaching from coniferous forest floor microcosms. Plant Soil 56: 301-322.

Cronan, C.S. 1985. Chemical weathering and solution chemistry in acid forest soils: Differential influence of soil type, biotic processes, and H^+ deposition. In J.I. Dreuer, ed., The Chemistry of Weathering. D. Reidel Publish. Co., New York, pp. 175-195.

Cronan, C.S. and C.L. Schofield. 1979. Aluminum leaching response to acid precipitation: Effects on high-elevation watersheds in the Northeast. Science 204: 304-306.

Cronan, C.S. and W.A. Reiners. 1983. Canopy processing of acidic precipitation by coniferous and hardwood forests in New England. Oecologia 59: 216-223.

Cronan, C.S., W.A. Reiners, R.C. Reynolds Jr., and G.E. Lang. 1978. Forest floor leaching: Contributions from mineral, organic and carbonic acids in New Hampshire subalpine forests. Science 200: 309-311.

Dochinger, L.S. and T.A. Seliga, eds. 1976. Proc. 1st Intl. Symposium on Acid Precipitation and the Forest Ecosystem. U.S.D.A. Forest Service, Gen. Tech. Rep. No. NE-23, Upper Darby, PA, 1074 pp.

Etherington, J.R. 1975. Environment and Plant Ecology. Wiley, New York, 347 pp.

Evans, L.S., N.F. Gmur, and F. DaCosta. 1978. Foliar response of a six clones of hybrid poplar. Phytopathology 68: 847-856.

Feller, M.C.1977. Nutrient movement through western hemlock-western red cedar ecosystems in southwestern British Columbia. Ecology 58: 1269-1293.

Fernandez, I.J. 1987. Vertical trends in the chemistry of forest soil microcosm following experimental acidification. Tech. Bull. No. 126, Maine Ag. Exp. Sta., Orono, ME, 19 pp.

Fernandez, I.J. and P. Kosian. 1986. Chemical response of soil leachate to alternative approaches to experimental acidification. Commun. Soil Sci. Plant. Anal. 17: 953-973.

Foster, N.W., J.A. Nicolson, and I. K. Morrison. 1983. Acid deposition and element cycling in eastern North American forests. Proc. Acid Rain and Forest Resources Conference. Quebec City, Canada, June, 1983.

Frink, C.R. and G.K. Voigt. 1976. Potential effects of acid precipitation on soils in the humid temperate zone. In L.S. Dochinger and T.A. Seliga, eds., 1st Intl. Symp. Acid Precipitation and the Forest Ecosystem, U.S.D.A. Forest Service, Gen. Tech. Rep No. NE-23, Upper Darby, PA, pp. 685-709.

Galloway, J.N. and E.B. Cowling. 1978. The effects of precipitation on aquatic and terrestrial ecosystems. A proposed precipitation chemisty network. J. Air Pollu. Control Assoc. 28: 229-235.

Galloway, J.N., G. E. Likens, and E.S. Edgerton. 1976. Acid precipitation in the northeastern United States: pH and acidity. Science 194:722-724.

Glatzel, V.G., M. Kazda, D. Grill, G. Halbwachs, and K. Katzensteiner. 1987. Nutritional disorders in spruce (Picea abies) as a consequence of damage to needle surfaces and deposition of atmospheric nitrogenous compounds: One of the mechanisms of forest decline? Allg. Forst. u. f. Ztg 158: 91-97.

Graustein, W.C., K. Cromack Jr., and P. Sollins. 1977. Calcium oxalate: Occurrence in soils and effect on nutrient and geochemical cycles. Science 198: 1252-1254.

Grennfelt, P., C. Bengtson, and L. Skärby. 1978. An estimation of the atmospheric input of acidifying substances to a forest ecosystem. Swedish Water and Air Pollution Res. Instit. No. B438, Gottenburg, Sweden, 12pp.

Grether, D.F. 1976. The effects of a high-stack coal-burning power plant on the relative pH of the superficial bark of hardwood trees. In L.S. Dochinger and T.A. Seliga, eds., Proc. 1st Intl. Symp. Acid Precipitation and the Forest Ecosystem. U.S.D.A. Forest Service, Gen. Tech. Rep. No. NE-23, Upper Darby, PA, pp. 913-918.

Grill, D., H. Pfeifhofer, G. Halbwachs, and H. Waltinger. 1987. Investigations on epicuticular waxes of different damaged spruce needles. Eur. J. For. Pathol. 17: 246-254.

Grodzinska, K. 1976. Acidity of tree bark as a bioindicator of forest pollution in southern Poland. In L.S. Dochinger and T.A. Seliga, eds., Proc. 1st Intl. Symp. Acid Precipitation and the Forest Ecosystem. U.S.D.A. Forest Service, Gen. Tech. Rep. No. NE-23, Upper Darby, PA, pp. 905-911.

Guderian, R., A. Klumpp, and K. Küppers. 1987. Gehalte and leaching von magnesium, calcium und kalium bei fichte (Picea abies Kant) nach einwirkung von ozon und schwefeldioxid. Verhandlungen der Gesellschaft für Okologie 16: 311-322.

Hitchocck, D.R. 1976. Atmospheric sulfates from biological sources. J. Air Pollu. Control Assoc. 26: 210-215.

Hüettl, R.F. 1986. Forest decline and nutritional disturbances. Proceedings Intl. Union of Forest Research Organizations (IUFRO) Congress. Ljubljana, Yugoslavia, Sept. 7–12, 1986.

Hüettl, R.F. and J. Wisniewski. 1987. Fertilization as a tool to instigate forest decline associated with nutrient deficiencies. Water Soil Air Pollu. 33: 265-276.

Hüettl, R.F., S. Fink, J.J. Lutz, M. Poth, and J. Wisiewski. 1987. Forest decline, nutrient supply and diagnostic fertilization in Southwestern Germany and in southern California. In Management of Water and Nutrient Relations to Increase Forest Growth, Intl. Union of Forest Research Organizations (IUFRO) Seminar. Canberra, Australia, Oct. 19-22, 1987.

Husar, R.B., J.P. Lodge Jr., and D. J. Moore. 1978. Sulfur in the Atmosphere. Proc. Intl. Symp., Dubrovnik, Yugoslavia, 7–14 Sept. 1977. Atmos. Environ. 12: 1-796.

Johnson, D.W. 1984. Sulfur cycling in forests. Biogeochemistry 1: 29-43.

Johnson, D.W., and D.W. Cole. 1976. Sulfate mobility in an outwash soil in western Washington. In L.S. Dochinger and T.A. Seliga, eds., 1st Intl. Symp. Acid Precipitation and the Forest Ecosystem. U.S.D.A. Forest Service, Gen. Tech. Rep. No. NE-23, Upper Darby, PA, pp. 827-835.

Johnson, D.W., J. Turner, and J.M. Kelly. 1982. The effects of acid rain on forest nutrient status. Water Res. 18: 449-461.

Johnson, D.W., D.D. Richeter, H. Van Miegroet, and D.W. Cole. 1983. Contributions of acid deposition and natural processes to cation leaching from forest soils: A review. J. Air Pollu. Control Assoc. 33: 1036-1041.

Johnson, N.M. 1979. Acid rain: Neutralization within the Hubbard Brook ecosystem and regional implications. Science 204: 497-499.

Johnson, N.M., R.C. Reynolds, and G.E. Likens. 1972. Atmospheric sulfur: Its effect on the chemical weathering of New England. Science 177: 514-515.

Joslin, J.D. and P.F. Brewer. 1986. Chemical interaction between cloud water and red spruce foliage. Third Annual Acid Rain Conf, Tenn. Valley Auth. Gatlinburg, TN, Nov. 1986, pp. 25-26.

Karhu, M. and S. Huttenen. 1986. Erosion effects of air pollution on needle surfaces. Water Air Soil Pollu. 31: 417-423.

Keller. T. 1986. The electrical conductivity of Norway spruce needle diffusate as affected by certain air pollutants. Tree Physiol. 1: 85-94.

Kelly, J.M. and R.C. Strickland. 1987. Soil nutrient leaching in response to simulated acid rain treatment. Water Air Soil Pollu. 34: 167-181.

Leonardi, S. and W. Flückiger. 1987. Short-term canopy interactions of beech trees; Mineral ion leaching and absorption during rainfall. Tree Physiol 3: 137-145.

Likens, G.E. 1975. Acid precipitation: Our understanding of the phenomenon. Proc. Conf. Emerging Environmental Problems: Acid Precipitation, May 1975, Renssalaerville, NY. EPA-902/9-75-001. U.S. Environmental Protection Agency, New York, 115 pp.

Likens, G.E. 1976. Acid Precipitation. Chem. Eng. News 54: 29-44.

Likens, G E., F.H. Bormann, and N.M. Johnson. 1972. Acid rain. Environment 14:33-40.

Likens, G.E., F.H. Bormann, R.S. Pierce, J.S. Eaton, and N.M. Johnson. 1977. Biogeochemistry of a Forested Ecosystem. Springer–Verlag, New York, 146 pp.

Liljestrand, H.M. and J.J. Morgan. 1978. Chemical composition of acid precipitation in Pasadena, California. Environ. Sci. Technol. 12: 1271-1273.

Lindberg, S.E.K., G.M. Lovett, D.D. Richter, and D.W. Johnson 1986. Atmospheric depositon and canopy interactions of major ions in a forest. Science 231: 141-145.

MacCracken, M.C. 1978. MAP3S: An investigation of atmospheric energy related pollutants in the northeastern United States. Atmos. Environ. 12: 649-660.

Mahendrappa, M.K. 983. Chemical characteristics of precipitation and hydrogen input in throughfall and stemflow under some eastern Canadian Forest stands. Can. J. For. Res. 13: 948-955.

Mäkelä, A. and S. Huttenen. 1987. Cuticular needle erosion and winter drought in polluted enviroments — A model analysis. Working Paper No. WP-87-48, Intl. Inst. for Applied Systems Analysis, Laxenburg, Austria, 25 pp.

Martin, J.T. and B.E. Juniper. 1970. The Cuticles of Plants. Edward Arnold Publishers, Edinburgh, UK, 347 pp.

Matzner, E. 1985. Deposition/canopy-interactions in two forest ecosystems of northwest Germany. In H. W. Georgii, ed., Atmospheric Pollutants in Forest Areas. D. Reidel Publ. Co. New York, pp. 247-262.

Mayer, R. and B. Ulrich. 1976. Acidity of precipitation as influenced by the filtering of atmospheric sulfur and nitrogen compounds — Its role in the element balance and effect on soil. In L.S. Dochinger and T.A. Seliga, eds., 1st Intl. Symp. Acid Precipittion and the Forest Ecosystem. U.S.D.A. Forest Service, Gen. Tech. Rep. No NE-23, Upper Darby, PA, pp. 737-743.

McClenahen, J.R. 1987. Effects of simulated throughfall pH and sulfate concentration on a deciduous forest soil. Water Air Soil Pollu. 35: 319-333.

McColl, J.G. and D.S. Bush. 1978. Precipitation and throughfall chemistry in the San Francisco Bay area. J. Environ. Qual. 7: 352-357.

Mies, V.E. and H.W. Zöttl. 1985. Chronological change in chlorophyll and element contents in the needles of a yellow-chlorotic spruce stand. Forstwissen-Schaftliches Centralblatt 104: 1-8.

Miller, H.G. 1984. Deposition-plant-soil interactions. Phil. Trans. Royal Soc. London B 305: 339-352.

Mollitor, A.V. and D.J. Raynal. 1982. Acid precipitation and ionic movements in Adirondack forest soils. Soil Sci. Soc. Am. J.46: 137-141.

Mollitor, A.V. and D.J. Raynal. 1983. Atmospheric deposition and ionic input in Adirondack forests. J. Air Pollu. Control Assoc. 33: 1032-1036.

Odén, S. 1976. The acidity problem — An outline of concepts. In L. S. Dochinger and T.A. Seliga, eds., Proc. 1st Intl. Symp. Acid Precipitation and the Forest Ecosystem. U.S.D.A. Forest Service, Gen. Tech. Rep. No. NE-23, Upper Darby, PA, pp. 1-36.

Olson, R.J., D.W. Johnson, and D.S. Shriner. 981. Regional assessment of potential sensitivity of soils in the eastern United States to acid precipitation. Office of Technology Assessment, U.S. Congress, Washington, DC.

Overrein, L.N. 1972. Sulfur pollution patterns observed; leaching of calcium in forest soil determined. Ambio 1: 145-147.

Ovington, J.D. 1962. Quantitative ecology and the woodland ecosystem concept. Adv. Ecol. Res. 1: 103-192.

Pack, D.H. 1980. Precipitation chemistry patterns: A two-network data set. Science 208: 1143-1145.

Parker, G.G. 1983. Throughfall and stemflow in the forest nutrient cycle. Adv. Ecol. Res. 13: 57-133.

Parker, G.G. 1987. Uptake and release of inorganic and organic ions by foliage: Evaluation of dry deposition, pollutant damage, and forest health with throughfall studies. National council of the Paper Industry for Air and Stream Improvement. Tech. Bull. No. 532, New York, 67 pp.

Percy, K E. and R.T. Riding. 1978. The epicuticular waxes of *Pinus strobus* subjected to air pollutants. Can. J. For. Res. 8: 474-477.

Perhac, R.M. 1978. Sulfate regional experiment in the northeastern United States: The SURE program. Atmos. Environ. 12: 641-648.

Rambo, D.L.1978. Interim Report: Acid precipitation in the United States, history, extent, sources, prognoses. U.S. Environmental Protection Agency, Contract No. 68-03-2650. Corvallis, OR, 24 pp.

Reiners, W.A., G.M. Lovett, and R.K. Olson. 1987. Chemical interactions of a forest canopy with the atmosphere. U.S. Department of Energy, Conf. No. CONF-85-10250, Washington, DC, pp. 111-146.

Richter, D.D., D.W. Johnson, and D.E. Todd. 1983. Atmospheric sulfur deposition, neutralization, and ion leaching in two deciduous forest ecosystems. J. Environ. Qual. 12: 263-270.

Robage, W.P., R.I. Bruck, and E.B. Cowling. 1987. Throughfall and stemflow measurements at Mt. Mitchell, NC during the summer of 1986. A preliminary report. G. Hertel, ed., Effects of Atmospheric Pollutants on the Spruce-Fir Forests of the Eastern United States and Federal Republic of Germany. U.S.D.A. Forest Service. Genl. Tech Rep. No 255, Broomall, PA.

Seufert, G., V. Arndt, J.J. Jäger, J. Bender, and B. Schweizer. Long-term effects of air pollutants on spruce (*Picea abies*) and fir (*Abies alba*) in open-top chambers. In G.D. Hertel, ed., Effects of Atmospheric Pollutants on the Spruce-Fir Forests of the Eastern United States and Federal Republic of Germany. U.S.D.A. For. Service, Genl. Tech. Rep. No. 255, Broomall, PA.

Staxäng, B. 1969. Acidification of bark of some deciduous trees. Oikos 20: 224-230.

Stroo, H.F. and M. Alexander. 1986. Available nitrogen and nitrogen cycling in forest soils exposed to simulated acid rain. Soil Sci. Soc. Am. J. 50: 110-114.

Tamm, C.O. 1951. Removal of plant nutrients from tree crowns by rain. Physiol. Plant 4: 184-188.

Tamm, C.O. 1976. Acid precipitation and forest soils. In L.S. Dochinger and Forest Ecosystem. U.S.D.A. Forest Service, Gen. Tech. Rep. No. NE-23, Upper Darby, PA, pp. 681-683.

Tamm, C.O. and E.B. Cowling. 1976. Acidic precipitation and forest vegetation. In L.S. Dochinger and T.A. Seliga, eds., Proc. 1st Intl. Symp. Acid Precipitation and the Forest Ecosystem. U.S.D.A. Forest Service, Gen. Tech. Rep. No. NE-23, Upper Darby, PA, pp. 845-855.

Tamm, C.O., G. Wiklander, and B. Popovic. 1976. Effects of application of sulphuric acid to poor pine forests. In L.S. Dochinger and T.A. Seliga eds., 1st Intl. Symp. Acid Precipitation and the Forest Ecosystem. U.S.D.A. Forest Service, Gen. Tech. Rep. No. NE-23, Upper Darby, PA, pp. 1011-1024.

Trudgill, S.T. 1977. Soil and Vegetation Systems. Clarendon Press, Oxford, 180 pp.

Tukey, H.B. Jr. 1970. The leaching of substances from plants. Annu. Rev. Pl. Physiol. 21: 305-324.

U.S.D.A. Forest Service. 1976. Workshop report on acid precipitation and the forest ecosystem. U.S.D.A. Forest Service, Gen. Tech. Rep. No. NE-26, U.S.D.A. Forest Service, Upper Darby, PA 18 pp.

U.S. Environmental Protection Agency. 1978. Simulation of Nutrient Loss from Soils Due to Rainfall Acidity. U.S.E.P.A. Pub. No. 600/3-78-053, Corvallis, OR, 44 pp.

U.S. Environmental Protection Agency. 1984. The Acidic Deposition Phenomenon and Its Effects. Vol. II. Effects Sciences. U.S.E.P.A. Publ. No. 600/8-83-0168F, Washington, DC.

U.S. Environmental Protection Agency. 1987. The Acidic Deposition Phenomenon and Its Effects Critical Assessment Document. U.S.E.P.A. Publ. No. 600/8-85/001 Washington, DC 159 pp.

Vogels, K., R. Guderian, and G. Masuch. 1986. Studies on Norway spruce (Picea abies Kant.) in damaged forest stands and in climatic chamber experiments. EDD Conf. Acidification and its Policy Implications. Amsterdam, Netherlands, May 5–9, 1986.

Voigt, G K. 1980. Acid precipitation and soil buffering capacity. Proc. Conf. Ecological Impacts of Acid Precipitation. SNSF Project. Oslo, Norway, pp. 53-57.

White, G.N., S.B. Feldman, and L.W. Zelazny. 1988. Rates of nutrient release by mineral weathering. Tech. Bull. No. 542, National Council of the Paper Industry for Air and Stream Improvement, New York, 66 pp.

Wilson, W.E. 1978. Sulfates in the atmosphere: A progress report on project MISTT. Atmos. Environ. 12: 537-548.

Wood, T., and F.H. Bormann. 1974. The effects of an artificial acid mist upon the growth of Betula alleghaniesis Bri. Environ. Pollu. 7: 259-268.

Wood, T. and F.H. Bormann. 1975. Increases of foliar leaching caused by acidification of an artificial mist. Ambio 4: 169-171.

Wood T. and F.H. Bormann. 1976. Short-term effects of a simulated acid rain upon the growth and nutrient relations of Pinus strobus L. In L.S. Dochinger and T.A. Seliga, eds., Proc. 1st Intl. Symp. Acid Precipitation and the Forest Ecosystem. U.S.D.A. Forest Service, Gen. Tech. Rep. No. NE-23, Upper Darby, PA, pp. 815-825.

Wood, T. and F.H. Bormann. 1977. Short-term effects of a simulated acid rain upon the growth and nutrient relations of Pinus strobus L. Water Air Soil Pollu. 7: 479-488.

Wright, R.F. and E.T. Gjessing. 1976. Acid prcipitation: Changes in the chemical composition of lakes. Ambio 5: 219-223.

Zoettl, H.W. 1987a. Responses of forests in decline to experimental fertilization. In T.C. Hutchinson and K.M. Meema, eds., Effects of Atmospheric Pollutants on Forests, Wetlands, and Agricultural Ecosystems. Springer–Verlag, New York, pp. 255-265.

Zoettl, H.W. and R.F. Huettl. 1986. Nutrient supply and forest decline in southwest Germany. Water Air Soil Pollu. 31: 449-462.

Zoettl, H.W. 1987b. Element turnover in ecosystems of the Black Forest. Forstweissenschaftliches Centralblatt 106: 105-114.

11

Forest Nutrient Cycling: Rhizosphere and Symbiotic Microorganisms

The rhizosphere and symbiotic microorganisms have roles of very great importance in nutrient relations in forested ecosystems. Forests frequently flourish in regions of low, marginal, or poor soil nutrient status. In addition to nutrient conservation and tight control over nutrient cycling, trees have evolved critically significant rhizosphere processes and symbiotic relationships with soil fungi and bacteria that enhance nutrient supply and uptake. The interaction between air contaminants, the rhizosphere, symbiotic microbes, and their relationship with host trees is of critical importance. An adverse impact on the rhizosphere (Smith, 1987) and on mycorrhizae (Sootka, 1968) by air pollution has been hypothesized.

A. Rhizosphere

The rhizosphere is the very restricted zone of soil immediately surrounding plant roots (Figure 11-1). It is the zone of root exudation, sloughed root cells, and the root mucigel complex. Saprophytic, parasitic, and symbiotic microorganisms abound in the rhizosphere. Roots exchange ions, gases, and moisture in the rhizosphere. The soil of the rhizosphere has greater bulk density, generally lower pH, and a more variable water potential gradient than bulk soil (soil devoid of roots) (see Table 9-10). The dominant feature and primary characteristic of rhizosphere soil relative to bulk soil is the stimulation of microbial numbers and activity in the former zone. Populations of viruses, bacteria, actinomycetes, yeasts, other fungi, insects, mites, other arthropods, nematodes, flagellates, and amoebas are all increased in rhizosphere ecosystems (Foster et al., 1983) (Table 11-1). Microbial stimulation in the soil close to roots is primarily due to the eu-

Figure 11-1. Scanning election microscope micrograph of cross section of mature red spruce root and associated rhizosphere soil.

trophic nature of the rhizosphere environment relative to the oligotrophic conditions characteristic of soil without roots. The rhizosphere region exerts dominant regulation over nutrient uptake by roots, root disease due to biotic infection and infestation, and microbial saphrophyte symbiont ecology. Six hypotheses have been presented regarding potential impacts of atmospheric deposition on the nature and functions of rhizosphere soil. The hypotheses are: (a) alteration of rhizosphere chemistry due to above-ground stress; and alterations to rhizosphere regulation of, (b) habitat of microbes controlling nutrient uptake, (c) pathogen and saphrophyte ecology, (d) nutrient uptake, (e) heavy metal uptake, and (f) aluminum uptake (Smith, 1987). Hypotheses e and f were reviewed in Chapter 9. Hypotheses a through d will be introduced in this chapter.

Table 11-1. Approximate Maximum Population Densities of Major Microbial Groups of the Rhizosphere. Bulk Soil Populations Generally Range from 10 to 100 Times Less.

Microbial group	Number per gram of rhizosphere soil
Bacteria	1×10^9
Actinomycetes	46×10^6
Fungi	$1 \times 10^{4-5}$
Algae	$1 \times 10^{3-5}$
Amoebas	18×10^2
Flagellates	50×10^2

Source: Foster et al. (1983), Foster (1985).

1. Alteration of Rhizosphere Chemistry

It is especially critical to understand how atmospheric deposition may influence root growth and exudation. Very few studies have examined the former and no studies have explored the latter. A variety of environmental stress factors have been demonstrated to influence root exudation. Mechanical damage to roots and above-ground plant parts appears to stimulate root exudation and to cause qualitative changes in root exudate patterns. Limited evidence suggests that drought conditions may generally increase organic material release into the rhizosphere.

The influence of pH on root exudation is particularly important with reference to the potential for forest soil acidification. Unfortunately the influence of hydrogen ion concentration on root exudation has not received detailed study. Studies with nonwoody species grown in solution culture have revealed conflicting evidence. McDougall (1970) recorded a decrease in exudation of ^{14}C from wheat seedlings when the pH of the collecting solution was increased from pH 5.9 to pH 7.0. Bonish (1973), on the other hand, found that a certain fraction of red clover root exudate increased as the pH was increased.

A large number of chemicals applied to foliage have been evaluated with respect to their potential to alter root exudation and/or rhizosphere microbes. Most of the chemicals screened include those otherwise employed in plant management practices, including fertilizers, growth regulators, and pesticides. The results of these applications are highly variable, but they indicate that changes in rhizosphere chemistry and biology can be mediated by materials applied to the foliage (Smith, 1987).

Ozone is an important regional-scale air pollutant, absorbed by foliage, with significant potential to directly and indirectly influence forest tree root metabolism. Ozone is known to cause a reduction in apparent photosynthesis (Chapter 12). A sustained reduction in photosynthesis will ultimately influence root physiology, alter root exudation, and change the microflora of the rhizosphere. In addition to depressing photosynthesis in the foliage of numerous species, ozone has been shown to inhibit the allocation and translocation of photosynthate (e.g., sucrose) form the shoots to the roots (Chapter 12). In addition to direct root influences, severe ozone or other pollutant exposure may result in acute foliar morbidity or mortality. This could represent the equivalent of defoliation. It is known that defoliation can influence both the quantity and quality of root exudation. Late season defoliation of sugar maple has been shown to alter patterns of carbohydrate, amino acid, and organic acid exudation (Smith, 1972). Defoliated maples released greater quantities of fructose, cystine, glutamine, lysine, phenylalamine, and tyrosine, whereas foliated maples exudated greater amounts of sucrose, glycine, homoserine, methionine, threonine, and acetic acid (Table 11-2).

2. Alteration of Nutrient Uptake

The rhizosphere mediates the transformation of nutrient elements from unavailable to available forms. Rhizosphere processes that help provide free ions for symplast uptake include pH regulation, protonation, reduction, and complexation. Hydrogen ions and organic acids produced by both the root and microorgan-

Table 11-2. The Influence of Defoliation on the Quantities of Compounds Exuded by Roots of Sugar Maples. Data are $\mu g \times 10^{-1}$ of Each Material Released During 14 Days mg^{-1} of Oven Dry Root.

	1969[a]		1970[b]	
Compound	Control	Defoliated	Control	Defoliated
Carbohydrates				
Fructose	2.6 ± 0.2	4.3 ± 0.1[c]	4.3 ± 1.2	6.1 ± 0.9
Glucose	0	0	trace	trace
Sucrose	73 ± 0.1	2.9 ± 0.2[c]	7.9 ± 0.9	6.0 ± 0.3
Amino acids/amides				
Alanine	0	0	1.8 ± 0.4	1.7 ± 0.1
Cystine	0.2 ± 0.1	0.5 ± 0.1	0	0
Glutamine	2.3 ± 0.3	3.4 ± 0.5	3.6 ± 0.7	4.3 ± 1.1
Glycine	0.5 ± 0.1	0.3 ± 0.1	1.9 ± 0.3	1.1 ± 0.3
Homoserine	1.1 ± 0.2	0.3 ± 0.1[c]	2.4 ± 0.6	1.8 ± 0.3
Lysine	0.8 ± 0.1	1.8 ± 0.2[c]	0	0.5 ± 0.2[c]
Methoinine	0	0	1.5 ± 0.1	1.4 ± 0.4
Phenylalanine	0	0	2.7 ± 0.6	3.1 ± 0.5
Threonine	trace	trace	3.4 ± 0.3	0[c]
Tyrosine	0	0	0.9 ± 0.7	1.1 ± 0.2
Organic acids				
Acetic	49.7 ± 10.1	24.3 ± 9.4	63.2 ± 11.1	58.1 ± 13.3
Malonic	0	0	trace	0

[a]Mean and standard error of three replicate determinations using one composite exudate sample from 19 and 20 roots of control and defoliated tree, respectively.
[b]Mean and standard error of three replicate determinations using one composite exudate sample from 17 and 23 roots of control and defoliated tree, respectively.
[c]Control and defoliated figures significantly different at the 95% level.
Source: Smith (1972).

isms of the rhizosphere play central roles in these processes. Nonhumidified low-molecular weight organic acids in root exudates may be particularly important because they form soluble complexes with nutrient ions (Chapter 9). Nutrients in these complexes are judged to be more available than nutrients in strong chelates formed by the humidified organic acids predominating in bulk soil. Unfortunately, there are no studies that demonstrate the effects of atmospheric deposition on rhizosphere levels of hydrogen ions and organic acids. Since phosphorus, potassium, iron, and other micronutrients may be significantly regulated by rhizosphere chemistry (Smith, 1987), it is critically important to design appropriate studies to explore the linkage between the uptake of these nutrients and atmospheric deposition.

3. Alteration of Soil Saprophyte and Root Pathogen Ecology

The rhizosphere is of special significance to root-infecting fungi, as root exudates provide the nutrition for fungal propagules to germinate and to support vegetative growth in the rhizosphere. Exudates may also provide energy for infection and may attract motile spores of Phycomycetes (Lockwood and Filonow, 1981). As in the case of symbionts, any significant change in the quantity or quality of root exudation or rhizosphere chemistry could have a profound influence on pathogen spore germination, infection, and growth in the rhizosphere. Unfortunately air pollution studies directly addressing this important topic for forest trees are not available.

More information is available on atmospheric deposition effects on saprophytic microbes that participate in important soil functions. Several of these studies have investigated pollutant effects on bulk soil processes such as total microbial respiration, rather than effects on individual microbial species (Chapter 9). Bewley and Stotzky (1983d) employed a continuous perfusion technique to investigate influences of simulated acid rain and cadmium, alone and in combination, on ammonification and nitrification in soil returned to the laboratory form Ossining, New York. Ammonification was relatively insensitive to cadmium and acidity, and occurred even in soils treated with pH 2. 0 or 1000 ppm cadmium. Nitrification, on the other hand, was more sensitive and was retarded in ammonium-nitrogen supplemented soils exposed to pH 2.5 and was inhibited in soil exposed to pH 2.0. In an effort to evaluate the effects of cadmium and zinc on carbon mineralization and the microflora, Bewley and Stotzky (1983a) applied these heavy metals to "stored" (8 yrs) glucose-amended soils. The test soils were also amended with 9% kaolinite or montmorillorite to evaluate mitigation of heavy metal toxicity by clay minerals. Results indicated that the threshold of an important inhibitory effect on microbial activity in soils by zinc may be at approximately 10,000 ppm, and the threshold for cadmium is approximately 5000 ppm. The clays did not reduce the toxicity of the heavy metals for the microbes. In a subsequent study (Bewley and Stotzky, 1983b), both metals were applied simultaneously to test soils. In spite of a gradual lengthening in the time before initiation of glucose degradation as the metal concentration increased, the combination of metals resulted in an additive, rather than a synergistic, response. Both of these studies and others have emphasized the importance of redundancy in microbial ecosystems. After a sufficient length of time, a microbial population tolerant of zinc and cadmium will be selected. The growth of *Aspergillus niger* in soil acidified to pH levels of 3.6–4.2 was reduced by the addition of either 100 or 250 ppm cadmium or 1000 ppm zinc (Bewley and Stotzky, 1983c). These studies led Babich et al. (1983) to propose the use of an "ecological dose 50 percent (EcD 50)" concept to qualify the inhibition of microbe mediated soil processes by pollutants. A similar approach is keenly needed for rhizosphere microbes and processes.

B. Fungal Symbionts

An enormous number of soil fungi infect fine roots of forest trees and form mycorrhizal or "fungus roots." It is widely held that these roots colonized by beneficial fungi are essential for the growth of essentially all woody plants in natural forest environments. Morphological differences conveniently place mycorrhizal associations into one of two groups: ectomycorrhizae and endomycorrhizae. The latter, formed by fungal species of the Endogonaceae (Phycomycete class), are the most common mycorrhizal associates of forest trees. Infection by these fungi does not alter the gross morphology of the root, but mycelia do emanate from the roots to form a loose network in the rhizosphere. The formation of vesicles and arbuscules on hyphae inside the root has led to the designation *VA mycorrhizae*. Ectomycorrhizae, while less abundant than endomycorrhizae, form on a variety of extremely important forest tree species including pine, hemlock, spruce, fir, oak, birch, beech, eucalyptus, willow, and poplar among others. Ectomycorrhizal fungi are taxonomically diverse and may be Basidomycetes, Ascomycetes, or family Endogonaceae of Phycomycetes. Infection in this case does morphologically alter the external appearance of fine roots by the development of a distinctive mass of external hyphae (Ruehle and Marx, 1979; Trappe, 1977) (Figure 11-2).

Mycorrhizal roots are presumed to confer multiple advantages to host trees. More efficient water uptake and increased resistance to infection by soil pathogens are two of the most important. Traditionally, however, the advantages associated with nutrient availability and uptake have been judged to be paramount. The uptake of nutrients from the forest floor (litter) and soil throughout the relatively large interroot distances of forest vegetation is achieved by the longevity of mycorrhizal roots and especially by the growth of hyphal strands into rhizosphere soil and beyond. Radiotracer investigations with essentially all the macronutrients and numerous micronutrients have invariably confirmed uptake by the fungus and translocation to tree hosts. The significance of these specialized roots is particularly great for relatively immobile ions such as phosphate, zinc, copper, molybdenum, and occasionally ammonium (Bowen,

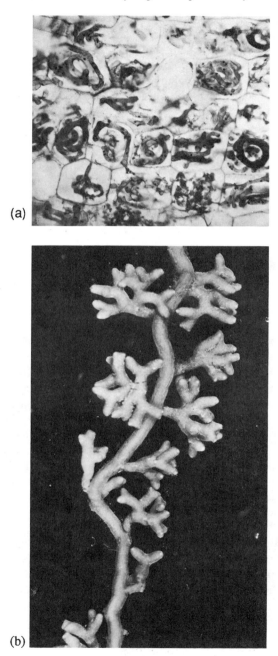

(a)

(b)

Figure 11-2. Mycorrhizal associations of yellow poplar and eastern white pine. (a) Fungal mycelium inside the endomycorrhizal roots of yellow poplar. (b) Mantle of fungal mycelium on the morphologically distinctive ectomycorrhizal roots of eastern white pine.

1973; Trappe and Fogel, 1977). Excellent reviews of mycorrhizal physiology and ecology are available (Hacskaylo, 1971; Harley, 1969; Lobanow, 1960; Marks and Kozlowski, 1973; Sanders et al., 1975; Ruehle and Marx, 1979).

1. Mycorrhiza and Toxic Metals

Ninety-five percent of the active ectomycorrhizae of mature Douglas fir and larch forests were found to be associated with the humus, decayed wood, and charcoal of the forest floor in Montana (Harvey et al., 1976). Concentration of mycorrhizal roots in the upper organic horizon of the soil profile places them in the zone of excessive heavy metal accumulation in those forest regions subject to trace metal input from the atmosphere (Chapter 5). Limited evidence for increased heavy metal uptake by mycorrhizal roots has been presented for zinc, copper, and manganese (Bowen, 1973). Unfortunately little is known concerning the relative tolerance of symbiotic fungi to heavy metal contamination. Fungi are known to accumulate metal ions and it has been hypothesized that they may combine these with oxalic acid and then dispose of them as insoluble oxalate salts (Cromack et al., 1975).

Under laboratory conditions, but with field collected root material, Bowen et al. (1974) have presented data indicating enhancement of zinc uptake from solution with both ectomycorrhizas of Monterey pine and with VA mycorrhizas of hoop pine compared with uninfected short roots. Zinc amendments of 45 and 135 $\mu g\ g^{-1}$ of soil decreased both nodulation and mycorrhizae of pinto bean when compared to amounts in nonamended soil (McIlveen et al., 1975). Smelter emissions and litter contaminated with heavy metals were added to intact forest microcosms, and the influence on litter-soil carbon metabolism was monitored by recording the daily efflux of carbon dioxide by Ausmus et al. (1978). Heavy metal contamination increased the rate of loss and the daily pattern of carbon dioxide efflux from the microcosms. Bacterial populations at the end of the experiment were greater in treated microcosms than in controls. The authors hypothesized, but did not provide specific evidence, that the disruption of mycorrhizal associations may have increased available substrate from fungal cells and stimulated carbon dioxide efflux.

McCreight and Schroeder (1982) tested (in vitro) nine ectomycorrhizal fungi for inhibition of growth by cadmium, lead, and nickel. All of the fungi were arrested by 350 $\mu g\ ml^{-1}$ (ppm) cadmium or less. Lead arrested five species at 200 ppm or less. Nickel arrested the growth of six fungi at 20 ppm or less (Table 11-3). Metal concentrations that did not arrest growth delayed growth for 1-3 weeks.

Hepper (1979) recorded decreased spore germination and growth of *Glomus caledonium* (VA mycorrhiza) with 2.6 $\mu M\ l^{-1}$ manganese in culture media. At 25.5 $\mu M\ l^{-1}$, germination was inhibited. Firestone et al. (1983) tested manganese and aluminum exposure against *Aspergillus flavus* spore metabolism. Spore growth was not inhibited in soil leachate by manganese at 470 $\mu M\ l^{-1}$, but was inhibited by aluminum at 600 $\mu M\ l^{-1}$. Thompson and Medve (1984) recorded that vegetative growth of ectomycorrhizal fungi, commonly associated with trees

Table 11-3. Cadmium, Lead, and Nickel Concentrations that Reduced the in vitro Doubling Rate of Growth by 50% (IC 50) Compared to Controls, and the in vitro Concentration Range at Which Growth was Arrested for 28 days for Nine Ectomycorrhizal Fungi.

Species	Metal (μg/ml)[a]					
	Cadmium		Lead		Nickel	
	IC50	Arrested	IC50	Arrested	IC50	Arrested
Amanita muscaria	–	150–200	70.0	100–150	28.2	30–40
Cenoccum graniforme	–	< 2	12,842	1,000–2,000	47.5	12–14[b]
Laccaria laccata	17.0	110–120	0.2	1,000–2,000	12.6	200–225
Pisolithus tinctorius	–	< 10	65.3	50–60	7.4	16–18
Rhizopogon roseolus	2.4	15–20	51.7	150–200	0.6	10–15
Suillus brevipes	1.5	20–25	101.2	150–200	0.1	10–15
Suillus grevellei	–	10–15	93.0	250–300	–	10–20
Suillus luteus	14.3	120[c]	527.8	1,000–2,000	64.1	150–175
Thelephora terrestris	15.2	300–350	82.5	150–175	19.2	12–14

[a]Mole equivalent, 100 μg/ml: cadmium, 0.890 mM; lead, 0.483 mM; nickel, 1.703 mM.
[b]Slight growth at 16 pp,
[c]Slight growth.
Source: McCreight and Schroeder (1982).

on acid sites, were more sensitive to aluminum than to manganese in laboratory media.

With current information, we cannot even approximate the threshold levels of any toxic metal that might exert an adverse influence on a mycorrhizal fungus in natural soil. Some data suggest toxicity to soil microbes from aluminum and manganese in soil solutions at levels lower than the toxicity thresholds for roots of forest trees. In view of the extraordinarily large number of fungi capable of forming mycorrhizal associations, estimated to be 2000 species for Douglas fir alone (Trappe, 1977), and general variation of fungi in response to heavy metal exposure, for example, the decrease in formation of ectomycorrhizae by some and the increase by others following application of a copper fungicide to seedlings (Göbl and Pümpel, 1973), it is assumed that the threshold range would be very broad, variable, and species specific.

2. Mycorrhizae and Acid Deposition

A soil pH of approximately 5 appears optimal for many mycorrhizal fungi (Meyer, 1973). Alkaline soil pH is known to be associated with poor mycorrhizal formation (Bowen and Theodorou, 1973). Could soil acidification and increased hydrogen-ion concentration resulting from acid precipitation have an adverse influence on mycorrhizal associations? Numerous ectomycorrhizal fungi are capable of reasonable in vitro growth at pH 3 (Hung and Trappe, 1983). Tolerance of ectomycorrhizal tree seedlings to soil acidity is demonstrated by the utility of selected trees for revegetating acid mine soil sites (Marx and Bryan, 1975).

Laboratory evidence regarding mycorrhizal activity is inconclusive. As rain simulant pH treatments were reduced from 5.6 to 3.0, pine infection (Shafer et al., 1985; Stroo and Alexander, 1985) and oak infection (Reich et al., 1985) were decreased. Treatment with pH 2.4 rain simulant, however, appeared to increase ectomycorrhizal infection (Schafer et al., 1985).

The influence of simulated acid rain on nitrogen uptake by the fungus *Glomus mosseae* endomycorrhizal with the roots of sweetgum seedlings has been studied by Haines and Best (1976). Natural forest soil profiles collected in North Carolina and supporting the growth of sweetgum seedlings (originating from planted seeds) were treated with solutions of pH 5.9 and 3.0, and artificial eastern United States rain simulant. Applications of the pH 5.9 and 3.0 solutions produced greater nitrate- and ammonium-nitrogen concentrations in the soil solution in columns with mycorrhizae relative to columns containing either soil alone or soil plus tree seedlings. When artificial acid rain acidified the top 5 cm of soil to a soil solution pH of 2.0, nitrate-nitrogen concentrations were unchanged by mycorrhizal columns while ammonia appeared excluded from soil exchange sites, presumably by hydrogen ions. The authors judged that if mycorrhizal roots are more subject to competitive inhibition of cation uptake by hydrogen ions than nonmycorrhizal roots, then acidification of soil solutions may result in decreased cation uptake by mycorrhizal roots.

A hypothesis associated with acid deposition impact on mycorrhizal associations suggests that symbiotic fungal activity may be restricted by nitrogen deposition. Decreased ectomycorrhizal colonization of spruce as a result of atmospheric nitrogen input has been recorded (Alexander and Farley, 1983; Meyer, 1985).

3. Mycorrhizae and Gaseous Pollutants

Unfortunately relatively little work has been done in this area. Mahoney et al. (1985) found that mycorrhizal infection in loblolly pine seedlings was not affected by chronic exposure to 70 ppb (137 μg m^{-3}) ozone. Reich et al. (1986) conducted five experiments with potted seedlings of eastern white pine and northern red oak exposed to low levels of ozone (20–140 ppb [39–274 μg m^{-3}], 7 hr day^{-1}, 3–7 weeks and acid deposition (pH 3.0–5.6, 1.25 hrs day^{-1}) for 2 weeks. No significant interaction between the pollutants was observed. Exposure to ozone for 3 days week^{-1} had no effect on white pine, but the same treatments for 5 days week^{-1} had significant effects on the mycorrhizae of both species (Figure 11-3). For seedlings of both species, increasing acidity of simulated rain resulted in significant decreases in mycorrhizal infection. The mechanism of ozone influence on mycorrhizae is not clear from this investigation. Reich et al. (1986) suggested, however, that the influence is presumably via an impact on seedling metabolism. They observed that soluble root carbohydrates are generally consid-

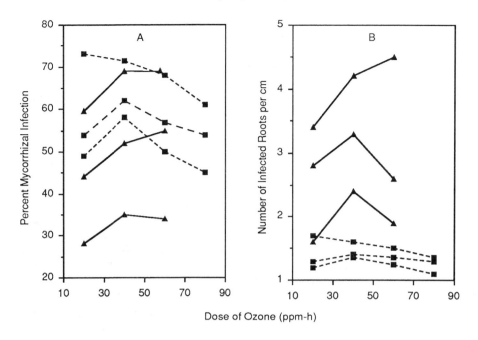

Figure 11-3. Percent of short roots infected with mycorrhizal (A) and number of infected short roots cm^{-1} of lateral root (B), for seedlings of white pine (squares) and red oak (triangles) (in three soil treatments each) in relation to the total exposure to ozone.
Source: Reiche et al. (1986).

ered a major factor in mycorrhizal infection and that any influence of ozone on the availability of root sugar would likely influence the carbohydrate metabolism of root fungi. Krupczak and Manning (1988) have investigated yellow birch seedlings inoculated (or not inoculated) with *Cenococcum geophilum* and exposed to low (10–20 ppb, 20–39 µgm^{-3}) or high (60–80 ppb, 118–157 µgm^{-3}) ozone for 8 hr day^{-1}, 5 days week^{-1} for 10 weeks. Between weeks 4 and 8, the rate of mycorrhizal infection was greater in the low ozone exposure. After 8 weeks, however, the rate of mycorrhizal infection was greater in the high ozone treatment.

C. Bacterial Symbionts

While generally much less abundant than fungal symbioses, mutually beneficial relationships with bacteria are important in nutrient relationships in certain forest ecosystems. The influence air contaminants may have on these microbes has been the topic of a very limited number of studies.

1. Rhizobia and Air Pollution Stress

Members of the *Rhizobium* genus can infect the root hairs of leguminous plants and initiate the formation of root nodules, within which they develop as intracellular symbionts and fix atmospheric nitrogen. The 12,000 leguminous species include trees, shrubs, and woody vines, of which less than 10% have been studied for nodulation (Postgate and Hill, 1979). Globally important tree genera include *Dalbergia, Pterocarpus, Albizzia, Peltogyne, Acacia, Cassia, Castanospermum, Haemotoxylon*, and *Cargana. Gleditsia* and *Robinia* are the two most important native American genera and, along with 16 other arborescent genera, contain 40 species.

The environmental factors that influence nitrogen fixation by rhizobia have been reviewed by Postgate and Hill (1979). It has been suggested by Tamm (1976) that microbial nitrogen fixation may be impaired by air pollution.

Ladino clover was treated with filtered air, 30 pphm (588 μg m^{-3}) ozone or, 60 pphm (1176 μg m^{-3}) ozone for two 2-hr exposures, 1 week apart in controlled environment chambers (Letchworth and Blum, 1977). Plants of various ages were tested. The influence of ozone varied with gas concentration and plant age. Ozone reduced the growth and nodulation of test plants. Nitrogenase activity per nodule and per plant was not significantly different from the control. Total nitrogen content per plant was correlated with plant biomass, but not with nodule number per plant (Table 11-4). Jones et al. (1985) investigated exposure of soybean cultivars, under greenhouse and field conditions, to ozone and sulfur dioxide in slight excess of concentrations found in polluted atmospheres during flowering and pod development. Total nodule activity, as quantified by acetylene reduction per plant, was significantly reduced in both studies. The authors felt that the ability of roots to fix nitrogen was mediated through foliar injury.

Heavy metals have been capable of altering nitrogen fixation in legumes (Döbereiner, 1966; Huang et al., 1974). Vesper and Weidensaul (1978) subjected soybeans to cadmium and nickel at 1, 2.5, and 5 ppm, and copper and zinc at 1, 5, and 10 ppm in sand culture. The degree of toxicity was cadmium > nickel > copper > zinc. Cadmium dramatically reduced nodule number, dry weight, and nitrogen fixation. Fixation by nickel treated plants was very low. Copper suppressed nodulation, but inhibited nitrogen fixation directly only at 5 and 10 ppm. Zinc reduced nodulation, but only slightly inhibited fixation (Table 11-5).

Unfortunately little information is available on the interaction of soil aluminum with legume-*Rhizobium* symbioses or even with *Rhizobium* alone. Wood and Cooper (1984) studied the response on six strains of *R. trifolii* to pH, calcium, aluminum, and phosphate in a defined medium that the authors claimed was a satisfactory model for the white clover rhizosphere. They observed that multiplication was inhibited by pH alone at 4.3 and by 50 μM aluminum at pH 5.5. Inhibition by aluminum was removed by increasing the pH > 6.0 or by increasing the phosphate concentration. Calcium concentration had no effect on the strains. The authors stressed that the aluminum toxicity was not due to a direct effect of monomeric aluminum nor an indirect effect of polymeric aluminum

Table 11-4. Effects of Ozone on Dry Weight, Nodule Number, and Plant Nitrogen of Ladino Clover.

	Top dry weight (g)	Root dry weight (g)	Nodule number	Nitrogen			
				Top (%)	Root (%)	Top (mg)	Root (mg)
Control	10.37(a)[a]	1.75(a)	483(a)	2.85(b)	2.66(b)	265.5(a)	44.5(a)
30 pphm ozone	9.65(b)	1.48(b)	317(b)	3.02(a)	2.79(ab)	247.5(ab)	38.0(b)
60 pphm ozone	8.15(c)	1.16(c)	309(b)	3.08(a)	2.87(a)	214.8(b)	30.8(c)
LSD (0.05)	0.650	0.17	41	0.14	0.14	18.5	4.3

Source: Letchworth and Blum (1977).
[a] Means followed by the same letter are not significantly different at the 5% level.

Table 11-5. Influence of Cadmium, Nickel, Copper, and Zinc on Soybean Nodule Size, Number, and Nitrogen-Fixing Ability.

Treatment (ppm)	Number of nodules per plant	Nodule size (mg)	Ethylene produced (nmoles plant^{-1} 0.5 h^{-1})
Cd			
1.0	12(g)[a]	25.31(d)	144(b)
2.5	12(g)	24.58(d)	87(c)
5.0	3(h)	13.47(e)	38(d)
Ni			
1.0	44(ab)	58.86(a)	824(e)
2.5	38(cd)	48.33(c)	611(f)
5.0	28(e)	47.41(c)	527(g)
Cu			
1.0	28(e)	54.09(b)	1509(h)
5.0	27(ef)	54.06(b)	991(i)
10.0	24(f)	47.85(c)	879(j)
Zn			
1.0	40(bc)	58.40(a)	1659(a)
5.0	38(c)	53.69(a)	1585(k)
10.0	36(d)	52.90(b)	1252(l)
Control	46(a)	62.30(f)	1640(a)

Source: Vesper and Weidensaul (1978).

[a] Values in a given column followed by the same letter are not significantly different ($P = 0.05$).

(see Chapter 9 for aluminum speciation), but rather was due to a direct effect of the polymeric aluminum. In subsequent studies, Wood et al. (1984a,b) employed an axenic solution culture system to investigate aluminum-pH stress on white clover-*R. trifolii* (strain HP3) symbioses. Their results suggested, in a soil of pH < 5.0 with high organic matter (low exchangeable aluminum), that nodulation of white clover could be limited by the lack of multiplication and modulating ability of *Rhizobium,* rather than by the growth of host roots. Establishment of "critical pH values" suggested that at pH 4.3–5.0 root elongation and root hair formation were limited by aluminum rather than by pH, whereas *Rhizobium* multiplication and nodule formation are limited by pH alone (Table 11-6). While recognizing the difficulty of extrapolating toxic aluminum concentrations in solution culture to aluminum levels in natural soil, Wood et al. (1984b) suggested that the poor response of white clover to nodulation on acid mineral soils may be due to aluminum stress. Under conditions of normal agricultural practice, where lime applications are made to optimize production, the effects of ambient levels of acid deposition in the eastern United States are unlikely to adversely impact leguminous crop nodulation in the field (Shriner and Johnston, 1981). The situation of leguminous woody species grow-

Table 11-6. Critical pH Values in the Absence and Presence of Aluminum with 10 μM Phosphate for Four Stages in the Development of the Symbiosis between White Clover and *R. trifolii.*

	Critical pH value	
Stage	No aluminum	50 μM aluminum
Root elongation	4.3	5.0
Root hair formation	4.3	5.0–5.5
Rhizobium multiplication	5.0	6.0
Nodule formation	5.0	6.0

Source: Wood et al. (1984a).

ing on selected forest soil types, however, may more closely approximate the acid soils investigated by Wood and colleagues rather than agricultural situations.

The clover exposures to ozone conducted by Letchworth and Blume (1977) employed ozone doses that were high relative to typical ambient levels. The experiments of Jones et al. (1985) were conducted at more realistic exposures. The heavy metal experiments of Vesper and Weidensaul (1978) used realistic metal concentrations of natural areas subject to modest trace metal pollution. The studies of Wood and collagues were conducted in laboratory solution systems rather than natural soil. The results of these investigations more than justify additional efforts to gather data from more legumes grown in natural environments. Unfortunately we were unable to find any studies concerned with nitrogen fixation and trees as affected by air pollution.

2. Alder-Type Endophytes (Frankia)

Although most species of *Alnus* do not attain a size nor form that makes them valuable for wood production, their nitrogen-fixing capabilities make them extremely valuable components of the extensive temperate and boreal forests where they naturally occur. The endophyte of *Alnus* nodules has been identified as *Frankia*, an actinomycete (Becking, 1974). The quantity of nitrogen fixed in alder nodules exceeds that of legume nodules and may range from 60 to 209 kg ha^{-1} yr^{-1} under favorable conditions. Growth of trees of the following genera have been shown to increase under the influence of associated alder: *Fraxinus, Liquidambar, Liriodendron, Picea, Pinus, Platanus, Populus,* and *Pseudotsuga* (Tarrant, 1968). Seven additional genera of woody, nitrogen-fixing angiospermous plants, including *Casuarina, Hippophal, Purshia, Caenthus, Myrica, Coriaria,* and *Dryas,* have "alder-type" nodules and are of varied importance and distribution throughout temperate wooded ecosystems.

Our ignorance of the ecology and physiology of the endophytes associated with woody plants with alder type nodules is very great. The limited number of investigators working in this area have been hampered by serious cultural limitations. Currently, however, many *Frankia* strains can be isolated from nodules

Table 11-7. Species with Sporulation in Actinorhizal Root Nodules as Observed in Nature.

Alnus rubra	*Caenothus velutinus*
Alnus crispa	*Hippopahaë rhamnoides*
Alnus incana, ssp. *rugosa*	*Casuarina cunninghamiana*
Alnus incana	*Casuarina equisetifolia*
Myrica gale	*Elaegnus umbellata*
Myrica cerifera	*Purshia tridentata*
Comptonia peregrina	

Source: Torrey (1986).

grown in pure culture and subsequently inoculated into host seedlings. At present, 24 host species from eight families are thought to form associations with different *Frankia* strains (Table 11-7) (Torrey, 1986). It is important to continue studies directed to improvement of our understanding of nutrient and cultural requirements of alder-type endophytes, since in-vitro cultivation experiments have shown actinomycete growth to be influenced by heavy metals in growth media. Cadmium has been shown to be toxic over a wide range of media concentrations. Aluminum and nickel are generally toxic above 10 ppm, while lead and vanadium appear to be relatively nontoxic (Waksman, 1967).

D. Other Nitrogen-Fixing Organisms

The relative importance of other nitrogen-fixing organisms, for example, free-living bacteria, algae, lichens and liverworts, mosses, and ferns with blue-green algal symbionts, in forest ecosystems is very poorly appreciated. The influence of air contaminants on these potentially significant organisms is, of course, equally poorly understood. Demison et al. (1976) have claimed that acid deposition has had an adverse impact on nitrogen-fixing lichens of western Washington coniferous forests, but the laboratory data they provided to support their contention were modest and quite variable. Hällgren and Huss (1975) have evaluated the influence of sodium hyposulfite on photosynthesis and nitrogen fixation of a lichen, collected from a Swedish pine forest, and the blue-green alga *Anabaena cylindrica*. Treatment with 5×10^{-4} M $NaHSO_3$ at pH 5.8 caused no reduction of photosynthesis in the lichen, while inhibition of nitrogen fixation was 97%. For the alga, the corresponding values were 50% and 75%, respectively. The authors speculated that sulfate may have some specific inhibitory action on nitrogenase enzyme activity.

E. Summary

The single most important zone of the soil ecosystem for forest tree health and growth is the rhizosphere. This zone, while only extending approximately 2 mm from the root surfaces, is an extremely unique environment and has extraordinary significance for nutrient dynamics and root metabolism. Microorganisms are stimulated in the rhizosphere because of the ready availability of organic nutrients from plant mucilages, mucigel, lysates, secretions, and root exudates. While bacteria and fungi exhibit the greatest stimulation, a large number of additional organisms are also enhanced.

Chemical and physical properties also differ between rhizosphere soil and soil devoid of roots. Typically, rhizosphere soil has lower pH, lower water potential, lower osmotic potential, lower redox potential, and higher bulk density than soil away from roots.

Due to its unique biological, physical, and chemical properties, the rhizosphere exerts dominant regulation over nutrient uptake by forest tree roots, root disease resulting from biotic agents, and microbial root symbiont and saphrophyte ecology. In view of this importance, it is essential to evaluate the potential interactions between pollutants deposited to forests from the atmosphere, and rhizosphere structure and function. A review of the literature has revealed that our understanding of this important topic is grossly deficient. Six hypotheses appear to describe the most important potential atmosphere–rhizosphere interactions. One involves alteration of rhizosphere chemistry by above-ground air pollution stress. The others involve deposition-induced alterations in rhizosphere regulation of nutrient uptake, heavy metal uptake, aluminum uptake, the habitat of microbes that mediate nutrient uptake, and the ecology of root pathogens and soil saphrophytes.

The research attention, given the interaction between mycorrhizal fungi, the specialized roots they form, and air pollution is grossly out of scale with the significance of this symbiotic relationship in forest ecosystems. The potential for heavy metal adverse impact on mycorrhizal associations appears particularly great due to the physical juxtaposition of the two entities in the forest floor. With current information, however, we cannot approximate the threshold levels of any toxic heavy metal that might exert an adverse influence on a mycorrhizal fungus in natural soil. Some data suggest aluminum may exert a toxic influence on selected mycorrhizal fungi at lower concentrations than those required to directly injure roots. Soil acidity is probably not directly restrictive to mycorrhizal development in natural forest soils. Gaseous pollutants, such as ozone, may have an adverse impact on fungal and bacterial symbionts by altering the quantity or quality of nutrient materials supplied to the roots by the foliage.

In acid soils, aluminum toxicity may play an important role in restricting symbioses involving *Rhizobium* endophytes. Evidence from forest systems is unfortunately not available.

Hopefully, research efforts to more adequately describe the influence of air contaminants on alder-type endophytes will expand as our application of isolation, cultivation, and inoculation techniques rapidly improve. Efforts directed toward clarification of air pollution influences on all forest organisms capable of nitrogen fixation are justified due to the potentially important role these organisms play in forest nutrient dynamics.

References

Alexander, I.J. and R.I. Fairley. 1983. Effects of N fertilization on populations of fine roots and mycorrhizas in spruce humus. Plant Soil 72: 49-53.

Ausmus, B.S., G.J. Dodson, and D.R. Jackson. 1978. Behavior of heavy metals in forest microcosms, III. Effects of litter-soil carbon metabolism. . 1 Water Air Soil Pollu. 10: 19-26.

Babich, H., R. J.F. Bewley, and G. Stotzky. 1983. Application of the "ecological dose" concept to the impact of heavy metals on some microbe-mediated ecologic processes in soil. Arch. Environ. Contam. Toxicol. 12: 421-426.

Becking, J.H. 1974. Frankiaceae Becking. In R .E. Buchanan and N.E. Gibbons, eds., Bergey's Manual of Determinative Bacteriology. Williams and Wilkens, Baltimore, MD, p., 701.

Bewley, R.J.F. and G. Stotzsky. 1983a. Effects of cadmium and zinc on microbial activity in soil: Influence of clay minerals. Part I: Metals added individually. Sci. Total Env. 31: 41-55.

Bewley, R.J.F. and G. Stotzsky. 1983b. Effects of cadmium and zinc on microbial activity in soil: Influence of clay minerals. Part II: Metals added simultaneously. Sci. Total Env. 31: 57-69.

Bewley, R.J.F. and G. Stotzsky. 1983c. Effects of combinations of simulated acid rain and cadmium or zinc on microbial activity in soil. Env. Res. 31: 332-339.

Bewley, R.J.F. and G. Stotzsky. 1983d. Effects of cadmium and simulated acid rain on ammonification and nitrification in soil. Arch. Env. Contam. Toxicol 12: 285-291.

Bonish, P.M. 1973. Cellulase in red clover exudates. Plant Soil 38: 307-314.

Bowen, G.D. 1973. Mineral nutrition in ectomycorrhizae. In G.C. Marks and T.T. Kozlowski, eds., Ectomycorrhizae. Their Ecology and Physiology. Academic Press, New York, pp. 151-205.

Bowen, G.D. and C. Theodorou. 1973. Growth of ectomycorrhizal fungi around seeds and roots. In G.C. Marks and T.T. Kozlowski, eds., Ectomycorrhizae. Their Ecology and Physiology. Academic Press, New York, pp. 107-150.

Bowen, G.D., M.F. Skinner, and D.I. Bevege. 1974. Zinc uptake by mycorrhizal and uninfected roots of Pinus radiata and Arucaria cunninghamii. Soil Biol. Biochem. 6: 141-144.

Cromack, K., R.L. Todd, and C.D. Monk. 1975. Patterns of Basidomycete nutrient accumulation in conifer and deciduous forest litter. Soil Bio. Biochem. 7: 265-268.

Demison, R., B. Caldwell, B. Bormann, L. Eldred, C. Swanberg, and S. Anderson. 1976. The effects of acid rain on nitrogen fixation in western Washington coniferous forests. In L.S. Dochinger and T.A. Seliga, eds., Proc. 1st Intl. Symp. Acid Precipitation and the Forest Ecosystem. U.S.D.A. Forest Service, Gen. Tech. Rep. No. NE-23, Upper Darby, PA, pp. 933-949.

Döbereiner, J. 1966. Manganese toxicity effects on nodulation and nitrogen fixation of beans (Phaseolus vulgaris L.) in acid soils. Plant Soil 24: 153-166.

Firestone, M.K., K. Killham, and J.G. McCall. 1983. Fungal toxicity of mobilized soil aluminum and manganese, Appl. Environ. Microbiol. 48: 556-560.

Foster, R.C. 1985. The biology of the rhizosphere. In C.A. Parker, A.D. Rovira, K.J. Moor, P.T. W. Wong, and J.F. Kollinorgen, eds., Ecology and Management of Soilborne Plant Pathogens. Am. Phytopathol Soc., St. Paul, MN, pp. 75-79.

Foster, R.C., A.D. Rovira, and T.W. Cook. 1983. Ultrastructure of the Root-Soil Interface. Am. Phytopathol. Soc., St. Paul, MN, 290 pp.

Göbl, F. and B. Pümpel. 1973. Einfluss von "grünkupfer linz" auf pflanzenausbildung mykorrhizabesatz sowie frosthärte von zirbenjungflazen. Eur. J. For. Pathol. 3: 242-245.

Hacskaylo, E. (ed.). 1971. Mycorrhizae. U.S.D.A. Forest Service, Misc. Publ. No. 1189, Washington, DC, 255 pp.

Haines, B. and G.R. Best. 1976. The influence of an endomycorrhizal symbiosis on nitrogen movement through soil columns under regimes of artificial throughfall and artificial acid rain. In L.S. Dochinger and T.A. Seliga, eds., Proc. 1st Intl. Symp. Acid Precipitation and the Forest Ecosystems. U.S.D.A. Forest Service, Gen. Tech. Rep. No. NE-23, Upper Darby, PA, pp. 951-961.

Hällgren, J.E. and K. Huss. 1975. Effects of SO_2 on photosynthesis and nitrogen fixation. Physiol. Plant 34: 171-176.

Harley, J.L. 1969. The Biology of Mycorrhiza. Leonard Hill, London, 334 pp.

Harvey, A.E., M.J. Larsen, and M.F. Jurgensen. 1976. Distribution of ectomycorrhizae in a mature Douglas-fir/larch forest soil in western Montana. For. Sci. 22: 393-398.

Hepper, C.M. 1979. Germination and growth of *Glomus caledonius* spores: The effects of inhibitors and nutrients. Soil Biol. Biochem. 11: 269-277.

Huang, C., F.A. Bazzaz, and L.N. Vanderhoef. 1974. The inhibition of soybean metabolism by cadmium and lead. Plant Physiol. 54: 122-124.

Hung, L.L. and J.M. Trappe. 1983. Growth variation between and within species of ectomycorrhizal fungi in response to pH *in vitro*. Mycologia 75: 234-241.

Jones, A.W., C.L. Mulchi, and W.J. Kenworthy. 1985. Nodule activity in soybean cultivars exposed to ozone and sulfur dioxide. J. Environ. Qual. 14: 60-65.

Krupczak, D.L. and W J. Manning. 1988. Reponse of mycorrhizal and nonmycorrhizal yellow birch seedlings to two ozone concentrations., Abstracts, Northeastern Div. Mtg, Am. Phytopath. Soc., September 29, 1988, Sturbridge, MA, p. 18.

Letchworth, M.B. and V.Blum. 1977. Effects of acute ozone exposure on growth, nodulation and nitrogen content of ladino clover. Environ. Pollu. 14: 303-312.

Lobanow, N.W. 1960. Mykotrophie der Holzpflanzen. Springer–Verlag, Berlin, 352 pp.

Lockwood, J.L. and A.B. Filonow. 1981. Responses of fungi to nutrient-limiting conditions and to inhibitory substances in natural habitats. Adv. Microb. Ecol. 5: 1-61.

Mahoney, M.J., B.I. Chevone, J.M. Skelly, and L.D. Moore. 1985. Influence of mycorrhizae on the growth of loblolly pine seedlings exposed to ozone and sulfur dioxide. Phytopathology 75: 679-682.

Marks, G.C. and T.T. Kozlowski (eds.). 1973. Ectomycorrhizae. Their Ecology and Physiology. Academic Press, New York, 444 pp.

Marx, D.H. and W.C. Bryan. 1975. The significance of mycorrhizae to forest trees. In B. Bernier and C.H. Winget, eds., Forest Soils and Forest Land Management. Laval Univ., Quebec, Canada, pp. 107-117.

McCreight, J.D. and D.B. Schroeder. 1982. Inhibition of growth of nine ectomycorrhizal fungi by cadmium, lead, and nickel *in vitro*. Environ. Experimen. Bot. 22: 1-7.

McDougall, B.M. 1970. Movement of ^{14}C-photosynthate into the roots of wheat seedlings and exudation of ^{14}C from intact roots. New Phytol. 69: 37-46.

McIlveen, W.D., R.A. Spotts, and D.D. Davis. 1975. The influence of soil zinc on nodulation, mycorrhizae, and ozone-sensitivity of Pinto bean. Phytopathology 65: 645-647.

Meyer, F.H. 1973. Distribution of ectomycorrhizae in native and man-made forests. In G.C. Marks and T.T. Kozlowski, eds., Ectomycorrhizae. Their Ecology and Physiology. Academic Press, New York, pp. 79-105.

Meyer, F.H. 1985. Effect of the nitrogen factor on the mycorrhizal complement of Norway spruce seedlings in humus from a damaged site. Allg. Forstz. 9/10: 208-219.

Postgate, J.R. and S. Hill. 1979. Nitrogen fixation. In J.M. Lynch and N.J. Poole, eds., Microbial Ecology: A Conceptual Approach. Wiley, New York, pp. 191-223.

Reich, P.B., A.W. Schoette, H.F. Stroo, J. Troiano, and R.G. Amundson. 1985. Effects of O_3 hr, SO_2, hr, and acidic rain on mycorrhizal infection in northern red oak seedlings. Can. J. Bot. 63: 2049-2055.

Reich, P.B., A.W. Schoettle, H.F. Stroo, and R.G. Amundson. 1986. Acid rain and ozone influence mycorrhizal infection in tree seedlings. J. Air. Pollu. Control Assoc. 306: 724-726.

Ruehle, J.L. and D.H. Marx. 1979. Fiber, food, fuel, and fungal symbionts. Science 206: 419-422.

Sanders, F.E., B. Mosse, and P.B. Tinker. 1975. Endomycorrhizas. Academic Press, New York, 626 pp.

Shafer, S.R., L.F. Grand, R.I. Brucks, and A.S. Heagle. 1985. Formation of ectomycorrhizae on Pinus taeda seedlings exposed to simulated acid rain. Can. J. For. Res. 15: 66-71.

Sheridan, R.P. 1979. Effects of airborne particulates on nitrogen fixation in legumes and algae. Phytopathology 69: 1011-1018.

Shriner, D.S. and J.W. Johnston. 1981. Effects of simulated, acidified rain on nodulation of leguminous plants by rhizobium spp. Environ. Exp. Bot. 21: 199-209.

Smith, W.H. 1972. Influence of artificial defoliation on exudates of sugar maple. Soil Biol. Biochem. 4: 111-113.

Smith, W.H. 1987. The Atmosphere and The Rhizosphere: Linkages with Potential Significance for Forest Tree Health. Tech. Bull. No. 527. National Council of the Paper Industry for Air and Stream Improvement. New York, pp. 30-94.

Sobotka, A. 1968. Wurzeln von Picea excelsa L. unter dem einfluss der industrieexhalate in gebiet des erzgebirges in der CSSR. Immissionene und waldöznoxen. cesk. Akad. Vd. Ustav pro Tvorlu a Ochr. Karn., Praha, p. 45.

Stroo, H.F. and M. Alexander. 1985. Effect of simulated acid rain on mycorrhizal infection of Pinus strobus L. . Water Air Soil Pollu. 25: 107-114.

Tamm, C.F. 1976. Acid precipitation — Biological effects on soil and forest vegetation. Ambio 5: 235-238.

Tarrant, R.F. 1968. Some effects of alder on the forest environment. In J.M. Trappe, J.F. Franklin, R.F. Tarrant, and G.M. Hansen, eds., Biology of Alder. U.S.D.A. Forest Service, Pacific Northwest Forest Range Exp. Sta., Portland, OR, pp. 193.

Thompson, G.W. and R.J. Medve. 1984. Effects of aluminum and manganese on the growth of ectomycorrhizal fungi. Appl. Environ. Microbiol. 48: 556-560.

Torrey, J. 1986. Personal communication. January 1986. Harvard Forest, Harvard University, Petersham, MA.

Trappe, J.M. 1977. Selection of fungi for ectomycorrhizal inoculation in nurseries. Annu. Rev. Phytopathol. 15: 203-222.

Trappe, J.M. and R.D. Fogel. 1977. Ecosystematic functions of mycorrhizae. In The Belowground Ecosystem: A Synthesis of Plant-Associated Processes. Range Sci. Dept., Science Ser. No. 26, Colorado State Univ., Fort Collins, CO, pp. 205-214.

Vesper, S.J. and T.C. Weidensaul. 1978. Effects of cadmium, nickel, copper, and zinc on nitrogen fixation by soybeans. Water Air Soil Pollu., 9: 413-422.

Waksman, S.A. 1967. The Actinomycetes. A Summary of Current Knowledge. Ronald Press, New York, 280 pp.

Wood, M. and J.E. Cooper. 1984. Aluminum toxicity and multiplication of *Rhizobium trifolii* in a defined growth medium. Soil Bio. Biochem. 16: 571-576.

Wood, M., J.E. Cooper, and A.J. Holding. 1984a. Aluminum toxicity and nodulation of *Trifobium repens*. Plant Soil 78: 381-391.

Wood, M., J.E. Cooper, and A.J. Holding. 1984b. Soil acidity factors and modulation of *Trifolium repens*. Plant Soil 78: 367-379.

12
Forest Tree Metabolism: Carbon Dynamics

Forest tree productivity, as reflected in the mass of carbon present, represents a dynamic balance between carbon fixed via photosynthesis and stored, and carbon lost to respiration, abscission, fine and woody root turnover, herbivory, infection, exudation, volatilization, and leaching. A comprehensive appreciation of air pollutant impact on tree metabolism will be realized only when we fully understand contaminant influence on carbon uptake, fixation, and allocation. Research on pollutant relationships with photosynthesis of agricultural and woody plants has been well developed for approximately 20 years. Fortunately recent research efforts have begun to explore stress and pollutant relationships with allocation (Lechowicz, 1987, Lorio and Sommers, 1986).

A. Photosynthesis

Photosynthesis is the most important metabolic process of forest ecosystems. In simple outline, the process amounts to the reduction of carbon dioxide to CH_2O and the oxidation of water to molecular oxygen, and results in the derivation of approximately 95% of the dry weight of plants. Photosynthesis is the process primarily responsible for forest productivity.

The few published reports of rates of net photosynthesis (gross carbon dioxide fixation less the losses due to respiration in the light and dark) of intact leaves of mature trees suggest that such rates are in general greater than 10 mg and less than 200 mg of carbon dioxide taken up per gram dry weight per day (Botkin, 1968; Kozlowski and Keller, 1966). Reported daily rates are generally greater for deciduous trees than for conifers on a leaf weight basis. Conifers, on the other hand, may be photosynthetically active for more months, even in northern latitudes. Linder and Troeng (1977) have indicated that Scotch pine may perform photosynthesis in mid-Sweden for approximately 10 months per year when supplied with adequate moisture. Polster (1950) reported carbon dioxide uptake per gram fresh weight for European birch to be 67 mg carbon dioxide g^{-1} day^{-1}, for beech 53 mg carbon dioxide g^{-1} day^{-1}, and for Norway spruce 14 mg carbon dioxide g^{-1} day^{-1}. In contrast, net photosynthesis for American white oak and scarlet oak are observed to range from 40 to 110 mg carbon dioxide g^{-1} day^{-1} (Botkin, 1968).

Differences among tree species are suggested in these reports but little data exist enabling direct comparative studies of species differences. Genetic differences in the rates of net photosynthesis have been found under certain environmental conditions for seedlings of white pine (Bordeau, 1963). Clonal differences have been observed for poplar by Huber and Polster (1955) and for larch by Polster and Weise (1962). Kruger and Ferrell (1965) observed differences for ecotypes of seedlings of Douglas fir from coastal and inland locations. Species differences have been found for seedlings of tree species under laboratory conditions. Kramer and Decker (1944) observed rates for seedlings of two oak species to be twice those of dogwood on a leaf area basis. Other similar reports are reviewed in Kramer and Kozlowski (1960), Kozlowski and Keller (1966), and Kramer and Kozlowski (1979), but it must be emphasized that too little is known to extrapolate from laboratory observations of seedlings to physiological reactions of mature trees under natural conditions.

From these studies, one would expect that differences exist among mature individuals of different tree species. Species differences have been found many times for herbaceous plants, as reported in Blackman and Black (1959), Blackman and Wilson (1951), and Hesketh and Moss (1963).

The situation for trees is complicated because differences have also been observed for leaves in different locations on the same tree. For example, shade leaves of European beech showed higher rates of net photosynthesis at low light intensities than sun leaves (Pisek and Tranquillini, 1954).

Rates of net photosynthesis for any tree are a function of time as well as location. Seasonal patterns have been studied particularly in conifers, and such differences have been observed for ponderosa pine (Fritts 1966) and Douglas fir (Hodges, 1962) among others. There are many reports of diurnal patterns in net photosynthesis, particularly in regard to a midday dip. These have been reported for coconut palm (McLean, 1920), ponderosa pine (Fritts, 1966), and other plants of the prairie and desert (Stocker, 1960), as well as for European beech and birch (Polster, 1950). The cause of this dip is not known and is the subject of controversy, attributed by some to environmental conditions, such as changes in carbon dioxide or water vapor concentrations in the air (Pisek and Tranquillini, 1954); to internal factors, such as the accumulation of net photosynthates (Thomas and Hill, 1949); and to the methods of measurement themselves (Bosian, 1965).

B. Constraints on Photosynthesis

Climate exerts profound control over photosynthetic rates (Zelitch, 1975; Govindjee, 1975). In addition to the obvious importance of solar radiation and ambient carbon dioxide (Anderson, 1973), moisture, temperature, and nutrition are critically important environmental variables. Water deficit decreases photosynthetic carbon dioxide uptake by restricting the transport of this gas to the chloroplasts in both the gaseous and liquid phase and by limiting the metabolic function of the chloroplasts themselves (Slavik, 1973). Carbon dioxide uptake generally increases with increasing temperature to some maximum and then decreases. Low temperature, chilling, and frost stress reduce photosynthetic rates (Bauer et al., 1973). Similar relationships between temperature and photosynthesis have been observed repeatedly under laboratory conditions, for example, between leaf temperature and carbon dioxide uptake of excised twigs of mature trees of Norway spruce and Swiss stone pine (Pisek and Winkler, 1958); between air temperature and carbon dioxide uptake by entire shoots of 35-day-old Douglas fir seedlings (Krueger and Ferrell, 1965); between air temperature and carbon dioxide uptake by four species of woody desert perennials, *Larrea divaricata, Hymenudea salsola, Encelia farinosa*, and *Chilopsis linearis* (Strain and Chase, 1966); and between air temperature and carbon dioxide uptake of excised apple leaves (Waugh, 1939). A second degree polynomial gave a significant regression of the logarithm of carbon dioxide uptake on air temperature for laboratory observations of seedlings of sand pine (Pharis and Woods, 1960). In a nonhomogeneous and mixed species forest, highly significant relationships were obtained between sunlight intensity and leaf temperature and rates of photosynthesis of individual oak leaves (Botkin, 1968). Sunlight and temperature accounted for more than 70% of the variation in rates of photosynthesis under these conditions.

There is abundant evidence demonstrating that the rate of photosynthesis may be decreased by a deficiency in any one of the essential nutrient elements (Nátr, 1973). Keller (1968, 1970) has shown a dramatic effect of nitrogen deficiency on the rate of photosynthesis of forest trees.

In addition to the restrictions imposed by these important environmental variables, it has become increasingly apparent that numerous materials released by human beings into their environment can reduce net photosynthesis of plants. These materials include radioisotopes (Nazirov, 1966), salt (Nellen, 1966), and numerous herbicides (Sassaki and Kozlowski, 1966a,b) and insecticides (Heinicke and Foott, 1966) among others. What is the evidence for adding air contaminants to this list?

C. Photosynthetic Suppression: Forest Tree Seedlings or Cuttings

Carbon dioxide uptake is a favored technique for monitoring photosynthetic rates. Since this measurement, along with pollutant fumigation, is most conve-

niently performed in a relatively small chamber, seedling size plants have been
the preferred research material for investigators interested in forest trees.

1. Sulfur Dioxide

Roberts et al. (1971) exposed 5-month-old maple seedlings to 1 ppm (2620 µg
m^{-3}) sulfur dioxide for 4–6 hr and observed a slight stimulation in carbon dioxide
exchange relative to unfumigated trees. At 6 ppm (15.7 x 10^3) sulfur dioxide depressed photosynthesis but the suppression was due to gross foliar necrosis.
Quaking aspen and white ash seedlings were exposed to 0.2 (524 µg m^{-3}), 0.5
(1310 µg m^{-3}), 1.0 (2620 µg m^{-3}), and 4.0 (10.5 x 10^3 µg m^{-3}) ppm sulfur dioxide for 2–4 hr by Jensen and Kozlowski (1974). No influence on photosynthetic
rate was detected at 0.2 or 0.5 ppm. One ppm caused a slight reduction and 4
ppm caused a substantial decrease in photosynthetic rate.

Theodore Keller of the Swiss Forest Research Institute in Birmensdorf,
Switzerland has done considerable research on the relationship between sulfur
dioxide and forest tree seedling metabolism. Exposure of Scotch pine to 0.05–
0.10 (131–262 µg m^{-3}) ppm sulfur dioxide for many weeks caused a reduction in
carbon dioxide assimilation of very approximately 40% of control rates (Keller,
1977a). Keller (1977b, 1978a,b) subsequently exposed genetically uniform 3-
year-old grafts of white fir, Norway spruce, and Scotch pine to 0.05 (131 µg m^{-3}), 0.1 (262 µg m^{-3}), and 0.2 (524 µg m^{-3}) sulfur dioxide for periods of up to 70
days. Prolonged exposure to 0.2 ppm caused a dramatic decrease in the photosyn-
thesis of all three species, especially the fir (Table 12-1). Visible symptoms of
foliar injury occurred at the end of the experiment only in Norway spruce subject
to the 0.2 ppm fumigation. Seasonal observations with the latter species have
revealed significant photosynthetic suppression at low doses during the spring
and has led Keller (1978a) to observe that the reduced annual increment may be
substantial because of this. Since suppressed root development may also be associated with high dose (0.1 and 0.2 ppm) exposure to sulfur dioxide, Keller
(1979) has judged that the impact of this gas on spruce may be substantial
growth reduction.

In a study of six early successional annual species exposed to zero or 0.25
ppm (655 µg m^{-3}) sulfur dioxide at 300 (5.4 x 10^5 µg m^{-3}), 600 (10.8 x 10^5 µg
m^{-3}), or 1200 (16.2 x 10^5 µg m^{-3}) ppm carbon dioxide, Carlson and Bazzaz
(1982) observed that the sensitivity (leaf area) to sulfur dioxide at elevated carbon
dioxide is lower for C$_3$ species and higher for C$_4$ species than that observed at
300 ppm carbon dioxide.

2. Ozone

In 1961, Taylor et al. presented evidence that the rate of carbon dioxide fixation
by potted lime seedlings was significantly inhibited by exposure to ozone at 0.6
ppm (1176 µg m^{-3}) for 1 hr without the development of leaf symptoms.

Table 12-1. Photosynthesis of 3-Year-Old Grafts of Tree Sedlings Subject to Continuous Exposure to Varying Concentrations of Sulfur Dioxide Under Controlled Environmental Conditions[a].

Season	Fumig. period (days)	Relative photosynthesis (%) at SO_2 conc.			
		Control	0.05 ppm	0.1 ppm	0.2 ppm
Silver fir					
Spring	1-15	40	28	27	10
	16-30	64	58	51	34
	31-60	93	74	49	24
Summer	1-15	109	108	114	100
	16-30	109	104	104	90
	31-60	96	87	77	58
Fall	1-15	122	94	91	87
	16-30	106	83	83	80
	31-60	100	72	70	62
Norway spruce					
Spring	1-15	56	52	58	53
	16-30	104	88	107	94
	31-60	110	89	88	72
Summer	1-15	105	113	108	95
	16-30	131	125	113	96
	31-60	127	120	101	80
Fall	1-15	113	134	114	96
	16-30	116	133	110	88
	31-60	100	100	84	68
Scotch pine					
Spring	1-15	75	77	70	38
	16-30	71	49	80	57
	31-60	86	64	83	49
Summer	1-15	130	132	124	111
	16-30	149	140	114	102
	31-60	136	116	83	44
Fall	1-15	110	113	118	116
	16-30	109	101	109	97
	31-60	100	80	82	72

Source: Keller (1977b).

[a] Carbon dioxide uptake rates are relative to uptake rate of control at bud break.

The first observations on the relationship between forest tree photosynthesis and oxidants was presented by Paul R. Miller in 1966. This investigator subjected 3-year-old ponderosa pine, grown from seed collected in the San Bernardino Mountains, California, to ozone fumigation 9 hr daily in a controlled environment chamber (Miller et al., 1969). A 60-day fumigation with 0.15 ppm (294 $\mu g\ m^{-3}$) resulted in a final 25% decrease in apparent photosynthesis and a 30-day fumigation with 0.30 ppm (588 $\mu g\ m^{-3}$) revealed a 67% depression (Figure 12-

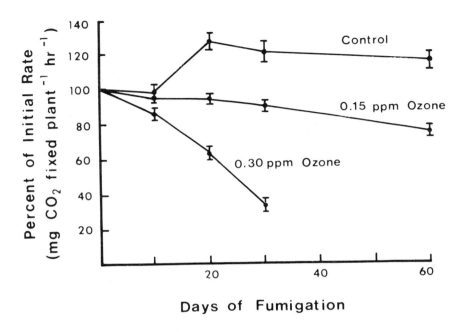

Figure 12-1. The influence of ozone on the apparent photosynthesis of ponderosa pine seedlings.
Source: Miller et al. (1969).

1). Soluble sugars and polysaccharides were observed to be significantly decreased in 1-year old needles following a 33-day exposure to 0.30 ppm ozone.

Table 12-2. Influence of Ozone on Photosynthesis and Respiration of Three Yellow Pine Seedlings.

	CO_2 exchange (mg g^{-1} dry wt hr^{-1})								
	Time of exposure (days)								
Concentration (pphm)	36			77			84		
	Slash	Pond	Loblolly	Slash	Pond	Loblolly	Slash	Pond	Loblolly
Photosynthesis									
0	2-6	3-5	3-7	0-9	0-9	1-8	1-9	3-4	4-3
5	3-1	4-2	4-0	1-6	1-8	1-8	1-7	3-8	4-0
15	2-0	3-3	3-2	0-9	1-6	1-7	1-4	3-3	4-1
Respiration									
0	0-7	1-0	0-7	0-4	0-6	0-6	0-5	1-0	1-0
5	0-6	0-9	0-7	0-4	0-5	0-6	0-5	1-2	1-0
15	1-3[a]	1-2	1-2[a]	0-6	0-5	0-5	0-8	1-2	1-2

Source: Barnes (1972).
[a] Significantly greater than control value ($P < 0.05$).

Barnes (1972) has examined the response of greenhouse-grown seedlings of four pine species to ozone fumigation. Seedlings of various ages, 2–8 months, were exposed to continuous fumigation with ozone in growth chambers at concentrations of 5 or 15 pphm (98–294 μg m^{-3}). Barnes recorded substantial variation in foliar response depending on seedling age. He also noted a consistent stimulation of respiration, in one instance 90%, at the 15 pphm fumigation dose. In younger seedlings of eastern white pine, which bore only primary needles, ozone had little influence on photosynthetic rate. In older seedlings with secondary needles, photosynthesis was slightly depressed. With seedlings of slash, pond, and loblolly pines, ozone exposure at 15 pphm (294 μg m^{-3}) had a relatively consistent depressing influence on photosynthesis of all species. At 5 pphm (98 μg m^{-3}), however, ozone appeared to have a stimulating influence on older secondary needles and a depressing effect on younger secondary needles (Table 12–2).

Reich (1983) exposed hybrid poplar cuttings to four ozone concentrations — 25, 50, 85, and 125 ppb (49, 98, 167, and 245 μg m^{-3}) — for approximately 5.5 hr day^{-1} for 62 days in environmental growth chambers. He observed that exposure to low concentrations of ozone had no immediate, direct effect on net photosynthesis, but chronic exposure to similar levels did have a number of effects on carbon dioxide exchange. He concluded that chronic exposure placed a stress on the photosynthetic and respiratory systems of the leaves, resulting in increased dark respiration and decreased leaf chlorophyll contents and net photosynthesis. Accelerated leaf aging due to ozone exposure was at least partially responsible for declining net photosynthetic capacity in treated leaves. In a related study, Reich and Lassoie (1984) concluded that ozone exposure also impaired stomatal function in hybrid poplar. Ozone exposure reduced the water-use efficiency and range of leaf conductance of individual leaves, and altered the relationship between the conductances of the two leaf surfaces (the ratio of abaxial to adaxial leaf conductance was increased). In summary, ozone exposure enhances desiccation stress. While ozone exposure had no effect on partitioning dry matter in hybrid poplar, chronic exposure to low concentrations did reduce growth and dry matter accumulation (Reich and Lassoie 1985).

3. Fluoride

Keller (1973) has evaluated the influence of particulate fluoride compounds — sodium fluoride, calcium fluoride, and natural and synthetic cryolite, on the photosynthetic rate of 3-year-old Scotch pine seedlings, 4-year-old Douglas fir seedlings, and 1-year-old birch grafts. Foliage was coated using fungicide applicators with a very thin whitish film under high and normal relative humidity in a greenhouse and in specially designed glass covers. Only treatment with sodium fluoride induced visible symptoms. In birch, all dusts caused a significant depression of photosynthesis. In conifers, the photosynthetic reduction was not significantly different from the controls in the absence of visible symptoms.

Seedlings of 12 forest tree species were potted and placed by Keller (1977c) at varying distances from an aluminum smelter. Photosynthesis was monitored for several months through summer and fall. Pine seedlings were observed to be particularly sensitive, as their photosynthetic rates dropped to below 60% of controls when foliar concentrations reached approximately 30 µg g^{-1} fluoride on a dry weight basis. Photosynthetic suppression decreased with both increasing distance from the fluoride source and increasing foliar tissue maturity. Simultaneous exposure of young spruce cuttings to sulfur dioxide and rooting substrate treated with sodium fluoride has revealed a synergistic suppression of photosynthesis (Keller, 1979, personal communication).

4. Heavy Metal Particulates

While the trace metal content was not determined, Auclair (1976) evaluated the influence of artificially applied coal (and cement dust) on 2-year-old Norway spruce seedlings. At all experimental light intensities employed, coal dust was found to significantly reduce photosynthesis. Whether this inhibition was due to stomatal blockage, attenuated light reception, or heavy metal toxicity was not clear from the experiment. Subsequent experiments examining the photosynthetic response of Scotch pine and poplar to coal dust led the investigator to conclude that the observed reduction in this metabolic process was due to reduced light (Auclair, 1977).

The influence of lead and cadmium on the photosynthesis of nursery grown 2- to 3-year-old seedlings of American sycamore has been studied by Carlson and Bazzaz (1977). These investigators planted seedlings in pots containing a silty clay loam field soil and treated the pots with 0, 50, 100, 250, 500, and 1000 µg g^{-1} cadmium as cadmium chloride and with an integrated treatment with each level of both metals combined. Figure 12–2 shows the relationship observed between lead and cadmium soil amendment and photosynthesis. While it can be seen that treatment with lead and cadmium both reduced the photosynthetic rate, no synergism was observed when the metals were combined.

5. Acid Deposition

While little evidence has been presented indicating that acid deposition can solely or directly influence the photosynthesis of forest trees, a variety of studies have examined the interaction of acid deposition with other stresses as related to carbon fixation.

Reich et al. (1986) exposed 2-year-old sugar maple and northern red oak seedlings to all combinations of several levels each of ozone and acid rain simulant. Ozone exposures were constant and ranged from 20 to 120 ppb (39–235 µg m^{-3}) applied 7 hr day^{-1} for 5 days week^{-1}. Rain simulants were normal for eastern North America and consisted of 12.5 mm week^{-1} with pH in the range of 3.0–5.6. Ozone exposure caused significant decreases in net photosynthesis in both species, with the greatest reductions recorded (30% in maple and 20% in oak)

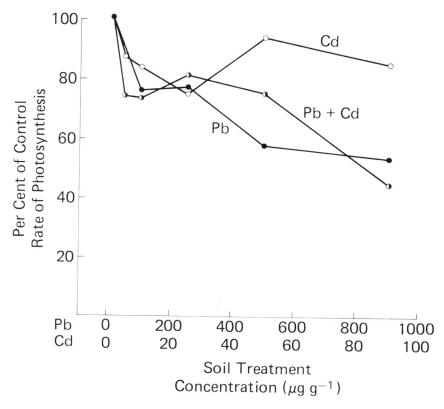

Figure 12-2. The influence of soil applied lead and cadmium on the photosynthesis of seedling American sycamore.
Source: Carlson and Bazzaz (1977).

after two months at the high (120 ppb) ozone treatment (Figure 12–3). Growth reductions occurred in sugar maple, but were not recorded in red oak. Chlorophyll contents were observed to increase in sugar maple leaves following ozone exposure. Acid rain applications had no effect on either net photosynthesis or growth in either species and no interactions of the two pollutants were observed.

Taylor et al. (1986) have investigated the interactive influence of ozone, mist chemistry, rain chemistry, and soil type on carbon dioxide assimilation and growth of 1-year-old red spruce seedlings over a 4-month period. The chemistry of the precipitation simulants and ozone exposure dynamics were based on reported characteristics of deposition in high-elevation forests of eastern North America. The soil types used were collected from Camels Hump Mountain in Vermont and Acadia National Park on the Maine coast. The rates of carbon dioxide assimilation and transpiration on a per gram needle dry weight basis were not influenced by any of the main treatment variables or their interactions (Table 12–

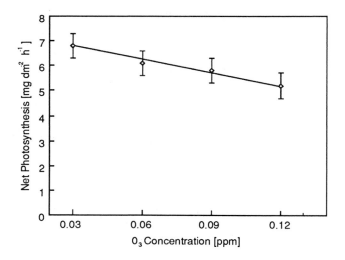

Figure 12-3. Mean net photosynthesis (± SE) of sugar maple seedlings in four ozone treatments.
Source: Reich et al. (1986).

3). As neither soil type nor precipitation chemistry influenced the efficiency of carbon dioxide assimilation, the authors speculated that the physiological mechanism responsible for the growth response of red spruce may have been due to a change in either whole-plant allocation of carbon resources or a direct toxic effect

Table 12–3. Statistical Summary of Treatment Effects on Indicators of Growth in Red Spruce.

Treatment	Growth parameter							
	Root	Needle	Stem	Shoot	Root/ Shoot	Total	Diameter	Height
Main								
Soil type	ns[a]	0.02	0.01	0.01	ns	0.01	ns	0.02
Rain chemistry	ns	ns	ns	ns	ns	ns	ns	ns
Mist chemistry	ns	ns	ns	ns	ns	ns	ns	ns
Ozone	0.03	ns	ns	ns	ns	ns	ns	ns
First-order interactions								
Soil × mist chemistry	0.05	ns	ns	ns	ns	ns	0.05	ns
All other	ns	ns	ns	ns	ns	ns	ns	ns
Second-order interactions								
All	ns	ns	ns	ns	ns	ns	ns	ns

[a]Probability of statistically significant treatment effect: ns = $P > 0.05$
Source: Taylor et al. (1986).

in the rhizosphere or root growth.

When working with eastern white pine, however, Reich et al. (1987) observed that interactions between soil properties, plant nutrition, and acid deposition were important in assessing the full impact of acid precipitation on pine seedlings. Several week-old seedlings were subjected to combined acid rain, ozone, and soil treatments. Acid rain treatments consisted of exposure to rain simulants of pH 3.0, 3.5, 4.0 and 5.6, while ozone exposures consisted of 20, 60, 100, or 140 ppb (40, 118, 196, or 274 μg m^{-3}) ozone. Seedlings were grown in five forest soils for a period of 4 months. Very minimal interaction between acid precipitation and ozone was observed with regard to photosynthesis or growth. Acid precipitation and soil type, however, had a strong interaction in determining pine response. In general, acid rain caused increased growth and net photosynthesis as a result of nitrogen fertilization from the rain simulant. The extent of the fertilization response was inversely correlated with nitrogen availability in each soil. Low-level ozone pollution caused decreases in photosynthesis of pine seedlings regardless of the soil.

Roberts (1988) exposed 1-year-old seedlings of yellow poplar to acid rain simulant, ozone, and drought for 5 months. Acid rain simulant was found to have a greater impact on physiological activity than ozone or drought. Seedlings treated with pH 3.0 rain simulant exhibited significant reductions in carbon exchange rate (29% percent), stomatal conductance (33%), and xylem pressure potential (17%) relative to trees treated with pH 5.5 rain.

D. Photosynthetic Suppression: Forest Tree Saplings

Investigations that have examined the response of photosynthetic rates of forest trees over the age of 5 years to air contaminant exposure have the advantage of avoiding metabolic pecularities that may be unique to the seedling stage of tree development. Because these studies are of necessity involved with relatively large plants, the monitoring of carbon dioxide flux and pollutant gas exposure is complicated. Experimental designs must monitor gas exchange in entire growth chambers, entire tree enclosures within growth chambers, in small enclosures placed on a portion of the foliage of experimental saplings, or deal with excised leaves.

1. Excised Leaf Study

Silver maple leaves were obtained by detaching the distal 30 cm of twigs from 8-year-old silver maple saplings by Lamoreax and Chaney (1978). Leaves of uniform size were excised from detached twigs in the laboratory. Some leaves were placed in solutions containing 0, 5, 10, or 20 ppm cadmium as cadmium chloride. Other leaves were treated with sulfur dioxide concentrations of 0, 1.0 (2620 μg m^{-3}), or 2.0 (5240 μg m^{-3}) ppm. Photosynthetic rates were determined after 45 hr of cadmium exposure and 30 min of sulfur dioxide treatment. Some leaves were exposed to the cadmium and sulfur dioxide treatments sequentially.

Cadmium at 20 ppm, which resulted in a petiole concentration of 4 μg cadmium g^{-1} petiole tissue and zero cadmium in leaf tissue, greatly reduced photosynthesis. The decrease caused by cadmium was greater when 1.0 or 2.0 ppm sulfur dioxide were present. In a peculiar deviation, treatment with 5 ppm cadmium precluded the synergistic influence of sulfur dioxide exposure.

2. Small Chamber Technique

The small chamber technique for the determination of photosynthetic rates was developed at Brookhaven National Laboratory, Upton, Long Island by Woodwell and Whittaker (1968). It is thoroughly described in Woodwell and Botkin (1970). D.B. Botkin, several graduate students, and the author have employed the small chamber technique to assess the influence of gaseous contaminants on tree sapling photosynthetic rates.

In this method a leaf, or group of leaves or needles, is surrounded by a small transparent plastic (polyvinyl chloride) cover supported on a lightweight aluminum frame and forming a cylindrical chamber approximately 28 cm long and 15 cm in diameter. Air is supplied to and sampled from this chamber. Carbon dioxide uptake by leaf tissue is determined by maintaining an empty chamber in addition to chambers containing leaves, and comparing the concentration of carbon dioxide in the air sampled from this chamber with that from a leaf-containing chamber. Gaseous pollutants, at a monitored concentration, are introduced to the chambers with the supply air.

Since the small-chamber technique completely encloses a leaf or group of leaves, it introduces some well appreciated artificialities. Air temperatures inside the chambers may become elevated in relation to ambient conditions outside the chambers. Temperatures within chambers are monitored with copper-constantin thermocouples shielded at the entrance to the air sampling line. The effect of temperature distortion has been investigated, and its influence on daily rates of carbon dioxide uptake by deciduous leaves in a coastal plain forest were found to be 15% or less (Botkin, 1968). Flow rates within the air of the chamber are unnatural in that it is impractical to simulate variable flow that mimics normal air movement within a forest canopy. Flow rates have a large effect on carbon dioxide uptake only when air flow is so slow that the photosynthetic rate is restricted by carbon dioxide availability. Our experimental flow rates were maintained at 5 l min^{-1} or better to avoid this problem.

A completely portable data monitoring system capable of recording photosynthetic rates, air pollutant levels, and pertinent environmental conditions was developed. The system recorded light intensity, temperature, and atmospheric concentrations of carbon dioxide, water vapor, and ozone or sulfur dioxide sequentially and continuously during experimental runs. Transducer measurements of all parameters, produced as a millivolt signal, were transmitted to a digital volt meter, and through interfacing electronics to a paper tape punch, where each signal was recorded as a voltage. A schematic outline of the system is presented in Figure 12–4.

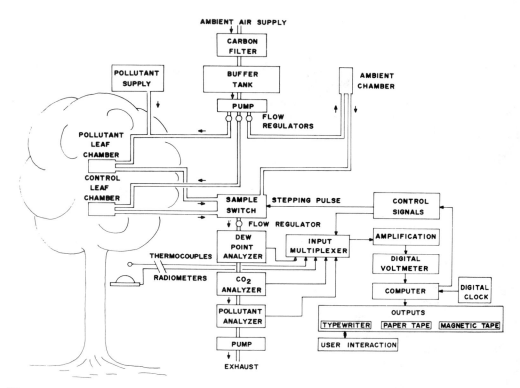

Figure 12-4. Outline of the data acquisition and recording system employed in the Botkin et al. (1971, 1972) small chamber technique for evaluating the influence of gaseous air pollutants on tree photosynthesis.

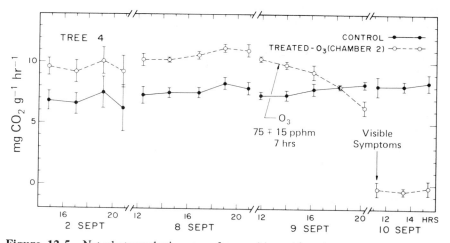

Figure 12-5. Net photosynthetic rates of a sensitive white pine exposed to ozone. Points represent means for 2-hr periods with 95% confidence intervals. Where not shown, confidence intervals were too small to graph.
Source: Botkin et al. (1972).

Our application of the small chamber technique to the study of eastern white pine photosynthesis as influenced by ozone has been presented in Botkin et al. (1971, 1972). All experimental trees were selected from suburban Connecticut and New York nurseries in locations subject to ozone pollution. Our monitoring system was applied to trees growing in large controlled environment facilities. Chamber temperature was maintained at 26°C daytime and 17°C nighttime with a photoperiod of approximately 13 hr. Lighting was mixed fluorescent and incandescent, giving a maximum of 0.23 g cal cm^{-2} min^{-1} at the top of the trees. Trees were exposed to ozone only during illumination.

The threshold of ozone suppression of white pine photosynthesis as indicated by this research was approximately 50 pphm (980 µg m^{-3}) for a minimum of 4 hr. Above this threshold, we distinguished three categories of ozone sensitivity: in sensitive trees dosages of 50–100 pphm (980–1960 µg m^{-3}) for 10 hr reduced net photosynthesis to essentially zero (Figure 12–5); in intermediate trees dosages within a similar range reduced net photosynthesis by approximately 50% (Figure 12–6); and in resistant trees these dosages had no effect on photosynthesis (Figure 12–7). In the intermediate trees, ozone-induced photosynthetic suppression was reversible if an ozone-free recovery period was made possible. Visible symptom expression (chlorosis or distal necrosis) on the foliage of the current year was not a good index of the timing or the severity of photosynthetic suppression of the needles following ozone exposure. Visible symptoms were induced by ozone exposure in some trees. Where both occurred, photosynthetic suppression preceded visible indication of injury.

Carlson (1979) employed the small chamber technique as described to evaluate the influence of sulfur dioxide, ozone, and combinations of both gases on 8– to 15–year–old saplings of black oak, sugar maple, and white ash collected from natural Connecticut forests and grown in containers filled with natural forest soil profiles. Treatments included fumigation at low (0.2–0.4), intermediate (0.5–0.6), and high (1.3–1.5 cal cm^{-2} min^{-1}) light intensity and low (22–43%) and high (55–92%) relative humidity. Following 1 week of fumigation at low humidity and low light intensity the rate of photosynthesis was 52, 46, and 80% of control for treatment with 50 pphm (1400 µg m^{-3}) sulfur dioxide; 52, 73, and 100% of control for treatment with 50 pphm (1100 µg m^{-3}) ozone; and 56, 59, and 62% of control for treatment with 50 pphm sulfur dioxide plus 50 pphm ozone, respectively, for black oak, sugar maple, and white ash. After 3 weeks of treatment, sulfur dioxide reduced photosynthesis to 26, 57, and 93% of control, while ozone reduced carbon dioxide uptake to 57, 45, and 94% of control, respectively, in black oak, sugar maple, and white ash. These results are summarized in Table 12–4. Simultaneous exposure to both sulfur dioxide and ozone caused a greater than additive reduction in net photosynthesis during the first 2 days of treatment in the case of sugar maple and white ash.

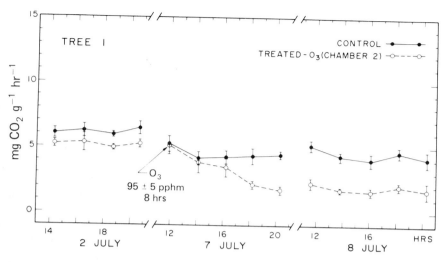

Figure 12-6. Net photosynthetic rates for a white pine of intermediate sensitivity to ozone.
Source: Botkin et al. (1972).

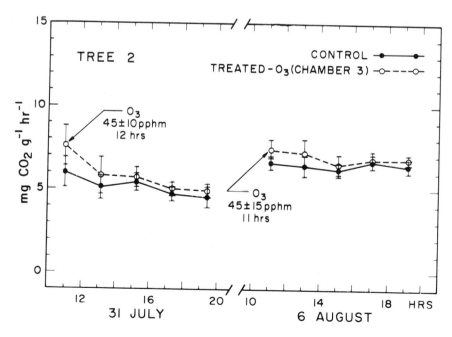

Figure 12-7. Net photosynthetic rates for a white pine resistant to ozone.
Source: Botkin et al. (1972).

3. Open Top Chamber Studies

Under field conditions, the most adequately tested exposure facility that allows control of gaseous pollutant concentrations is the open top chamber (Mandl et al., 1973, Heagle et al., 1973, Heagle et al. 1979) (Figure 12–8). The chambers are cylinders constructed with aluminum frames covered with clear plastic film. They are approximately 3 m in diameter by 2.4 m high and are equipped with blowers to introduce approximately 60 m^3 of air per minute (approximately three changes min[-1]) through a manifold surrounding the lower 1.2 m of the chamber. The air entering "control" chambers passes through both particulate and charcoal filters. Nonfiltered chambers permit the use of ambient pollutants as a baseline for developing additional exposures. Different exposures are obtained by either adding pollutants at various concentrations to the ambient pollution load or by partial filtration to reduce the ambient pollution load (Heck et al., 1979). The open-top chamber methodology provides a realistic strategy for experimentally manipulating air pollutant exposure in the field and for measuring the response of plants larger than seedlings.

Growth of woody plants in open-top chambers has not been extensive, but recent applications of the technique have demonstrated considerable utility for large seedling- and sapling-size plants (Duchelle et al., 1982).

Open-top chambers have provided the first quantitative, statistically significant data linking productivity declines in a forest species to ambient exposure to ozone in a rural area. Working with open-top chambers in Millbrook, New York (20 km northeast of Poughkeepsie), Wang et al. (1985a) presented evidence that ambient ozone, at levels below the United States air quality standard, caused a significant reduction(19%) in the growth of sapling hybrid poplars. Growth re-

Table 12-4. Influence of Sulfur Dioxide, Ozone, and a Mixture of Both Gases on the Net Photosynthesis of Forest Saplings following 1–2 days, 1 week, and 3 weeks of Intermittent Fumigation[a].

Species	Treatment (pphm)	% expected rate of net photosynthesis ± 1 SD		
		1-2 days	1 week	3 weeks
Black oak	50 SO_2	60 ± 15	52 ± 11	26 ± 7
	50 O_3	70 ± 10	52 ± 5	57 ± 5
	50 SO_2 + 50 O_3	55 ± 8	56 ± 9	—
Sugar maple	50 SO_2	78 ± 5	46 ± 9	57 ± 10
	50 O_3	79 ± 10	73 ± 16	45 ± 11
	50 SO_2 + 50 O_3	26 ± 10	59 ± 10	—
White ash	50 SO_2	73 ± 13	80 ± 12	93 ± 5
	50 O_3	100 ± 18	100 ± 15	94 ± 12
	50 SO_2 + 50 O_3	54 ± 8	62 ± 13	—

Source: Carlson (1979).
[a]The small chamber technique was employed along with environmental conditions of varying temperature, humidity, and light conditions.

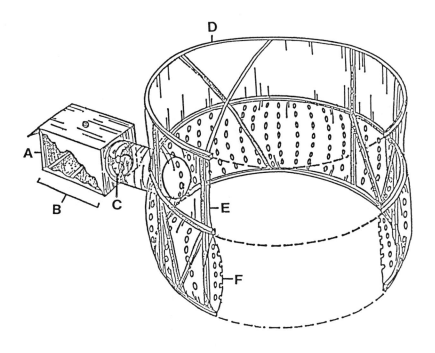

OPEN TOP FIELD CHAMBER

Figure 12-8. Open top field exposure chamber for air pollution studies. A, air filter, B. fan box, C, fan, D, lightweight metal frame, E, plastic cover, F, perforated plenum.

ductions for eastern cottonwood and black locust, examined at the same time, were not significant. Over a 3-year period at the same Millbrook site, Wang et al. (1986b) examined ozone sensitivities of four clones of trembling aspen. Ambient ozone was found to significantly reduce above ground dry-matter production (12–24 percent), and to modify tree morphology, root/shoot ratios, and rates of leaf senescence. Unfortunately no metabolic measurements were taken during these open-top chamber studies.

Reich and Amundson (1985) employed open-top chambers to study the influence of 23, 48, and 68 ppb (45, 94, and 133 µg m^{-3}) ozone 7 hr d^{-1}, 7 days week^{-1} for 9 weeks on net photosynthesis of red oak. Exposure to any increase in ozone was found to reduce photosynthesis.

In reviewing this study and several other investigations examining the ozone and acid precipitation influence on photosynthesis, the authors made the following summary conclusion (Reich and Amundson, 1985). In contrast to ozone, acid precipitation had no negative effect on photosynthesis in four tree species tested (sugar maple, eastern white pine, hybrid poplar, and northern red oak), and no interaction between ozone and acid rain was detected. Also since elevated levels of ozone occur in ambient air more frequently and over larger areas than do elevated levels of sulfur dioxide, it is probable that ozone, but not sulfur dioxide, is involved in significant reductions in tree metabolism. Ozone-induced reductions in photosynthesis were related to declines in growth or yield. Species with higher stomatal conductances and thus higher potential for pollutant uptake exhibited greater negative responses to similar ozone treatments. Reich and Amundson judged, that reductions in photosynthesis may be occurring over much of the eastern United States, since ozone exposure to concentrations typical of levels of pollution observed in ambient air reduced the rates of net photosynthesis in all species tested.

Using the results from field fumigation studies of unenclosed hybrid poplar cuttings exposed to ozone, Reich et al. (1984) used a percent reduction model (Heck et al., 1982) to predict poplar growth reductions in polluted atmospheres relative to a "clean" atmospheres with a mean daily ozone concentration of 25 ppb (49 μg m^{-3}). The model arbitrarily assumed that exposures would be for 5.9 hr day^{-1} on 16 days during a 1-month period. The predicted reductions in poplar growth and dry matter (Table 12–5) are approximately similar or smaller than predicted reductions for yield of a variety of agricultural species (Heck et al., 1982; Reich and Amundson, 1984). The model predicted that poplar trees exposed to 60 ppb (118 μg m^{-3}) and 100 ppb (196 μg m^{-3}) ozone would have dry weights approximately 12% and 25% percent lower, respectively, than plants grown in unpolluted air. These predictions are consistant with the open-top evidence provided by Wang et al. (1986a,b) and previously discussed.

E. Photosynthetic Suppression: Large Forest Trees

The influence of air contaminants on photosynthesis of large trees growing under field conditions is very poorly understood. Experimental designs necessary for this research are complicated and expensive. Soil moisture, relative humidity, temperature, and light characteristics all influence photosynthetic rates under natural conditions, and all must be continuously and carefully monitored. Access to mid- and upper-crown areas of large trees is generally by elaborate scaffolding, which is awkward, risky, and expensive to erect.

Legge and Jacques (1977) have conducted field studies of photosynthesis in lodgepole pine-jack–pine-hybrids, white spruce, and aspen that were potentially influenced by sulfur dioxide and hydrogen sulfide from a natural gas processing plant near Whitecourt, Alberta, Canada. Small assimilation chambers, or cuvettes, were placed around portions of tree branches and were used to monitor carbon dioxide flux. A 15-m high scaffold was used to gain access to midcrown

Table 12-5. Percent Reduction Models for Hybrid Poplar as a Function of Ozone Concentration (for 5.9 hr day^{-1} on 17 day in a 1-month period).

Parameter	Linear function[a]	Predicted percent reduction at	
		0.06 ppm	0.10 ppm
Experiment No. 1			
Stem height	$Y = 5.47 - 218.90c$	7.7	16.4
No. of leaves	$Y = 15.27 - 610.86c$	21.4	45.8
Total above ground dry mass	$Y = 15.27 - 402.88c$	14.1	30.2
Experiment No. 2			
Stem height	$Y = 4.24 - 169.81c$	5.9	12.7
No. of leaves	$Y = 14.9 - 576.06c$	19.7	42.7
Total above ground dry mass	$Y = 7.01 - 280.41c$	9.8	21.0

[a]Y is the percent reduction in a parameter in comparison with the estimated value at 0.025 ppm: c is the mean 5.9-hr O_3 concentration (parts per million).
Source: Reich et al. (1984).

foliage of the pine hybrids (20 m ht). The spruce (3 m ht.) and aspen (2 m ht.) were reached from the ground. Efforts to correlate trends in photosynthetic rates of these trees with episodic release of sulfur gases from the processing plant were frustrated by reductions in light quantity coincident with peak sulfur gas exposure and the fact that the measurements were made in September and the authors judged that foliar senescence may have been a factor in the metabolic responses observed. Nevertheless, ambient sulfur dioxide concentrations ranging from 5–10 pphm (131–262 μg m^{-3}) and persisting for several hours in the study forest canopy were recorded. It was the authors' judgment that all the sampled trees exhibited less than the expected photosynthetic rates. The maximum rates for monitored trees were pine, 3.28 mg dm^{-2} hr^{-1}; spruce, 2.3 mg dm^{-2} hr^{-1}; and aspen 3 mg dm^{-2} hr^{-1}. Additional data would be required to more specifically relate ambient sulfur gas concentrations to photosynthetic performance of the forest surrounding the Windfall gas plant.

We have made an effort to monitor the photosynthetic response of mature deciduous species growing in natural Connecticut forests to artificially applied ozone by employing the portable small chamber technique described in Section D (Figure 12–9). Preliminary and unpublished results indicated that thresholds of suppression in red maple and white ash were comparable to the threshold determined for young white pine.

Skärby et al. (1987) investigated ozone uptake, transpiration, net photosynthesis, and dark respiration in a 20 year-old stand of Scots pine in Sweden by employing branch chambers. Current shoots were treated with ozone concentrations ranging from 60 ppb (120 μg m^{-3}) to 200 ppb (400 μg m^{-3}) during 1 month. No significant impact on net photosynthetic performance was detected.

Figure 12-9. Field installation employed to obtain information on the photosynthetic response of mature Connecticut deciduous trees to artificially applied ozone using the small chamber technique. Scaffolding employed to gain upper crown access is visible to the left. The trailer to the right housed all data monitoring and recording equipment while the polyethylene "balloon" was used to stabilize air to be supplied to the chambers against short-term fluctuations in carbon dioxide concentration.

At the highest exposure, ozone appeared to decrease daytime transpiration. Dark respiration was increased by ozone exposure.

A very ambitious open fumigation study has been initiated in the Liphook forest, on the Sussex-Hampshire border, U.K., by the Central Electricity Research Laboratories (Shaw 1986) (Figure 12–10). The outdoor fumigation system employed on these experimental pots is based on a technology described by McLeod et al. (1985). Initial results will describe the extended term (several years) interaction of young Corsican pine, Scots pine, Sitka spruce, and Norway spruce to sulfur dioxide and ozone exposures.

F. Photosynthetic Response to Air Contaminants: Mechanisms of Suppression

As is evident from this review, the typical method for monitoring the photosynthetic rate of forest species is carbon dioxide flux. When used to evaluate the influence of an air contaminant, this method tells us little about the physiological

Figure 12-10. Aerial view of Liphook forest (U.K.) experimental site where open fumigation of forest trees has been initiated.
Source: Central Electricity Generating Board, Surrey, UK.

and biochemical mechanisms(s) of alterations in rates that may be observed. Fortunately this topic has been approached by numerous investigators, commonly employing nonwoody species, and several hypotheses have been offered. While a thorough discussion of these is considered to be beyond the scope of this book, the most important suggestions have been as follows. Evidence has been provided to suggest that air pollutants may alter stomatal opening, interfere with chloroplast membranes, influence chlorophyll concentrations; affect pH; alter redox reactions, electron flow, and phosphorylation, or attack critical proteins or enzymes involved in photosynthesis. Reviews addressing the relationship between photosynthetic metobolism and specific air pollutants are available for sulfur dioxide (Hällgren, 1978; Ziegler, 1975; Keller and Schwager, 1977, Winner et al., 1985), ozone (Evans and Ting, 1974; Verkroost, 1974; Malhotra and Khan 1984, Reich, 1987b), fluoride (McCune and Weinstein, 1971;

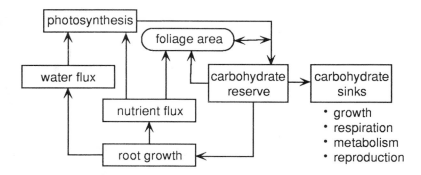

Figure 12-11. Physiological process model for forest trees. Photosynthetic decrease and reduction in carbohydrate reserve would reduce the amount and alter the allocation of carbon going to roots, shoots, foliage, and stemwood! Source: Ford (1982).

McCune, 1986), and heavy metals (Arndt, 1974; Höll and Hampp, 1975; Flückiger et al., 1978).

G. Carbon Allocation

In addition to fixation (photosynthesis), the utilization and allocation of carbon, along with water and mineral nutrients, within the plant is a critical regulator of plant health. The concepts of resource allocation within plants and its significance for yield and competitiveness are well developed (Waring and Patrick, 1975, Wilson, 1972, Mooney, 1972, McLaughlin and Shriner, 1980).

The concepts of whole-plant allocation are especially significant to the comprehension of the effects of pollutants on tree metabolism, as both sources and sinks may be impacted by pollutant stress (McLaughlin 1987) (Figure 12–11). Almost all sustained stresses act to reduce canopy integrity, to suppress photosynthesis, and to impact storage reserves. Shade, drought, mechanical abrasion, and nutrient stress cause distinctive alterations in how photosynthate is allocated along the stem and to the roots (Waring, 1987; Waring and Schlesinger, 1985).

A large number of studies with nonwoody species have examined the influence of gaseous pollutants, particularly sulfur dioxide and ozone, on biomass partitioning into leaves, shoots, and roots. While generalizations are difficult, substantial evidence suggests that exposure to these gases can reduce root biomass more than leaves or shoots, and can result in reduced root/shoot ratios (Miller 1987).

McLaughlin et al. (1982) studied the growth trends of nine 25-year-old white pine trees, in varying stages of apparent vigor, growing in a plantation in Oak Ridge, Tennessee. Tolerant, intermediate, and sensitive trees were examined for differences in patterns of photosynthate allocation which may have been due to ambient exposure to air pollutants, especially ozone. Higher retention of [14]C -

photosynthate by foliage and branches of sensitive trees indicated that photosynthate export to boles and roots was reduced. The ratio of respiratory to photosynthetic activity was significantly higher for foliage of sensitive trees.

Reich and Lassoie (1985) exposed cuttings of hybrid poplar in chambers to daily 5.5-hr exposures to ozone at 25 ppb (49 μg m^{-3}), 50 ppb (98 μg m^{-3}), 85 ppb (167 μg m^{-3}), and 125 ppb (245 μg m^{-3}) for 10 weeks. At the end of the study, dry weights of plants in the 85 and 125 ppb treatments were 10–15% lower than in the 25 and 50 ppb treatments, but ozone did not have any influence on dry matter partitioning.

Dr. Peter Sharpe and colleagues have exposed loblolly pine seedlings to 120 ppb (235 μg m^{-3}) ozone (7 hr d^{-1}, 5 days week^{-1}) for 12 weeks in controlled environment chambers. Treated and control seedlings were exposed to $^{11}CO_2$. Preliminary results suggest that transport of photosynthate to roots and shoots was substantially reduced by chronic exposure to ozone prior to ^{11}C treatment. Ozone exposure during ^{11}C experiments appeared to have little impact on photosynthate transport (NCASI, 1988).

H. Respiration

Allocation of carbon resources to respiration is central to whole plant carbon dynamics. Plant growth, in effect, is the balance of photosynthetic gains and respiratory losses. Models of plant growth and productivity must describe respiration as adequately as photosynthesis (Amthor, 1984). Similarly, air pollution studies must evaluate the impact of pollutant exposure on respiration as well as photosynthesis — they have not done so!

A useful, contemporary model of dark respiratory losses partitions the losses into two functional components, a growth component and a maintenance component. The growth component is that associated with the synthesis of new biomass, while the maintenance component is that associated with sustaining existing biomass (Amthor, 1984, 1986a).

White pine seedlings grown in a growth chamber and exposed to 50 ppb (98 μg m^{-3}) ozone for five weeks exhibited increased respiration (Barnes, 1972). Treatment of slash, pond, loblolly, and white pine with 150 ppb (294 μg m^{-3}) for several weeks also resulted in increased respiration by second year needles. McLaughlin et al. (1982) recorded higher rates of respiration by branches of "sensitive" relative to "intermediate or tolerant" white pine growing in the field. This increase may have been due to ambient ozone. Reich (1983) has reported an increase in respiration by leaves of hybrid poplar fumigated with low levels, 85 ppb (167 μg m^{-3}) and 125 ppb (245 μg m^{-3}), of ozone. This response was most dramatic in younger leaves.

Yang et al. (1983) exposed 2-year-old "sensitive" white pine seedlings to 100 ppb (196 μg m^{-3}), 200 ppb (392 μg m^{-3}), and 300 ppb (588 μg m^{-3}) ozone for 50 days. These high exposures resulted in a decrease in needle respiration. Intermediate and tolerant clones did not exhibit alterations in respiration rate following ozone exposure.

Working with small chambers placed on 20-year-old Scots pine and exposed to ozone ranging from 60 ppb (120 µg m⁻³) to 200 ppb (400 µg m⁻³) for 1 month, Skärby et al. (1987) observed that dark respiration increased throughout the treatment period, and the accumulated respiration was approximately 60% higher for the ozone-exposed shoots at the end of the experiment. Kupers and Klumpp (1987) exposed Norway spruce foliage to ozone, sulfur dioxide and ozone plus sulfur dioxide under chamber conditions. They observed an elevation in dark respiration in all treatments in current year needles. Ozone alone induced the greatest increase.

Amthor (1986b) exposed pinto bean unifoliate leaves to 90 ppb (176 µg m⁻³) for 6 hr day⁻¹ in a growth chamber. He observed a 10–15% increase in maintenance respiration and no change in growth respiration. In open-top chamber experiments in the field, employing pinto bean, Amthor also recorded an increase in maintenance respiration and no change in growth respiration. The increase in maintenance respiration was 10–30% over three ozone exposures (zero ozone, ambient, and 2x ambient). Amthor concluded that an important mechanism for low level ozone inhibition of plant growth and productivity may be the increase in maintenance respiration.

I. Summary

Photosynthesis is the most fundamental metabolic process of forest ecosystems and is the primary determinant of growth and biomass accumulation. The rate of net photosynthesis of mature trees frequently is within the range of 10–200 mg of carbon dioxide taken up per gram of dry weight per day. The rate is extremely variable, however, and is influenced by genetic, clonal, and provenance differences, season of the year, time of day, position within the crown of the tree, age of foliage, climate, and edaphic factors.

Studies with a wide variety of agricultural and herbaceous species, under controlled environmental conditions, have indicated that air contaminants must be added to the list of environmental variables that can potentially alter the rate of photosynthesis.

Because of ease of handling and experimental design, investigators studying the relationship betwen air pollutants and tree photosynthesis have primarily employed tree seedlings for research material and controlled environmental facilities for growth. Evidence has been provided, under the above circumstances, for photosynthetic suppression caused by sulfur dioxide, ozone, fluoride, heavy metals, and coal dust.

The thresholds of photosynthetic toxicity for tree seedlings vary with individual species, individual pollutants, length of exposure, and other experimental conditions. For several seedlings, the threshold of sulfur dioxide photosynthetic influence may approximate 1000 ppb (2620 µg m⁻³) for exposure of several hours or 100 ppb (262 µg m⁻³) for continuous exposure for several weeks. For ozone, the threshold of photosynthetic response may approximate 100 ppb (196 μg m^{-3}) if the exposure occurs for several hours per day and persists for several weeks (Table 12–6).

Considerable risk is associated with extrapolation of seedling photosynthetic data accumulated in controlled environment facilities to older trees in natural forests. Excised-leaf and small-chamber techniques, therefore, have been employed to assess the air pollutant influence on photosynthetic rates of trees 5-years-old and older. The use of sapling-age experimental material avoids the unique characteristics of seedling metabolism. Evidence for forest tree sapling photosynthetic suppression has been presented for sulfur dioxide, ozone, and cadmium. For sulfur dioxide and ozone exposure, the sapling evidence suggests that the threshold of photosynthetic reduction may approximate 500 ppb sulfur dioxide for many hours for 2 or 3 days, and 500 ppb ozone for similar length exposures. Chronic, many week, exposures to ozone as low as 25 ppb (49 μg m^{-3}) for several hours per day, however, may represent an extended term threshold (Table 12–7).

With regard to sulfur dioxide and agricultural species, Roberts (1984) has generalized as follows: 19–38 ppb (50-100 μg m^{-3}) for several months has produced beneficial as well as detrimental effects on yield, 38–76 ppb (100–200 μg m^{-3}) has produced yield losses in some studies, but not all, and 76–150 ppb (200–393 μg m^{-3}) generally produces significant yield losses in the selected crops studied. Based on the limited woody plant studies available, a significant impact of sulfur dioxide on forest trees may fall only in the 100–150 ppb chronic exposure range. With regard to ozone and agricultural productivity, the National Crop Loss Assessment Program has indicated yield losses in numerous crops where the growing season 7-hr mean for ozone is 40 ppb (78 μg m^{-3}) (Miller, 1987). Woody plant evidence available for selected species suggests forest trees may also experience important growth reductions at the ambient exposures that adversely impact agricultural species. Reich (1987b) has provided a comprehensive and integrated review of ozone impact on plant photosynthesis and productivity, and has reached numerous important conclusions. At low ozone concentration (approximately 50 ppb, 98 μg m^{-3}), evidence from agricultural crops, hardwood trees, and conifers all suggest linear reductions in net photosynthesis and growth with respect to ozone uptake. When uptake is the same, agricultural crops are more sensitive than hardwoods, which in turn are more sensitive than conifers. These differences in sensitivity are presumed to be due to several factors, with some of the most important being that conifers generally have lower diffusive conductances than crop species and require longer exposure to ozone to have comparable uptake, conifer foliage is less productive per unit time than crop foliage, and conifer needles have a higher capacity to resist stress (low nutrient supply, herbivory, infection abiotic stress) than crop foliage. Hardwoods are judged to be intermediate with regard to the above characteristics. Ambient air over large portions of eastern North America has an average of 50–70 ppb (98–1372 μg m^{-3}) ozone on clear summer days (natural background 20–30 ppb, 39–59 μg m^{-3}), with frequent peak concentrations of 80–110 ppb, 159–216 μg m^{-3} (Chapter 2). Evidence available suggests that following 1–2 weeks of typical

Table 12-6. Threshold Dose for Photosynthetic Suppression of Selected Forest Tree Seedlings by Air Contaminants.

Pollutant	Concentration	Time	Experiment Duration	Species	Reference
SO_2	6 ppm $(15.7 \times 10^3 \mu g\ m^{-3})$	4–6 hr	Single treatment	Red maple	Roberts et al. (1971)
	1 ppm $(2620 \mu g\ m^{-3})$	2–4 hr	Single treatment	Quaking aspen White ash	Jensen and Kozlowski (1974)
	0.1 ppm $(262 \mu g\ m^{-3})$	Continuous	2 weeks	White fir	Keller (1977b)
	0.2 ppm $(524 \mu g\ m^{-3})$	Continuous	2 weeks	Norway spruce Scotch pine	Keller (1977b)
O_3	0.30 ppm $(588 \mu g\ m^{-3})$	9 hr day^{-1}	10 days	Ponderosa pine	Miller et al. (1969)
	0.15 ppm $(294 \mu g\ m^{-3})$	Continuous	19 days	White pine	Barnes (1972)
	0.15 ppm $(294 \mu g\ m^{-3})$	Continuous	84 days	Slash pine Pond pine Loblolly pine	Barnes (1972)
	0.085 ppm $(167 \mu g\ m^{-3})$	5.5 hrs day^{-1}	62 days	Hybrid Poplar	Reich (1983)
	0.120 ppm $(235 \mu g\ m^{-1})$	7 hr day^{-1} 5 day wk^{-1}	60 days	Sugar maple	Reich et al. (1986)
F	30 $\mu g\ g^{-1}$ d.w. basis foliar tissue			Pine (various)	Keller(1977c)
Pb	< 10 $\mu g\ g^{-1}$ d.w. basis foliar tissue			American sycamore	Carlson and Bazzaz (1977)
Cd	< 10 $\mu g\ g^{-1}$ d. w. basis foliar tissue			American sycamore	Carlson and Bazzaz (1977)

Table 12-7. Threshold Dose for Photosynthetic Suppression of Selected Forest Tree Saplings by Air Contaminants.

Pollutant	Concentration	Time	Experiment Duration	Species	Reference
SO_2	1 ppm $(2620 \mu g\ m^{-3})$	30 min	Single treatment	Silver maple (excised leaves)	Lamoreaux and Chaney (1978)
	0.5 ppm $(1310 \mu g\ m^{-3})$	7–11 hr	1–2 days	Black oak	
	0.5 ppm $(1310 \mu g\ m^{-3})$	7–11 hr	1–2 days	Sugar maple	Carlson (1979)
	0.5 ppm $(1310 \mu g\ m^{-3})$	7–11 hr	1–2 days	White ash	
O_3	0.5 ppm $(980 \mu g\ m^{-3})$	4 hr	Single treatment	White pine	Botkin et al. (1972)
	0.5 ppm $(980 \mu g\ m^{-3})$	7–11 hr	1–2 days	Black oak	Carlson (1979)
	0.5 ppm $(980 \mu g\ m^{-3})$	7–11 hr	1–2days	Sugar maple	Carlson (1979)
	0.023 $(45 \mu g\ m^{-3})$	7 hrs day^{-1} 7 day wk^{-1}	9 weeks	Red oak	Reich and Amundson (1985)
Cd	$\approx$100 $\mu g\ g^{-1}$ (?)	45 hr	Single treatment	Silver maple (excised leaves)	Lamoreaux and Chaney (1978)

growing season pollution, mean daytime concentration of 50–70 ppb (exposure above a background of 3–7 ppm-hr, agricultural crops will exhibit significant declines in net photosynthesis and growth. Hardwood forest trees will begin to exhibit similar decreases following several additional weeks of elevated concentrations, that is, with an exposure above background of approximately 10–20 ppm-hr). The threshold for conifer impact may be several months of elevated ozone concentrations resulting in exposures approximating 25–100 ppm-hr.

A variety of cautions must be considered when considering air contaminants and photosynthesis. Most of the evidence available has been generated by seedling, sapling, or small chamber studies. Direct evidence from closed-canopy forests is very meager. Much of the seedling and sapling evidence suggests that the photosynthetic inhibition caused by sulfur dioxide and ozone is reversible if the pollutant stress is removed. Under the circumstance of variable pollutant concentration in ambient atmospheres, photosynthetic recovery might be common. Synergism, or greater stress resulting from simultaneous pollutant exposure relative to either pollutant alone, appears frequently in the seedling and sapling literature. Evidence for synergistic photosynthetic suppression by sulfur dioxide and ozone, fluoride and cadmium has been presented. Studies with short-term exposure periods, have not provided evidence for an interaction with ozone and acid deposition on net photosynthesis or growth (Chappelka et al., 1985; Chappelka and Chevone, 1986l Reich et al., 1986, 1987a). Almost all the studies report photosynthetic depression in the absence of, or at least prior to, the appearance of visible foliar symptoms.

Despite the cautions, we concur with Reich and colleagues that current levels of ambient ozone are causing declines in net photosynthesis and growth in natural forests over significant portions of North America.

In addition to the influence on photosynthesis, exposure to ozone may influence carbon allocation within the plant. While all evidence is not consistent, several studies have suggested reduced movement of carbon to roots following exposure. In trees, between 15% and 50% of photosynthate is allocated to produce, maintain, and replace mycorrhizal fine-root systems annually (Marx, 1988). A reduction in root growth could increase the risk of drought stress. Root growth reduction could also exacerbate nutrient deficiencies. In addition, a variety of studies have indicated an increased allocation to respiration following ozone exposure. This increase may be especially significant in the maintenance component of respiratory function. Finally several studies have indicated that a variety of pollutants, including ozone and sulfur dioxide, suppress reproduction.

References

Amthor, J.S. 1984. The role of maintenance respiration in plant growth. Plant Cell Environ. 7: 561-569.

Amthor, J.S. 1986a. Evolution and applicability of a whole plant respiration mode. J. Theor. Biol. 122: 473-490.

Amthor, J.S. 1986b. Ozone-induced increase in bean leaf maintenance respiration. Unpublished Ph.D. Thesis. School of Forestry and Environ. Studies, Yale Univ., New Haven, CT, 166 pp.

Andersin, M.C. 1973. Solar radiation and carbon dioxide in plant communities — Conclusions. In J.P. Cooper, ed., Photosynthesis and Productivity in Different Environments. Cambridge Univ. Press, New York, pp. 234-354.

Arndt, V. 1974. The Kaulsky-effect: A method for the investigation of the actions of air pollutants in chloroplasts. Environ. Pollu. 6: 181-194.

Auclair, D. 1976. Effects of dust on photosynthesis. I. Effects of cement and coal dust on photosynthesis of spruce. Annu. Sci. For. 33: 274-255.

Auclair, D. 1977. Effects of dust on photosynthesis. II. Effects on particulate matter on photosynthesis of Scots pine and poplar. Annu. Sci. For. 34: 47-57.

Barnes, R.L. 1972. Effects of chronic exposure to ozone on photosynthesis and respiration of pines. Environ. Pollut. 3: 133-138.

Bauer, H., W. Larcher, and R. B. Walker. 1973. Influence of temperature stress on CO_2-gas exchange. In J.D. Cooper, ed., Photosynthesis and Productivity in Different Environments. Cambridge Univ. Press, New York, pp. 557-586.

Blackman, G.E. and J.N. Black. 1959. Physiological and ecological studies in the analysis of plant environment. XII. The role of the light factor in limiting growth. Annu. Bot. N.S. 23: 131-145.

Bosian, G. 1965. The controlled climate in the plant chamber and its influence upon assimilation and transpiration. In Methodology of Plant Ecophysiology. Proc. Montpellier Symp., UNESCO, New York, pp. 225-232.

Botkin, D.B. 1968. Observed and predicated rates of carbon dioxide uptake for oak leaves in a coastal plain forest. Ph.D. Thesis. Rutgers Univ., New Brunswick, New Jersey, 171 pp.

Botkin, D.B., W.H. Smith, and R. W. Carlson. 1971. Ozone suppression of white pine net photosynthesis. J. Air Pollu. Control Assoc. 21: 778-780.

Botkin, D.B., W.H. Smith, R.W. Carlson, and T.L. Smith. 1972. Effects of ozone on white pine saplings: Variation in inhibition and recovery of net photosynthesis. Environ. Pollu. 3: 273-289.

Bourdeau, P.F. 1963. Photosynthesis and respiration of *Pinus strobus* seedlings in relation to provenance and treatment. Ecology 44: 710-716.

Carlson, R.W. 1979. Reduction in the photosynthetic rate of *Acer, Quercus*, and *Fraxinus* species caused by sulphur dioxide and ozone. Environ. Pollu. 18: 159-170.

Carlson, R.W. and F.A. Bazzaz. 1977. Growth reduction in American sycamore (*Plantanus occidentalis* L.) caused by Pb-Cd interaction. Environ. Pollu. 12: 243-253.

Carlson, R.W. and F.A. Bazzaz, 1982. Photosynthetic and growth response to fumigation with SO_2 for CO_3 and C_4 plants. Oecologia 54: 50-54.

Chappelka, A.H. and B.I. Chevone, 1986. White ash seedling growth response to ozone and simulated acid rain. Can. J. For. Res. 16: 786-790.

Chappelka, A.H., B.I. Chevone, and T.E. Burk. 1985. Growth rsponse of yellow poplar (Liriodendron tulipifera L.) seedlings to ozone, sulfur dioxide and simulated acidic precipitation, alone and in combination. Environ. Exp. Bot. 25: 233-244.

Duchelle, S.F., J.M. Skelly, and B.I. Chevone. 1982. Oxidant effects on forest tree seedling growth in the Appalachian Mountains. Water Soil Air Pollu. 18: 363-373.

Evans, L.S. and I.P. Ting. 1974. Ozone sensitivity of leaves: Relationship to leaf water content, gas transfer resistance, and anatomical characteristics. Am. J. Bot. 61: 592-597.

Flückiger, W., H. Flückiger-Keller, and J.J. Oerti. 1978. Biochemische veränderungen in jungen birken im nachbereich einer autobahn. Eur. J. For. Pathol. 8: 154-163.

Ford, E.D. 1982. Physiological process model for mature forest trees. Scottish For. 36: 9-24.

Fritts, H.C. 1966. Growth-rings in trees: Their correlation with climate. Science 154: 973-979.

Govindjee. 1975. Bioenergetics of Photosynthethesis. Academic Press, New York, 687 pp.

Hällgren, J.E. 1978. Physiological and biochemical effects of sulfur dioxide on plants. In J. E. Nriagu, ed., Sulfur in the Environment. Part II. Ecological Impacts. Wiley, New York, pp. 163-209.

Heagle, A.S., D.E. Body, and W.W. Heck. 1973. An open-top field chamber to assess the impact of air pollution on plants. J. Environ. Qual. 2: 365-368.

Heagle, A.S., R.B. Philbeck, and M.B. Letchworth. 1979. Injury and yield responses of spinach cultivars to chronic doses of ozone in open-top field chambers. J. Environ. Qual. 8: 268-373.

Heck, W.W., S.V Krupa, and S.N. Linzon. 1979. Handbook of Methodology for the Assessment of Air Pollution Effects on Vegetation. Air Pollu. Control Assoc., Pittsburgh, PA.

Heck, W.W., O.C. Taylor, R. Adams, G. Bingham, J. Miller, E. Preston, and L. Weinstein. 1982. Assessment of crop loss from ozone. J. Air Pollu. Control Assoc. 32: 353-363.

Heinicke, D.R. and J.W. Foott. 1966. The effect of several phosphate inscticides on photosynthesis of red delicious apple leaves. Can. J. Plant Sci. 46: 589-591.

Helms, J.A. 1965. Diurnal and seasonal patterns of net assimilation in Douglas fir, Pseudotsuga menziesii (Mirb.) Franco, as influenced by environment. Ecology 46: 698-708.

Hesketh, J.D. and D. N. Moss. 1963. Variation in the response of photosynthesis to light. Crop Sci. 3: 107-110

Hodges, J.D. 1962. Photosynthetic efficiency and patterns of photosynthesis of seven different conifers under different natural environmental conditions. M. F. Thesis, Univ. Washington, Seattle, WA, 99 pp.

Höll, W. and R. Hampp. 1975. Lead and plants. Residue Rev. 54: 79-111.

Huber, B. and H. Polster. 1955. Zur frage der physiologischen ursachen der unterschiedlichen stofferzeugung von pappelklonen. Bio. Zentralbl. 74: 370-420.

Keller, T. 1968. The influence of mineral nutrition on gaseous exchange by forest trees. In Phosphorus in Agriculture. Intl. Superphosphate Manufact. Assoc., Ltd., Agricul. Comm. Bull. No. 50, June 1968, Paris, pp 1-11.

Keller, T. 1970. Gaseous exchange — A good indicator of nutritional status and fertilizer response of forest trees. Proc 6th Intl. Colloq. Plant Analysis and Fertilizer Problems. ISHS, Tel Aviv, pp. 669-678.

Keller, T. 1973. On the phytotoxicity of dust-like fluoride compounds. Staub Reinhal. Luft 33: 379-381.

Keller, T. 1977a. Definition and importance of latent injury by air pollution. Allg. Forst-u. Jagdztg. 148: 115-120.

Keller, T. 1977b. The effect of long term low SO_2 concentration upon photosynthesis of conifers. 4th Intl. Clean Air Congress, pp. 81-83.

Keller, T. 1977c. The influence of air pollution by fluorides on photosynthesis of forest tree species. In W. Bosshard, ed., Mitt. Schweiz. Anst. Forstl. Ver'wes 53: 163-198.

Keller, T. 1978a. Influence of low SO_2 concentrations upon CO_2 uptake of fir and spruce. Photosynthetica 12: 316-322.

Keller, T. 1978b. The influence of SO_2 treatment at different seasons on the spruce trees intake of CO_2 and its yearly growth pattern. Schweier. Zeitsch. Fortwesen 129: 381-393.

Keller, T. 1979. The influence of SO_2 gasing on the growth of spruce tree roots. Schweizer. Zeitsch. Fortwesen 130: 429-435.

Keller, T. and H. Schwager. 1977. Air pollution and ascorbic acid. Eur. J. For. Pathol. 7: 338-350.

Kozlowski, T.T. and T. Keller. 1966. Food relations of woody plants. Bot. Rev. 32: 293-383.

Kramer, P.J. and J.P. Decker. 1944. Relation between light intensity and rate of photosynthesis of loblolly pine and certain hardwoods. Plant Physiol. 19: 350-358.

Kramer, P.A. and T.T. Kozlowski. 1960. Physiology of Trees, McGraw-Hill, New York, 642 pp.

Kramer, P.J. and T.T. Kozlowski. 1979. Physiology of Woody Plants. Academic Press, New York, 811 pp.

Krueger, K.W. and W.K. Ferrell. 1965. Comparative photosynthetic and respiratory responses to temperature and light by Pseudotsuga menziesii var. menziesii and var. glauca seedlings. Ecology 46: 797-801.

Küppers, K. and G. Klumpp. 1987. Effects of ozone, sulfur dioxide, and nitrogen dioxide on gas exchange and starch economy in Norway spruce (Picea abies L. Karsten). Proc. XIV Intl. Bot. Cong. Berlin, FRG, 24 July-1 Aug., 1987.

Lamoreaux, R.J. and W.R. Chaney. 1978. Photosynthesis and transpiration of excised silver maple leaves exposed to cadmium and sulphur dioxide. Environ. Pollu. 17: 259-268.

Lechowicz, M.J. 1987. Resource allocation by plants under air pollution stress: Implications for plant-pest-pathogen interactions. Bot. Rev. 53: 281-300.

Legge, A.H. and D.R. Jaques. 1977. Field studies of pine, spruce and aspen periodically subjected to sulfur gas emissions. Water Soil Air Pollu. 8: 105-129.

Linder, S. and E. Troeng. 1977. The seasonal course of net photosynthesis and stem respiration in a 20-year-old stand of Scots pine (*Pinus silvestris* L.) V.K. Sci. Comm., 4th Intl. Congr. Photosynthesis, London, p. 221.

Lorio, P.L. and R.A. Sommers. 1986. Evidence of competition for photosynthates between growth synthesis in *Pinus taeda* L. Tree Physiol. 2: 301-306.

Malhatra, S.S. and A.A. Kahn. 1984. Biochemical and physiological impact of major pollutants. In M. Treshow, ed., Air Pollution and Plant Life. John Wiley and Sons, New York, pp. 113-157.

Mandl, R.H., L.H. Weinstein, D.C. McCune, and M. Keverny. 1973. A cylindrical open-top chamber for exposure of plants to air pollutants in the field. J. Environ. Qual. 2: 371-376.

Marx, D.H. 1988. Personal Communication. U.S.D.A Forest Service. Southeastern Forest Experiment Station. Athens, GA.

Miller, J.E. 1987. Effects of ozone and sulfur dioxide stress on growth and carbon allocation in plants. Paper No. 10807, J.S.N.C.A.R. Service, Raleigh, NC.

Miller, P.R. 1966. The relationship of ozone to suppression of photosynthesis and to the cause of chlorotic decline of ponderosa pine. Diss. Abstr. 26: 3574-3575.

Miller, P.R., J.R. Parmeter Jr., B.H. Flick, and C.W. Martinez. 1969. Ozone dosage response of ponderosa pine seedlings. J. Air Pollu. Control Assoc. 19: 435-438

Mooney, H.A. 1972. The carbon balance of plants. Annu. Rev. Ecol. Syst. 3: 315-346.

McCune, D.C. 1986. Hydrogen fluoride and sulfur dioxide. In A.H. Legge and S.V. Krupa, eds., Air Pollutants and Their Effects on the Terrestrial Ecosystem. John Wiley and Sons, New York, pp. 305-324.

McLaughlin, S.B. 1987. Carbon allocation as an indicator of pollutant impacts on forest trees. M. Cammell and D.L. Lavender, eds., Proc. I UFRO Symposium. Woody Plant Growth in a Changing Chemical and Physical Environment. July 1987, Vancouver, Canada.

McLaughlin S.B. and D.S. Shriner. 1980. Allocation of resources to defense and repair. Chapter 22 in Plant Disease, Vol.V. Academic Press, New York, pp. 407-431.

McLaughlin, S.B., R.K. McConathy, D. Duvick, and L.K. Mann. 1982. Effects of chronic air pollution stress on photosynthesis, carbon allocation, and growth of white pine trees. For. Sci. 28:60-70.

McLean, F. T. 1920. Field studies of the carbon dioxide absorption of coconut leaves. Annu. Bot 34: 367-389.

McLeod, A.R., J.E. Fackrell, and K. Alexander. 1985. Open air fumigation of field crops: Criteria and design for a new experimental system. Atmos. Environ. 19: 1639-1649.

National Council of the Paper Industry for Air and Stream Improvement. 1988. NCASI-sponsored carbon traces study nears completion at Texas A & M. Air Quality/Forest Health Program News 3: 1.

Nátr, L. 1973. Influence of mineral nutrition on photosynthesis and the use of asimilates. In J.P. Cooper, ed., Photosynthesis and Productivity in Different Environments. Cambridge Univ. Press, New York, pp. 537-555.

Nzirov, N.N. 1966. Deistvie ioniziruy-uschei radiatsu na fatozintez u razlichnykk po radioustoichivasti sortov khlopchatnika. Uzbeksku Biol. Zh. 10: 3-8.

Nellen, V.R. 1966.Über den Einfluss des Salzgehaltes auf die photosynthetische leistung verschiedener Standardformen von Delesseria sanguinea und Fucus serratus. Hergoländer Wiss. Meeresunters 13: 288-313.

Pharis, R.P. and F.W. Woods. 1960. Effects of temperature upon photosynthesis and respiration of Choctawatchee sand pine. Ecology 41: 797-799.

Pisek, A. and W. Tranquillini. 1954. Assimilation und kohlenstoffhaushalt in der krone von fichten (Picea excelsa Link) und rotbuchenbäumen (Fagus silvatica L.) Flora (Jena) 141:237-270.

Pisek, A. and E. Winkler. 1958. Assimilationsvermögen und respiration der fichte (picea excelsa Link) in verschiedener höhenlage und der zirbe (pinus cembra L.) der alpinen waldgrenze. Planta 51:518-543.

Polster, H. 1950. Die physiologischen grundlagen der stofferzeugung im walde. Untersuchengen über assimilation, respiration und transpiration unserer hauptholzarten. Bayrischer land wirtschaftsverlag, G. m. b. H., München, 96 pp.

Polster, H. and G. Weise. 1962. Vergleichende assimilation—suntersuchungen an klonen verschiedener lärchenherkünfte (Larix decidua and Larix leptolepis) unter freiland und klimaraumbedingungen. Zuchter 32: 103-110.

Reich, P.B. 1983. Effects of low concentrations of O_3 on net photosynthesis, dark respiration, and chlorophyll contents in aging hybrid poplar leaves. Plant Physiol. 73: 291-296.

Reich, P.B. 1987b. Quantifying plant response to oxone: A unifying theory. Tree Physiol. 3: 63-91.

Reich, P.B. and R.G. Amundson. 1984. Low level ozone and/or SO_2 exposure causes a linear decline in soybean yield. Environ. Pollu. Ser. A. 34: 345-355.

Reich, P.B. and R.G. Amundson. 1985. Ambient levels of ozone reduce net photosynthesis in tree crop species. Science 230: 566-570.

Reich, P.B. and J.P. Lassoie. 1984. Effects of low level O_3 exposure on leaf diffusive conductance and water-use efficiency in hybrid poplar. Plant Cell. Environ. 7: 661-668.

Reich, P.B. and J.P. Lassoie. 1985. Influence of low concentrations of ozone on growth, biomass partitioning and leaf senescence in young hybrid poplar plants. Environ. Pollu. 39:39-51.

Reich, P.B., J.P. Lassoie, and R.G. Amundson. 1984. Reduction in growth of hybrid poplar following field exposure to low levels of O_3 and (or) SO2. 62: 2835-2841.

Reich, P.B., A.W. Schoettle, and R.G. Amundson. 1986. Effects of ozone and acidic rain on photosynthesis and growth in sugar maple and northern red oak seedlings. Environ. Pollu. 40: 1-15.

Reich, P.B., A.W. Schoettle, H.F. Stroo, J. Troiano and R.G. Amundson. 1987. Effects of ozone and acid rain on white pine (*Pinus strobus*) seedlings grown in five soils. I. Net photosynthesis and growth. Can. J. Bot. 65: 977-987.

Roberts, B.R. 1988. Effect of atmospheric deposition and drought on physiological activity and root growth potential in yellow poplar. Abstract. 1988. Am. Soc. Pl. Physiol. Annual Meeting, Reno, Nevada, July 10-14, 1988.

Roberts, T.M. 1984. Long-term effects of sulfur dioxide on crops: An analysis of dose-response relations. Phil. Trans. Royal Soc. London B 305: 299-316.

Sasaki, S. and T.T. Kozlowski. 1966a. Variable photosynthetic responses of *Pinus resinosa* seedlings to herbicides. Nature 209: 1042-1043.

Sasaki, S. and T.T. Kozlowski. 1966b. Effects of herbicides on carbon dioxide uptake by pine sedlings. Can. J. Bot. 45: 961-971.

Shaw, P.J.A. 1986. The Liphook forest fumigation experiment: Description and project plan. Central Electricity Research Laboratories. Technology Planning and Research Division. Publ. No. TPRD-L-2985-R86. Surrey, UK.

Skärby, L., E. Troeng, and C-Å. Boström. 1987. Ozone uptake and effects on transpiration, net photosynthesis, and dark respiration in Scots pine. For. Sci. 33: 801-808.

Slavik, B. 1973. Water stress, photosynthesis and the use of photosynthates. In J. P. Cooper, ed., Photosynthesis and Productivity in Different Environments. Cambridge Univ. Press, New York, pp. 511-536.

Stocker, O. 1960. Die photosynthetischen leistungen der steppen und wüstenpflanzen. In W. Ruhland, ed., Handbuch der Pflanzenphysiologie, Vol. 5, pp. 460-491.

Strain, B.R. and V.C. Chase. 1966. Effect of past and prevailing temperatures on the carbon dioxide exchange capacities of some woody desert perenials. Ecology 47: 1043-1045.

Taylor, O.C., W.M. Dugger Jr., M.D. Thomas, and C.R. Thompson. 1961. Effect of atmospheric oxidants on apparent photosynthesis in citrus trees. Plant Physiol. (Suppl.) 36: 26.

Taylor, G.E. Jr., R.J. Norby, S.B. McLaughlin, A.H. Johnson and R.S. Turner. 1986. Carbon dioxide assimilation and growth of red spruce (*Picea rubens* Larg.) seedlings in response to ozone, precipitation chemistry, and soil type. Oecologia 70: 163-171.

Thomas, M.D. and G.R. Hill. 1937. Relation of sulphur dioxide in the atmospere to photosynthesis and respiration of alfalfa. Plant Physiol. 12: 309-383.

Thomas, M.D. and G.R. Hill. 1949. Photosynthesis under field conditions. In J. Frank and W.F. Loomis, ed., Photosynthesis in Plants. Iowa State Univ. Press, Ames, IA, pp. 19-52.

Todd, G.W. 1958. Effect of ozone and ozonated 1-hexene on respiration and photosynthesis of leaves. Plant Physiol. 33: 416-420.

Todd, G.W. and B. Propst. 1963. Changes in transpiration and photosynthetic rates of various leaves during treatment with ozonated hexene. Plant Physiol. 16: 57-65.

Verkroost, M. 1974. The effect of ozone on photosynthesis and respiration of *Scendesmus obtusiusculus* Chod., with a general discussion of effects of air pollutants in plants. Mededelingen Landbouwhogeshool Wageningen 19: 1-78.

Wang, D., F.H. Bormann, and D.F. Karnosky. 1986a. Regional tree growth reductions due to ambient ozone: Evidence from field experiments. Environ. Sci. Technol. 20: 1122-1125.

Wang, D., D.F. Karnosky, and F.H. Bormann. 1986b. Effects of ambient ozone on the productivity of *Populus tremuloides* Michx. grown under field conditions. Can. J. For. Res. 16: 47-55.

Wareing, P.F. and J. Patrick. 1975. Source-sink relations and the partition of assimilates in the plant. In H.P. Cooper, ed., Photosynthesis and Productivity in Different Environments. Cambridge Univ. Press, New York pp 481-499.

Waring, R.H. 1987. Characteristics of trees predisposed to die. BioScience 37: 569-574.

Waring, R.H. and W.H. Schlesinger. 1985. Forest Ecosystems: Concepts and Management. Academic Press, Orlando, FL.

Waugh, J.G. 1939. Some investigatins on the assimilation of apple leaves. Plant Physiol. 14: 436-477.

White, K.L., A.C. Hill, and J.H. Bennett. 1974. Synergistic inhibition of apparent photosynthetic rate of alfalfa by combinations of sulfur dioxide and nitrogen dioxide. Environ. Sci. Technol. 8: 575-576.

Wilson, W.J. 1972. Control of crop processes. In A.R. Rees, K.E. Cockshull, D.W. Hand, and R.G. Hurd, eds., Crop Processes in Controlled Environments. Academic Press, New York, pp. 7-30.

Winner, W.E., H.A. Mooney, K. Williams and S. von Caemmerer. 1985. Measuring and assessing SO_2 effects on photosynthesis and plant growth. In W.E. Winner, H.A. Mooney and R.A. Goldstein, eds., Sulfur Dioxide and Vegetation. Stanford Univ. Press, Stanford, CA, pp. 118-132.

Woodwell, G.M. and D.B. Botkin. 1970. Metabolism of terrestrial ecosystems by gas exchange techniques: The Brookhaven appraoach. In D.E. Reichle, ed., Analysis of Temperate Forest Ecosystems. Springer–Verlag, New York, pp. 73-85.

Woodwell, G.M. and R.H. Whittaker. 1968. Primary production in terrestrial ecosystems. Am. Zool.. 8: 19-30.

Yang, Y.-S., J.M. Skelley, B.I. Chevone, and J.B. Birch. 1983. Effects of long-term ozone exposure on photosynthesis and dark respiration of eastern white pine. Environ Sci. Technol. 17:371-373.

Zelitchl, I. 1975a. Environmental and biological control of photosynthesis: General assessment. In R. Marcelle, ed., Environmental and Biological Control of Photosynthesis. Dr. W. Junk, The Hague, pp. 251-262.

Ziegler, I. 1975. The effect of SO_2 pollution on plant metabolism. Residue Rev. 56: 79-105.

13

Forest Biotic Agent Stress: Air Pollutants and Phytophagous Forest Insects

Arthropods have roles of enormous importance in the structure and function of terrestrial ecosystems. Forest ecosystems, in particular, typically have large and diverse arthropod populations. The importance of pollinating (Chapter 8) and litter metabolizing (Chapter 9) species has already been introduced. The damaging influence of high population densities of certain insects can be very visible and can cause widespread forest destruction; witness contemporary or recent North American situations involving the Douglas fir tussock moth, the gypsy moth, the eastern and western spruce budworms, and southern and western bark beetles. It is critically important, however, to keep in perspective that there is substantial evidence to support the notion that forest insects, even those that cause massive destruction in the short run, may play essential and beneficial roles in forest ecosystems in the long run. These roles involve regulation of tree species competition, species composition and succession, primary production, and nutrient cycling (Huffaker, 1974; Mattson and Addy, 1975; Seastedt and Crossley, 1984; Smith, 1986).

There is increasing indication that a variety of particularly damaging forest insects detect and respond to stress induced alterations in host tree physiology (Jones and Coleman, 1989). The stresses are variable and may include microbial infection, climatic extremes, edaphic factors, and age. Massive insect infestations are characteristically initiated in middle-aged to mature forests typified by reduced productivity rates. Localized and scattered insect outbreaks are associated with forests of all ages, but are generally associated with the least vigorous trees with slow growth rates. Some investigators judge that insect population growth is inversely related to host plant vigor (Mattson and Addy, 1975).

It is essential, therefore, to appreciate the interactions between air pollutants and forest insects because of the critical importance of these animals to forest ecosystem structure and function, and because air contaminants may be an additional environmental stress factor capable of predisposing forest tree species to detrimental arthropod influence.

Air pollutants may directly affect insects by influencing growth rates, mutation rates, dispersal, fecundity, mate finding, host finding, and mortality. Indirect effects may occur through changes in host age structure, distribution, and acceptance. Research dealing with these possible interactions, however, is not extensive, despite the fact that insect-air pollution relations has a research history that extends over 50 years. European literature dealing with this topic is substantially larger than the North American literature.

A variety of studies has presented data indicating that species composition or population densities of insect groups are altered in areas of high air pollution stress, for example, roadside (Przybylski, 1979) or industrial (Lebrun, 1976; Novakova, 1969; Sierpinski, 1967) environments. Specific information is further available on the general influence of polluted atmospheres on population characteristics of forest insects (Charles and Villemant, 1977; Boullard, 1973; Hay, 1975; Schnaider and Sierpinski, 1967; Sierpinski, 1970, 1971, 1972 a,b, 1981; Sierpinski and Chlodny, 1977; Templin, 1962; Wiackowski and Dochinger, 1973). Johnson (1969) has reviewed much of the literature dealing with air pollutants and insect pests of conifers. Comprehensive literature reviews available concerning forest insects and air contaminants have been presented by Alstad and Edmunds (1982), Flückiger (1987), Flückiger et al. (1986, 1987), Führer (1985), Hughes and Laurence (1984), and Villemant (1979).

A. Sulfur Dioxide

While primarily concerned with honeybees, Hillman (1972) reviews several references concerned with the interaction of insects with sulfur dioxide. Additional references are contained in Ginevan and Lane (1978). As is generally true with insects from a variety of ecosystems, forest insect population densities appear to be both increased and decreased by exposure to sulfur dioxide depending on the species. In an inventory of eastern white pine stems for white pine weevil deformity, Linzon (1966) recorded fewer deformed stems near to, relative to far from, sulfur dioxide sources. These results suggest an adverse impact on the weevil. Several European researchers, on the other hand, have associated increased population densities of the European pine shoot moth, a serious pest of red, Scotch, and Austrian pine, with increased ambient levels of sulfur dioxide and smoke (Sierpinsky and Chlodny, 1977).

Hughes et al. (1981) reported that exposure of bean plants to low-level sulfur dioxide fumigation increased feeding preference and reproduction of the Mexican bean beetle. Increased levels of free amino acids and certain sugars in foliage in response to fumigation may have been partially responsible for insect stimulation (Chiment et al., 1986). In a related experiment, soybeans were exposed to 200 ppb (524 μg m⁻³) sulfur dioxide (just above NAAQS, but below threshold of acute foliar injury) and Mexican bean beetle responses were evaluated (Hughes et al., 1983). Larvae developed faster and grew larger when fed fumigated leaves. Adult females showed significant feeding preference for fumigated leaves and were more fecund when fed treated foliage.

O. L. Gilbert of the University of Sheffield, England has been primarily concerned with lichen distribution as influenced by sulfur dioxide in urban and industrial areas. In one of his surveys of the Newcastle upon Tyne area he recorded arthropod numbers occurring 1–2 m above the ground on the bark of European ash trees. Numbers of herbivorous insects were significantly reduced in regions with high sulfur dioxide concentrations, while carnivorous insects did not show a significant correlation with sulfur dioxide levels (Gilbert, 1971). Annual surveys of the black bean aphid abundance in southeast England have suggested higher infestations on bean and soybean downwind of London. Dohmen et al. (1984) have provided evidence that ambient sulfur and nitrogen dioxide concentrations alter amino acid profiles of foliar tissue and enhance the growth rates of black bean aphids (Table 13-1).

A few studies have suggested that sulfur dioxide may be involved in the predisposition of woody plants to insect infestation (Strubel and Johnson, 1964). Anderson (1970) has reported an association between abnormal Christmas tree growth and an eriophytid mite on white pine subject to sulfur dioxide exposure from a power plant in a West Virginia–Maryland site. Saunders (1972) has also recorded increased mite infestations in Christmas tree plantations damaged by power plant effluent.

B. Oxidants

There has been surprisingly little research concerned with the direct influence of oxidants on insects. Concentrations of 10 pphm (196 μg m⁻³) ozone for prolonged exposures have been found to cause mortality of adult house flies (Beard, 1965). Beard also observed reduced egg laying by females following high ozone exposure. Low levels of ozone appeared to have a favorable influence on adult flies. Levy et al. (1972) exposed three Diptera species to high ozone levels and found slightly reduced egg hatch in two species. Adult response, however, included a dramatic stimulation of oviposition with subsequent increase in adult populations. In subsequent trials with ozone exposure of cockroach species and the red imported fire ant, Levy et al. (1974) observed that several-day exposure to 30 pphm (588 μg m⁻³) did not produce any obvious deleterious effect to adult or immature stages. These species, in fact, appeared even more tolerant than house flies.

Jeffords and Endress (1984) used third-instar larvae of gypsy moth to evaluate their feeding preference for seedling white oak foliage, which had been exposed to three levels of ozone. Insects preferred to feed on foliar material that had been exposed to the highest ozone concentration (150 ppb - 295 μg m⁻³)7 hr day⁻¹ for 11 exposure days. When ozone concentration was reduced to 60 ppb (118 μg

Table 13-1. Mean Relative Growth Rate[a] over 3 Days of *Aphis fabae* Subjected to SO_2 or NO_2 Either Directly on Artificial Diet or Indirectly Via Prefumigated Plants.

		Pollutant	Control	Significance
SO_2	Plant[b]	0.540	0.507	$P < 0.01$
	Artificial diet[c]	0.395	0.394	NS
NO_2	Plant[d]	0.575	0.530	$P < 0.01$
	Artificial diet[e]	0.369	0.370	NS

[a]Mean relative growth rate $\mu g \ g^{-1} \ day^{-1}$ = In (final weight) - In (initial weight)/ no. days growth. NS, not significant.
[b]Mean of five experiments with 15 replicates per treatment.
[c] Mean of two experiments with 15 replicate sachets per treatment, each with a mean of 10 aphids.
[d]Mean of four experiments with 15 replicates per treatment.
[e]One experiment with 15 replicate sachets per treatment, each with a mean of 10 aphids.
Source: Dohmen et al. (1984).

m^{-3}), treated leaves were less preferred than control leaves. The authors speculated that changes in the physical or chemical properties of the leaf, caused by ozone exposure, were probably regulating palatability of the foliage to the gypsy moth.

1. Oxidants and Aphids

As in the case of sulfur dioxide, the influence of oxidants on aphid dynamics has received some research attention. In France, Villemant (1981) has emphasized that Scots pine forest regions with elevated pollution loads are characterized by increases in some aphid populations and decreases in others. Braun and Flüchiger (1988) have studied the effects of ambient air (7 km west of Basel, Switzerland) with supplemented ozone concentration and acid mist simulant on the development of two aphid populations. Treatments inhibited the growth of *Aphis fabae* on bean and stimulated the growth of *Phyllaphis fagi* on European beech seedlings.

In roadway environments, aphid populations have been recorded as larger, when located closer to the road than at several meters distance, on hawthorn (Flückiger and Oertli, 1978) and viburnum (Bolsinger and Flückiger, 1987) (Figure 13-1). Increased nitrogen content of foliage, perhaps related to elevated concentrations of nitrogen oxides in the roadway environments, may improve the nutritional quality of the foliage for the aphids (Bolsinger and Flückiger, 1987, 1988).

Coleman and Jones (1988a) have investigated the effect of acute ozone exposure of eastern cottonwood on the survivorship, reproduction, and development of the cottonwood aphid *(Chaitophorus populicola)*. Cottonwood cuttings were exposed to 200 ppb (397 $\mu g \ m^{-3}$) ozone for 5 hr or filtered air. Aphid performance was not significantly altered in plants exposed to ozone relative to fil-

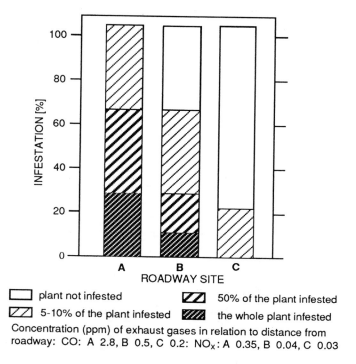

Figure 13-1. Infestation of hawthorn by aphids in relation to distance from roadway in mid-July. A dividing strip, $n = 25$; B area at a distance of 8–10 m, $n = 79$; C, area at a distance of 50–70 m $n = 83$.
Source: Fluckiger and Oertli (1978).

larvae of the imported willow leaf beetle (*Plagiodera versicolora*) preferred to feed on, and consumed more, cottonwood material that had been previously exposed to ozone (Figure 13-2). Females, however, preferred to oviposit on unexposed controls. The authors concluded that ozone exposure probably increased foliage consumption as a compensation for decreased foliage quality and that fecundity reduction was due to decreased oviposition rates (Coleman and Jones, 1988b).

2. Ozone Predisposition to Bark Beetle Infestation

Bark beetles are the single most damaging and economically significant insect pest of commercially important conifers in the United States (Figure 13-3). These insects are among the few native insects capable of killing large numbers of trees in 1 year. Most bark beetles breed only in trees that are severely diseased or dead. The insects deposit their eggs in galleries excavated in the phloem, cambium, and outer sapwood, and successful brood production requires mortality of

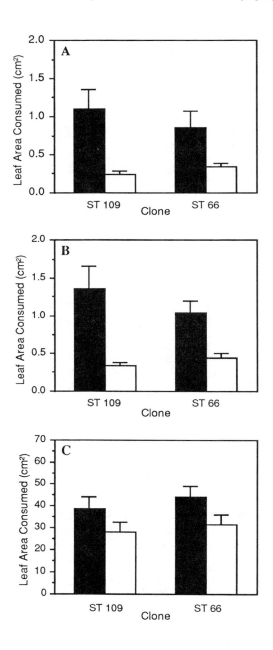

Figure 13-2. Mean leaf area consumed (± 1 S.E.) by adult *Plagiodera versicolora* in choice tests with cottonwood of two clones (ST109, ST66), treated with ozone (■) or charcoal filtered air (□) in different trials: A, Consumption after 7 days in a dual-choice disc assay, 1984. B, consumption after 7 days in a dual choice disc assay, 1985; C, consumption after 14 days in a multiple-choice whole plant assay, 1986.

Figure 13-3. Adult bark beetles. The larger beetle on right is *Dendroctonus valens* and the smaller on the left is *Ips pilifrons* .
Source: Fred B. Knight, University of Maine.

infested tissues. A few aggressive bark beetle species attack healthy trees (Christiansen et al., 1981).

Beetle outbreaks in western forests are associated with weakening caused by microbial infection; for example, root disease initiated by *Heterobasidion annosum* or *Verticicladiella wagenerii* fungi in ponderosa pine (Stark and Cobb, 1969); and insect defoliation, for example, pine looper stripping of ponderosa pine (Dewey et al., 1974); or various climatic stresses including drought and windthrow (Rudinsky, 1962). In the latter 1960s, California investigators added ozone to the list of biotic and environmental stresses that predispose ponderosa pine to bark beetle infestation. This is perhaps the most completely documented example of insect damage enhancement caused by air pollution in North America (Miller and Elderman, 1977).

During the summer of 1966 a survey of ponderosa pines was carried out in the San Bernadino Mountains of California. These forests are subject to elevated atmospheric oxidants from the Los Angeles urban complex to the west. Over 1000 trees were examined for degree of ozone damage and infestation and mortality from either the western pine beetle or mountain-pine beetle or both. Trees with the greatest pollution injury were found to be most commonly supporting populations of one or both bark beetle species (Figure 13-4). As the degree of oxidant damage increased, the live crown ratio decreased and the occurrence of bark beetle infestation increased (Stark et al., 1968).

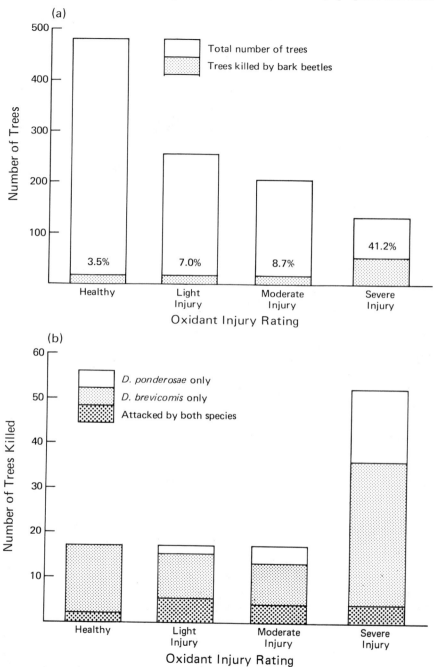

Figure 13-4. Relationship between the degree of photochemical air pollution injury and beetle attack (a) and mortality (b) in the San Bernadino Mountains of California.
Source: Stark and Cobb (1969).

Table 13-2. Comparison of Resin and Sapwood Characteristics of Ponderosa Pine
With and Without Symptoms of Oxidant Stress.

Characteristic	Without symptoms	With symptoms
Oleoresin exudation pressure	Greater	Lesser
Yield/rate of resin flow	Greater	Lesser
Crystallization rate	Lesser	Greater
Sapwood moisture content	Greater	Lesser
Phloem thickness	Greater	Lesser
Oleoresin chemistry	Same	Same
Resin acid chemistry	Same	Same
Soluble sugars	Greater	Lesser
Reserve polysaccharides	Greater	Lesser
Phloem pH	Same	Same

Source: Cobb et al. (1968a); Miller et al. (1968).

In an effort to appreciate ponderosa pine resistance mechanisms for bark beetle infestation, the investigators embarked on an ambitious attempt to examine
resin and sapwood characteristics of this species. The goal was to elucidate specific hypotheses regarding the apparent predisposition to insect attack by oxidants. A variety of field and laboratory tests was performed on 65- to 85-year-old
ponderosa pine growing in an all-age, second growth stand. Several physiological and chemical differences were recorded between trees exhibiting various degrees of oxidant symptoms (Table 13-2). In addition to the examination of terpene fractions of stem xylem, Cobb et al. (1972) have also examined the essential oils from foliage of stressed and healthy ponderosa pine. No significant differences were observed in the monoterpenoids of foliage, as in the xylem observations, but methyl chavicol was found to be much lower in the injured trees.

Cobb et al. (1986), based on their various observations, advanced the hypothesis that reduction of oleoserin exudation pressure, quantity, rate of flow,
and increased propensity of oleoserin to crystallize, and a reduction in phloem
and sapwood moisture content, all correlated with photochemical atmospheric
pollution in ponderosa pine, increased the damage caused by the western pine
beetle and mountain beetle in California. The investigators concluded that the
changes induced in resin and sapwood characteristics facilitated bark beetle activity particularly in the concentration and establishment phases. Air pollution
stressed trees were judged to be less suitable for beetle breeding than nonstressed
trees. As a result, symptomatic trees may act to "trap" beetles. The authors observed that the latter may have accounted for their failure to observe obvious increases in bark beetle infestations in their study sites. Christiansen et al. (1987)
have emphasized that the capacity of trees to withstand bark beetle attacks is associated with the amount of carbohydrates that can be utilized for defensive reactions. Any stress factor that restricts the size of the canopy or photosynthetic efficiency can reduce the resistance of a tree. Ozone can be capable of the latter and
can predispose trees to bark beetle infestation.

In the southern United States, longleaf, loblolly, shortleaf, and slash pines are all subject to economically significant damage by the southern pine beetle. Loblolly and shortleaf are especially susceptible. Hodges et al. (1979) have recently completed measurements on physical and chemical parameters of the resin of 50 trees each of the four southern pine species. As in the California study, susceptibility was strongly correlated with physical properties of the resin. If these properties are influenced by any southern air contaminants, for example, ozone, an atmospheric predisposition factor for beetle infestation may occur in the Southeast as well as in the West.

C. Fluoride

The tendency for fluoride to accumulate in biota has precipitated studies concerned with fluoride uptake and its metabolic influence on insects. Bees have received the greatest research attention (Lezovic, 1969; Caparrini, 1957; Guilhon et al., 1962; Marier, 1968; Maurizio and Staub, 1956), but other species including the Mexican bean beetle (Boyce Thompson Institute, 1971), desert locust, yellow meal worm (Outram, 1970), and others (Weismann and Svatarakova, 1974), have been investigated. A limited number of studies dealing with fluoride and forest insects are available.

1. Glacier National Park and Flathead National Forest Studies

Clinton Carlson and co-workers of the Forest Insect and Disease Branch, U.S.D.A. Forest Service, Missoula, Montana have been involved in a comprehensive study of the influence of fluorides on the forest ecosystem surrounding an aluminum reduction plant in Columbia Falls, Montana (Carlson and Dewey, 1971). Specific attention has been given to the level and consequences of fluoride contamination of trees and insects. Aluminum is produced by the electrolytic reduction of alumina. The process releases to the atmosphere approximately equal amounts of fluoride in the particulate and gaseous forms. Particulates consist of sodium fluoride (NaF) and aluminum fluoride (AlF_3), while gaseous emissions include hydrogen fluoride (HF) along with small amounts of carbon tetrafluorude (CF_4). Symptoms of fluoride injury occur on trees and shrubs throughout large areas of the Glacier National Park and Flathead National Forest in the vicinity of Columbia Falls. High but asymptomatic levels of fluoride are associated with other portions of the forests.

A wide variety of insects was collected within approximately 1 km of the aluminum plant and analyzed for fluoride (Dewey, 1973). Control insects collected from areas not subject to plant emissions had fluoride body burdens ranging from 3.5 to 16.5 μg g^{-1} (dry-weight basis). Analysis of forest insects within the 1-km zone revealed the following ranges for major insect groups: 58.0–585.0 μg g^{-1} pollinators, 6.1–170.0 μg g^{-1} predators, 21.3–225 μg g^{-1} foliar feeders, and 8.5–52.5 μg g^{-1}, cambial region feeders. Among the phytophagous species, the foliage feeders had the highest mean fluoride concentration. This is presumably due to the injestion of particulate fluoride associated with leaf surfaces as well as the fluoride contained within the leaves. The relatively high body burdens of the predatory osmotids, dragonflies, and damselflies suggested to Dewey that these insects may obtain some fluoride via respiratory processes or food chain accumulation. Dewey also observed that the elevated fluoride in the bark beetles examined may indicate fluoride contamination of host tree vascular tissue as well as foliage.

Lodgepole pine occupies a position of prominence in the forest ecosystems surrounding the aluminum reduction plant. It forms nearly pure stands over roughly 50% of the study area under the influence of fluoride deposition. Over 50 years ago, Keen and Evenden (1929) observed that defoliators, bark beetles, and wood borers of lodgepole pine may have contributed to the mortality of numerous trees that were severely defoliated by fume damage in the vicinity of a smelter at Northport, Washington. Since much of the lodgepole pine in the vicinity of the Columbia Falls aluminum plant was dying or symptomatic, Carlson et al. (1974) initiated efforts to correlate foliar and ambient fluoride concentrations with insect infestation. Stepwise multiple regression techniques were employed to statistically analyze the relationship between damage caused by the pine needle sheath miner, a needle miner, sugar pine tortix, and the fluoride parameters mentioned. Foliar fluoride concentration was significantly related to needle miner damage and to damage caused by the pine needle sheath miner. The authors concluded that the data strongly suggest that fluoride contamination may be a contributing factor in predisposing lodgepole pines to damage by these insects. While not specifically given by the authors, it may be judged that the foliar fluoride threshold level predisposing to needle miner and pine-needle sheath miner may approximate 30 µg g^{-1}.

2. Other Forest Insect-Fluoride Studies

Foliar fluoride concentrations can become substantial in woody vegetation. McClenahen and Weidensaul (1977) have mapped the distribution of fluorides in black locust leaves in the vicinity of a point source in southeast Ohio (Figure 13-5).

Elevated fluoride concentrations have been found associated with the European pine shoot moth in the vicinity of an aluminum production facility (Mankovska, 1975). An infestation of black pine leaf scale on ponderosa pine near Spokane, Washington from 1948 to 1950, originally thought to be associated with fluoride released from an aluminum facility, was ultimately judged to be correlated with cement and silicon dust (Johnson, 1950).

D. Acid Deposition

Few researchers have investigated the effects of acidic deposition on insects. Some studies relative to acidity effects on aquatic insects are available

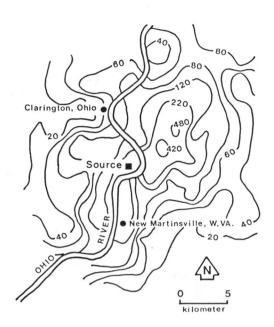

Figure 13-5. Isopleths of black locust foliar fluoride (μg g^{-1} dry wt basis) surrounding an Ohio point source
Source: McClenahen and Weidensaul (1977).

(Bornstrom and Hendrey, 1986). Terrestrial arthropods, on the other hand, have been the subject of very few studies. Hagvar et al. (1976) concluded that acidic precipitation, probably from western and central Europe, increases the susceptibility of Scotch pine forests to the pine bud moth (*Exoteleia dodecella*). The evidence available to support this conclusion was modest.

Two insect situations may be especially vulnerable to impact by acidic deposition: (a) foliar and other surface-feeding phytophagous insects and (b) litter- and soil-inhabiting microarthropods. Additional associations of high potential vulnerability are those involving predaceous and parasitoid insects and arachnids (including phytophages) via direct deposit of pollutants on cuticle and eggs, and internal feeders in wood, galls, and leaf mines, if there are significant alterations induced in host metabolism, carbon balance, or in secondary plant chemicals. Very little is known at present of the consequences of acidic precipitation for leaf-feeding insects. Montgomery and Meyer (1985) applied rain simulants of pH 3 and 5 to white oak seedlings for 2 weeks. Subsequent feeding by gypsy moth larvae demonstrated that caterpillars had higher survival and weight on plants receiving pH 3 rain than control plants receiving no rain. It would be desirable to expose seedling oak to greater exposure by increasing the length of the treatment period. It would further be important to apply treatments under field conditions. It is especially important to determine if acidic precipitation can induce alterations in leaf chemistry under field conditions. Recent evidence suggests drought stress increases foliar polyphenol levels in trees. Ethylene, produced by injured leaves, may increase polyphenol levels in adjacent leaves. Ethylene production modification as a result of acidic deposition has not been examined. Leaf tannin increases apparently make oak leaves less palatable to gypsy moth larvae. Defoliation appears to elevate tannin levels in succeeding growing seasons. Alterations of assimilable forms of foliar nitrogen by acid deposition is also of key interest. Nitrogen availability is fundamental to successful insect herbivory (Strong et al., 1984; White, 1984).

Due to great differences in mouthpart styles, studies will need to include significant tests of the differences between phytophages with chewing mouthparts (primarily larvae of lepidoptera, sawflies, and chrysomelid beetles, all stages of Arcrididae and Tettigoniidae, and adults of beetle families, notably Scarabeidae, Meloidae, and Chrysomelidae), and those with piercing and sucking mouthparts that allow them to bypass compounds on plant surfaces (primarily Hemiptera, Thysanoptera, and phytophagous mites). The former all ingest entire leaves and sometimes other plant structures; therefore, any deposits on plant surfaces enter the alimentary tract. The latter ingest only the internal fluids, and a reasonable working hypothesis is that aphids, scales, mirid bugs, thrips, and tetranychid and eriophyid mites are relatively unaffected by leaf-surface pollutants.

E. Trace Metals

In consideration of the potential non-soil-inhabiting phytophagous insects have for accumulation of heavy metals from hosts growing in roadside, industrial, or urban environments, it is surprising to find relatively little published data on this topic.

Giles et al. (1973) have examined lead burdens of several insect species in the roadside environment of Interstate 83, 4 km north of Baltimore, Maryland, an area with an average daily traffic of 13,000 vehicles. Japanese beetles did not show significant increases in lead relative to nonroadside control samples. Damselflies did not exhibit elevated lead burdens consistently over the sampling period. All collections of the European mantids, however, did exhibit higher concentrations of lead than control samples, but the levels were not excessive and averaged approximately 8 $\mu g\ g^{-1}$.

Price et al. (1974) have also sampled insects in the roadside environment from locations having average daily traffic values ranging from < 10 to 12,900 vehicles in the vicinity of Urbana, Illinois. In areas of higher lead emission, insect species that suck plant juices, chew plant parts, or prey on other insects had approximately 10, 16, and 25 $\mu g\ g^{-1}$ lead, respectively. In low lead emission areas the same feeding categories had 5, 3, and 3 $\mu g\ g^{-1}$, respectively.

F. Summary

Air pollutants may directly affect insects by influencing growth rates, mutation rates, dispersal, fecundity, mate finding, host finding, and mortality. Indirect effects may occur through changes in host age structure, distribution, vigor, and acceptance. The literature addressing the relationship between air pollutants and temperate forest insects is very meager and extremely disproportionate to the importance of arthropods in forest ecosystesm. Much of the research deals with relationships in a superficial manner. Alteration of population densities had been related to general ambient air quality. The specific pollutants present are neither inventoried nor measured. The insect surveys are restricted to one or a few species and do not constitute comprehensive surveys. Very limited information is available on the interaction of specific air contaminants with specific behavioral and physiological functions of insects. Information on fecundity, stage specific morbidity and mortality, and influence on sex ratios, for example, are needed. Meaningful studies must compare the effects on a few well-chosen arthropods with (a) different phyletic ancestry, (b) different feeding mechanisms, and (c) different body size (both early vs. late instar effects and tiny vs. large species). In addition, several developmental stages must be examined.

Ginevan and Lane (1978) have provided an experimental protocol ideally suited for initial information needed for important insects. These investigators found that laboratory exposure of developing larvae of fruit flies to sulfur dioxide at 700 ppb (1834 µg m^{-3}) for 10 days resulted in large increases in developmental time and decreased survival. Fecundity was not influenced by adult fumigation at 700 ppb for 4 days. Fumigation of prepupae and pupae at 400 ppb (1048 m^{-3}) for 4 days resulted in decreased survival that averaged 17%. The authors observed that factors that cause approximately 5–10% mortality in the pupal stage have been judged to have potentially significant effects on natural insect population dynamics. While the doses employed were high, they are not unknown in certain ambient situations.

The limited data available do suggest that insect-air pollutant interactions are likely to be very variable. The same dose of sulfur dioxide may stimulate or inhibit arthropods depending on the species. A given insect may be variously influenced by ozone depending on the gas concentration and time of exposure.

The capability that various air pollutants have to alter tree host susceptibility to insect influence is one of the most important relationships between the two potential stresses. Convincing evidence in the case of ponderosa pine has been presented for predisposition to bark beetle infestation by oxidants. Additional evidence supports the potential importance of sulfur dioxide and fluoride on mite and lepidopterous infestation, respectively, in other pine species. The threshold exposures required for significant predisposition under field conditions are not adequately described, but may preliminarily be judged to be relatively low in the ponderosa and lodgepole pine cases described.

The limited data available concerning insect body burdens of recalcitrant material, for example, fluoride and heavy metals, suggest that fluoride accumulation may be excessive only very close (within 0.5–1 km) to a primary source and that heavy metals may not accumulate to levels characteristic of floral components, even in ecosystems subject to excessive loading, for example, roadside environments.

Priority research attention should be given to selected United States insect species, most notably, the eastern and western spruce budworms, gypsy moth, tussock moth, and the bark beetles. Air pollution interaction with insects predatory on important forest insects should be studied more intensively. Bark- and wood-boring insects, shoot and bud-tunneling insects, and scales and aphids should receive more research attention, as they frequently respond favorably to increased host stress and are insulated from the atmosphere for portions of their life cycle.

References

Alstad, D.N. and G.F. Edmunds. 1982. Effects of air pollutants on insect populations. Annu. Rev. Entomol. 27: 369-384.

Anderson, R.F. 1970. Relation of insects and mites to the abnormal growth of the Christmas trees in Mt. Storm, West Virginia–Gorman, Maryland vicinity. U.S. Environmental Protection Agency Report, Durham, NC, 31 pp.

Baker, G.M. and E.A. Wright. 1977. Effects of carbon monoxide on insects. Bull. Environ. Contam. Toxicol 17: 98-104.

Beard, R.L. 1965. Observation on house flies on high-ozone environments. Annu. Entomol. Soc. Am. 58: 404-405.

Bolsinger, M. and W. Flückiger. 1987. Enhanced aphid infestation at motorways: The role of ambient air pollution. Entomol. Exp. Appl. 45: 237-243.

Boullard, B. 1973. Interactions entre les pollutants atmosphériques et certains parasites des essences forestières (champignons et insectes). For. Privée 94: 31, 33, 35, 37.

Bolsinger, M. and W. Flückiger. 1988. Ambient air pollution induced changes in amino acid pattern of phloem sap in host plant-relevance to aphid infestation. Environ. Pollu. 28: 158-160.

Borstrom, F. and G.R. Hendrey. 1976. pH tolerance of the first larva stages of Lepidurus arcticus (Pollas) and adult Gammarus lacustris. G O. Sars. Report, Zoological Museum, Oslo University, Oslo, Norway.

Boyce Thompson Institute. 1971. Annual Report of the Boyce Thompson Institute for Plant Research. Yonkers, New York, pp. 4-7.

Braun, S. and W. Flückiger. 1988. Effect of ambient ozone and acid mist on aphid development. Environ. Pollu. 28: 89-92.

Caparrini, W. 1957. Fluorine poisoning in domestic animals (cattle) and bees. Zooprofilass 12: 249-250.

Carlson, C.E. and J.E. Dewey. 1971. Environmental Pollution by Fluorides in Flathead National Forest and Glacier National Park. U.S.D.A. Forest Service. Forest Insect and Disease Branch, Missoula, MN, 57 pp.

Carlson, C.E., W.E. Bousfield, and M. D. McGregor. 1974. The relationship of an insect infestation on lodgepole pine to fluorides emitted from a nearby aluminum plant in Montana, Report No. 74-14, U.S.D.A. Forest Service, Div. State Private For., Missoula, MN, 21 pp.

Charles, P.J. and C. Villemant. 1977. Modifications des niveaux de population d'insectes dans les jeunes plantations de pins sylvestres de la fôret de Roumare (Seine-Maritime) soumises à la pollution atmosphérique. C.R. Acad. Agric. Fr. 63: 502-510.

Chiment, J.J., R. Alscher, and P.R. Hughes. 1986. Glutathione as an indicator of SO_2-induced stress in soybean. Environ. Exper. Bot. 26: 147-152.

Christiansen, E., R.H. Waring, and A.A. Berryman. 1987. Resistance of conifers to bark beetle attacks: Searching for general relationships. For. Ecol. Manage. 22: 89-106.

Cobb, F.W. Jr., D.L. Wood, R.W. Stark, and P.R. Miller. 1968a. II. Effect of injury upon physical properties of oleoserin, moisture content, and phloem thickness. Hilgardia 39: 127-134.

Cobb, F.W. Jr., D.L. Wood, R.W. Stark, and J R. Parmeter Jr. 1968b. Theory on the relationship between oxidant injury and bark beetle infestation. Hilgardia 39: 141-152.

Cobb, F.W. Jr., E. Zavarin, and J. Bergot. 1972. Effect of air pollution on the volatile oil from leaves of *Pinus ponderosa*. Phytochemistry 11: 1815-1818.

Coleman, J.S. and C.G. Jones. 1988a. Acute ozone stress on eastern cottonwood (*Populus deltoides* Bartr.) and the pest potential of the aphid *Chaitophorus populicola* Thomas (Homoptera: Aphididae). Environ. Entom. 17: 207-212.

Coleman, J.S. and C.G. Jones. 1988b. Plant stress and insect performance: Cottonwood, ozone, and a leaf beetle. Oecologia 76: 57-61.

Dewey, J.E. 1973. Accumulation of fluorides by insects near an emission source in western Montana. Environ. Entomol. 2: 179-180.

Dewey, J.E., W.M. Ciesla, and H.E. Meyer. 1974. Insect defoliation as a predisposing agent to a bark beetle outbreak in eastern Montana. Environ. Entomol. 3: 722.

Dohmen, G.P., S. McNeill, and J.N.B. Bell. 1984. Air pollution increases *Aphis fabae* pest potential. Nature 307: 52-53.

Flückiger, W. 1987. Effect of pollution on natural communities.In V. Delucchi, ed., Parasitis 86, pp. 331-349.

Flückiger, W., S. Braun, H. Flückiger–Keller, S. Leonardi, N. Asche, V. Bühler, and M. Lier. 1986. Untersuchungen über Waldschäden in festen Buchenbeobachtungsflächen der kabtone Basel-Landeschaft Basel-Stadt, Aargau Solothurn, Bern, Zürich and Zug. Aeitschrift für Forstwesen 137: 917-1010.

Flückiger, W., S. Braun, and M. Bolsinger. 1987. Air pollution: Effect on host plant-insect relationships.In 2nd Intl. Symp. on Air Pollution and Plant Metabolism. Gesellschaft für Shahlen-und Umwellforschung. Neuherberg, FRG, April 6–9, 1987.

Flückiger, W. and J.J. Oertli. 1978. Observations of an aphid infestation on hawthorn in the vicinity of a motorway. Nalurfwissencshaften 65: 39-40.

Gilbert, O.L. 1971. Some indirect effects of air pollution on bark-living invertebrates. J. Appl. Ecol. 8: 77-84.

Giles, F.E., S.G. Middleton, and J.G. Grau. 1973. Evidence for the accumulation of atmospheric lead by insects in areas of high traffic density. Environ. Entomol. 2: 299-300.

Ginevan, M.E. and D.D. Lane. 1978. Effects of sulfur dioxide in air on the fruit fly, *Drosophila melanogaster*. Environ. Sci. Technol. 12: 828-831.

Guilhon, J., R. Truhaut, and J. Bernuchon. 1962. Studies on the variations in fluorine levels in bees with respect to industrial atmospheric air pollution in a Pyrenean village. Acad. d'Agr. de France, Compt. Rendt. 48: 607-615.

Hagvar, S., G. Abrahamsen, and A. Bokhe. 1976. Attack by the pine bud moth in southern Norway: Possible effect of acid pollution. For. Abstr. 37: 694.

Hay, C.J. 1975. Arthropod stress.In W.H. Smith and L.S. Dochinger, eds., Air Pollution and Metropolitan Woody Vegetation. U.S.D.A. Forest Service, Publ. No. PIEFR-PA-1, Upper Darby, PA, pp. 33-34.

Hillman, C. 1972. Biological effects of air pollution on insects emphasizing the reactions of the honey bee (Apis mellifera L.) to sulfur dioxide. Ph.D. Thesis. PA State Univ., State College, PA, 170 pp.

Hodges, J.D., W.W. Elam, W.F. Watson, and T.E. Nebeher. 1979. Oleoserin characteristics and susceptibility of four southern pines to southern pine beetle (Coleoptera: Scolytidae) attacks. Can. Entomol. 111: 889-896.

Huffaker, C. B. 1974. Some implications of plant-arthropod and higher-level arthropod-arthropod food links. Environ. Entomol. 3: 1-9

Hughes, P.R. and J.A. Laurence. 1984. Relationships of biochemical effects of air pollutant of plants to environmental problems: Insect and microbial interactions.In M.J. Koziol and F.R. Whatley, eds., Gaseous Air Pollutants and Plant Metabolism. Butterworths, London, pp. 261-377.

Hughes, P.R., J.E. Potter, and L.H. Weinstein. 1981. Effects of air pollutants on plant/insect interactions. Reactions of the Mexican bean beetle to SO_2 fumigated pinto beans. Environ. Entomol. 7: 29-34.

Hughes, P.R., A.I. Dickie, and M A. Penton. 1983. Increased success of the Mexican bean beetle on field-grown soybeans exposed to sulfur dioxide. J. Environ. Qual. 12: 565-568.

Jeffords, M.R. and A.G. Endress. 1984. Possible role of ozone in tree defoliation by the gypsy moth (Lepidoptera: Lymantrüdae) Environ. Entom. 13: 1249-1252.

Johnson, P.C. 1950. Entomological aspects of the ponderosa pine blight study, Spokane, Washington. Unpubl. Report, U.S.D.A. Bur. Entomol. and Plant Quar., Forest Insect Laboratory, Coeur d'Alene, ID, 15 pp.

Johnson, P.C. 1969. Atmospheric pollution and coniferophagous invertebrates. Proc. 20th Annu. Western For. Insect Work Conf., Coeur d'Alene, ID.

Jones, C.G. and J.S. Coleman. 1988. Plant stress and insect behavior. Cottonwood, ozone and the feeding and oviposition preference of a beetle. Oecologia 76: 51-56.

Jones, C.G. and G.S. Coleman. 1989. Plant "stress" and insect herbivory: Toward an integrated perspective.In H.A. Mooney, W. Winner, and E. Pell, eds., An Integrated Approach to the Study of Environmental Stress on Plant Growth. Stanford Univ. Press, Stanford, CA.

Keen, F.P. and J.C. Evenden. 1929. The role of forest insects in respect to timber damage and smelter fume area near Northport, Washington. Unpubl. Report, U.S.D.A. Bur. Entomol., Stanford Univ., Stanford, CA, 12 pp.

Lebrun, P. 1976. Effects écologiques de la pollution atmosphériques sur les populations et communautés des microarthropods corticoles (Acariens, Collemboles et Ptérygotes.) Bull. Soc. Ecol. 7: 417-430.

Levy, R., Y.J. Chiu, and H.L. Cromroy. 1972. Effects of ozone on three species of Diptera. Environ. Entomol. 1: 608-611.

Levy, R., D.P. Jouvenaz, and H.L. Cromroy. 1974. Tolerance of three species of insects to prolonged exposures to ozone. Environ. Entomol. 3: 184-185.

Lezovic, J. 1969. The influence of fluorine compounds on the biological life near an aluminum factory. Fluoride Q. Rev. 2: 1.

Linzon, S. N. 1966. Damage to eastern white pine by sulfur dioxide, semimature-tissue needle blight, and ozone. J. Air Pollu. Control Assoc. 16: 140-144.

Mankovska, B. 1975. Influence of fluorine emissions from an aluminum factory plant on the content in different developmental stages of European pine shoot moth, Rhyacionia buoliana Schiff. Biologia (Bratislava) 30: 355.

Marier, J.R. 1968. Fluoride research. Science 159: 1494-1495.

Mattson, W.J. and N.D. Addy. 1975. Phytophagous insects as regulators of forest primary production. Science 190: 515-522.

Maurizio, A. and M. Staub. 1956. Poisoning of bees with industrial gases containing fluorine in Switzerland. Schweiz. Bienen Ztg. 79: 476-484.

McClenahen, J.R. and T.C. Weidensaul. 1977. Geographic Distribution of Airborne Fluorides Near a Point Source in Southeast Ohio. Ohio Agricultural Research and Development Center, Res. Bull. No. 1093, Wooster, OH, 29 pp.

Miller, P.R. and M.J. Elderman (eds.). 1977. Photochemical Oxidant Air Pollutant Effects on a Mixed Conifer Forest Ecosystem. EPA-600/3-77-104. U.S.E.P.A., Corvallis, OR, 338 pp.

Miller, P.R., F.W. Cobb Jr. and E. Zavarin. 1968. III. Effect of injury upon oleosresin composition, phloem carbohydrates and phloem pH. Hilgardia 39: 135-140.

Montgomery, M.E. and G.A. Meyer. 1985. Effect of acidified rain on gypsy moth host plant relationships. In Proc., 4th Intl. Union of Forestry Research Organizations, Conference: Resistance in Trees to Harmful Agents. Curitiba, Brazil.

Novakova, E. 1969. Influence des pollutions industrielles sur les communautes animals et l'utilisation des animaux comme bioindicateurs. Proc. 1st Eur. Cong. Influence of Air Pollution on Plants and Animals, Wageningen, 1968, pp. 41-48.

Outram, I. 1970. Some effects of fumigant sulphryl fluoride on the gross metabolism of insect eggs. Fluoride Q. Rep. 3: 2.

Price, P.W., B.J. Rathcke, and D. A. Gentry. 1974. Lead in terrestrial arthropods: Evidence for biological concentration. Environ. Entomol. 3: 370-372.

Przybylski, Z. 1979. The effects of automobile exhaust gases on the arthropods of cultivated plants, meadows and orchards. Environ. Pollu. 19: 157-161.

Rudinsky, J.A. 1962. Ecology of Scolytidae. Annu. Rev. Entomol. 7: 327-348.

Saunders, J.L. 1972. Disease and insect pests of Christmas trees. School for Christmas Tree Growers. College of Agriculture, Proc. Cornell Univ., Ithaca, New York, pp. 88-90.

Schnaider, Z. and Z. Sierpinski. 1967. Dangerous condition for some forest tree species from insects in the industrial region of Silesia. Prace Instytut Badawczy Tesnictwa (Warsaw), Bull. No. 316, pp. 113-150.

Seastedt, T.R. and D.A. Crossley. 1984. The influence of arthropods on ecosystems. BioScience 34: 157-161.

Sierpinski, Z. 1967. Influence of industrial air pollutants on the population dynamics of some primary pine pests. Proc. 14th Congr. Intl. Union For. Res. Organiz. 5(24): 518-531.

Sierpinski, Z. 1970. Economic significance of noxious insects in pine stands under the chronic impact of the industrial air pollution. Sylwan 114: 59-71.

Sierpinski, Z. 1971. Secondary noxious insects of pine in stands growing in areas with industrial air pollution containing nitrogen compounds. Sylwan 115: 11-18.

Sierpinski, Z., 1972a. The economic importance of secondary noxious insects of pine on territories with chronic influence of industrial air pollution. Mitt. Forstl. Bundesversuchsanst Wien 97: 609-615.

Sierpinski, Z. 1972b. The occurrence of the spruce spider (*Paratetranychus (Oligonychus) ununquis* Jacoby) on Scotch pine in the range of the influence of industrial air pollution.In Institute Badawczego Lesnictwa (Warsaw), Bull. No. 433–434, pp. 101-110.

Sierpinski, Z. 1981. Rücksgang der tanne (*Abies albal* Mill.) in Polen. Eur. J. For. Pathol. 11: 153-162.

Sierpinski, Z. and J. Chlodny. 1977. Entomofauna of forest plantations in the zone of disastrous industrial pollution.In J. Woldk, ed., Relationship Between Increase in Air Pollution Toxicity and Elevation Above Ground. Institute Badawczego Lesnictwa (Warsaw), pp. 81-150.

Smith, W.H. 1986. Role of phytophagous insects, microbial, and other pathogens in forest ecosystem structure and function.In B.L. Bedford, ed., Modification of Plant-Pest Interactions by Air Pollutants. Ecosystems Research Center, Publ.. No. ERC-117, Cornell Univ., Ithaca, NY, pp. 34-52.

Stark R.W. and F.W. Cobb Jr. 1969. Smog injury, root diseases beetle damage in ponderosa pine. Calif. Agric. Sept., 1969: 13-15.

Stark, R.W., P.R. Miller, F. W. Cobb Jr., D.L. Wood, and J.R. Parmeter Jr. 1968. I. Incidence of bark beetle infestation in injured trees. Hilgardia 39: 121-126.

Strong, D.R., J.H. Lawton, and R. Southwood. 1984. Insects on Plants. Harvard University Press, Cambridge, MA. 313 pp.

Struble, G.R. and P.C. Johnson. 1964. Black pine leaf scale. U.S.D.A. Forest Serv., Forest Pest Leaflet No. 91, Washington, DC, 6 pp.

Templin, E. 1962. On the population dynamics of several pine pests in smoke-damaged forest stands. Wissenschafthce Zeitschrift der Techniscen Universität, Dresden 113: 631-637.

Villemant, C. 1979. Modifications de l'enlomocenose due pin sylvestre en liaison avec la pollution atmosphérique en fôret de Roumare (Seine-Maritime). Doctoral Dissertation, Pierre and Marie Curie University, Paris, 161 pp.

Villemant, C. 1981. Influence de la pollution atmospherique sur les populations d'aphides du pin sylvestre en foret de Roumare (Seine-Maritime). Environ. Pollu. 24: 245-262.

Weisman, L. and L. Svatarakova. 1974. Toxicity of sodium fluoride on some species of harmful insects. Biologia (Bratislava) 19: 847-852.

White, T.C.R. 1984. The abundance of invertebrate herbivores in relation to the availability of nitrogen in stressed food plants. Oecologia 63: 90-105.

Wiackowski, S.K. and L.S. Dochinger. 1973. Interactions between air pollution and insect pests in Poland. 2nd Intl. Congr. Plant Pathol., Univ. of Minnesota, Minneapolis, MN, Abstr. No. 0736, p. 1.

14

Forest Biotic Agent Stress: Air Pollutants and Disease Caused by Microbial Pathogens

Abnormal physiology, or disease, in woody plants follows infection and subsequent development of an extremely large number and diverse group of microorganisms internally or on the surface of tree parts. All stages of tree life cycles and all tree tissues and organs are subject, under appropriate environmental conditions, to impact by a heterogeneous group of microbial pathogens including viroids, viruses, mycoplasmas, bacteria, fungi, and nematodes. The influence of a specific disease on the health of an individual tree may range from innocuous to mild to severe. Over extended time periods, the interaction of native pathogens with natural forest ecosystems is significant, and frequently beneficial, in terms of ecosystem development and metabolism. As in the instance of insect interactions (Chapter 13), microbes and the diseases they cause, play important roles in forest succession, species composition, density, competition, and productivity. In the short term, the effects of microbial pathogens may conflict with forest management objectives and may assume a considerable economic or managerial as well as ecologic significance (Smith, 1970).

The interaction between air pollutants and microorganisms in general is highly variable and complex. Considerable attention has already been given to soil microorganisms (Chapter 9) and symbiotic microbes (Chapter 11). Babich and Stotzky (1974) have provided a comprehensive overview of the relationships between air contaminants and microorganisms. Microbes may serve as a source as well as a sink for air pollutants. A specific air pollutant, at a given dose, may be stimulatory, neutral, or inimical to the growth and development of a particular virus, bacterium, or fungus. In the latter, fruiting body formation, spore production, and spore germination may be stimulated or inhibited. Microorganisms that normally develop in plant surface habitats may be especially subject to air pollutant influence. These microbes have received considerable research attention and have been the subject of review (Saunders, 1971, 1973, 1975). The author has employed a strategy analogous to this volume in an attempt to summarize the interaction between air contaminants and plant-surface microbial ecosystems (Smith, 1976). Class I, II, and III interactions are identifiable for these ecosystems as well as for forest ecosystems. As in the latter case, the variable physiological responses of individual elements of the biota translate into increased, no change, or decreased biomass and biological activity at the ecosystem level (Table 14-1).

Microorganisms that function as plant pathogens are, of course, no exception to Table 14-1 generalizations. As a consequence, it is of no surprise that the apparent influence of individual pollutants and combinations of pollutants on microbial plant parasites is to both *increase* and *decrease* their activities. The actual impact of air pollution stress on disease expression is especially complicated, however, as the air contaminants not only influence the metabolism and ecology of the microbe but also influence the physiology of the host plant. Even under "unpolluted atmospheric conditions," disease in plants is a complex integration of pathogen physiology, host plant physiology, and ambient environmental conditions. The addition of an air pollutant stress has the effect of adding an additional complexing variable to an already elaborate and complicated interaction. Numerous comprehensive reviews have summarized the interactions between air contaminants and plant diseases. In 1973, Allen S. Heagle, U.S.D.A., Agricultural Research Service, Raleigh, North Carolina summarized nearly 100 references and found that sulfur dioxide, ozone, or fluoride had been reported to increase the incidence of 21 diseases and to decrease the occurrence of nine diseases in a variety of nonwoody and woody hosts. Michael Treshow of the Department of Biology, University of Utah, Salt Lake City, Utah, has provided a detailed review concerning the influence of sulfur dioxide, ozone, fluoride, and particulates on a variety of plant pathogens and the diseases they cause (Treshow, 1975). Treshow lamented the fact that most of the data available deal with in-vitro or laboratory accounts of microbe-air-pollutant interactions, while only a few investigations have been performed that have examined the influence of air pollutants on disease development under field conditions. In a review provided by William J. Manning, Department of Plant Pathology, University of Massachusetts, Amherst, Massachusetts, it was pointed out that most research attention has been directed to fungal-pathogen–air-pollutant interactions (Manning, 1975). John A. Laurence, Boyce Thompson Institute, Ithaca, New York, and Karin Kvist, Swedish University of Agricultural Sciences, Uppsala, Sweden, both emphasize in their reviews that most research efforts have been focused on fungi and on agricultural crops (Laurence, 1981; Kvist, 1986). Laurence (1981) further suggested that diseases caused by obligate fungal parasites usually have been found to be reduced in development by air pollutant exposure. Greater research perspective is needed concerning air pollution influence on viruses, bacteria, nematodes, and the diseases they cause. Macroscopic agents of disease, most importantly true and dwarf mistletoes, should also be examined relative to

Table 14-1. Influence of Air Pollution on Plant-Surface Microbial Ecosystems.

Air pollution dose	Response of microbe	Impact on microbial ecosystem	Reaction of host plant
Class I Low	1. Act as a source of air contaminants 2. Act as a sink for air contaminants	1. No effect or potentially some allelopathic influence 2. No or minimal physiological alteration or potentially some fertilization, stimulation	

Class II Intermediate	1. Abnormal metabolism altered pigmentation, morphology, enzyme activity
	2. Reduced reproduction (reduced competitiveness)
	  (a) lessened spore production or dispersal
	  (b) reduced or delayed spore germination
	3. Reduced growth (reduced productivity and competitiveness)
	  (a) vegetative retardation
	  (b) vegetative inhibition
		1. No significant or very minor perturbation
		2. Altered species composition and succession
		
		3. Reduced microbial biomass, altered structure and function (energy flow, nutrient cycling, competition, succession)
			Altered surface microflora, changed relationship with saprophytes, increased/decreased disease caused by parasites

Class III High	1. Stimulation of individual species 2. Acute morbidity of individual species 3. Mortality of individual species	1. Increased microbial biomass, altered structure and function 2. Reduced microbial biomass, altered structure and function 3. Simplification	Altered surface microflora, changed relationship with saprophytes, increased/decreased disease caused by parasites

Source: Smith (1976).

air pollution impact, especially in the western part of North America, where the latter are extremely important agents of coniferous disease.

Specific Air Pollutants and Forest Tree Disease

Forest trees, because of their large size, extended lifetime and widespread geographic distribution, are subject to multiple microbially induced diseases frequently acting concurrently or sequentially. The reviews of Heagle (1973), Treshow (1975), Manning (1975), and Laurence (1981) included consideration of a variety of pollutant–woody plant pathogen interactions, but were not specifically concerned with forest tree disease. In their review of the impact of air pollutants on fungal pathogens of forest trees of Poland, Gizywacz and Wazyn (1973) referenced literature citations indicating that air pollution stimulated the activities of at least 12 fungal tree pathogens, while restricting the activities of at least 10 others.

A. Sulfur Dioxide

Elemental sulfur has been long appreciated for its toxic influence on fungi. At extremely high concentrations, 900–2500 ppm (24–66 x 10^5 µg m^{-3}), sulfur dioxide itself has been employed as a fungicide (Treshow, 1975). Even though ambient concentrations may only approximate 1% of this extreme fungicidal dose, it has been observed that sulfur dioxide appears to have the ability to adversely influence pathogens directly (Manning, 1975). The observation that numerous forest pathogens appear to be restricted in regions subject to high ambient concentrations from point sources supports this generalization. The pioneering research of Scheffer and Hedgcock (1955) provides a classic example. In their intensive observations of the forest ecosystems surrounding metal smelters in Washington and Montana, these pathologists included surveys of parasitic fungi. In those areas subjected to elevated sulfur dioxide, a large number of fungi, particularly those parasitic on foliage, appeared to be suppressed. This observation appeared especially true for species of *Cronartium, Coleosporium, Melampsora, Peridermium, Pucciniastrum, Puccinia, Lophodermium, Hypoderma,* and *Hypodermella.* Gradients of rust infection, *Melampsora albertensis* on quaking aspen and *M. occidentalis* on black cottonwood, were observed coincident with ambient sulfur dioxide. These fungi were not found in the zone of greatest sulfur dioxide tree injury, were sparse in the zone of moderate injury, and were most abundant where the injury was least. Similar observations were made with *Pucciniastrum pustulatum* on grand and subalpine firs; *Coleosporium solidaginis* on lodgepole pine; and *Cronartium harknessii, C. comandrae,* and *Lophodermium pinastri* on ponderosa and lodgepole pines. *Hypodermella laricis,* parasitic on larch needles, was absent from areas affected by sulfur dioxide, while it was more common in regions free of sulfur gas stress. *Cronartium ribicola,* the causal agent of white pine blister rust, has been observed to be almost absent in forests to distances of 40 km (25 miles) northeast of the Sudbury, Ontario, smelters in the direction of the prevailing wind. With increasing distance from the Sudbury sulfur dioxide source, white pine blister rust incidence invariably increases (Linzon, 1978).

A variety of additional fungi that infect tree foliage has been shown to be variously impacted by sulfur dioxide. *Microsphaera alni*, the causal agent of oak powdery mildew, has been observed to be absent from the vicinity of a paper mill in Hinterburg, Austria (Koeck, 1935). Although unable to implicate a specific pollutant, Hibben and Walker (1966) have determined that lilacs growing in New York City appear to have substantially less powdery mildew caused by *Microsphaera alni* than lilacs in nonurban areas. At 0.3–0.4 ppm (785–1048 μg m^{-3}) sulfur dioxide exposure for 24–72 hr, these investigators recorded decreased *M. alni* spore germination of 50–60% on leaf discs and decreased disease development beyond the appressorium (infection peg) stage (Hibben and Taylor, 1975). At a sulfur dioxide acute dose of 1 ppm (2620 μg m^{-3}) for 1, 2, 4, or 6 hr, no effect on detached spore metabolism was noted. Chronic exposure was apparently necessary for a suppressive effect. Ham (1971) has indicated that *Scirrha acicola*, the causal agent of brown spot disease of loblolly pine foliage, grew normally and produced viable spores when exposed in vitro to 1 ppm (2620 μg m^{-3}) sulfur dioxide for 4 hr. It is of interest to speculate on what the response of this fungus would have been if it had also been subjected to chronic, low dose exposure. Additional evidence for reduced foliar disease in areas of high ambient sulfur dioxide, has been presented for the following fungal pathogens: *Lophodermium juniperi* and *Rhytisma acerinum* (Barkman et al., 1969), *Hysterium policore* (Skye, 1968), and *Venturia inaequalis* (Przybylski, 1967).

In contrast to these examples of suppression, the significance of some-foliage infecting fungi has been shown to be enhanced under conditions of elevated ambient sulfur dioxide. In an effort to explain the increased incidence of pine needle blight on Japanese red pine in central Japan caused by *Rhizosphaero kalkhoffii*, Chiba and Tanka (1968) exposed inoculated and uninoculated seedlings to 2 ppm (5240 μg m^{-3}) sulfur dioxide for 1-, 2-, 3-, and 4-hr fumigations. Generally disease was most severe on those seedlings receiving the greatest sulfur dioxide dose. Jancarik (1961) has recorded a higher incidence of *Lophodermium piceae* on spruce needles damaged by sulfur dioxide exposure. Fungal bark pathogens on larch and oak species have been recorded as more numerous in industrial regions of Poland with high pollution exposure (Kowalski, 1982, 1983). Beech bark disease, on the other hand, has been reported to decrease in the presence of increasing sulfur dioxide (Decourt et al. 1980).

Fungi that cause wood decay have also received some examination relative to interaction with sulfur-dioxide-stress. Jancarik (1961) conducted a survey of macroscopic fruiting bodies of wood decay producing Basidiomycete fungi in northern Czechoslovakia in areas with healthy and sulfur dioxide damaged conifers. Of 40 decay-producing species recorded, 12 were present in regions of slight injury but absent from areas of severe sulfur dioxide damage. Six decay fungi were recorded only on severely damaged trees. These included *Glocophyllum abietinum*, *Trametes serialis*, and *Trametes hetermorpha*. *Poria* sp., *Mycena* spp., *Schizophyllum commune*, and *Polyporus versicolor* were found exclusively in areas of minor sulfur dioxide injury. Scheffer and Hedgcock (1955), on the other hand, were unable to find any influence of sulfur dioxide on decay incidence in Montana conifers. Close to the Anaconda sulfur gas point source, approximately 7% of the mature lodgepole pine and 72% of the Douglas fir were infected in various degrees by *Polyporus schweinitsii* or by *Fomes pini*. These percentages were comparable to the incidence in surrounding forests not under the influence of elevated sulfur dioxide.

Soil-inhabiting fungi have exhibited a variable response to sulfur dioxide deposition. Killham and Wainwright (1981) have stressed that air pollution deposits to the soil in heavily polluted regions can provide microbial nutrients, especially carbon and sulfur. Bewley and Parkinson (1984, 1985) have characterized the soil microflora in pine forest soils at three sites (2.8, 6.0, and 9.6 km) downwind of a sour gas plant (sulfur dioxide emissions) in Alberta, Canada. They found a reduction in total microbial biomass in the forest floor at 2.8 km relative to 6.0 and 9.6 km, but no differences between sites in the mineral soil. Fewer total numbers of bacteria and a greater proportion of spore forming bacteria characterized forest floor samples closest to the point source. The predominant fungi of the forest floor were *Aureobasidium pullulans* associated with fresh needle litter and *Trichoderma viride*, associated with forest floor material. Sulfur dioxide deposition appeared to have little influence on the distribution of either *A. pullulans* or *T. viride* in the forest floor.

Fungi that induce root disease are among the most important pathogens of managed forest ecosystems. An example of enormous importance to temperate forest ecosystems is the ubiquitous *Armillaria* genus (Figure 14-1). These fungi are very widespread, have an extremely broad host range, and can function as aggressive killers of healthy trees, secondary pathogens of stressed trees, and saprophytic decayers of dead trees (Wargo and Shaw, 1985). Sinclair (1969) has pointed out that the relationship between *A. mellea* infection and air pollution stress may be a classic example of disease predisposition caused by air contamination. Scheffer and Hedgcock (1955) did indeed find that the association of *A. mellea* with pine roots was greatest inside the zone of sulfur dioxide pine damage. Additional evidence has been provided indicating predisposition to *A. mellea* infection by trees stressed by sulfur dioxide exposure (Donaubauer, 1968; Jancarik, 1961; Kudela and Novakova, 1962). We have investigated the role of *Armillaria* infection in red spruce dieback/decline (Section D4, this chapter).

The information available frustrates attempts to generalize concerning the influence of elevated ambient sulfur dioxide on forest tree disease induced by biotic agents. Substantial observations of fungal disease incidence suggest that the activities of some pathogens are suppressed while others are enhanced. Much of this information, however, stems from relatively simple disease surveys in stressed and nonstressed environments. In some of the studies, it is not obvious that appropriate attention has been given to factors other than sulfur dioxide that also could have accounted for altered disease incidence. Little attention has been given to specific mechanisms of sulfur-dioxide–plant-pathogen interaction that

Figure 14-1. Fruiting bodies of *Armillaria* developing at the base of a yellow birch in Vermont. This fungus causes one of the most significant and common root diseases in temperate forests.

may account for increased or decreased disease. It has been hypothesized that increased stomatal aperture may facilitate foliar infection, but the evidence is not extensive (Unsworth et al., 1972; Williams et al., 1971). It is probable in natural forest ecosystems subject to sulfur dioxide stress that disease incidence may be altered by sulfur-gas influence on the pathogen and the host depending on the relative susceptibilities of the organisms and the nature of the sulfur dioxide exposure. Before and after fumigation of tomato and bean plants with less than 20 ppm (524 μg m^{-3}) sulfur dioxide for several days and inoculation with the fungal agents causing early blight and bean rust, respectively, an influence on the latter disease only was revealed (Weinstein et al., 1975). In this case, however, the decreased incidence and severity of the disease was judged to have resulted from alterations in both the pathogen and the host. This study is representative of the considerable evidence indicating that spore production and germination, vegetative development of some microbial pathogens, and metabolism and physiology of foliar tissues of some hosts are proportional to sulfur dioxide exposure (National Academy of Sciences, 1978).

The results of the fine study of air pollution impact on fungal pathogens of Polish forests conducted by Grzywacz and Wazny (1973) are consistent with this exposure generalization. These investigators found substantial quantitative differences in the occurrence of *Armillaria mellea, Ophodermium pinastri, Fomitopsis annosa (Heterobasidion annosum), Cronartium flaccidum, Melampsora pinitorqua, Phellinus pini, Cerangium abietis,* and *Microsphaera alphitoides* relative to air quality. In all these cases, high exposure to industrial sulfur dioxide acted to destroy or inhibit the growth of these fungi. Low exposure acted to stimulate their activities. Plots of disease incidence against distance from point sources revealed curves of similar shape for all fungi examined (for example, Figure 14–2).

B. Ozone

As in the instance of sulfur dioxide, ozone has a research history as a potential microbial pesticide. In the ozone case, its proposed use as a fungicide (Hartman, 1924) predates its recognition as an air pollutant by several decades. Also, as in the case of sulfur dioxide, an extensive literature is available concerning the interaction of this gas with microbial development under laboratory conditions. Unlike sulfur dioxide, however, extensive field correlations of plant disease incidence and ambient ozone concentration are lacking. Heagle (1973) has correctly indicated that this is primarily due to the absence of point sources of ozone and the lack of distinct gradients of ozone concentrations in natural environments.

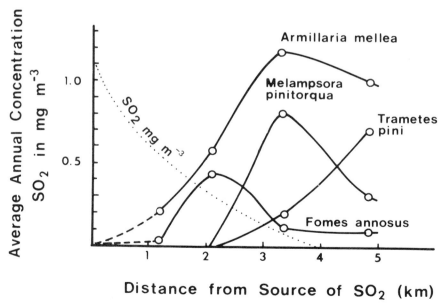

Figure 14-2. Average percentage of trees infected by various fungal pathogens in Polish forests at varying distances from sulfur dioxide point sources.
Source: Gryzywacz and Wazny (1973).

Excellent reviews of the interactions between ozone, plant pathogens, and agricultural crops have been provided by Heagle (1973), Treshow (1975), and Laurence (1981). These reviews reveal considerable evidence from laboratory and greenhouse studies to indicate that ozone can decrease infection, invasion, and spore production of fungal pathogens and that this can inhibit parasitism. The majority of evidence indicates that ozone alters parasitism via effects on host plants. Ozone can also apparently stimulate the growth and development of microbes on plant surfaces.

Ozone doses required for direct impact on microbial metabolism may be quite high. The data of Hibben and Stotsky (1969) are illustrative. These investigators examined the response of detached spores of 14 fungi to 10-100 pphm (196–1960 μg m^{-3}) ozone for 1-, 2-, and 6-hr periods. The large pigmented spores of *Chaetomium* sp., *Stemphylium sarcinaeforme*, *S. loti*, and *Alternaria* sp. were uninfluenced by 100 pphm (1960 μg m^{-3}). Germination of *Trichoderma viride*, *Aspergillus terreus*, *A. niger*, *Penicillium egyptiacum*, *Botrytis allii*, and *Rhizopus strolonifera* spores were reduced by ozone exposure, but only by concentrations above 50 pphm (980 μg m^{-3}). The small colorless spores of *Fusarium oxysporum*, *Colletotrichum largenarium*, *Verticillium albo-atrum*, and *V. dahliae* had germination percentages reduced by 50 pphm (980 μg m^{-3}) and occasionally by doses of 25 pphm (490 μg m^{-3}) for 4–6 hr. Concentrations less than this latter dose stimulated spore germination in some cases. Rist and Lorbeer (1984) studied the interaction of onion plants exposed to ozone and leaf infection by *Botrytis cinerea* and *B. squamosa*. Plants were subjected to chronic exposure with ozone to a maximum of 140 ppb (274 μg m^{-3}) for 5 hr day^{-1} for 5 days. Under these conditions, sporulation of *B. cinerea* and *B. squamosa* occurred only occasionally and was not different in frequency, extent, or density from plants not exposed.

The literature addressing the interaction of ozone with woody plant disease is modest. The reduced incidence of powdery mildew disease on lilac documented by Hibben and Walker (1966) in urban areas could have been related to ozone. Laboratory exposure of conidia of *Microsphaera alni* , however, to 0.9–1.0 ppm (1760–1970 μg m^{-3}) ozone for 1, 2, 4, and 6 hr; 0.5 ppm (980 μg m^{-3}) ozone for 6 hr; 0.25 ppm (490 μg m^{-3}) ozone for 72 hr; or 0.1–0.15 ppm (200-290 μg m^{-3}) ozone for 72 hr had essentially no influence on germination and early fungal development (Hibben and Taylor, 1975).

There is some indication that ozone may enhance disease development by pathogens that normally infect stressed or senescent plant parts or invade nonliving woody plant tissues. *Lophodermia pinastri* and *Pullalaria pullulans* were most commonly associated with eastern white pine foliage injury when inoculated in conjunction with tree exposure to 7 pphm (137 μg m^{-3}) ozone for 4.5 hr (Costonis, 1968; Costonis and Sinclair, 1972).

Weidensaul and Darling (1979) inoculated Scotch pine seedlings with *Scirrhia acicola* 5 days before or 30 min following fumigation for 6 hr with 0.20 ppm (533 μg m^{-3}) sulfur dioxide, 0.20 (332 μg m^{-3}) ozone, or both gases combined. Significantly more brown spot lesions were formed on seedlings fumigated with sulfur dioxide alone or combined with ozone than on controls when inoculation was done 5 days before fumigation. When inoculation was done 30 min after gas exposure, seedlings exposed to sulfur dioxide alone had more lesions than those exposed to ozone alone or combined with sulfur dioxide, but no significant differences were noted between treated seedlings and controls. The authors judged that ozone-induced stomatal closure may have been responsible for the latter observation.

Heterobasidion annosum is another Basidiomycete fungus capable of causing widespread and significant root disease and decay in a variety of coniferous hosts throughout temperate forests (Ross 1973, Shaw 1981). A comprehensive examination of oxidant stress on southern California forest ecosystems (Chapter 16) has included investigation of ozone influence on this fungus and the disease it causes in ponderosa and Jeffrey pines (Miller 1977). Laboratory studies suggested that *H. annosum* conidia were not influenced at exposures similar to ambient in the San Bernardino National Forest. At 100 ppb (196 μg m^{-3}) for 8 hr germination was reduced less than 10%, and at 100 ppb for 1 hr it was actually increased. Germ tube elongation and branching appeared more sensitive than germination and at 90 ppb (176 μg m^{-3}) for 12 hrs both were reduced approximately 50% (James et al. 1982). Basidiospores rather than conidia, however, are judged to be the major inoculum for stump infection in the forest. Unfortunately basidiospore response to ozone has not been examined. James (1977) found no correlation between *H. annosum* basidiospore deposition rate and air quality in the San Bernardino Mountains. Artificial root inoculation was conducted with trees exhibiting various degrees of oxidant stress. Pine seedlings were also artificially inoculated following fumigation with ozone. In light of the importance of freshly cut stump surfaces in the spread of this fungus, trees of various suceptibility classes were cut and their stump surfaces were inoculated with *H. annosum*. Laboratory exposures of pure cultures of *H. annosum* to ozone were also performed. Preliminary results have indicated the following. Field inoculation of roots of both ponderosa and Jeffrey pines did not reveal any correlation with the degree of oxidant damage. Stump inoculation, however, did suggest that air pollution injury may have increased the susceptibility of pine stumps to colonization by *H. annosum* (James et al., 1980). The percentage of infection of fumigated seedlings was also greater than that of nonfumigated seedlings (Table 14–2).

Heagle (1975) has determined that ozone, at doses comparable to ambient conditions in certain areas, was capable in inhibiting various phases of development of two rust fungi and a powdery mildew fungus of cereal crops. Since it is of extreme interest to have comparable information on the fungi responsible for the large number of economically significant forest tree rust diseases, we initiated rust studies in 1984. The interaction of an acute ozone dose, plant genotype, and leaf ontogeny on the development of leaf rust on eastern cottonwood was investigated (Coleman et al., 1987). A rust-resistant and a rust-susceptible clone were exposed to charcoal-filtered air or were fumigated with 200 ppb (393 μg m^{-3}) ozone for 5 hr. Forty hours after fumigation, leaf material of different developmental ages was inoculated with urediospores of *Melampsora medusae*, and

Table 14-2. Infection of Ozone Fumigated and Unfumigated Jeffrey and Ponderosa Pine Seedlings by *Heterobasidion annosum*[a].

Pine species	Seedling number	Ozone fumigation concentration (μg m^{-3})	Infection (%)
Jeffrey	32	0	53.1
Jeffrey	16	431.2	75.0
Jeffrey	16	882.0	81.0
Ponderosa	32	0	62.0
Ponderosa	16	431.2	81.0
Ponderosa	16	882.0	75.0

Source: Miller (1977).

[a] Seedlings at each concentration were exposed for a period ranging between 58 and 87 days.

uredia production was measured after 10 days. Ozone fumigation of cottonwoods significantly reduced uredia production by *M. medusae* on both clones and all leaf ages, without causing visible leaf injury or measurable changes in cottonwood height growth, leaf production, leaf length, or root/shoot biomass. (Figure 14–3). Uredia production was strongly affected by ozone treatment, cottonwood genotype, and leaf age, but interactions among these three factors did not occur.

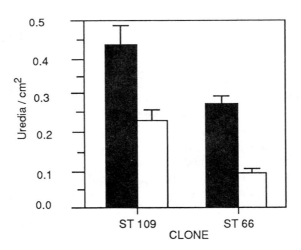

Figure 14-3. Mean number of *M. medusae* uredia produced cm^2 ($\pm$ 1 SE) on leaf plastochron index 5 for two cottonwood clones 10 d after inoculation which was done 40 hr after ozone (200 ppb, 393 μg m^{-3}) for 5 hr (white bars) or charcoal-filtered air (dark bars) exposure.
Source: Coleman et al. (1987).

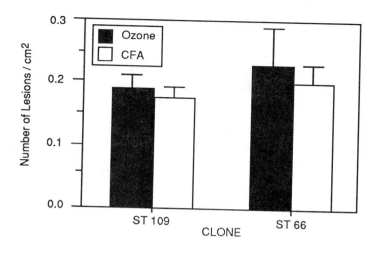

Figure 14-4. Mean number (± 1 SE) of lesions produced cm² by *Marssonina brunnea* on leaf plastochron index 5 for two cottonwood clones treated with ozone at 200 ppb (393 µg m⁻³) for 5 hr or with charcoal-filtered air.
Source: Coleman et al. (1988).

In a companion experiment, a similar acute ozone exposure was evaluated with regard to lesion production of the leaf spot fungus *Marssonina brunnea* (Coleman et al., 1988). Again two cottonwood clones were exposed to 200 ppb (393 µg m⁻³) ozone for 5 hr and then were inoculated with conidia of *M. brunnea* in leaf-disk and whole-plant arrays on leaves of different ages. There was no significant difference in lesion production on ozone-treated or control foliage (Figure 14-4). Our cottonwood studies suggest that disease response to ozone exposure is clearly pathogen dependent. In nature, ozone exposure of cottonwoods may have the potential to alter pathogen community dynamics by differentially altering host susceptibility to different pathogens.

In addition to fungal agents of disease, recent research attention has been given to other important microorganisms responsible for plant disease. Papers dealing with air contaminant interaction with bacterial disease agents (Laurence and Wood, 1987ab; Howell and Graham, 1977; Pell et al., 1977) and viral disease agents (Brennan, 1975; Bisessar and Temple, 1977; Davis and Smith, 1975, 1976; Moyer and Smith, 1975; Reinert and Gooding, 1978; Vargo et al., 1978) are available for nonwoody host species. Similar research on woody plant viruses and bacteria, and in addition on nematodes, would provide important perspective to those attempting to evaluate the importance of pathogen–plant–air pollutant interactions.

C. Fluoride

The influence of plant accumulated fluoride on disease development has received only very limited research attention. Some data have been provided concerning the microbial response to elevated fluoride provided in laboratory media. As expected, various fungi respond differently to sodium fluoride incorporated into agar. *Pythium debaryanum* has a lower threshold of inhibition than *Verticillium alboatrum* and *Helminthosporium sativum.* The growth of *Botrytis cinerea* and two *Colletotrichum* species was enhanced by low concentrations of fluoride (Heagle, 1973).

There is some evidence that foliar fluoride may reduce agricultural plant disease. Bean plants exposed to hydrogen fluoride at 7–10 μg m^{-3} and with foliar fluoride concentrations approximating 400 μg g^{-1} fluoride were less severely infected with powdery mildew than control plants (McCune et al., 1973). Treshow (1975) reported similar protection from powdery mildew for chrysanthemum plants that had foliar concentrations between 350 and 400 μg g^{-1} following exposure to hydrogen fluoride at 2 μg m^{-3} for 4 hr per day for several days. Elevated fluoride has further been correlated with reduced bacterial disease, but apparently tobacco mosaic virus symptoms of bean can be reduced or enhanced depending on the fluoride concentration (Manning, 1975).

Unfortunately almost no information is available concerning fluoride and forest tree disease. Barkman et al. (1969) have recorded that *Melampsoridium betulinum,* the causal agent of a birch leaf rust and that is common in Norway, was absent from birches growing near a fluoride source.

D. Particulates

A large number of particulates, including coarse dust, trace metals, and acid precipitation, have been implicated in alterations of plant disease.

1. Coarse Dust

Accumulation of coarse dust particles has been demonstrated to increase foliar disease. Lime dust significantly enhanced *Cercospora beticola* infection of sugar beet leaves following artificial application (Schönbeck, 1960). Manning (1971) has detailed the influence of dust on foliar disease in a forest ecosystem (Jefferson National Forest) surrounding quarries and limestone-processing facilities in a mountain valley in southwestern Virginia. Grape and sassafras foliage with moderate dust deposits generally had more fungal infection than leaves lacking dust deposits. Dusty leaves also had increased numbers, but not kinds, of bacteria and fungi relative to clean leaves. On dusty hemlock leaves, bacterial numbers were greatly reduced, while fungal incidence was increased (Table 14–3). Sassafras and grape were judged to be predisposed to leaf spot disease caused by the fungi *Guignardia bidwellii* and *Gloeosporium* sp. when dusty.

Table 14-3. Influence of Limestone Dust from Quarries and Rock-Processing Plants on the Occurrence of Fungi on the Foliage of Woody Plants in the Jefferson National Forest, Virginia.

	Leaf prints[a]						
	No. isolates per genus expressed as percent of total no. colonies isolated						
	Wild grape			Sassafras		Hemlock	
Genera of fungi isolated	Heavy crust	Moderate dust	No dust	Moderate dust	No dust	Heavy dust	No dust
Alternaria		6.3		2.8			28.0
Candida	20.0		15.6				
Cladosporium	26.6	2.3				26.9	19.4
Colletotrichum				31.8	21.6	3.5	
Cryptococcus		13.2	25.2	6.0	14.9	28.2	22.8
Curvularia							
Fusarium				8.1			
Penicillium		22.0	42.5				
Periconiella						12.8	
Pestalotia		9.0			5.5		12.3
Piricauda		2.5	9.2		12.7		
Rhodotorula	26.6	33.6	4.1		17.7		
Saccharomyces		5.8				17.8	
Black mycelium	6.8	5.3	4.2			6.8	
Brown mycelium	20.0			23.8	24.9		
Gray mycelium			3.3	15.4		6.8	17.5
White mycelium				8.0	2.7		
Total no. isolates	5	10	6	9	7	7	5
Total no colonies	15	534	237	362	179	78	57
	Dilution plates[b]						
Alternaria			4.2	8.1	4.0		
Candida	97.6		12.2			2.3	6.8
Cladosporium	1.2	16.4	18.5	32.5	8.7	11.7	13.6
Colletotrichum			8.4		6.2		
Cryptococcus						10.1	13.7
Curvularia					6.2	3.3	5.1
Fusarium					2.0		
Penicillium		2.7	4.2				
Periconiella				6.4	31.2		
Pestalotia							
Piricauda		4.1	1.7	4.2	6.2	3.3	4.0
Rhodotorula	0.6	56.1		12.1	12.6	69.3	56.8
Saccharomyces		2.7	34.6	28.4	16.6		

Table 14-3 (continued)

	Dilution plates						
	No. isolates per genus expressed as percent of total no. colonies isolated						
	Wild grape			Sassafras		Hemlock	
Genera of fungi isolated	Heavy crust	Moderate dust	No dust	Moderate dust	No dust	Heavy dust	No dust
Black mycelium				4.2			
Brown mycelium	0.6						
Gray mycelium		18.0	16.2	4.1	6.2		
White mycelium							
Total no. isolates	4	6	8	8	9	7	6
Total no. colonies	21.5	365	49	123	48	295	58

Source: Manning (1971).

[a] Total no. colonies isolated (upper and lower surfaces) for leaf prints.

[b] No. colonies g^{-1} leaf tissue in 1000s for dilution plates.

2. Smoke

Smoke consists of aerosol particles resulting from combustion processes. The average diameter of smoke particles is very approximately 0.075 μm. Commonly smoke consists of carbonaceous compounds, particularly hydrocarbons and resins (Spedding 1974).

The microbial toxicity of smoke is not well appreciated. Zagory and Parmeter (1984) investigated the fungitoxicity of smoke by exposing representative phycomycetes, ascomycetes, and basidiomycetes to gaseous smoke and liquid smoke condensates from burning wheat or barley straw. Generally, low exposures were fungistatic rather than fungicidal. Fungi appeared capable of becoming tolerant of smoke. Smoke has been shown to be capable of reducing the inoculum of and/or plant infection caused by several pathogens (Melching et al., 1974; Zagory and Parmeter, 1974).

In a novel experiment, unredospores of *Puccinia graminis tritici* and *P. striiformis* and conidia of *Pyricularia oryzae* and an *Alternaria* sp. failed to germinate on water agar when exposed to 6000 μl of cigarette smoke l[-1] of air in an incubation chamber (Melching et al., 1974). Whatever component of the 1200 identified compounds known to occur in tobacco smoke was responsible for the inhibition was not determined. The authors judged that it was probably not nicotine, carbon monoxide, pyridine, phenol, or hydrogen cyanide acting alone.

Spore germination or mycelial growth of several fungi, including some forest tree pathogens, was recently shown to be reduced on cellophane previously ex-

Figure 14-9. Excessive smoke resulting from the Poverty Flat wildfire, Payette National Forest, Idaho.
Source: U.S.D.A. Forest Service.

posed to smoke from burning pine needles. Prior exposure of Monterey pine seedlings to smoke reduced the amount of gall rust following inoculation (Parameter and Uhrenholdt, 1975). These authors noted that forest burning, particularly wildfires, may result in dense clouds of smoke that drift for many kilometers through forest ecosystems (compare Figure 14–9). Smoke deposits on dead branches, stubs, exposed wounds, and other tree surfaces might reduce the activities of important forest fungi, including pathogens, if smoke deposits are toxic on plant surfaces in nature as well as on cellophane in the laboratory.

In view of the important functions of natural and managed fires in forest ecosystems, the role of smoke in forest pest population dynamics is worthy of additional research.

3. Trace Metals

The potential importance of trace metals associated with particles and microbial metabolism has been previously stressed in this volume (Chapter 9). In view of the considerable capacity of foliage to accumulate trace metal particles (Chapter 6), there is substantial interest to evaluate the interaction of metal cations with those microorganisms of the phyllosphere capable of causing disease or influencing those that do function as pathogens.

Gingell et al. (1976) positioned cabbage plants and 5- to 6-year old Austrian pine saplings 0.6 km (0.4 mile) northeast of a smelter complex in Avonmouth, England and monitored changes in foliar microbes at this site relative to plants located 7 km (4 miles) southeast of the industry. The zinc, lead, and cadmium levels of the cabbages and pine needles were much higher in the test plants located at 0.6 km. Isolations on Martin's rose-bengal streptomycin agar and tryptic-soy agar revealed significantly fewer microbes from the polluted relative to the control site for both pine needles and cabbage. The reduction in the number and diversity of organisms on the contaminated cabbage was primarily due to a significantly lower population of bacteria and pigmented yeasts.

Specific information on the interaction of trace metals with foliar pathogens is not extensive. Leaves artificially contaminated with zinc, lead, and cadmium were shown to be less infected by *Botyritis cinerea* than uncontaminated leaves (Gingell, 1975).

In view of the excessive contamination of urban trees with trace metals in the roadside environment, the author's laboratory has attempted to explore the interactions between pollutant metals and leaf inhabiting fungi. Leaf washing and impression techniques were employed to isolate phylloplane fungi from the leaves of mature, roadside London plane growing in New Haven, Connecticut. Numerous fungi were consistently isolated from various crown positions and at different times during the growing season. Those existing primarily saprophytically included *Aureobasidium pullulans, Chaetomium* sp., *Cladosporium* sp., *Epicoccum* sp., and *Philaphora verrucosa*. Those existing primarily parasitically included *Gnomonia platani, Pestalotiposis* sp., and *Pleurophomella* sp. The following cations were tested in vitro for their ability to influence the growth of these fungi: cadmium, copper, manganese, aluminum, chromium, nickel, iron, lead, sodium, and zinc. The results of this effort indicated variable fungal response with no correlation between saprophytic or parasitic activity and sensitivity to trace metals. Both linear extension and dry weight data indicated that the saprophytic *Chaetomium* sp. was very sensitive to numerous metals. *Aureobasidium pullulans, Epicoccum* sp., and especially *P. verrucosa*, on the other hand, appeared to be much more tolerant. Of the parasites, *Gnominia platani* appeared to be more tolerant than *Pestalotiopsis* sp. and *Pleurophomella* sp. Metals exhibiting the broadest spectrum growth suppression were iron, aluminum, nickel, zinc, manganese, and lead (Smith, 1977).

Because of the important anthracnose disease caused by *Gnomonia platani* on *Platanus* species (Figure 14-5), we have been particularly interested in the tolerance of this organism to trace metals. In vitro linear extension of mycelial growth was significantly inhibited by aluminum, iron, and zinc. These three cations, in addition to cadmium, chronium, manganese, and nickel, also significantly suppressed spore formation (Figure 14-6). Dry weight determinations following growth in liquid culture indicated that mycelial growth was significantly reduced by aluminum, iron, zinc, nickel, and copper. Amendment of shake cul-

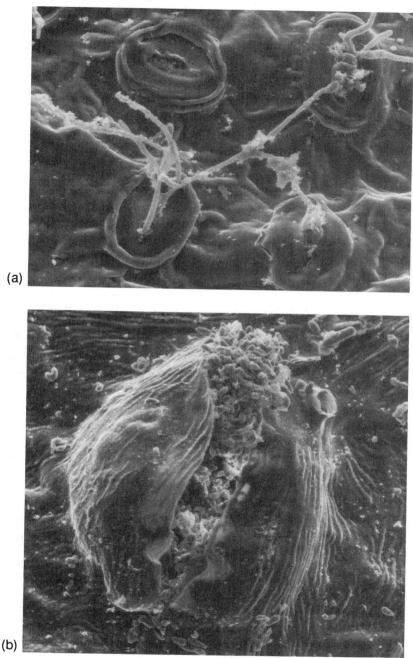

Figure 14-5. Scanning electron micrograph of *Gnomonia platani,* causal agent of anthracnose disease of *Platanus* species, growing on the leaf surface of American sycamore: (a) mycelium, (b) acervulus with conidia. Scale, 10 μm.

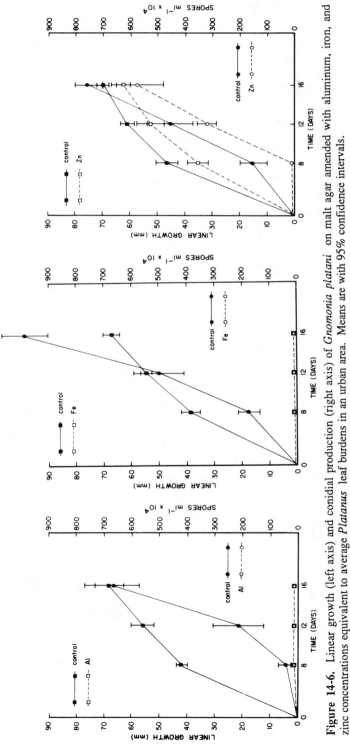

Figure 14-6. Linear growth (left axis) and conidial production (right axis) of *Gnomonia platani* on malt agar amended with aluminum, iron, and zinc concentrations equivalent to average *Platanus* leaf burdens in an urban area. Means are with 95% confidence intervals. Source: Staskawicz and Smith (1977).

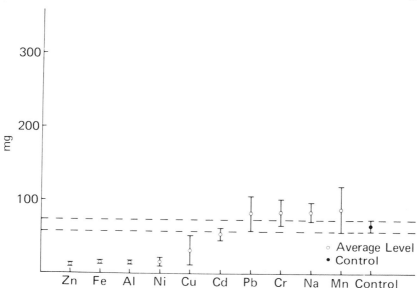

Figure 14-7. Dry weight of *Gnomonia platani* after 7 days growth in malt broth amended with concentrations of trace metals equivalent to average leaf burdens in an urban area. Means with 95% confidence intervals.
Source: Staskawicz and Smith (1977).

tures with lead, chromium, sodium, or manganese did not cause significant growth stimulation or inhibition (Figure 14-7). When condia were placed on a medium containing aluminum, iron, zinc, or nickel, spore germination was significantly suppressed (Figure 14-8). Reduction of mycelial growth, spore formation, and spore germination in nature would lessen the competitive capability of *Gnomonia plantani* and may lessen the ability of this fungus to cause foliar disease in *Platanus* (Staskawicz and Smith, 1977). Unfortunately it is difficult to extrapolate from our observations in vitro to the natural environment. The cation concentrations employed, while approximating measured field burdens, are arbitrary. It is most difficult to judge the actual concentration a specific fungus will encounter on a particular leaf surface. A dose-response test of metals and fungi discussed in this section revealed that only zinc was toxic to *Chaetomium* under very low dose conditions (Smith et al., 1978). Our in vitro efforts have generally employed nitrate salts in order to supply a common anion and completely soluble compounds. In nature, trace metals probably occur on leaf surfaces as less-soluble oxides, halides, sulfates, or phosphates (Koslow et al., 1977). The use of any natural product medium in in vitro efforts will presumably cause alteration of available metal concentrations due to binding by media components (Ko et al., 1976; Romamoorthy and Kushner, 1975). Since the metals were reacted with the fungi individually, it is possible that important antagonistic, additive, or synergistic interactions were overlooked. Because the phyllosphere has a complex microflora (Last and Deighton, 1965) with much interaction between para-

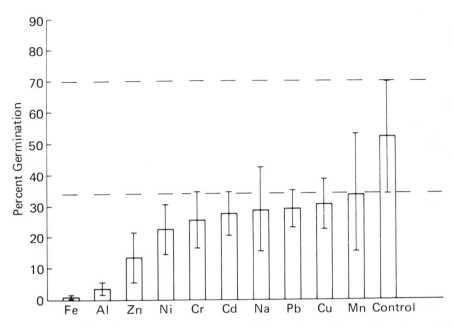

Figure 14-8. Percent germination of *Gnominia platani* conidia transferred to malt agar amended with concentrations of trace metals equivalent to average leaf burdens in an urban area. Means with 95% confidence intervals.
Source: Staskawicz and Smith (1977).

sitic and nonparasitic microbes (Fokkema and Lorbeer, 1974; Last and Warren, 1972), it is possible that the trace element effect on other organisms that influence the fungi examined in our studies may be more significant in nature than the direct toxic effect on our test organisms. In spite of these limitations, we feel that our data support the general suggestion that foliar fungi, including pathogens, respond differentially to foliar metal contamination and in this sense are very similar to the relationships recorded for other air contaminants and pathogenic microorganisms.

Bewley (1981) employed a leaf washing method to compare the numbers and species composition of the phylloplane microflora of Corsican pine on saplings uncontaminated, exposed under field conditions to zinc, lead and cadmium contaminants from a smelter in the UK, and artificially exposed to these heavy metal oxides to simulate emissions in the absence of other pollutants. Despite in vitro differences in tolerance between certain groups of microbes, Bewley's results suggest marked resistance in vivo of most leaf surface microbes to the effects of heavy metals.

4. Acid Deposition

The influence of acid deposition on disease agents and processes is not well appreciated. Only a small number of studies have addressed this topic in agricultural and woody species.

Lacy et al. (1981) examined the response of two epiphytic bacteria, *Erwinia herbicola* and *Pseudomonas syringae*, to exposure to water acidified to pH 2 for 50 min. No bacteria could be recovered on agar following the treatment. When treated at pH 3, bacteria could be recovered but colony forming units were significantly depressed. Shafer et al. (1985) have studied the effects of simulated rain solutions (pH 5.6, 4.0, 3.2, or 2.4) on *Phytophthora cinnamomi* infection of blue lupine seedlings potted in Lakeland sand. Sporangium production was suppressed by simulated rain solutions at acidities similar to that of ambient precipitation for parts of the eastern United States. Release of zoospores into soil extracts prepared with similar solutions, however, was unaffected. The number of zoospore-induced infection sites on lupine roots was suppressed by a single 1 hr exposure to simulated rain at pH 2.4. The authors concluded that although simulated acid rain affects epidemiologically important steps in the life cycle of *P. cinnamomi*, gradual deposition of H ions in rain probably has little short-term effect on root rot disease. The long-term effects of acid deposition could not be assessed. Martin et al. (1987) have examined the interaction of acid deposition with *Phytophthora infestans*, the causal agent of late blight disease of potato. Laboratory and greenhouse studies indicated that highly acidic rain could reduce disease severity. When rain simulants were applied in field studies, however, disease reduction at low precipitation pH was not recorded.

David S. Shriner, Environmental Sciences Division, Oak Ridge National Laboratory, Tennessee, has examined the effects of simulated rain acidified with sulfuric acid on several host-parasite systems under greenhouse and field conditions (Shriner, 1974, 1975, 1977). The simulated rain he employed had a pH of 3.2 or 6.0 (representing extremes of natural precipitation). The application of simulated rain of pH 3.2 resulted in (a) an 86% restriction of tilia production by *Cronortium fusiforme* (fungus) on willow oak, (b) a 66% inhibition of *Meloidogyne hapla* (root-knot nematode) on kidney bean, (c) a 29% decrease in percentage of leaf area of kidney bean affected by *Uromyces phaseoli* (fungus), and (d) either stimulated or inhibited development of halo blight of kidney bean caused by *Pseudomonas phaseolicola* (bacterium). In the latter case, the influence of acid precipitation varied and depended on the particular stage of the disease cycle when the exposure to acid precipitation occurred. Simulated sulfuric acid rain applied to plants prior to inoculation stimulated the halo blight disease by 42%. Suspension of inoculum in acid rain decreased inoculum potential by 100%, while acid rain applied to plants after infection had occurred inhibited disease development by 22% (Table 14-4).

Examination of the willow oak and bean leaves using the scanning electron microscope revealed distinct erosion of the leaf surface by rain of pH 3.2. This may suggest that altered disease incidence may be due to some change in the

Table 14-4. Effects of Simulated Rain Acidified with Sulfuric Acid on Selected Host-Parasite Interactions under Greenhouse and Field Conditions.

Host-pathogen system	Acidity of simulated rain (pH)			Disease measure	
	Preinoculation	Inoculation	Postinoculation	Infected leaves/plant[a]	Telia/infected leaf[a]
Greenhouse studies					
Willow oak	3.2	3.2	3.2	3.8[b]	15[b]
Cronartium fusiforme oak-pine-rust of oak	6.0	6.0	6.0	6.5	115
				Dead leaflets/plant[c,d] (no.)	
Phaseolus vulgaris- *Pseudomonas phaseolicola* halo blight of bean	3.2	3.2	3.2	0.0a[e]	
	3.2	3.2	6.0	0.0a	
	3.2	6.0	3.2	4.2d	
	3.2	6.0	6.0	4.8d	
	6.0	3.2	3.2	0.0a	
	6.0	3.2	6.0	0.0a	
	6.0	6.0	3.2	2.0b	
	6.0	6.0	6.0	3.2c	
			LSD 0.05	0.77	
				Eggs/plant[f]	% root galled[f,g]
Field studies					
Phaseolus vulgaris- *Meloidogyne hapla* root knot nematode on beans	3.2		3.2	74[b]	26
	6.0		6.0	217	50

structure or function of the cuticle. Shriner has also proposed that the low pH rain may have increased the physiological age of exposed leaves. Shriner (1978) concluded his initial experiments by suggesting that he had not established threshold pH levels at which significant biological ramifications to pathogens occur from acid precipitation. He did suggest, however, that artificial precipitation of extremely low pH probably alters infection and disease development of a variety of microbial pathogens.

Bruck et al. (1981) exposed two half-sib families of loblolly pine to rain simulants at four pH levels for 1 hr on each of 2 days before and after inoculation with basidispores of *Cronartium quercuum* f. sp. *fusiforme*. Significantly fewer galls formed on trees treated with rain simulants at pH 4.0 or less than on trees treated with simulants at pH 5.6.

Manion (1983) studied the influence of pH 3.5 rain simulant on *Gremmeniella abietina* infection (scleroderris canker disease) on 3-0 red pine seedlings in New York. Acid rain treatment did not alter the natural infection rates relative to neutral rain or lime treatments. Kvist and Barklund (1984) have reported that acid precipitation favors spore germination of *G. abietina* in Sweden.

Red spruce dieback/decline intensity (Chapter 18) is positively correlated with elevation in the eastern United States and is most prominent in the transitional and montane boreal forests. These high elevation sites are characterized by very acid precipitation events, especially cloudwater deposition, and by highly organic forest floors that can accumulate high levels of potentially toxic metals, including aluminum and a variety of heavy metals (Chapter 9). As a result, we initiated a series of studies to explore the role of *Armillaria* infection in red spruce dieback/decline and to evaluate the tolerance of *Armillaria* to high elevation soil chemistry. Roots of 288 red spruce trees in mixed hardwood, transitional, and montane boreal forests in New England and New York were excavated and examined for colonization by *Armillaria* (Carey et al., 1984). The fungus was found associated with declining and dead spruce in all locations. The percentage of roots colonized by the fungus, however, increased with increasing severity of decline symptoms but decreased with the increasing elevation (Figure 14-10). In high-elevation boreal forests, where the decline has been documented to be most severe, 75% of the recently dead and severely declining trees were not colonized by *Armillaria* (Figure 14-11). In a companion study, the occurrence of rhizomorphs of *Armillaria* in soil around dead red spruce trees was determined (Wargo et al., 1987). Rhizomorph incidence and density were significantly lower in the higher elevation transition and montane boreal forest types (Table 14-5). We concluded, therefore, that infrequent colonization of declining red spruce at high elevations was due to low levels of inoculum of *Armillaria* in these forest soils. High lead concentration and low pH of forest floors in the higher elevation spruce-fir stands in the Northeast were correlated with low levels of inoculum. These factors alone, however, are not thought to fully explain the variation of the fungus. We further concluded that although *Armillaria* is involved in the red spruce

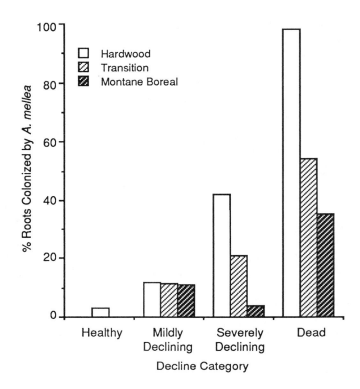

Figure 14-10. Percentage of roots of red spruce colonized by *Armillaria* in relation to decline symptoms in three forest communities. Roots were examined to at least 1 m from the bole.
Source: Carey et al. (1984).

dieback/decline phenomenon, it is not the major cause of the current morbidity and mortality.

E. Summary

Several conclusions are possible concerning the interaction between air pollutants and microbial pathogens of forest trees. The response of individual pathogens and the diseases they cause to atmospheric components is extremely variable, and in a given host-parasite-environment system exposure to a particular pollutant may increase, decrease, or have no apparent influence on a given disease situation. In those instances in which air contaminants do alter disease occurrence or severity, the primary mechanism may be a direct influence on the causal microorganism, an indirect influence on the causal microorganism via a direct influence on an associated microbe, or an indirect influence on the pathogen via an alteration in host physiology or metabolism. Microorganisms that cause foliar disease, or that infect plants through the leaves, may be expected

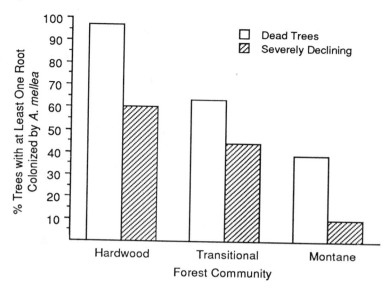

Figure 14-11. Percentage of dead and severely declining red spruce colonized by *Armillaria* in three forest communities.
Source: Carey et al. (1984).

to be especially subject to influence by air pollutants. This is true for at least three reasons: (a) these microbes may grow and develop vegetatively and reproductively in environments with relatively high levels of ambient pollutant concentrations (for example, leaf surface); (b) foliar tissue is known to be the site of primary accumulation of recalcitrant materials from the atmosphere, for example fluoride and heavy metals; and (c) foliar tissue is the primary site of direct damage to the plant occasioned by air pollution, and leaf tissue may be expected to be predisposed to infection if physical or metabolic resistance mechanisms are adversely influenced. Most of the evidence is consistent with these three observations. It is also true, however, that most investigators have concentrated their attention on foliar infecting groups such as rust fungi, powdery mildew fungi, and other bacteria and fungi that cause foliar symptoms. Unfortunately, there is less information available concerning the response to air pollution of microorganisms that infect trees via the root, stem, fruit, seed, or branch.

As in the instance of bark beetle infestation, it might be expected that those pathogens that normally infect woody plants and cause significant disease when the host is under stress would be facilitated in environments of poor air quality. Perhaps the best example of this are the disease situations involving *Armillaria* species. This fungus is frequently most severe under conditions of host predisposition by some insect, microbial, edaphic, or environmental stress. While some evidence has been provided to indicate air quality can act to predispose northeastern United States hardwoods to *Armillaria* infection, no such evidence is available for significant *Armillaria* involvement in the contemporary northeastern red spruce decline.

Table 14-5. Frequency and Density of *Armillaria* Rhizomorphs in Samples of Forest Floor from Red Spruce Stands in Hardwood, Transition, and Montane Boreal Forest Types in Five Northern and Two Southern Sites.

Rhizomorph measurement	Forest type	Northern sites[a]						Southern sites[a]		
		WF	CH	HB	MW	WC	%	GK	MM	%
Frequency[b]	Hardwood	6/6	5/6	5/6	5/6	5/6	87	1/6	1/6	17
	Transition	2/5	1/6	3/6	3/6	1/4	37	0/6	1/6	8
	Montane boreal	2/6	1/6	...	5/6	2/6	42	0/6	0/6	8
							Av.			Av.
Density[c]	Hardwood	6.8	29.4	17.5	12.4	23.3	17.9	9.7	2.7	6.2
	Transition	2.0	0.8	0.6	5.7	0.0	1.8	0.0	0.5	0.25
	Montane boreal	0.0	1.9	...	2.9	2.1	1.7	0.0	0.0	0.0

[a]Northern sites from west to east: WF = Whiteface Mountain, New York; CH = Camel's Hump, Vermont; HB = Hubbard Brook, New Hampshire; MW = Mt. Washington, New Hampshire; and WC = Wildcat Mountain, New Hampshire. Southern sites: GK = Gaudineer Knob, West Virginia, and MM = Mt. Mitchell, North Carolina.
[b]Number of plots out of six in which rhizomorphs occurred (three samples per plot).
[c]Density is centimeters of rhizomorphs X 10^{-3}/cm^3 of forest floor. n = 3 x Number of plots per forest type.
Source: Wargo et al. (1987).

    Unfortunately we are deficient in our appreciation of the interaction of air pollutants and important tree disease groups caused by viruses and mycoplasma, nematodes, dwarf mistletoes, and fungal species responsible for root disease, vascular disease, stem and branch disease, seed disease, and wood decay.

    The pollutants that have received the greatest research effort include sulfur dioxide, ozone, fluoride, dust, and heavy metals. There is substantial circumstantial evidence to indicate that a variety of forest tree diseases appear to be less abundant in areas of grossly elevated ambient sulfur dioxide, for example, within a few kilometers of a major point source such as a smelter. Under these conditions, it is presumed that sulfur compounds may be exerting a directly toxic influence on the microbial pathogen.

    The direct influence of ozone on microbial pathogens may occur at substantially higher ambient concentrations than is the general case with sulfur dioxide. The influence of this gas on disease development, however, may be typically via an alteration in host metabolism, for example, through a change in some resistance mechanism. Accumulation of fluoride to levels approximately 300–400 µg g^{-1} of leaf tissue (dry weight basis) appears, in several cases, to reduce the incidence of microbial disease. Lime dust contamination of leaves, on the other hand, appears to increase the incidence of foliar infection.

Trace metal pollutants have a high potential to interact with tree pathogens on vegetative surfaces. In vitro evidence indicates variable influence of metals on microbes depending on the cation and the specific pathogens. Generally, however, most fungi are probably not directly influenced unless the surface contimination on the plant is extremely high as might occur in the immediate vicinity of a smelter point source. Metal accumulation in forest floors, however, may have long-term impacts on microbial pathogen populations in soil profiles.

Evidence is not available to suggest that acid deposition plays an important role in short-term forest tree pathogen ecology or disease processes. Research on extended-term interactions between acid precipitation, other stresses, and tree pathogens, on the other hand, is fully justified.

References

Babich, H. and G. Stotzky. 1974. Air pollution and microbial ecology. Crit. Rev. Environ. Cont. 4:353-420.

Barkman, J.J., F. Rose, and V. Westhoff. 1969. The effects of air pollution on non-vascular plants. Section 5 discussion. Proc. Eur. Congr. Influence Air Pollu. Plants. Wageningen, The Netherlands, pp. 237-241.

Bewley, R.J.F. 1981. Effects of heavy metal pollution on the micro-flora of pine needles. Holarctic Ecol. 4: 215-220.

Bewley, R. J. F. and D. Parkinson. 1984. Effects of sulphur dioxide pollution on forest soil microorganisms. Can. J. Microbiol. 30: 179-185.

Bewley, R. J. F. and D. Parkinson. 1985. Bacterial and fungal activity in sulphur dioxide polluted soils. Can. J. Microbiol. 31: 13-15.

Bisessar, S. and P.J. Temple. 1977. Reduced ozone injury on virus-infected tobacco in the field. Plant Dis. Reptr. 61: 961-963.

Brennan, E. 1975. On exclusion as the mechanism of ozone resistance in virus-infected plants. Phytopathology 65: 1054-1055.

Bruck, R.I., S.R. Shafer, and A.S. Heagle. 1981. Effects of simulated acid rain on the development of fusiform rust on loblolly pine. Photopathology 71: 864.

Carey, A.C., E.A. Miller, G.T. Geballe, P. M. Wargo, W. H. Smith, and T. G. Siccama. 1984. *Armillaria mellea* and decline of red spruce. Pl. Dis. 68: 794-795.

Chiba, O. and K. Tanaka. 1968. The effect of sulfur dioxide on the development of pine needle blight caused by *Rhizosphaera kalkhoffii*. Bubak. J. Jpn. For. Soc. 50: 135-139.

Coleman, J.S., C.G. Jones, and W. H. Smith 1987. The effect of ozone on cotton-wood-leaf rust interactions: Independence of abiotic stress, genotype, and leaf ontogeny. Can. J. Bot. 65: 949-953.

Coleman, J.S., C.G. Jones, and W. H. Smith. 1988. Interactions between an acute ozone dose, eastern cottonwood, and *Marssonina* leaf spot: Implications for pathogen community dynamics. Can. J. Bot. 66: 963-868.

Costonis, A.C. 1968. Relationship of ozone, *Lophodermium pinastri* and *Pullularia pullulans* to needle blight of eastern white pine. Ph.D. Thesis, Cornell Univ., Ithaca, NY. 176 pp.

Costonis, A.C. and W.A. Sinclair. 1972. Susceptibility of healthy and ozone-injured needles of Pinus strobus to invasion by *Lophodermium pinastri* and *Aureobasidium pullulans*. Eur. J. For. Path. 2: 65-73.

Davis. D.D. and S.H. Smith. 1975. Bean common mosaic virus reduces ozone sensitivity of pinto bean. Environ. Pollu. 9: 97-101.

Davis, D.D. and S.H. Smith. 1976. Reduction of ozone sensitivity of pinto bean by virus-induced local lesions. Plant Dis. Reptr. 60: 31-34.

Decourt, N., C.B. Malphettes, R. Perrin, and D. Caron. 1980. Does sulfur-pollution limit the development of beech bark disease? Annales des Sciences Forestiéres 37: 135-145.

Donaubauer, E. 1968. Skundärschäden in österreichischen rauchschadensbebieten. schwierigkeiten der diagnose und bewertung. In Materialy VI Miedzynarodowej Konferencji, Katowice, Poland, Sept. 9–14, 1968. Polaska Akademia Nauk, pp. 277-284.

Fokhema, N.J. and J.W. Lorbeer. 1974. Interactions between *Alternaria porri* and the saprophytic mycoflora of onion leaves. Phytopathology 64: 1128-1133.

Gingell, S.M. 1975. The effect of heavy metal pollution on the leaf surface microflora. B.Sc. Thesis, Univ. of Bristol, England.

Gingell, S.M., R. Campbell, and M.H. Martin. 1976. The effect of zinc, lead and cadmium pollution on the leaf surface microflora. Environ. Pollu. 11: 25-37.

Grzywacz, A. and J. Wazny. 1973. The impact of inductrial air pollutants on the occurrence of several important pathogenic fungi of forest trees in Poland. Eur. J. For. Path. 3: 129-141.

Ham, D.L. 1971. The biological interactions of sulfur dioxide and *Scirrhia acicola* on loblolly pine. Ph.D. Thesis, Duke University, Durham, NC.

Hartman, F.E. 1924. The industrial application of ozone. J. Am. Soc. Heat. Vent. Engin. 30: 711-727.

Heagle, A.S. 1973. Interactions between air pollutants and plant parasites. Annu. Rev. Phytopathol. 11: 365-388.

Heagle, A. S. 1975. Response of three obligate parasites to ozone. Environ. Pollu. 9: 91-95.

Hibben, C.R. and G. Stotsky. 1969. Effects of ozone on the germination of fungus spores. Can. J. Microbiol. 15: 1187-1196.

Hibben, C.R. and M.P. Taylor. 1975. Ozone and sulphur dioxide effects on the lilac powdery mildew fungus. Environ. Pollu. 9: 107-114.

Hibben, C.R. and J.T. Walker. 1966. A leaf roll-necrosis complex of lilacs in an urban environment. Proc. Am. Soc. Hort. Sci. 89: 636-642.

Howell, R.K. and J. H. Graham. 1977. Interaction of ozone and bacterial leaf-spot on alfalfa. Plant Dis. Reptr. 61: 565-567.

James, R.L. 1977. The effects of photochemical air pollution on the epidemiology of *Fomes annosus* . Unpublished Ph.D. dissertation. University of California, Berkeley, CA, 200 pp.

James, R.L., F.W. Cobb Jr., W.W. Wilcox, and D.L. Rowney. 1980. Effects of photochemical oxidant injury of ponderosa and Jeffrey pines on susceptibility of sapwood and freshly-cut stumps to *Fomes annosus* . Phytopathology 70: 704-708.

James, R.L., F.W. Cobb Jr. and J.R. Parmeter Jr. 1982. Effects of ozone on sporulation, spore germination, and growth of *Fomes annosus*. Phytopathology 72: 1205-1208.

Jancarik, V. 1961. Vyskyt drovokaznych hub v kourem poskozovani oblasti. Krusnych hor. Lesnictvi 7: 667-692.

Killam, K. and M. Wainwright. 1981. Microbial release of sulphur ions from atmospheric pollution deposits. J. Appl. Ecol. 18: 889-896.

Ko, W.H., J.T. Kliejunas, and J.T. Shimooka. 1976. Effect of agar on inhibition of spore germination by chemicals. Phytopathology 66: 363-366.

Koeck, G. 1935. Mildew on oak trees and flue-gas damage. Z. Pflanzenkr. Pflanzensch. 45:1-2.

Koslow, E.E., W.H. Smith, and B.J. Staskawicz. 1977. Lead-containing particles on urban leaf surfaces. Environ. Sci. Technol. 11: 1019-1021.

Kowalski, Von T. 1982. Fungi occurring in forest injured by air pollutants in the Upper Silesia and Cracow industrial regions. VIII. Mycoflora on *Larix decidua* located in a zone with medium-high air pollution damage. Eur. J. For. Path. 12: 262-272.

Kowalski, Von T. 1983. Fungi occurring in forest injured by air pollutants in the Upper Silesia and Cracow industrial regions. IX. Mycoflora on *Quercus robur* L. and *Q. rubra* L. located in a zone with medium-high air pollution damage. Eur. J. For. Pathol. 13: 46-59.

Kudela, M. and E. Novakova. 1962. Lesni skudci a slpdu zvero v lesich poskozovanych Kourem. Lesnictvi 6: 493-502.

Kvist, K. 1986. Fungal pathogens interacting with air pollutants in agricultural crop production. In Proceedings, Effects of Air Pollution on Terrestrial and Aquatic Ecosystems. EC Workshop, Uppsala, Sweden, June 1986, pp. 67-78.

Kvist, K. and P. Barklund. 1984. Air pollution problems in Swedish forests. Forstwissen. Central. 103: 74-82.

Lacy, G.H., B.I. Chevone, and N.P. Cannon. 1981. Effects of simulated acidic precipitation on *Erwinia herbicola* and *Pseudomonas* populations. Phytopathology 71: 888.

Last, F.T. and F.C. Deighton. 1965. The nonparasitic micro-flora on the surfaces of living leaves. Trans. B. Mycol. Soc. 48: 83-99.

Last, F.T. and R.C. Warren. 1972. Non-parasitic microbes colonizing green leaves: Their form and functions. Endeavour 31: 143-150.

Laurence, J.A. 1981. Effects of air pollutants on plant-pathogen interactions. J. Pl Dis and Protect. (Stuttgart) 87: 156-172.

Laurence, J.A. and F.A. Wood. 1987a. Effects of ozone on infection of soybean by *Pseudomonas glycinea*. Phytopathology 68: 689-692.

Laurence, J.A. and F.A. Wood. 1978b. Effect of ozone on infection of wild strawberry by *Xanthomonas fragariae*. Phytopathology 68: 689-692.

Linzon, S.N. 1978. Effects of airborne sulfur pollutants on plants. In J.O. Nriagu, ed., Sulfur in the Environment: Part II. Ecological Impacts. Wiley, New York, pp. 109-162.

Manion, P. 1983. Effects of acid precipitation on scleroderris canker. U.S.D.A. Forest Service, Scleroderris Canker Study Progress Report No. 7, Washington, DC.

Manning, W.J. 1971. Effects of limestone dust on leaf condition, foliar disease incidence, and leaf surface microflora of native plants. Environ. Pollu. 2: 69-76.

Manning, W.J. 1975. Interactions between air pollutants and fungal, bacterial and viral plant pathogens. Environ. Pollu. 9: 87-90.

Martin, S.B., C.L. Campbell, and R.I. Bruck. 1987. Influence of acidity level in simulated rain on disease progress and sporangial germination, infection efficiency, lesion expansion, and sporulation in the potato late blight system. Phytopathology 77: 969-974.

McCune, D.C., L.H. Weinstein, J.F. Mancini, and P. Van Leuken. 1973. Effects of hydrogen fluoride on plant-pathogen interactions. Proc. Intl. Clean Air Congr., Dusseldorf, Germany.

Melching, J.S., J.R. Stanton, and D.L. Koogle. 1974. Deleterious effects of tobacco smoke on germination and infectivity of spores of *Puccinia graminis tritici* and on germination of spores of *Puccinia striiformis*, *Pyricularia oryzae*, and an *Alternaria* species. Phytopathology 64: 1143-1147.

Miller, P. R. 1977. Photochemical Oxidant Air Pollutant Effects on a Mixed Conifer Ecosystem. A Progress Report. U.S. Environmental Protection Agency, Publica. Report No. EPA-600/3-77-104, Corvallis, OR, 338 pp.

Moyer, J.W. and S.H. Smith, 1975. Oxidant injury reduction on tobacco induced by tobacco etch virus infection. Environ. Pollu. 9: 103-106.

National Academy of Sciences. 1978. Sulfur Oxides. NAS, Washington, DC., pp. 80-129.

Parmeter, J. R. and B. Uhrenholdt. 1975. Some effects on pine-needle or grass smoke on fungi. Phytopathology 65: 28-31.

Pell, E.J., F.J. Lukezic, R.G. Levine, and W.C. Weissberger. 1977. Response of soybean foliage to reciprocal challenges by ozone and a hypersensitive response-inducing Pseudomonad. Phytopathology 67:1342-1345.

Przylbyiski, Z. 1967. Results of observations of the effect of SO_2 hr, SO_3, and H_2SO_4 on fruit trees, some harmful insects near the sulfur mine and sulfur processing plant at Machow near Tarnobrzeg. Postepy Nauk Roin 2:111-118.

Reinert, R.A. and G.V. Gooding Jr. 1978. Effect of ozone and tobacco streak virus alone and in combination on Nicotiana tabacum. Phytopathology 68: 15-17.

Rist, D.L. and J.W. Lorbeer. 1984. Moderate dosages of ozone enhance infection of onion leaves by Botrytis cinerea but not by B. squamosa. Phytopathology 74: 761-767.

Romamoorthy, S. and D.J. Kushner. 1975. Binding of mercuric and other heavy metal ions by microbial growth media. Microb. Ecol. 2: 162-176.

Ross, E.W. 1973. Fomes annosus in the Southeastern United States: Relation of Environmental and Biotic Factors to Stump Colonization and Losses in the Residual Stand. Forest Service, Tech. Bull. No. 1459, U.S. Dept of Agriculture, Washington, DC., 26 pp.

Saunders, P.J.W. 1971. Modification of the leaf surface and its environment by pollution. In   T.F. Preece and C.H. Dickinson, eds., Ecology of Leaf Surface Microorganisms. Academic Press, New York, pp. 81-89.

Saunders, P.J.W. 1973. Effects of atmospheric pollution on leaf surface micro-flora. Pestic. Sci. 4: 589-595.

Saunders, P.J.W. 1975. Air pollutants, microorganisms and interaction phynomena. Environ. Pollu. 9: 85.

Scheffer, T.C. and G.G. Hedgecock. 1955. Injury to Northwestern Forest Trees by Sulfur Dioxide from Smelters. U.S.D.A. Forest Service. Tech. Bull. No. 1117, Washington, DC., 49 pp.

Schönbeck, H. 1960. Beobachtungen zur frage des einflusses von industriellen immossionen auf die krankheitsbereitschaft der pflanze. Ber. Landesanst. Bodennutzungsschutz 1:89-98.

Shafer, S.R., R.I. Bruck, and A.S. Heagle. 1985. Influence of simulated acidic rain on Photophthora   cinnamomi and Phytophthora root rot of blue lupine. Phytopathology 75:996-1003.

Shaw, C.G. III. 1981. Infection of western hemlock and Sitka spruce thinning stumps by Fomes annosus and Armillaria mellea in Southeast Alaska. Pl. Dis. 65: 967-971.

Shriner, D.S. 1974. Effects of simulated rain acidified with sulfuric acid on host-parasite interactions. Ph.D. Thesis, North Carolina State Univ., Raleigh, NC, 79 pp.

Shriner, D.S. 1975. Effects of simulated rain acidified with sulfuric acid on host-parasites interactions. In   L.S. Dochinger and T.A. Seliga, eds., The 1st Intl. Symp. on Acid Precipitation and the Forest Ecosystem. U.S.D.A. For. Serv. Genl. Tech. Rep. No. NE-23, Upper Darby, PA, p. 919-925.

Shriner, D.S. 1977. Effects of simulated rain acidified with sulfuric acid on host-parasite interactions. Water Air Soil Pollu. 8: 9-14.

Shriner, D.S. 1978. Effects of simulated acidic rain on host-parasite interactions in plant diseases. Phytopathology 68: 213-218.

Sinclair, W.A. 1969. Polluted air: Potent new selective force in forests. J. For. 67: 305-309.

Skye, E. 1968. Lichens and air pollution. Acta Phytogeogr. Suec. 52: 1-23.

Smith, W.H. 1970. Tree Pathology — A Short Introduction. Academic Press, New York, 309 pp.

Smith, W.H. 1976. Air pollution — effects on the structure and function of plant-surface microbial-ecosystems. In C.H. Dickinson and T.F. Preece, eds., Microbiology of Aerial Plant Surfaces. Academic Press, New York, pp. 75-105.

Smith, W.H. 1977. Influence of heavy metal leaf contaminants on the *in vitro* growth of urban tree phylloplane-fungi. Microb. Ecol. 3:231-239.

Smith, W.H., B.J. Staskawicz, and R.S. Harkov. 1978. Trace-metal pollutants and urban-tree leaf pathogens. Trans. B. Mycol. Soc. 70: 29-33.

Spedding, D.J. 1974. Air Pollution. Oxford Chemistry Series. Oxford University Press, London, 76 pp.

Staskawicz, B.J. and W.H. Smith. 1977. Trace-metal leaf-pollutants suppress *in vitro* development of *Gnomonia platani*. Eur. J. For. Pathol. 7: 51-58.

Treshow, M. 1975. Interaction of air pollutants and plant diseases. In J. B. Mudd and T. T. Kozlowski, eds., Responses of Plants to Air Pollution. Academic Press, New York, pp. 307-334.

Unsworth, M.H., P.V. Biscal, and H.R. Pinckey. 1972. Stomatal response to sulfur dioxide, Nature 239: 458-459.

Vargo, R.H., E.J. Pell, and S.H. Smith. 1978. Induced resistance to ozone injury of soybean by tobacco ringspot virus. Phytopathology 68: 715-719.

Wargo, P.M and C G. Shaw III. 1985. *Armillaria* root rot: The puzzle is being solved. Pl. Dis. 69: 826-832.

Wargo, P.M., A.C. Carey, G.T. Geballe, and W.H. Smith. 1987. Occurrence of rhizomorphs of *Armillaria* in soils from declining red spruce stands in three forest types. Pl. Dis. 71: 163-167.

Weidensaul, T.C. and S.L. Darling. 1979. Effects of ozone and sulfur dioxide on the host-pathogen relationship of Scotch pine and Scirrhia acicola. Phytopathology 69: 939-941.

Weinstein, L.H., D.C. McCune, A.L. Aluisio, and P. Van Leuken. 1975. The effect of sulfur dioxide on the incidence and severity of bean rust and early blight of tomato. Environ. Pollu. 9: 145-155.

Williams, R.J.H., M.M. Lloyd, and G.R. Ricks. 1971. Effects of atmospheric pollution on deciduous woodland I. Some effects on leaves of *Quercus petraea* (Mattuscha) Leibl. Environ. Pollu. 2: 57-68.

Zagory, D. and J.R. Parmeter Jr. 1984. Fungitoxicity of smoke. Phytopathology 74: 1027-1031.

15

Forest Abiotic Agent Stress: Symptomatic Foliar Damage Directly Caused by Air Contaminants

Under conditions of sufficient dose, air pollutants directly cause visible injury to forest trees. The accumulation of particulate contaminants on leaf surfaces or the continued uptake of gaseous pollutants through leaf stomata will eventually result in cell and tissue damage that will be manifest in foliar symptoms obvious to the trained, but unaided eye. This direct induction of disease in trees by air pollutants is the most dramatic and obvious individual tree response of all Class II interactions. It is the only Class II interaction that can be detected in the field by casual observation. Unlike altered reproductive strategy, nutrient cycling, tree metabolism, or insect and disease relationships; the degree of foliar symptoms induced by air pollutants can be relatively easily observed, inventoried, and quantified. In the presence of a sufficient dose, tree damage may be of sufficient severity to cause mortality. Tree death directly induced by ambient air pollution exposure is considered a Class III interaction and is treated in Chapter 17.

This chapter will consider each of the 10 most important air contaminant groups capable of causing direct tree morbidity. For each pollutant summary, a perspective will be presented including (a) foliar symptoms, (b) physiological-biochemical mechanism(s) of toxicity, (c) threshold dose, and (d) relative woody plant susceptibility. There is, of course, an enormous literature on the above topics for a large variety of plants. It is well beyond the scope of this book to provide a comprehensive review of these areas. The author has found the following publications especially useful for general information and perspective: Centre for Agricultural Publishing and Documentation (1969), Darley and Middletown (1968), Guderian (1977), Heggestad (1968), Hindawi (1970, Jacobson and Hill (1970), Lacasse and Moroz (1969), Legge and Krupa (1986) Linthurst (1984), Mudd and Kozlowski (1975), Naegele (1973), Stern et al.

(1973), Treshow (1984a), U.S. Environmental Protection Agency (1976), VDI Commission for Air Pollution Prevention (1983), and Winner et al. (1985).

A. Limitations on Generalizations Concerning Direct Air Pollutant Influence on Trees

Before presenting the discussion of the influence of individual pollutants on forest species, it is enormously important to realize the limitations associated with these general comments. Readers are cautioned that published information concerning symptoms, toxicology, threshold doses, and susceptibility is not absolute, but rather is quite arbitrary and variable. Substantial variations in the parameters discussed in this chapter occur for a variety of reasons. The most important reasons include variation in inherent characteristics and age of trees, along with variation caused by environmental conditions, all of which have been the subject of review (Davis and Wood, 1973a, b; Heck, 1968; Heggestad and Heck, 1971; Legge and Krupa, 1986;Treshow, 1984).

1. Plant Factors

Evidence has been provided, some with woody plant species, that vegetative response to air contaminants is genetically controlled (Townsend, 1975). Genetic considerations must be evaluated, therefore, when attempting to generalize about varietal or species reaction to a particular air pollutant. David F. Karnosky, Cary Arboretum, Millbrook, New York, has provided several reviews of this general topic (Karnosky, 1974, 1978a,b, 1985). Intraspecific variation in the response to a variety of pollutants was recorded for 11 forest trees by Karnosky in 1974. Variable responses of 32 cultivars of eight forest tree species, widely employed in urban plantings, have been recorded for exposure to sulfur dioxide and ozone (Karnosky, 1978a) (Table 15–1). Karnosky et al. (1986) collected deciduous tree material from 10 national parks. The trees were clonally propagated and exposed to ozone. Fumigations were conducted on 79 quaking aspen, 77 red maple, 39 black cherry, 29 flowering dogwood, and 15 paper birch clones. Significant variation in ozone tolerance was found among clones in almost all populations examined. Henry D. Gerhold, Forest Resouces Laboratory, Pennsylvania State University, University Park, Pennsylvania, has performed pioneering research on the genetic control of *Pinus* species response to air pollution in the United States. In his 1975 review, Gerhold tabulated the variable response of 20 northeastern forest tree species to sulfur dioxide, nitrogen oxides, ozone, and fluoride. In a comprehensive review of North American and European literature, Gerhold (1977) cited references to intraspecific variation in eight conifers and seven deciduous species or hybrid groups within which significant genetic variation in response to one or more air contaminants has been found (Table 15–2). Additional evidence for genetically controlled variable response has been provided for pines and sulfur dioxide and ozone (Genys and Heggestad, 1978), trembling aspen and sulfur dioxide and ozone (Karnosky, 1977), ash and

Table 15-1. Response of 32 Forest Tree Cultivars, Widely Employed in United States Urban Plantings, to Ozone and Sulfur Dioxide, Alone and in Combination, as Determined by Chamber Tests and to Oxidants (Primarily Ozone) as Determined by Field Tests.

| | | Sensitivity[a] to: | | | |
| | | | Sulfur | Ozone plus | |
Species	Cultivar	Ozone	dioxide	sulfur dioxide	Oxidants
Norway maple	Cleveland	R	I	R	R
	Crimson King	R	R	R	–
	Crimson Sentry	R	I	R	R
	Columnar	I	R	S	R
	Emerald Queen	I	R	I	R
	Green Mountain	R	R	I	–
	Jade Glen	I	I	I	R
	Schwedler	I	I	I	R
	Summershade	R	R	R	–
Red maple	Autumn Flame	R	I	R	R
	Bowhall	I	R	S	–
	Red Sunset	R	R	R	R
	Tilford	I	S	I	–
Sugar maple	Goldspire	I	R	S	–
	Temple's Upright	R	R	R	–
European beech	Rotundifolia	R	R	R	–
White ash	Autumn Purple	S	S	S	I
European ash	Hessei	I	S	I	R
Green ash	Marshall's Seedless	I	S	S	I
	Summit	I	I	I	R
Ginko	Autumn Gold	R	I	I	–
	Fairmont	R	I	I	–
	Fastigiate	R	I	R	–
	Sentry	R	R	R	–
Honey locust	Emerald Lace	S	I	I	R
	Imperial	S	R	R	S
	Majestic	S	S	R	I
	Shademaster	S	R	I	I
	Skyline	S	S	R	R
	Sunburst	S	S	I	R
London plane	Bloodgood	S	S	S	S
English oak	Fastigiate	S	R	I	–

Source: Karnosky (1978a).
[a] S = sensitive, I = intermediate, R = tolerant.

sulfur dioxide and ozone (Steiner and Davis, 1979, Karnosky and Steiner, 1981), and loblolly pine and ozone (Kress et al., 1982).

Age is another plant factor that must be considered when reviewing tree response to air contaminants. Evidence from pine species indicates foliar age is

Table 15-2. Forest Tree Species That Have Exhibited Genetically Variable Response to One or More Air Pollutants.

Species	Differences among	Pollutant
Acacia	Populations	CO
Red maple	Populations	O_3
Sugar maple	Seedlings	O_3
European and Japanese larch	Families	SO_2
Norway spruce	Families	SO_2, HF
	Clones	
	Populations	
	Families	
Lodgepole pine	Populations	SO_2
	Populations	SO_2
	Populations	SO_2
Ponderosa pine	Populations	O_x, HF
	Populations	SO_2, O_3
White pine	Clones	O_3, SO_2
	Clones	HF, O_3, SO_2
Scotch pine	Clones	HF, SO_2
	Populations	SO_2
	Clones, families	SO_2
Platanus	Families	O_3, SO_2
Poplar	Clones	SO_2
	Clones	SO_2
	Clones	O_3
Trembling aspen	Clones	O_3, SO_2
Douglas fir	Populations	SO_2
American elm	Families	O_3, SO_2
	Clones	O_3, SO_2

Source: Gerhold (1977).

important in tree reaction to sulfur dioxide (Berry, 1974; Smith and Davis, 1977) and ozone (Davis and Coppolino, 1974; Davis and Wood, 1973b). A unique age associated factor is encountered in air pollution studies involving trees. Laboratory, greenhouse, growth chamber, and field studies are most conveniently performed with seedling-or sapling-size plants. As a result, most acute response data has been accumulated from studies using relatively young trees. The response of mature and overmature trees is invariably extrapolated from these studies. In awareness of differences in physiological processes of mature versus juvenile individuals, it should be recognized that risks are associated with this extrapolation.

Tree health is a final factor that may influence the response of a tree to air pollution stress. Chapters 13 and 14 have clearly shown that insect and microbial stress may interact with air contaminants in a complex way. Reduced foliar gas exchange occasioned by microbial infection or arthropod feeding may reduce gaseous air pollutant uptake and may reduce acute air pollution injury.

Accumulation of persistent air contaminants, for example, heavy metals and fluoride, on the other hand, may exacerbate impaired tree health that is already abnormal by virtue of insect and disease impact.

2. Environmental Factors

Woody plant studies that have examined acute plant responses to an air pollutant are typically performed under controlled environment conditions. Most data have been produced from experiments conducted in growth chambers or greenhouses. A long list of environmental variables strongly mediate the response of a plant to air pollution exposure (Davis and Wood, 1973a; Davis, 1975a; Tingey and Taylor 1982; Kozlowski, 1980). This realization has necessitated the use of controlled facilities in order to manage confusing and complexing environmental responses. Controlled facilities themselves introduce well-appreciated artificialities. Closed chambers, for example, typically have higher temperature and lower light than ambient conditions. Developments in open-top chamber design partially address this problem (Heagle et al., 1979). Environmental variables of primary significance in plant responses to pollutant stress include temperature (Davis, 1975a), light intensity and quality (Dunning and Heck, 1973), humidity (Leone and Brennan, 1969; Otto and Daines, 1969; Wilhour, 1970), and wind velocity (Brennan and Leone, 1968; Heagle et al., 1971).

In addition to the large number of above-ground environmental variables that influence plant responses to air pollution, considerable evidence suggests that numerous below ground or edaphic variables are also important (Smith, 1975). The greatest amount of information relates to the influence of soil moisture and soil nutrient content on plant responses. The general observation has been that plants are generally more resistant to damage from gaseous air contaminants when they are under moisture stress. Presumably the drought-induced reduced stomatal aperture permits less uptake of pollutants from the ambient atmosphere and therefore occasions less plant injury (cf., Fuhrer and Erismann, 1980). Stomatal conductance and sulfur dioxide uptake in river birch seedlings were reduced 40% and 45% by flooding, respectively, and sulfur dioxide exposure caused less visible injury and less growth inhibition in flooded than in unflooded seedlings (Norby and Kozlowski, 1983).

The influence of soil nutrient status on plant susceptibility to air contaminants, especially oxidants, has also been investigated. The nonwoody plant data are variable however, and generalizations are difficult (Heagle, 1979; Jager and Klein, 1976; Leone, 1976). Some research consideration has been given to the response of trees to soil nutrients and air pollution damage. Bjorkman (1970) concluded that optimal nutrient concentrations reduced sulfur dioxide damage to Scotch pine. Applications of lime and fertilizer, however, failed to protect eastern white pine from air pollution damage (Dochinger, 1964; Dochinger and Seliskar, 1970). Following fertilization of field-grown white pine, Berry and Hepting (1964) did not detect any change in tree susceptibility to air pollution injury. Applications of N,P,K fertilizer to air pollution sensitive clones of eastern white pine, however, did increase their tolerance to air pollution stress (Cotrufo and Berry, 1970). Cotrufo (1974) has treated clonal ramets of eastern white pine with factorial combinations of N, P, and K and exposed the trees to ambient sulfur dioxide pollution. The addition of N at all levels of P and K significantly increased sulfur dioxide susceptibility. The addition of P at all levels of N and K reduced susceptibility, while the ramets did not respond significantly to the addition of K.

In addition to soil moisture and nutrient content, other soil variables may influence plant responses to air pollution. Composition of the soil atmosphere, for example, oxygen or carbon dioxide concentrations, may be important. Soil temperature is another variable that may have some significance in controlling plant responses.

3. Pollutant Interactions

In natural environments, trees may be exposed to more than one air pollutant concurrently or sequentially within a relatively short period of time. A large number of investigations cited in this volume and elsewhere (Miller and Davis, 1981; Ormrod et al., 1984; Reinert, 1975, 1984) have indicated that contaminants may interact to produce a unique plant response. In general, three kinds of interactions have been recognized. When the influence of exposure to one pollutant is merely added onto the influence of another or several other pollutants, the interaction is termed *additive*. In this case an individual pollutant does not either increase nor decrease the influence of another contaminant. When exposure to two or more air pollutants simultaneously produces a plant response more severe than any induced by individual contaminant stress, then the interaction is termed synergistic. If two or more pollutants interact to reduce the impact of a given pollutant, then the interaction is termed *antagonistic*.

Investigations specifically designed to explore these interactions have been conducted and have demonstrated all of these responses in trees: additive (Matsushima and Brewer, 1972; Tingey and Reinert, 1975); synergistic (Dochinger et al., 1970; Krause and Kaiser, 1977); and antagonistic (Davis, 1977; Nielson et al., 1977). The size and longevity of forest tree species makes considerations of potential pollutant interactions especially important.

4. Dose-Response Considerations

Dose measurement of air pollutant exposure involves a measured concentration of a toxicant for a known duration of time (Weinstein, 1975). A large volume of literature reports only single dose exposures. It is frequently most difficult to predict tree response over a range of contaminant concentrations for varying time periods. Cultural treatments performed preceding or following pollutant exposure can modify dose-response data. The large variability in dose administration itself frequently makes comparison between laboratories and experiments difficult (cf., Chapter 2, section C2).

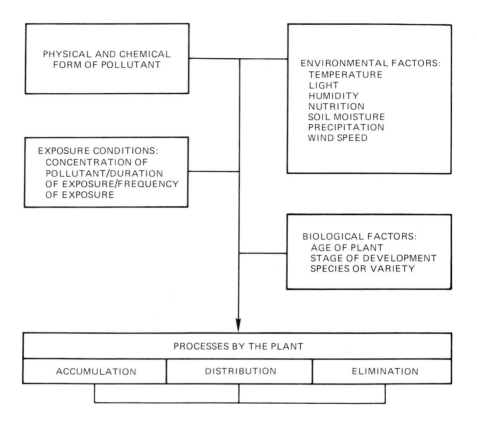

Figure 15-1. Factors affecting the response of plants to air pollutants .
Source: Weinstein and McCune (1979).

Ideally investigations should include doses that will induce effects that range from none to extreme and data that can be employed to estimate a median effective dose (compare ED_{50} or effective dose at which 50% of the test plants are influenced). The dose associated with the lowest probable threshold for the occurrence of an effect is also needed (Weinstein, 1975).

5. Summary

A very impressive set of variables influence the response of plants to air pollutant exposure (Figure 15–1). Because of the size and age of trees, research artificialities introduced by experimental design are more severe for forest vegetation than other plant types. As a result, severe limitations are placed on our ability to extrapolate much of the wealth of data available in the literature to natural forest environments. It is further important to recognize that the symptoms of acute foliar response to air pollutants are not unlike the symptoms induced in tree foliage by a host of other abiotic and biotic stress factors (Treshow, 1984b). The diagnosis of specific tree response to air pollution stress will involve the same care, knowledge, and perspective that all woody plant diagnoses involve, that is, a good appreciation of all stress factors and the symptoms they cause for a given plant in a given location. It will further be necessary to have some appreciation of the historical stresses that have impacted the tree in order to make an accurate diagnosis. Excellent manuals for diagnosis of damage caused by air pollution, other abiotic stresses, insects, and pathogens of forest trees in North America have been provided (Malhotra and Blauel, 1980; Skelly et al., 1987; Sinclair et al., 1987; Johnson and Lyon, 1988).

B.   Sulfur Dioxide

Due to the long history of awareness of the adverse impact of sulfur dioxide on vegetative health and the extensive global distribution of point sources of sulfur dioxide, an enormous literature has accumulated on the acute influence of this gas on plants. A variety of excellent reviews is available, including those dealing with general vegetation (Linzon, 1978; National Academy of Sciences, 1978a; Tamm and Aronsson, 1972), agricultural crops (Heitschmidt and Altman, 1978), and forest trees (Forestry Commission, 1971; Davis and Wilhour, 1976).

1.   Foliar Symptoms

In angiosperms, the formation of marginal and interveinal necrotic areas, which are dark green or dull in color and have a water-soaked appearance are the initial symptoms. Eventually these necrotic zones dry and bleach to a light (ivory-white-hyaline) color. Necrotic spots result as mesophyll cells collapse in those areas near stomata. Lamina portions immediately bordering veins rarely become necrotic, as stomata are few and intercellular space is limited in these areas. Necrotic zones extend through the leaf and are visible on adaxial (upper) as well as abaxial (lower) surfaces. Leaves are most susceptible to damage when they have just reached their fully expanded size. Older leaves are less sensitive and very young expanding leaves are the most resistant. Gymnosperm needles develop a water-soaked appearance and typically turn reddish-brown in color. As in the case of angiosperms, recently fully expanded needles are the most sensitive (U.S. Environmental Protection Agency, 1976).

Excellent color photographs of sulfur dioxide symptoms are available in Jacobson and Hill (1970), Loomis and Padgett (1973), Hindawi (1970), U.S.D.A. Forest Service (1973), U. S. Environmental Protection Agency (1976), and Skelly et al. (1987).

2.   Mechanism of Toxicity

Excellent reviews are available concerning the phytotoxicity of sulfur dioxide (Alscher et al., 1987; Rothermel and Alscher, 1985). Toxicity results when the

Table 15-3. Toxic Species Formed in Chloroplasts as a Result of Oxidative Stress and Potential Metabolic Toxicity.

Species	Toxicity
Superoxide (O_2^-)	Formation of reactive species
Singlet oxygen ($O_2[^1\Delta_g]$)	Lipid peroxidation
Hydroxyl radical (OH•)	Macromolecule reactions
Hydrogen peroxide (H_2O_2)	SH enzyme inactivation
Glutathione (oxidized form GSSG)	Enzyme inactivation
Peroxy radical (O_2^-)	Memberane dysfunction

Source: Alsher et al. (1988).

rate of sulfur dioxide uptake exceeds the plant's capacity to detoxify sulfite. Following absorption of sulfur dioxide through stomatal pores, the gas is dissolved in free space moisture in leaf interiors. In solution, sulfur dioxide exists as bisulfite and sulfite. The latter species is presumed to dominate at the site of sulfite transformation in chloroplasts. Metabolic transformations of sulfite include both oxidation to sulfate and eventual incorporation into the sulfate assimilation pathway, and reduction to hydrogen sulfide and eventual release as a gas to the atmosphere (Rothermel and Alscher, 1985). Since photoreduction of free sulfite involves sulfite reductase and utilizes reduced ferredoxin from photosynthetic electron transport, carbon dioxide fixation and photosynthesis may be inhibited by this reductant removal (Rothermel and Alscher 1985). Any disruption of chloroplast metabolism involving inhibition of carbon fixation or electron transport can increase oxidative stress and lead to the formation of numerous toxic species. These toxins can adversely impact metabolism in a variety of ways, including enzyme and membrane dysfunction (Alsher et al. 1988, Table 15–3). Sulfite also directly adds to ketones and aldehydes to form a-hydroxysulfonates. It reacts with pyrimidine bases in DNA and RNA. Sulfite may interfere with a variety of enzymes by competing with phosphate and carbonate groups at binding sites, by reaction with disulfide bonds and altering structure, or by forming inhibiting compounds (Wallace and Spedding, 1976). Interference at the physiological level, resulting from biochemical alterations induced by sulfite, may involve abnormalities in stomatal opening, amino acid metabolism, photosynthesis, chlorophyll levels, ATP levels, and enzyme levels (Mudd, 1973, 1975; Jäger et al., 1985; Keller 1984, Keller and Häsler, 1986; Harvey and Legge, 1979).

3.   Threshold Dose

The National Academy of Sciences (1978) has provided a comprehensive review of sulfur dioxide concentrations that cause acute foliar damage to plants under experimental laboratory and field conditions as well as following ambient exposures. The variation in threshold doses is extensive but not surprising in view of the experimental limitations previously discussed. General ranges of sulfur

Table 15-4. Sensitivity Groupings of Trees Based on Visible Injury at Different SO_2 Exposures[a].

Sensitivity grouping	SO_2 concentration, $\mu g/m^3$ (ppm), and duration time, hr			
	Peak[b]	1 hr	3 hr	Plants
Sensitive	2620–3930 $\mu g/m^3$ (1.0–1.5 ppm)	1310–2620 $\mu g/m^3$ (0.5–1.0 ppm)	790–1570 $\mu g/m^3$ (0.3–0.6 ppm)	Southern pines Red and black oaks White ash Sumacs
Intermediate	3930–5240 $\mu g/m^3$ (1.5–2.0 ppm)	2620–5240 $\mu g/m^3$ (1.0–2.0 ppm)	1570–2100 $\mu g/m^3$ (0.6–0.8 ppm)	Maples Locust Sweetgum Cherry Elms Tulip tree
Resistant	> 5240 $\mu g/m^3$ (> 2.0 ppm)	> 5240 $\mu g/m^3$ (> 2.0 ppm)	> 2100 $\mu g/m^3$ (> 0.8 ppm)	White Oaks Dogwood Peach

[a]Based on observations over a 20-year period of visible injury occurring on over 120 species growing in the vicinities of coal-fired power plants in the southeastern United States.
[b]Maximum 5 min. concentration.
Source: after U.S.E.P.A.(1987), Jones et al. (1974).

Table 15-5. Conifer Effects From Long-Term Sulfur Dioxide Exposures.

SO_2 annual mean ($\mu g\ m^{-3}$)	Species	Effects
25–50	Picea abies	needle loss
50–100	Picea abies	moderate damage
100–180	Picea abies	severe damage
< 100	Pinus sylvestris	no impact
> 100	Pinus sylvestris	slight damage, needle loss

Source: International Electric Research Exchange (1981).

dioxide concentrations injuring forest trees at peak, 1-hr, 3-hr, and annual mean exposures are presented in Tables 15–4 and 15–5. The thresholds of visible injury are likely to be reduced by simultaneous exposure to other phytotoxic gases such as ozone, nitrogen oxides, or hydrogen fluoride.

4. Relative Susceptibility

A large number of forest tree rankings of relative susceptibility to acute sulfur dioxide injury are available. Specialty lists for Australia (O'Connor et al., 1974) and Europe (Tamm and Aronsson, 1972) have been provided. The most comprehensive and useful list for North American woody plants has been provided by Davis and Wilhour (1976) and is reproduced in Table 15–6.

C. Nitrogen Oxides

The most comprehensive and recent reviews of the influence of oxides of nitrogen on plants have been provided by Taylor et al. (1975) and the National Academy of Sciences (1977b). Nitric oxide is less phytotoxic than nitrogen dioxide, and both are less damaging to plants than sulfur dioxide, gaseous fluoride, or oxidants.

1. Foliar Symptoms

Initial symptoms involve the development of diffuse discolored spots of gray-green or light brown color. These spots eventually weather and dry, and become bleached, as in the instance of sulfur dioxide injury. In angiosperms, the necrotic discolored spots that form in interveinal portions of the leaf frequently combine to form stripes. Marginal necrosis may appear, especially in locust, oak, and maple. High dose exposure may cause intense net necrosis and leave only "fingers of green" along the veins. Gymnosperm symptoms typically initially involve red-brown or fuchsia discoloration in the distal portion of the needle. The discoloration may eventually extend to the needle base. A distinct boundary between necrotic and healthy needle tissue may occur in pine and fir. A general chlorosis of older leaves may occur after long-term exposure to low concentrations (National Academy of Sciences, 1977b). These symptoms are generally not distinctive, are infrequent, and field diagnosis based on symptoms alone would be extremely difficult.

2. Mechanism of Toxicity

Following absorption of nitrogen dioxide through the stomata, the gas will react with water on the moist surfaces of mesophyll cells and form nitrous (HNO_2) and nitric (HNO_3) acids. Toxicity may partially result from a pH decrease (Zeevaart, 1976). Deamination reactions may be induced in amino acids and nucleic acid bases. The acids may react with unsaturated compounds and cause isomerization and free radical formation. Excessive nitrate anions may react with amines to form nitrosamines. At the physiological level cellular pH is altered and pollutant interactions with cellular components lead to altered metabolism, for example, carbon dioxide fixation may be inhibited (nitrite is much more toxic than nitrate in this regard), acetate metabolism may be inhibited, and the final result is that growth may be suppressed (Mudd, 1973; Taylor et al., 1975; Zeevaart, 1976).

Table 15-6. Relative Sensitivity of North American Woody Plants to Acute Damage by Sulfur Dioxide.

Sensitive	Intermediate	Tolerant
	Angiosperms	
Thinleaf alder	Mountain alder	Buck-brush
Large-toothed aspen	Chinese apricot	Buffalo-berry
Trembling aspen	Basswood	Redstem ceanothus
Green ash	Water birch	Forsythia
European birch	Box elder	Black hawthorn
Gray birch	Bitter cherry	Kinnikinnick
Western paper birch	Chokecherry	Littleleaf linden
White birch	Black cottonwood	Cure-leaf mountain mahogany
Yellow birch	Eastern cottonwood	Norway maple
Lowbush blueberry	Narrowleaf cottonwood	Silver maple
Bitter cherry	Sticky currant	Sugar maple
Chinese elm	Red osier dogwood	Gambel oak
Beaked hazel	Blueberry elder	Pin oak
California hazel	American elm	Northern red oak
Manitoba maple	Wild grape	Oregon grape
Rocky Mountain maple	Red hawthorne	London plane
Lewis mock-orange	Tatarian honeysuckle	Western poison ivy
Sitka mountain-ash	Witch hazel	Carolina poplar
Texas mulberry	Hydrangea	Squawbush
Pacific ninebark	Common lilac	Smooth sumac
Ocean-spray	Mountain mahogany	
Lombardy poplar	Douglas maple	
Creambush rockspirea	Rocky Mountain maple	
Low serviceberry	Red maple	
Saskatoon serviceberry	Coronarius mock-orange	
Utah serviceberry	Virginalis mock-orange	
Staghorn sumac	European mountain-ash	
Black willow	Western mountain-ash	
	Mountain laurel	
	White oak	
	Balsam poplar	
	Big sagebrush	
	Mountain snowberry	
	Columbia snowberry	
	Van Houts spirea	
	Shiny leaf spirea	
	Gymnosperms	
Western larch	Douglas fir	Arborvitae (white cedar)
Eastern white pine	Balsam fir	Western red cedar
Jack pine	Grand fir	Silver fir
Red pine	Western hemlock	White fir

Table 15-6 (continued)

Sensitive	Intermediate	Tolerant
	Gymnosperms	
	Austrian pine	Ginkgo
	Lodgepole pine	Common juniper
	Ponderosa pine	Rocky Mountain juniper
	Western white pine	Utah juniper
	Engleman spruce	Western juniper
	White spruce	Limber pine
		Pinyon pine
		Blue spruce
		Pacific yew

Source: Davis and Wilhour (1976).

3. Threshold Dose

Nitrogen oxides are much less phytotoxic than ozone or peroxyacylnitrates. Vegetative injury would be anticipated only in localized regions immediately adjacent to excessive industrial sources. Leaf symptoms would be expected at doses approximating 1.6–2.6 ppm (3000–5000 µg m^{-3}) for periods of up to 48 hr. The threshold for leaf injury may require exposure to 20 ppm (38 X 10^3 µg m^{-3}) if the exposure is only for 1 hr, while a concentration of 1 ppm (1900 µg m^{-3}) might require up to 100 hr to produce symptoms (National Academy of Sciences, 1977b). Nitric oxide injury to trees in the field is unknown while nitrogen dioxide injury would be expected only in the vicinity of an excessive industrial source.

4. Relative Susceptibility

Information on relative susceptibility of forest trees to nitrogen dioxide is extremely limited and is entirely based on experimental fumigations. A relative ranking provided by Davis and Wilhour (1976) is reproduced in Table 15–7.

D. Ozone

Since acute ozone damage to vegetation was first observed in the mid 1940s an enormous literature has developed. Reviews have been provided by Heath (1975), Treshow (1970), the National Academy of Sciences (1977c), Guderian (1985), Runeckles (1986), and the U.S. Environmental Protection Agency (1986).

Table 15-7. Relative Sensitivity of Woody Plants to Acute Damage by Artificial Exposures to Nitrogen Oxides, Primarily Nitrogen Dioxide.

Sensitive	Intermediate	Tolerant
	Angiosperms	
Weeping birch	Norway maple	Black locust
Showy apple	Fan maple	Hornbeam
Wild pear	Winter lime	European beech
Rose	Summer lime	Common elder
Azalea	Rhododendron	Mountain elm
Pyracantha	Ligustrum	Common oak
	Gymnosperms	
European larch	Blue spruce	Yew
Japanese larch	White spruce	Black pine
	Lawson's cypress	Gingko
	Japanese fir	Shore juniper
	Common silver fir	

Source: Davis and Wilhour (1976).

1. Foliar Symptoms

In angiosperms the classic symptoms are necrotic spots on the adaxial surface typically termed *stipples* or *flecks*. Unlike sulfur and nitrogen oxide impact on mesophyll cells, ozone symptoms result from palisade cell destruction. Eventually the entire upper leaf surface may assume a bleached appearance (coalescence of flecks). In other cases, palisade cells may accumulate dark alkaloid pigments and present the stipple symptom. Ultimately ozone influence may extend to the spongy cells and produce abaxial necrotic spots. High exposure may induce dark, water-soaked areas to form. Chlorosis may occur at low level exposure (Davis and Skelly, 1988). In gymnosperms, eastern white pine elongating needles undergo distal necrosis, tip dieback, and browning when subject to ozone exposure. Chlorotic flecks and mottling may be associated with tip necrosis. In ponderosa pine, ozone exposure causes premature needle shed, dwarfed and chlorotic needles, and a general decline syndrome (National Academy of Sciences, 1977c).

Color photographs of ozone symptoms are available in Jacobson and Hill (1970), Loomis and Padgett (1973), Hindawi (1970), U.S.D.A. Forest Service (1973), U.S. Environmental Protection Agency (1976), and Skelly et al. (1987).

2. Mechanism of Toxicity

Ozone enters foliage through stomata. Once in the substomatal cavity the gas is quickly dissolved in the moisture film on cell surfaces. Ozone or its decomposition products rapidly react with cellular components. With a sufficient dose, cellular damage may be reflected as visible foliar injury, premature senescence, reduced yield or growth, reduced vigor, or mortality.

As ozone enters the liquid phase, it undergoes transformations that yield a variety of free radicals, including superoxide and hydroxyl radicals (Tingey and Taylor, 1982). Plant defenses are mobilized in response to superoxide and hydrogen peroxide and involve the enzymes superoxide dismutase, peroxidase and catalase (Tandy et al., 1987). With a sufficient dose, however, defensive strategies are overcome. Ozone, or its products, then react in cells to alter metabolism or structure. Ozone can oxidize a number of biological molecules, including reduced nicotinamide adenine dinucleotide (NADH), DNA, RNA, purine, pyrimidenes, indole acetic acid, some amino acids, many proteins (including several enzymes), and numerous lipids (U.S. Environmental Protection Agency, 1986). Numerous adverse physiological responses have been associated with ozone exposure. Ozone alters the function (permeability and others) of plasma and organelle membranes (Tingey and Taylor, 1982). Ozone interferes with photosynthesis by causing chloroplast dysfunction and/or reduced carbon dioxide uptake. Ozone reduces the carbon allocation and translocation from shoots to roots (Jacobson, 1982). Numerous questions remain concerning the relative importance of ozone itself or its decomposition or induced products and specific sites of adverse biochemical or physiological response. Numerous cellular effects of oxidative stress, however, are clear. Increases are generally recorded in superoxide dismutase, reduced glutathione, detoxification mediated by gentathione, and maintenance and repair. Decreases are generally recorded in growth, carbon dioxide uptake, reductive pentose phosphate pathway, and intergrowth, carbon dioxide uptake, reductive pentose phosphate pathway, and intercellular sucrose export (Alscher, 1988). The phytotoxicity of this highly reactive gas is amply documented!

3.   Threshold Dose

With regard to visible symptom production in woody plants, a variety of reports support the establishment of the following thresholds; 200–510 ppb (392–1000 μg m^{-3}) for 1 hr., 100–250 ppb (196–490 μg m^{-3}) for 2 hrs., and 60–170 ppb (118–333 μg m^{-3}) for 4 hr (U.S. Environmental Protection Agency, 1986; National Academy of Sciences, 1977c). Individual tree responses, however, are highly variable. Resistant species, for example, Norway spruce, failed to exhibit visible symptoms following ozone exposure of 155 ppb (300 μg m^{-3}) for 1215 hr (Keller and Häsler, 1987). Several studies suggest that the threshold of visible injury of eastern white pine approximates 150 ppb (294 μg m^{-3}) for 5 hr (Costonis, 1976). Ponderosa pine in the San Bernardino Mountains of California has a threshold of symptomatic injury approximating 80 ppb (157 μg m^{-3}) for 12–13 hr (Taylor, 1973).

With regard to threshold exposures that result in growth or yield declines, as opposed to visible symptom development, numerous studies suggest a range approximating 50–450 ppb (98-882 μg m^{-3}) for exposure periods of several weeks (Table 15–8).

Table 15-8. Effects of long-term, controlled ozone exposures on growth, yield, and foliar injury in selected forest trees.

Tree species	Ozone concentration $\mu g/m^3$ (ppm)	Exposure time	Tree response, % reaction from control
Pine, ponderosa	290 (0.15)	9/day, 10 days	4, photosynthesis
	290 (0.15)	9/day, 20 days	25, photosynthesis
Pine, ponderosa	290 (0.15)	9/day, 30 days	25, photosynthesis
	290 (0.15)	9/day, 60 days	34, photosynthesis
	588 (0.30)	9/day, 10 days	12, photosynthesis
	588 (0.30)	9/day, 20 days	50, photosynthesis
	588 (0.30)	9/day, 30 days	72, photosynthesis
	880–588 (0.30)	9/day, 30 days	85, photosynthesis
Poplar, yellow	588–880 (0.45)	9/day, 30 days 13 weeks	82, leaf drop; 0, height
Maple, silver	588 (0.30)	8/day, 5 days/wk 13 weeks	50, leaf drop; 78, height
Ash, white	588 (0.30)	8/day, 5 days/wk 13 weeks	66 leaf drop; 0, height
Sycamore	588 (0.30)	8/day, 5 days/wk 13 weeks	0, leaf drop; 22, height
Maple, sugar	588 (0.30)	8/day, 5 days/wk 13 weeks	28, leaf drop; 64, height
Pine, ponderosa[a]	196 (0.10)	6/day, 64 days	45, 25, 35 for same responses
	196 (0.10)	6/day, 126 days	12, root length 21, stem dry wt; 26, root dry wt
Pine, western white[a]	196 (0.10)	6/day, 126 days	13, foliage dry wt 9, stem dry weight
Poplar, hybrid	290 (0.15)	8/day, 5 days/wk 6 weeks	50, shoot dry wt; 56, leaf dry wt; 47, root dry wt
Pine, eastern white	196 (0.10)	4/day, 5 d ays/wk 4 weeks (mixture of O_3 and CO_2 for same periods)	3, needle mottle (over 2-3 days of exposure) 16, needle mottle

[a]Studies conducted under field conditions, except that plants were enclosed to ensure controlled pollutant doses. Plants grown under conditions making them more sensitive.
Source: U.S.E.P.A. (1986).

4. Relative Susceptibility

Davis and Wilhour (1976) have reviewed the considerable literature dealing with woody plant susceptibility, and their summary listing is provided in Table 15–9.

Table 15-9. Relative Sensitivity of North American Woody Plants to Acute Damage by Ozone.

Sensitive	Intermediate	Tolerant
	Angiosperms	
Tree-of-heaven (ailanthus)	Chinese apricot	Apricot
Green ash	Box elder	Arborvitae (white cedar)
White ash	Incense cedar	Chinese azalea
Quaking aspen	Lambert cherry	Avocado
Campfire azalea	Northern black currant	European beech
Hinodegiri azalea	Black bead elder	European birch
Korean azalea	Chinese elm	Japanese box
Snow azalea	Lynwood gold forsythia	Gray dogwood
Bridalwreath	Sweetgum	White dogwood
Bing cherry	Blue-leaf honeysuckle	Dwarf winged Euonymus
Rock cotoneaster	Common lilac	Laland's firethorne
Spreading cotoneaster	Sweet mock-orange	Black gum
Concord grape	Black oak	American holly
Honey locust	Pin oak	English holly
Chinese lilac	Scarlet oak	Hetz Japanese holly
European mountain-ash	Common privet	Mountain laurel
Gambel oak	Eastern redbud	American linden
White oak	Rhododendron	Little-leaf linden
Hybrid poplar	Vaccinioides snowberry	Black locust
Tulip poplar	Linden viburnum	Trailing mahonia
Londense privet	Tea viburnum	Bigtooth maple
Saskatoon serviceberry		Norway maple
Alba snowberry		Sugar maple
Fragrant sumac		Red maple
American sycamore		Bur oak
English walnut		English oak
		Northern red oak
		Shingle oak
		Pachiystima
		Japanese pagoda
		Barlett pear
		Japanese pieris
		Poison ivy
		Amur north privet
		Carolina rhododendron
		Rose woods
		Common sagebrush
		Korean spice viburnum
		Burkwoodii viburnum
		Black walnut
	Gymnosperms	
European larch	Japanese larch	Balsam fir
Austrian pine	Eastern white pine	Douglas fir

Table 15-9 (continued)

Sensitive	Intermediate	Tolerant
	Gymnosperms	
Coulter pine	Knobcone pine	White fir
Jack pine	Lodgepole pine	Eastern hemlock
Loblolly pine	Scotch pine	Digger pine
Monterey pine	Shortleaf pine	Red pine
Ponderosa pine	Slash pine	Redwood
Virginia pine	Sugar pine	Giant sequoia
	Torrey pine	Black Hills spruce
		Colorado blue spruce
		Norway spruce
		White spruce
		Dense yew
		Hatifield's yew

Source: Davis and Wilhour (1976).

5. Field Survey

Very unfortunately, few field surveys have been conducted over wide areas for air pollutant induced forest tree symptomology. Anderson et al. (1988) did conduct such a survey for needle tip necrosis and chlorotic mottling of eastern white pine in the southern Appalachian Mountains. Symptoms were found in approximately 23% of sampled stands in Virginia, North and South Carolina, Georgia, Tennessee, and Kentucky (Figure 15–2). Plantations had a higher percentage of symptomatic trees than did natural stands. The authors concluded that the percentage of trees impacted was generally low and that the pattern of incidence was not uniform.

E. Peroxyacetylnitrate

Peroxyacetylnitrate is one of a series of photochemically produced oxidant homologues. Its chemical formula is $CH_3 CO \cdot O_2 NO_2$ and the gas is commonly referenced using the abbreviation PAN. Peroxyacetylnitrate, along with peroxypropionylnitrate (PPN) and peroxybutyrylnitrate (PBN) are phytotoxic. Only PAN is found in ambient air at concentrations sufficient to be potentially phytotoxic. Excellent reviews of vegetative stress by peroxyacetylnitrate have been provided by Mudd (1975b), the National Academy of Sciences (1977c), and the U.S. Environmental Protection Agency (1986).

1. Foliar Symptoms

Descriptions of ambient peroxyacetylnitrate symptoms come primarily (if not exclusively) from nonwoody species. Laboratory induced woody plant symp-

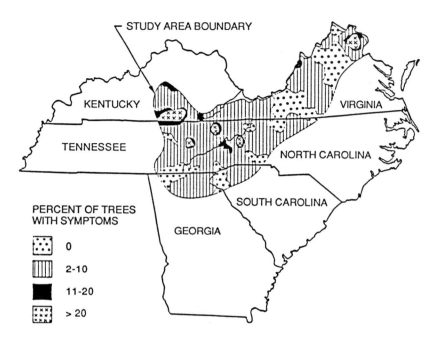

Figure 15-2. Eastern white pine exhibiting tip necrosis and chlorotic mottling symptoms in a six-state survey area.
Source: Anderson et al. (1988).

toms, however, are generally consistent with classic field symptoms resulting from exposure to this gas. Peroxyacetylnitrate preferentially influences the spongy mesophyll cells, and deciduous species typically exhibit symptoms initially on the abaxial surface. The classic syndrome is a glaze appearance followed by bronzing of the lower leaf surface. High dose exposure may result in upper leaf surface injury. Premature senescence and leaf abscission may occur. Low dose exposure may cause chlorosis only. Symptoms associated with peroxypropionylnitrate and peroxybutyrylnitrate are similar. Young expanding leaves are normally most sensitive to injury (National Academy of Sciences, 1977c). Adequate differentiation of peroxyacetylnitrate symptoms in gymnosperms has not been provided.

2. Mechanism of Toxicity

Following stomatal uptake of peroxyacetylnitrate, the gas dissolves in the aqueous layer of substomatal cells as in the case of ozone. PAN is less soluble than ozone and is highly unstable in solution. Decomposition products include acetate, nitrite, oxygen, and water (Nicksic et al., 1967). As in the case of ozone, phytotoxicity targets may include cell and organelle membranes, chloroplasts, or the cytoplasm. Several enzymes are inhibited by the gas. Peroxyacetylnitrate may react with sulfhydryl groups and may inhibit fatty acid synthesis. The gas is capable of oxidizing a large number of sulfur compounds. Eventually, as in the case of ozone, leaf cell plasmolysis occurs and cell death results. The specific sequence and relative importance of various biochemical events remain unclear. Physiological abnormalities may be caused in carbohydrate or hormone metabolism and photosynthesis (Mudd, 1975b; National Academy of Sciences, 1977c, U.S. Environmental Protection Agency, 1986).

3. Threshold Dose

Peroxyacetylnitrates are the most phytotoxic of the various photochemical oxidants described. Insufficient data are available to develop reliable dose-response curves for peroxyacetylnitrate and homologues, especially for woody plants. Laboratory evidence with agricultural species indicate a sensitive plant threshold dose of approximately 10–20 ppb (49–99 μg m^{-3}) for 4 hr (U.S. Environmental Protection Agency, 1976). Davis (1975b) could not injure cotyledons or primary needles of ponderosa pine with peroxyacetylnitrate concentrations of 80, 200, and 400 ppb (396, 989, or 1979 μg m^{-3}) for 8-hr exposures. Variable symptoms were reported in maple, ash, oak, and honey locust when exposed to 200–300 ppb (989–1484 μg m^{-3}) for 8 hr (Drummond, 1971).

4. Relative Susceptibility

In their review, Davis and Wilhour (1976) do not record any North American woody plants sensitive or intermediate in resistance to damage by peroxyacetylnitrate. They list 36 species as tolerant of laboratory exposures (Table 15–10).

F. Fluoride

The impact of fluoride on vegetation has an extended research history and a very substantial literature. Numerous reviews of this research have been provided (Chang, 1975; Keller, 1975a,b; McCune and Weinstein, 1971; Weinstein and McCune, 1971). An extremely comprehensive and very useful review was recently provided by Leonard H. Weinstein, Boyce Thompson Institute for Plant Research, Ithaca, New York (Weinstein, 1977).

1. Foliar Symptoms

In angiosperms the initial symptom is leaf tip chlorosis. The chlorotic condition eventually expands around the leaf margin and inward along the midvein. With continued exposure chlorotic areas become necrotic. Necrotic tips and margins may ultimately fall from the leaf. Occasionally premature leaf abscission precedes development of intensive necrosis (Weinstein, 1977).

Table 15-10. North American Woody Plants Tolerant of Peroxyacetylnitrate Exposure under Laboratory Conditions.

Angiosperms	Gymnosperms
Green ash	Arborvitae (white cedar)
White ash	Balsam fir
Basswood	Douglas fir
European white birch	White fir
White dogwood	Eastern hemlock
Sweet gum	European larch
Honey locust	Japanese larch
Common lilac	Austrian pine
Norway maple	Eastern white pine
Silver maple	Pitch pine
Sugar maple	Ponderosa pine
American mountain-ash	Red pine
English oak	Scotch pine
Northern red oak	Virginia pine
Pin oak	Hybrid poplar
White oak	Tulip poplar
	Black Hills spruce
	Blue spruce
	Norway spruce
	White spruce

Source: Davis and Wilhour (1976).

The typical response of gymnosperms to fluoride exposure is the development of distal necrosis or "tipburn." A well-defined boundary usually occurs between live and necrotic portions. The boundary zone may become darker brown than the necrotic tip, which is typically some shade of reddish-brown (U.S. Environmental Protection Agency, 1976). Unfortunately these symptoms are not highly specific and may be induced by a large number of tree stresses. Color photographs of fluoride injury are provided in Weinstein (1977), the U.S. Environmental Protection Agency (1976), Loomis and Padgett (1973), and Jacobson and Hill (1970).

2. Mechanisms of Toxicity

Trees may accumulate fluoride through stomata, the cuticle, or via other exterior surfaces. The specific biochemical ramifications of fluoride contamination remain incompletely appreciated. A variety of enzymes sensitive to fluoride stimulation or inhibition is known. Fluoride may alter cell nutrient status by complexing with a variety of metal cations. The interaction of fluoride with enzymes and nutrients is translated into effects on metabolite levels, metabolic reactions, for example, oxygen uptake, cell wall formation, starch synthesis, and growth, all thoroughly reviewed by Weinstein (1977).

3. Threshold Dose

Gaseous compounds containing fluoride are more toxic than particulate forms. The phytotoxicity of the latter is largely related to solubility. Ammonium fluoride, for example, is much more toxic to vegetation than low solubility cryolite (Na_3AIF_6). In general fluoride is considerable more phytotoxic than the previously discussed gaseous pollutants. Susceptible plants may be injured at ambient concentrations that are 10–1000 times lower than sulfur dioxide, ozone, peroxyactylnitrate, and nitrogen oxides. Since fluorides accumulate in foliar tissue, toxicities are generally expressed in terms of tissue concentrations on a dry-weight basis. For most plants designated as susceptible, this concentration is typically less than 100 μg g^{-1}. Plants classified as intermediate or tolerant can probably tolerate concentrations in excess of 200 μg g^{-1} without the development of visible symptoms (Weinstein, 1977). Selected woody plants may accumulate in excess of 4000 μg g^{-1} without visible stress. For a large number of trees the fluoride dose for symptomatic injury ranges from approximately 0.75 to 50 μg m^{-3} fluoride for several hours to 10 or more days.

4. Relative Susceptibility

Plants exhibit a broad range of tolerances to fluoride. A comprehensive listing of relative susceptibility provided by Weinstein (1977) is reproduced in Table 15–11.

G. Trace Metals

A very large number of trace metals may be associated with higher plants. Many of these are heavy metals, that is, they have a density greater than 5. Some of these metals are required micronutrient elements, for example, iron, manganese, copper, and zinc. Other metals have no known metabolic function, for example, lead, cadmium, nickel, and tin. All these trace metals have the potential to be toxic to trees if present in sufficient concentrations. Trace metals identified as having especially high potential to acutely injury trees, because of widespread distribution or intensive local release as a result of anthropogenic activities, include cadmium, cobalt, chromium, copper, lead, mercury, nickel, thallium, vanadium, and zinc. These metals are the subject of an enormous literature. Comprehensive general reviews include Energy Research and Development Administration (1975), Kathny (1973), Oehme (1978), Purves (1977), and Smith et al. (1986). Particular element reviews include cadmium (Commission of the European Communities, 1978; Friberg et al., 1974), chromium (National Academy of Sciences, 1974), copper (National Academy of Sciences, 1977a; Nriagu, 1979), lead (Peterson, 1978), mercury (Friberg and Vostal, 1972; National Academy of Sciences, 1978b), nickel (National Academy of Sciences, 1975), and zinc (Nriagu, 1980).

Table 15-11. Relative Sensitivity of Woody Plants to Acute Damage by Fluoride.

Sensitive	Intermediate	Tolerant
	Angiosperms	
Chinese apricot	Apple	Alder
Blueberry	Green ash	Japanese barberry
Box elder	Trembling aspen	Black birch
European grape	Azalea	White birch
Sheep laurel	Warty barberry	Cutleaf birch
Bradshaw plum	Wintergreen barberry	Blackberry
Italian prune	Common boxwood	Black-haw
	Bing cherry	Cotoneaster
	Royal Ann cherry	Currant
	Choke cherry	Flowering dogwood
	Flowering cherry	Red osier dogwood
	Concord grape	Elderberry
	Mountain laurel	American elm
	Lilac	Chinese elm
	American linden	Siberian elm
	Littleleaf linden	Winged euonymus
	Hedge maple	Pyracantha
	Norway maple	Forsythia
	Red maple	Honeysuckle
	Silver maple	Black locust
	Sugar maple	Honey locust
	European mountain ash	Oak
	Mulberry	Pear
	Peach	London plane
	Flowering plum	Balsam poplar
	Rhododendron	Carolina poplar
	Serviceberry	Lombardy poplar
	Smooth sumac	Silver-leaved poplar
	Staghorn sumac	Privet
	Double fill viburnum	Russian olive
	Black walnut	Spirea
	English walnut	Sweetgum
	Yew	Sycamore
		Tree-of-heaven
		Tuliptree
		Arrowwood viburnum
		Leatherleaf viburnum
		Scibold viburnum
		Goat willow
		Laurel leaf willow
		Weeping willow
	Gymnosperms	
Douglas fir	Balsam fir	Arborvitae (white cedar)
Western larch	Grand fir	Eastern red cedar

Table 15-11 (continued)

Sensitive	Intermediate	Tolerant
	Gymnosperms	
Eastern white pine	Ginkgo	Western red cedar
Mugho pine	Western white pine	Cypress
Loblolly pine	Jack pine	Andorra juniper
Lodgepole pine	Austrian pine	Creeping juniper
Ponderosa pine	Blue spruce	Pfitzen juniper
Scotch pine	Birds nest spruce	
	Black spruce	
	Engelman spruce	
	White spruce	

Source: Weinstein (1977).

1. Foliar Symptoms

Specific symptom descriptions for forest vegetation exposed to excessive trace metals are not abundant. The syndrome of a large number of plants acutely injured by trace metal accumulation, however, involves development of interveinal chlorosis, tip and margin necrosis, and premature leaf abscission. Mitchell and Fretz (1977) have described symptoms of cadmium and zinc toxicity to forest tree seedlings grown in solution culture amended with heavy metals. Symptoms of cadmium toxicity appeared on the youngest foliage of red maple in the form of interveinal chlorosis. Foliage size was stunted in most cases. High cadmium levels resulted in loss of leaf turgor, wilting, and ultimately death. Initial symptoms of cadmium stress of white pine were inhibition of needle expansion, and on Norway spruce, chlorotic tips of new growth. Ultimately wilting progressed basipetally, followed by necrosis. Low zinc levels induced interveinal chlorosis in red maple. High zinc concentrations resulted in wilting of new growth and eventually necrosis of leaf and stem tissue. Zinc produced symptoms on white pine and Norway spruce similar to those described for cadmium.

2. Mechanisms of Toxicity

Specific mechanisms of toxicity of trace metals are presumed to be quite variable. It is probable, however, that trace metal injury to woody plants involves one or more of the following biochemical abnormalities: (a) metal interferes with enzyme function, (b) metal serves as an antimetabolite, (c) metal forms a stable precipitate or chelate with an essential metabolite, (d) metal catalyzes the decomposition of an essential metabolite, (e) metal alters the permeability of cell membranes or (f) the metal replaces important structural or electrochemically important elements in the cell (Bowen, 1966). Each of these abnormalities is capable of inducing a substantial adverse impact on a variety of critical physiological functions.

Table 15-12. Baseline (uncontaminated) Trace Metal Concentrations (ash weight basis) for United States Forest Foliage.

Element	Baseline concentration (mean and range) (ppm)
Cadmium	7.0 (.05-60)
Chromium	8.0 ($<$ 2-150)
Cobalt	6.2 ($<$ 1-10,000)
Copper	128 ($<$ 10-3000)
Lead	135 ($<$ 10-3000)
Mercury	25 ($<$ 25-50)
Nickel	37 ($<$ 2-1300)
Thallium	4 (2-100)
Vanadium	7.7 ($<$ 5-70)
Zinc	740 (100-7400)

Source: Calculated from Connor et al. (1975) and Shacklette et al. (1978).

3. Threshold Dose

It is extremely difficult to generalize concerning threshold doses required for acute injury to woody plants. Temple and Hill (1979) have presented threshold levels of contamination for vegetation for a variety of trace elements. Some of the trace metals are essential plant nutrients and are required by various species in differing amounts. Numerous plants have the capacity to evolve substantial tolerance to high environmental levels of certain heavy metals (Antonavics et al., 1971). Some elements, for example, cadmium, nickel, and thallium, are especially mobile in plants. Other elements have considerable potential to accumulate, for example, mercury and vanadium. Table 15–12 presents a general suggestion of baseline trace metal content for the 10 metals judged to have especially high potential to injure trees.

Mitchell and Fretz (1977) have suggested that toxicity thresholds (foliar symptoms) for red maple, white pine, and Norway spruce seedlings were 23, 61, and 8 μg g^{-1} cadmium, respectively, in leaf tissue (dry weight basis). In the instance of zinc, foliar symptoms resulted when leaf concentrations reached 421 μg g^{-1} zinc in red maple, 1006 μg g^{-1} in white pine, and 596 μg g^{-1} in Norway spruce. There is insufficient information available to suggest threshold levels of these various trace metals required to cause acute injury to sensitive woody plants. These thresholds are surely very variable and are most probably within the ranges contained in Table 15–12.

4. Relative Susceptibility

Relative susceptibility of forest trees to acute injury by trace metal contamination cannot be established with current information.

H. Acid Deposition

Acute woody plant response to acid deposition has not been documented under field conditions. Symptomology has been investigated with seedling trees grown in managed environments and subjected to precipitation simulants.

1. Foliar Symptoms

Necrotic spots (lesions) on deciduous foliage and distal needle necrosis and/or chlorosis on coniferous foliage are the primary symptoms recorded. Foliar galls have been recorded in pin oak and poplar. Foliar galls in poplar consisted of hyperplasia and hypertrophy of both palisade and spongy parenchyma cells. In pin oak, galls resulted from abnormal cell functions in spongy cells only (Evans and Curry, 1979).

2. Mechanisms of Toxicity

Cell necrosis may result primarily from accumulation of toxic ion species, most probably hydrogen ion (H^+) or sulfate (SO_4^{2-}). Malformation or malfunction of guard cells may be involved. Enhancement of foliar leaching (Chapter 10) associated with cuticular erosion may be associated with chlorotic symptomology.

Jacobson et al. (1988) have emphasized that intermittent mist exposures are more injurious to conifer seedlings than continuous exposures per unit amount of acidity deposited, especially when intervals between mist exposures are sufficient to allow evaporation of liquid from treated foliage. Unsworth (1984) has hypothesized that measurements of acidity of collected cloudwater underestimates the effective concentration of acids on foliar surfaces due to solute concentration increase resulting from evaporation.

3. Threshold Dose

Chevone et al. (1986) exposed yellow poplar seedlings to ozone (100 ppb, 200 $\mu g\ m^{-3}$), sulfur dioxide (80 ppb, 210 $\mu g\ m^{-3}$), and simulated rain solutions (pH 5.6, 4.3, and 3.0) for 6 wks. Dry or wet deposition of sulfate at these levels, which approximate ambient levels, did not induce a visible symptom response. The phytotoxicity of ozone and sulfur dioxide, however, were altered by rain treatments. Skeffington and Roberts (1985) failed to detect any increase in foliar leaching when 3-year-old Scots pine seedlings were simultaneously exposed to ozone (up to 150 ppb, 300 $\mu g\ m^{-3}$) and acid mist (pH 3.0). Foliar symptoms have been recorded on yellow birch seedlings exposed to rain simulant at pH 3.0 (Wood and Bormann, 1974). Weinstein et al. (1987) have investigated the response of 3-year-old balsam fir and 4-year-old red spruce seedlings to acidity and anion content (sulfate or nitrate) of mist simulants. Three levels of acidity were employed, pH 2.5, 3.5, and 4.5; and three anion levels of sulfate or nitrate at 3.2, 0.32, and 0.032 equiv l^{-1} for the three levels of acidity respectively. Mist treatments were initiated after needle expansion and 32 exposures averaging 13.9 hr day[-1] were given over a 57-day period. Seedlings were exposed to mist on 56% of the days and 35% of the hours. Visible symptoms (needle chlorosis or necrosis) were recorded only at the pH 2.5 treatment (Table 15–13). The severity of symptomology was greater when the anion was sulfate than when it was nitrate or a combination of sulfate and nitrate (Figure 15–3). No bud necrosis or needle loss was recorded during the spring following treatment. Continuation of comparable exposures over three growing seasons did not meaningfully alter the results (Jacobson and Lassoie, 1987). Visible necrosis of needles occurred at pH 2.6 and below, but not at pH 2.8 and above. The combination of high acidity and high sulfate in mist was the most toxic.

4. Relative Susceptibility

Systemic screening of forest species to acute acidic deposition exposure has not been conducted. Nevertheless, it is reasonable to presume that the species response would be highly variable. Cuticle thickness, foliar wettability, leaf surface diffusion rates, foliar buffering, and detoxification capacities all vary greatly in forest species.

I. Other Air Contaminants

Under certain circumstances localized forest areas or ornamental woody plants may be subject to *acute* injury by a minor list of gaseous pollutants including ammonia, chlorine, hydrocarbons, and hydrogen sulfide. Typically these injuries result following accidental or unusual industrial, commercial, or transportation release, or in the presence of extremely atypical climatic conditions.

Ammonia may induce the formation of marginal leaf spots and cause leaves to assume a dull and dark green color. Eventually leaves turn brown or black. In the presence of excessive chlorine, angiosperms typically develop interveinal leaf spots while gymnosperms generally exhibit distal necrosis. Exposure to excessive ethylene may cause trees to develop chlorosis and necrosis of older leaves followed by leaf abscission. Conifers may exhibit dwarfed needle growth and premature cone shed. Hydrogen sulfide typically causes interveinal white to tan discoloration.

The approximate threshold dose required for acute injury to sensitive plants is quite variable for miscellaneous pollutants. Agricultural species may be acutely injured by ammonia at 1000 ppm (70×10^4 μg m^{-3}) for 3 min or 55 ppm (38×10^3 μg m^{-3}) for 1 hr. For chlorine, injury thresholds approximate 0.5–1.5 ppm (1400–4350 μg m^{-3}) for 0.5–3 hr. Nonwoody vegetation susceptible to ethylene injury may become symptomatic at ambient concentrations as low as 0.04–0.1 ppm (4–115 μg m^{-3}) for 8 hr or 0.002–0.02 ppm (2.3–23 μg m^{-3}) for 24 hr. Hydrogen sulfide at 86 ppm (12×10^4 μg m^{-3}) will injure the most sensitive plant species in 5 hr (U.S. Environmental Protection Agency, 1976). Continuous fumigation of Douglas-fir seedlings with hydrogen sulfide caused

Table 15-13. Foliar Injury[a] of a Red Spruce and Balsam Fir Seedlings After 57 Days of Intermittent Exposure to Acidic Mist.

pH	Anion	Red spruce I Spring	Fall	Red spruce II Spring	Fall	Balsam fir Spring	Fall
Untr.	Control	0.3	0.3	0.3	0.9	0.8	0.3
	Sulfate	0.6	0.4	0.3	0.9	0.7	0.4
	Nitrate	0.4	0.3	0.2	0.8	0.7	0.5
	S plus N	0.4	0.5	0.5	1.0	0.9	0.3
	Sufate	0.5	0.5	0.5	0.9	0.7	0.4
3.5	Nitrate	0.3	0.4	0.4	1.0	0.6	0.4
	S plus N	0.6	0.4	0.6	1.3	1.0	0.3
	Sulfate	2.1	1.0	2.6	1.4	5.0	1.1
2.5	Nitrate	0.8	0.8	0.8	1.1	1.5	0.5
	S plus N	1.2	0.8	0.9	1.6	3.0	1.0

[a]Foliar injury scale: 0, no needles injured; 1, less than 5% injured; 2, 5–10%; 3, 15–30%; 4, 35–65%; 5, 65–85%; 6, 85–95%; 7, over 95% injured.
[b]I and II represent two red spruce populations.
Source: Weinstein et al. (1987).

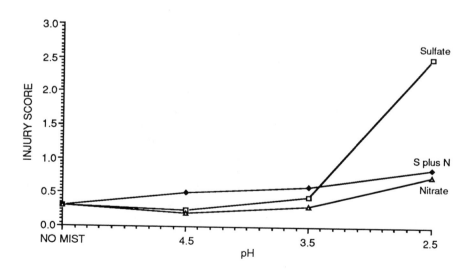

Figure 15-3. Enhanced influence of sulfuric acid mist on injury to needles of red spruce seedlings. Injury score is described in Table 15-12.
Source: Weinstein et al. (1987).

Table 15-14. Relative Sensitivity of Woody Plants to Acute Damage by Chlorine.

Sensitive	Intermediate	Tolerant
	Angiosperms	
Box elder		Azalea
Crab apple		
Horse chestnut		
Pin oak		
Sugar maple		
Sweetgum		
Tree-of-heaven		
	Gymnosperms	
		Balsam fir
		Pine

Source: U.S. Environmental Protection Agency (1976).

slight symptoms at 100 ppb (139 μg m^{-3}) and extensive foliar damage at 300 ppb (417 μg m^{-3}) (Thompson and Kats, 1978).

The relative susceptibility of several forest tree species to chlorine is presented in Table 15–14.

J. Summary

Acute foliar disease may be caused in forest vegetation by widespread air contaminants including sulfur dioxide, nitrogen oxides, ozone, peroxyacetylnitrates, fluoride, and several trace metals, and by localized air contaminants including acid deposition, ammonia, chlorine, hydrocarbons, and hydrogen sulfide. The response of woody plants to these atmospheric pollutants is extremely variable and is dramatically controlled by genetic factors, plant age and health, and environmental conditions. Field symptoms of air pollution injury are not highly specific, are mimicked by symptoms caused by a wide variety of other tree stress factors, and are useful only to experienced observers familiar with the range of edaphic, entomological, and pathological stress factors characteristic of a given flora in a given location. The dose required to produce acute injury varies widely with the pollutant and vegetative type. There has been sufficient work done to enable a generalized ranking of relative forest tree sensitivity to the most important air pollutants. A summary treatment of general symptoms and injury thresholds for the major contaminants is contained in Table 15–15.

Table 15-15. Acute Foliar Disease of Forest Trees Caused by Major and Minor Air Pollutants.

Pollutants	Symptoms[a]	Threshold dose
A. Major		
1. Sulfur dioxide	Angiosperms: interveinal necrotic blotches Gymnosperms: red-brown dieback or banding	0.50 ppm (1310 μg m^{-3}) 0.18 ppm (468 μg m^{-3}) for 8 hr; 0.008–0.017 ppm (21–44 μg m^{-3}) for growing season
2. Nitrogen dioxide	Angiosperms: interveinal necrotic blotches similar to SO_2 injury Gymnosperms: red-brown distal necrosis	20 ppm (30 x 10^3 μg m^{-3}) for 1 hr; 1.6–2.6 ppm (3000–5000 μg m^{-3}) for 48 hr; 1 ppm (1900 μg m^{-3}) for 100 hr
3. Ozone	Angiosperms: upper surface flecks Gymnosperms: distal necrosis, stunted needles	0.20–0.30 ppm (392–588 μg m^{-3}) for 2–4 hrs; some conifers 0.10 ppm (196 μg m^{-3}) chronic, extended term (wks, months) exposure
4. Peroxyacetyl-nitrate	Antiosperms: lower surface bronzing Gymnosperms: chlorosis, early senescence	0.20–0.80 ppm (989–3958 μg m^{-3}) for 8 hr
5. Fluoride	Angiosperms: tip and margin necrosis Gymnosperms: distal necrosis	50–100 μg g^{-1} fluoride, dry wt. basis
6. Trace metals	Angiosperms: interveinal chlorosis, tip and marginal necrosis Gymnosperms: distal necrosis	Variable, undetermined
B. Minor		
1. Acid rain	Angiosperms: necrotic spots Gymnosperms: distal necrosis	pH < 3.0 pH < 2.8
2. Ammonia	Angiosperms: interveinal necrotic blotches similar to SO_2 injury Gymnosperms: distal necrosis	55 ppm (38,280 μg m^{-3}) for 1 hr
3. Chlorine	Angiosperms: chlorosis, upper surface fleck similar to O_3 Gymnosperms: distal necrosis	0.5–1.5 ppm (1400–4530 μg m^{-3}) for 0.5–3 hr
4. Ethylene	Angiosperms: chlorosis, necrosis, abscission Gymnosperms: dwarfing, premature defoliation	Variable, undetermined
5. Hydrogen sulfide	Angiosperms: interveinal necrotic blotches Gymnosperms: distal necrosis	100 ppm (14 x 104 μg m^{-3}) for 5 hr

[a]Symptoms and dose thresholds are for the most sensitive species.

References

Alscher, R.G. 1988. Personal communication. Boyce Thompson Institute for Plant Research, Cornell Univ., Ithaca, NY.

Alcher, R., M. Franz, and C.W. Geske. 1987. Sulfur dioxide and chloroplast metabolism. In J.A. Saunders, L. Kosak-Channing, and E.E. Conn, eds., Phytochemical Effects of Environmental Compounds. Plenum Publishing Corp., New York, pp. 1-28.

Anderson, R.L., H.D. Brown, B.I. Chevone, and T.C. McCartney. 1988. Occurrence of air pollution symptoms (needle tip necrosis and chlorotic mottling) on eastern white pine in the southern Appalachian Mountains. Pl. Dis. 72: 130-132.

Antonovics, J., A.D. Bradshaw, and R.G. Turner. 1971. Heavy metal tolerance in plants. Adv. Ecol. 7: 1-85.

Berry, C.R. 1974. Age of pine seedlings with primary needles affects sensitivity to ozone and sulfur dioxide. Phytopathology 64: 207-209.

Berry, C.R. and G.H. Hepting. 1964. Injury to eastern white pine by unidentified atmospheric constituents. For. Sci. 10: 2-13.

Bjorkman, E. 1970. The effect of fertilization on sulfur dioxide damage to conifers in industrial and built-up areas. Stud. For. Suec. 78:1-48.

Bowen, H.J.M. 1966. Trace Elements in Biochemistry. Academic Press, New York, 241 pp.

Brennan, E. and I.A. Leone. 1968. The response of plants to sulfur dioxide or ozone-polluted air supplied at varying flow rates. Phytopathology 58: 1661-1664.

Centre for Agricultural Publishing and Documentation. 1969. Air Pollution Proc. First European Congress on the Influence of Air Pollution on Plants and Animals. Wageningen, The Netherlands, April 22-27, 1968, 415 pp.

Chang. C.W. 1975. Fluorides. In J.B. Mudd and T.T. Kozlowski, eds., Responses of Plants to Air Pollution. Academic Press, New York, pp. 57-95.

Chevone, B.I., D.E. Herzfeld, S.V. Krupa, and A.H. Chappelka. 1986. Direct effects of atmospheric sulfate deposition on vegetation. J. Air Pollu. Control Assoc. 36: 813-815.

Commission of the European Communities. 1978. Criteria Dose-Effect Relationships for Cadmium. Pergamon Press, New York, 202 pp.

Connor, J.J., H.T. Shacklette, R.J. Ebens, J.A. Erdman, A.T. Miesch, R.R. Tidball, and H.A. Bourtelot. 1975. Background Geochemistry of Some Rocks, Soils, Plants and Vegetables in the Conterminous United States. U.S. Geological Survey, Professional Paper No. 574-F, Washington, DC, 168 pp.

Costonis, A.C. 1976. Criteria for evaluating air pollution injury to forest trees. IUFRO Congress, Oslo, Norway, June 21–26, 1976.

Cotrufo, C. 1974. The sensitivity of a white pine clone to air pollution as affected by N, P, and K. U.S.D.A. Forest Service, Research Note No. SE-198, Southeastern Forest Exp. Sta., Asheville, NC, 4 pp.

Cotrufo, C. and C.R. Berry. 1970. Some effects of a soluble NPK fertilizer on sensitivity of eastern white pine to injury from SO_2 air pollution. For. Sci. 16: 72-73.

Darley, E.F. and J.T. Middleton. 1966. Problems of air pollution in plant pathology. Annu. Rev. Phytopath. 4: 103-118.

Davis, D.D. 1975a. Variable tree response due to environmental factors — climate. In W.H. Smith and L.S. Dochinger, eds., Air Pollution and Metropolitan Woody Vegetation. U.S.D.A. Forest Service. PIEFR-PA-1, Upper Darby, PA, pp. 14-16.

Davis, D.D. 1975b. Resistance of young ponderosa pine seedlings to acute doses of PAN. Plant Dis. Reptr. 59: 183-184.

Davis, D.D. 1977. Response of ponderosa pine primary needles to separate and simultaneous ozone and PAN exposures. Plant Dis. Reptr. 61: 640-644.

Davis, D.D. and J.B. Coppolino. 1974. Relationship between age and ozone sensitivity of current needles of ponderosa pine. Plant Dis. Reptr. 58:660-663.

Davis, D.D. and J.M. Skelly. 1988. Relative sensitivity of eight eastern hardwood tree species to ozone and/or acidic precipitation. Annual Report 1978–88, Eastern Hardwood Research Cooperative, National Acid Precipitation Assessment Program, PA State Univ., University Park, PA, 21 pp.

Davis, D.D. and R.G. Wilhour. 1976. Susceptibility of Woody Plants to Sulfur Dioxide and Photochemical Oxidants. U. S. Environmental Protection Agency Publ.. No. EPA-600/3-76-102, Corvallis, OR, 71 pp.

Davis, D.D. and F.A. Wood. 1973a. The influence of environmental factors on the sensitivity of Virginia pine to ozone. Phytopathology 63: 371-376.

Davis, D.D. and F.A. Wood. 1973b. The influence of plant age on the sensitivity of Virginia pine to ozone. Phytopathology 63: 381-388.

Dochinger, L.S. 1964. Effects of nutrition on the chlorotic dwarf disease of eastern white pine. Plant Dis. Reptr. 48: 107-109.

Dochinger, L.S. and D.E. Seliskar. 1970. Air pollution and the chlorotic dwarf disease of eastern white pine. For. Sci. 16: 46-55.

Dochinger, L.S., F.W. Bender, F.L. Fox, and W.E. Heck. 1970. Chlorotic dwarf of eastern white pine caused by an ozone and sulphur dioxide interaction. Nature 225:476.

Drummond, D.B. 1971. Influence of high concentrations of peroxyacetylnitrate on woody plants. Phytopathology 61: 178.

Dunning, J.A. and W.W. Heck. 1973. Response of pinto bean and tobacco to ozone as conditioned by light intensity and/or humidity. Environ. Sci. Technol. 7: 824-826.

Energy Research and Development Administrtion. 1975. Biological Implications of Metals in the Environment. ERDA Symposium Series No. 42, Washington, DC, 682 pp.

Evans, L.S. and T.M. Curry. 1979. Differential responses of plant foliage to simulated acid rain. Am. J. Bot. 66: 953-962.

Forestry Commission (England). 1971. Fume Damage to Forests. Research and Development Paper No. 82, London, 50 pp.

Friberg, L. and J. Vostal. 1972. Mercury in the Environment. Chemical Rubber Co. Press, Cleveland, OH, 215 pp.

Friberg, L., M. Piscator, G. Nordberg, and T. Kjellstrom. 1974. Cadmium in the Environment. Chemical Rubber Co. Press, Cleveland, OH, 248 pp.

Fuhrer, J. and K.H. Erismann. 1980. Uptake of NO_2 by plants grown at different salinity levels. Experientia 36: 409-410.

Genys, J.B. and H.E. Heggestad. 1978. Susceptibility of different species, clones and strains of pines to acute injury caused by ozone and sulfur dioxide. Plant Dis. Reptr. 62: 687-691.

Gerhold, H.D. 1975. Resistant varieties. In W.H. Smith and L.S. Dochinger, eds., Air Pollution and Metropolitan Woody Vegetation. U.S.D.A. Forest Service PIERR-PA-1, Upper Darby, PA, pp. 45-49.

Gerhold, H.D. 1977. Effect of Air Pollution on *Pinus strobus* L. and Genetic Resistance. U.S. Environmental Protection Agency, Publ.. No. EPA-600/3-77-002. Corvallis, OR, 45 pp.

Guderian, R. 1977. Air Pollution Phytoxicity of Acidic Gases and Its Significance in Air Pollution Control. Ecological Studies No. 22, Springer–Verlag, New York, 122 pp.

Guderian, R. (ed.). 1985. Air Pollution by Photochemical Oxidants: Formation, Transport, Control, and Effects on Plants. Springer–Verlag, New York, 346 pp.

Harvey, G.W. and A.H. Legge. 1979. The effect of sulfur dioxide upon the metabolic level of odenosine triphosphate. Can. J. Bot. 57: 759-764.

Heagle, A.S. 1979. Effects of growth media, fertilizer rate and hour and season of exposure on sensitivity of four soybean cultivars to ozone. Environ. Pollut. 18: 313-322.

Heagle, A.S., W.W. Heck, and D. Body. 1971. Ozone injury to plants as influenced by air velocity during exposure. Phytopathology 61: 1209-1212.

Heagle, A.S., R.B. Philbeck, H.H. Rogers, and M.B. Letchworth. 1979. Dispensing and monitoring ozone in opentop field chambers for plant-effects studies. Phytopathology 69: 15-20.

Heath, R.L. 1975. Ozone. In J.B. Mudd and T.T. Kozlowski, eds., Responses of Plants to Air Pollution. Academic Press, New York, pp. 23-55.

Heck, W.W. 1968. Factors influencing expression of oxidnt damage to plants. Annu. Rev. Phytopathol. 6: 165-188.

Heggestad, H.E. 1968. Diseases of crops and ornamental plants incited by air pollutants. Phytopathology 58: 1089-1097.

Heggestad, H.E. and W.W. Heck. 1971. Nature, extent, and variation of plant response to air pollutants. Adv. Agronomy 23: 111-145.

Heitschmidt, R.K. and J. Altman. 1978. Probable Effects of SO2 on Agricultural Crops. Experiment Station, Colorado State University, Tech. Bull. No. 133, Fort Collins, CO, 7 pp.

Hindawi, I.J. 1970. Air Pollution Injury to Vegetation. U.S. Dept. Health, Education and Welfare, National Air Pollution Control Administration, Raleigh, NC, 44 pp.

International Electric Research Exchange. 1981. Effects of SO_2 and its Derivatives on Health and Ecology. Vol. 2. Natural Ecosystems Agriculture Forestry and Fisheries. Research Reports Center, Electric Power Research Institute, Palo Alto, CA.

Jacobson, J.S. 1982. Ozone and the growth and productivity of agricultural crop. In M.H. Unsworth and D.F. Ormrod, eds., Effects of Gaseous Air Pollution in Agriculture and Horticulture. Butterworth Scientific, London, pp. 293-304.

Jacobson, J.S. and A.C. Hill. 1970. Recognition of Air Pollution Injury to Vegetation: A Pictorial Atlas. Air Pollu. Control Assoc., Pittsburgh, PA.

Jacobson, J.S. and J.P. Lassoie. 1988. Response of red spruce to sulfur- and nitrogen-containing contaminants in simulated acidic mist. In G. Hertel, ed., The Effects of Atmospheric Pollution on Spruce and Fir Forests in the Eastern United States and the Federal Republic of Germany. U.S.D.A. Forest Service Genl. Tech. Publ.. No 255, Northeastern Forest Exper. Stat., Broomall, PA.

Jacobson, J.S., L.I. Heller, K.E. Yamada, J.F. Osmeloski, and T. Bethard. 1988. Foliar injury and growth response of red spruce to sulfate and nitrate acidic mist. Can. J. For. Res. (in press).

Jäger, H.J and H. Klein. 1976. Studies on the influence of nutrition on the susceptibility of plants to SO_2. Eur. J. For. Pathol. 6: 347-353.

Jäger, H.J., J. Bender, and L. Gruunhage. 1985. Metabolic responses of plants differing in SO_2 sensitivity towards SO_2 fumigation. Environ. Pollu. 39: 317-335.

Johnson, W.T. and H.H. Lyon. 1988. Insects that Feed on Trees and Shrubs. Cornell University Press, Ithaca, NY, 464 pp.

Jones, H.C., D. Weber, and D. Balsillie. 1974. Acceptable limits for air pollution dosages and vegetation effects: Sulfur dioxide. Proc. 67th Annual Meeting, Air Pollu. Control. Assoc., Paper No. 74-225. Denver, CO.

Karnosky, D.F. 1974. Implications of genetic variation in host resistance to air pollutants. Proc. 9th Central States Forest Tree Improvement Conference, Ames, IA, pp. 7-20.

Karnosky, D.F. 1977. Evidence for genetic control of response to sulfur dioxide and ozone in *Populus tremuloides* . Can. J. For. Res. 7: 437-440.

Karnosky, D.F. 1978a. Selection and testing programs for developing air pollution tolerant trees for urban areas. Proc. IUFRO Air Pollution Meeting, Sept. 18–23, 1978, Ljubljana, Yugoslavia.

Karnosky, D.F. 1978b. Genetics of air pollution tolerance of trees in the Northeastern Forest Tree Improvement Conf., July 25-27, 1978, PA State Univ. State College. PA, p. 161-178.

Karnosky, D.F. 1985. Genetic variability in growth responses to SO_2. In W.E. Winner, H.A. Mooney, and R.A. Goldstein, eds., Sulfur Dioxide and Vegetation. Stanford University Press, Stanford, CA, pp. 346-356.

Karnosky, D.F. and K.C. Steiner. 1981. Provenance and family variation in response of *Fraxinus americana* and F. pennsylvanica to ozone and sulfur dioxide. Phytopathology 71: 804-807.

Karnosky, D.F., P. Berrang, and R. Mickler. 1986. A Genecological Evaluation of Air Pollution Tolerances in Hardwood Trees in Eastern Forest Parks. Final Report, Contract No. CX-001-4-0057, Air Quality Division, National Park Service, Denver, CO, 93 pp.

Kathny, E.L. (ed.). 1973. Trace Elements in the Environment. American Chemical Soc., Adv. in Chem. Series No. 123, Washington, DC, 149 pp.

Keller, T. 1984. The influence of SO_2 on CO_2 uptake and peroxidase activity. Eur. J. For. Path 14: 354-359.

Keller, T. 1975a. On the phytotoxicity of fluoride immissions for woody plants. Mitt. Eidg. Anst. Forstl. Vers'wes 51: 303-331

Keller, T. 1975b. On the translocation of fluoride in forest trees, Mitt. Eidg. Anst. Forstl. Vers'wes 51: 335-356.

Keller, T. and R. Häsler. 1986. The influence of a prolonged SO_2 fumigation on the stomatal reaction of spruce. Eur. J. For. Path. 16: 110-115.

Keller, T. and R. Häsler. 1987. Lame effects of long-term fumigations on Norway spruce. Trees 1: 129-133.

Kozlowski, T.T. 1980. Impacts of air pollution on forest ecosystems. BioScience 30: 88-93.

Krause, G.H. and H. Kaiser. 1977. Plant response to heavy metals and sulphur dioxide, Environ. Pollu. 12: 63-71.

Kress, L.W., J.M. Skelly, and K.H. Hinkelmann. 1982. Relative sensitivity of 18 full -sib families of *Pinus taeda* to O_3 hr. Canad. J. For. Res. 12: 203-209.

Lacasse, N.L. and W.J. Moroz. 1969. Handbook of Effects Assessment-Vegetation Damage, Center for Air Environment Studies, PA State University, University Park, PA.

Legge, A.H. and S.V. Krupa (eds.). 1986. Air Pollutants and Their Effects on the Terrestrial Ecosystem. John Wiley and Sons, New York, 662 pp.

Leone, I.A. 1976. Response of potassium deficient tomato plants to atmospheric ozone. Phytopathology 66: 734-736.

Leone, I.A. and E. Brennan. 1969. The importance of moisture in ozone phyto-toxicity. Atmos. Environ. 3: 399-406.

Linthurst, R.A. (ed.). 1984. Direct and Indirect Effects of Acidic Deposition on Vegetation. Vol. 5. Acid Precipitation Series. Butterworth Publishers, Boston, 117 pp.

Linzon, S.N. 1978. Effects of airborne sulfur pollutnts on plants. In J.O. Nriagu, ed., Sulfur in the Environment: Part II, Ecological Impacts. Wiley, New York, pp. 109-162.

Loomis, R.C. and W.H. Padgett. 1973. Air Pollution and Trees in the East. U.S. D.A. Forest Service, State and Private Forestry, Atlanta, GA, 28 pp.

Malhotra, S.S. and R.A. Blauel. 1980. Diagnosis of Air Pollutant and Natural Stress Symptoms on Forest Vegetation in Western Canada. Publ. No. NOR-X-228. Northern Forest Research Centre, Canadian Forestry Service, Environment, Canada, Edmonton, Alberta, Canada, 84 pp.

Matsushima, J. and R.F. Brewer. 1972. Influence of sulfur dioxide and hydrogen fluoride as a mix or reciprocal exposure on citrus growth and development. J. Air Pollu. Control Assoc. 22: 710-713.

McCune, D.C. and L.H. Weinstein. 1971. Metabolic effects of atmospheric fluorides on plants. Environ. Pollu. 1: 169-174.

Miller, C.A. and D.D. Davis. 1981. Response of pinto bean plants exposed to O_3 hr, SO_2 hr, or mixtures at varying temperatures. Hort. Sci. 16: 548-550.

Mitchell, C.D. and T.A. Fretz. 1977. Cadmium and zinc toxicity in white pine, red maple and Norway spruce. J. Am. Soc. Hort. Sci. 102: 81-84.

Mudd, J.B. 1973. Biochemical effects of some air pollutants on plants. In J.A. Naegele, ed., Air Pollution Damage to Vegetation. Adv. Chem. Series No. 122, Am. Chem. Soc., Washington, DC, pp 31-47.

Mudd, J.B. 1975a. Sulfur dioxide. In J.B. Mudd and T.T. Kozlowski, eds., Responses of Plants to Air Pollution. Academic Press, New York, pp. 9-12.

Mudd, J.B. 1975b. Peroxyacetyl nitrates. In J.B. Mudd and T.T. Kozlowski, eds., Responses of Plants to Air Pollution. Academic Press, New York, pp. 97-119.

Mudd, J.B. and T.T. Kozlowski. 1975. Responses of Plants to Air Pollution. Academic Press, New York, 383 p.

Naegele, J.A. 1973. Air Pollution Damage to Vegetation. Adv. in Chem. Series No. 122, Am. Chem. Soc., Washington, DC, 137 p.

National Academy of Sciences. 1974. Chromium, NAS, Washington, DC, 155 pp.

National Academy of Sciences. 1975. Nickel. NAS, Washington, DC, 277 pp.

National Academy of Sciences. 1977a. Copper, NAS, Washington, DC, 115 pp.

National Academy of Sciences. 1977b. Effects of nitrogen oxides on vegetation. In Nitrogen Oxides, NAS, Washington, DC, pp. 147-158.

National Academy of Sciences. 1977c. Oxone and Other Photochemical Oxidants. NAS, Washington, DC, 789 pp.

National Academy of Sciences. 1978a. Effects of atmospheric sulfur oxides and related compounds on vegetation. In Sulfur Oxides. NAS, Washington, DC, pp. 80-129.

National Academy of Sciences. 1978b. An Assessment of Mercury in the Environment, NAS, Washington, DC, 192 pp.

Nicksic, S.W., J. Harkins, and P.K. Mueller. 1967. Some analyses for PAN and studies of its structure. Atmos. Environ. 1: 11-18.

Nielsen, D.G., L.E. Terrell, and T.C. Weidensaul. 1977. Phytotoxicity of ozone and sulfur dioxide to laboratory fumigated Scotch pine. Plant Dis. Reptr. 61: 699-703.

Norby, R.J. and T.T. Kozlowski. 1983. Flooding adn SO_2 stress interaction in *Betula papyrifera* and *B. nigra* seedlings. For. Sci. 29: 739-750.

Nriagu, J.O., (ed.). 1980. Zinc in the Environment. Part I. Ecological Cycling Wiley-Interscience, Somerset, NJ, 464 pp.

O'Connor, J.A., D.G. Parbery, and W. Strauss. 1974. The effects of phytotoxic gases on native Austrialian plant species. Part I. Acute effects of sulphur dioxide. Environ. Pollu. 7: 7-23.

Oehme, F.W., (ed.). 1978. Toxicity of Heavy Metals in the Environment. Dekker, New York, Part I, 515 pp.; Part II, 970 pp.

Ormrod, D.P., D.T. Tingey, M.L. Gumpertz, and D.M. Olszyk. 1984. Utilization of a response-surface technique in the study of plant responses to ozone and sulfur dioxide mixtures. Pl. Physiol. 75: 43-48.

Otto, H.W. and R.H. Daines. 1969. Plant injury by air pollutants. Influence of humidity on stomatal apertures and plant response to ozone. Science 163: 1209-1210.

Purves, D. 1977. Trace element contamination of the environment. Elsevier, New York, 260 pp.

Reinert, R.A.   1975.   Pollutant interactions and their effect on plants. Environ. Pollut. 9: 115-116.

Reinert, R.A.   1984.   Plant response to air pollutant mixtures. Annu. Rev. Phytopathol. 22: 421-442.

Rothermel, B. and R. Alscher. 1985. A light-enhanced metabolism of sulfite in cells of *Cucumis sativus* L. cotyledons. Planta 166: 105-110.

Runeckles, V.C. 1986. Photochemical oxidants. In A.H. Legge and S.V. Krupa, eds., Air Pollutants and Their Effects on the Terrestrial Ecosystem. John Wiley and Sons, New York, pp. 265-303.

Shacklette, H.T., J. A. Erdman, T.F. Harms, and C.S.E. Pupp. 1978. Trace elements in plant foodstuffs. In F. W. Oehme, ed., Toxicity of Heavy Metals in the Environment. Dekker, New York, pp. 25-68.

Sinclair, W.A., H.H. Lyon, and W.T. Johnson. 1987. Diseases of Trees and Shrubs. Cornell University Press, Ithaca, NY, 574 pp.

Skeffington, R.A. and T.M. Roberts. 1985. The effects of ozone and acid mist on Scots pine saplings. Ocecologia 65: 201-206.

Skelly, J.M., D.D. Davis, W. Merrill, E.A. Cameron, H.D. Brown, D.B. Drummond, and L. S. Dochinger, (eds.). 1987. Diagnosing Injury to Eastern Forest Trees. Forest Response Program. U.S.D.A. Forest Service PA State University, University Park, PA, 122 pp.

Smith, H.J. and D.D. Davis. 1977. The influence of needle age on sensitivity of Scotch pine to acute doses of SO_2 hr. Plant Dis. Reptr. 61: 870-874.

Smith, W.H. 1975. Variable tree response due to environmental factors — edaphic. In W.H. Smith and L.S. Dochinger, eds., Air Pollution and Metropolitan Woody Vegetation. U.S.D.A. Forest Service. PIEFR-PA-1, Upper Darby, PA, pp. 17-18.

Smith, W.H., T.G. Sicama, and S.L. Clark. 1986. Atmospheric deposition of heavy metals and forest health: An overview and a ten-year budget for the input/output of seven heavy metals to a northern hardwood forest. Publ. No. FWS-87-02, Distinguished Lectureship Program, Virginia Polytechnic Institute and State University, Blacksburg, VA.

Steiner, K.C. and D.D. Davis. 1979. Variation among *Fraxinus* families in foliar response to ozone. Can. J. For. Res. 9: 106-109.

Stern, A.C., H.C. Wohlers, R. W. Boubel, and W.P. Lowry. 1973. Fundamentals of Air Pollution. Academic Press, New York 492 pp.

Tamm, C.O. and A. Aronson. 1972. Plant Growth as Affected by Sulphur Compounds in Polluted Atmosphere. A Literature Survey. Royal College of Forestry, Dept. Forest Ecology and Forest Soils, Research Note No. 12, Stockholm, Sweden, 53 pp.

Tandy, N.E., R.T. DiGuilio, and C.J. Richardson. 1987. Isozymes of superoxides dismutase in red spruce and their importance in protecting against oxidative stress. In G. Hertel, ed., The Effects of Atmospheric Pollution on Spruce and Fir Forests in the Eastern United States nd the Federal Republic of Germany. U.S.D.A. Forest Service Genl. Tech. Publ. No. 255, Northeastern Forest Exper. Stat., Broomall, PA.

Taylor, O.C. 1973. Oxidant Air Pollutant Effects on a Western Coniferous Forest Ecosystem. Task C Report No. EP-R3-73-043B, Statewide Air Pollut. Res. Center, Riverside, CA, 189 pp.

Taylor, O.C., C.R. Thompson, D.T. Tingey, and R.A. Reinert. 1975. Oxides of nitrogen. In J.B. Mudd and T.T. Kozlowski, eds., Response of Plants to Air Pollution. Academic Press, New York, pp. 121-139.

Temple, P.J. and R. Wills. 1979. Sampling and analysis of plants and soils. In W.W. Heck, S.V. Krupa, and S.N. Linzon, eds., Methodology for the Assessment of Air Pollution Effects on Vegetation. Air Pollu. Control Assoc., Pittsburgh, PA, pp. 1-23.

Thompson, C.R. and G. Kats. 1978. Effects on continuous H_2 S fumigation on crop and forest plants. Environ. Sci. Technol. 12: 550-553.

Tingey, D.T. and R.A. Reinert. 1975. The effect of ozone and sulphur dioxide singly and in combination on plant growth. Environ. Pollu. 9: 117-125.

Tingey, D.T. and G.E. Taylor Jr. 1982. Variation in plant response to ozone: A conceptual model of physiological events. In MH. Unsworth and D.F. Ormrod, eds., Effects of Gaseous Air Pollution in Agriculture and Horticulture. Buterworth Scientific, London, pp. 113-138.

Townsend, A.M. 1975. Variable tree response due to genetic factors. In W.H. Smith and L.S. Dochinger, eds., Air Pollution and Metropolitan Woody Vegetation. U.S.D.A. Forest Service. PIEFR-PA-1, Upper Darby, PA, pp. 18-19.

Treshow, M. 1970. Ozone damage to plants. Environ. Pollu. 1: 115-161.

Treshow, M. 1984a. Air Pollution and Plant Life. John Wiley and Sons, New York, 486 pp.

Treshow, M. 1984b. Diagnosis of air pollution effects and mimicking symptoms. In M. Treshow, ed., Air Pollution and Plant Life. John Wiley and Sons, New York, pp. 97-112.

U.S.D.A. Forest Service. 1973. Air Pollution Damages Trees. Stte and Private Forestry, Upper Darby, PA, 32 pp.

U.S. Environmental Protection Agency. 1976. Diagnosing Vegetation Injury Caused by Air Pollution. U.S.E.P.A., Washington, DC.

U.S. Environmental Protection Agency. 1982. Air Quality Criteria for Particulate Matter and Sulfur Oxides. Vol. III. U.S.E.P.A. Publ. No. 600-8-82-029c. Research Triangle Park, NC.

U.S. Environmental Protection Agency. 1986. Air Quality Criteria for Ozone and Other Photochemical Oxidants. Vol. III. Publ.. No. EPA-600-8-84-020cF. Research Triangle Park, NC.

Unsworth, M.H. 1984. Evaporation from forests in cloud enhances the effects of acid deposition. Nature 312: 262-264.

UDI—Commission for Air Pollution Control. 1987. Acidic Precipitation. Formation and Impact on Terrestrial Ecosystems. Verein Deutscher Ingenieure, Düsseldorf, FRG, 281 pp.

Wallace, R.G. and D.J. Spedding. 1976. The biochemical basis of plant damage by atmospheric sulphur dioxide. Clean Air 10: 61-64.

Weinstein, L. H. 1975. Dose-response relationships. In W.H. Smith and L.S. Dochinger, eds., Air Pollution and Metropolitan Woody Vegetation. U.S.D.A. Forest Service. PIEFRA-PA-1, Upper Darby, PA, pp. 11-13.

Weinstein, L.H. 1977. Fluoride and plant life. J. Occup. Med. 19: 49-78.

Weinstein, L.H. and D.C. McCune. 1971. Effects of fluoride on agriculture. J. Air Pollu. Control Assoc. 21: 410-413.

Weinstein, L.H. and D.C. McCune. 1979. Air pollution stress. In H. Mussell and R. Staples, eds., Stress Physiology in Crop Plants. Wiley, New York, pp. 328-341.

Weinstein, L.H., R.J. Kohut, and J.S. Jacobson. 1987. Research at Boyce Thompson Institute on the Effects of Ozone and Acidic Precipitation on Red Spruce. Proc. 80th Annual Meeting, Air Pollu. Conrol Asoc., June 21–26, 1987, New York.

Wilhour, R.G. 1970. The influence of temperature and relative humidity on the response of white ash to ozone. Phytopathology 70: 579.

Winner, W.E., H.A. Mooney, and R.A. Goldstein. 1985. Sulfur Dioxide and Vegetation. Stanford University Press, Stanford, CA, 593 pp.

Wood, T. and F.H. Bormann. 1974. The effects of an artificial acid mist upon the growth of *Betula alleghaniensis* Br. Environ. Pollu. 7: 259-267.

Zeevaart, A.J. 1976. Some effects of fumigating plants for short periods with NO_2 hr. Environ. Pollu. 11: 97-108.

16

Class II Summary: Forest Responds by Exhibiting Alterations in Growth, Biomass, Species Composition, Disease, and Insect Outbreaks

With sufficient exposure to an air pollutant, forest trees will be adversely impacted. When this occurs the threshold between Class I and Class II interactions is crossed. At an intermediate dose, the specific contaminant concentration and time of exposure varying greatly with the specific pollutant and forest situation, the influence on individual forest components may range from extremely subtle to visibly dramatic. Chapters 8 through 15 have reviewed the evidence available to support the hypotheses that intermediate air pollution loads may alter or inhibit forest tree reproduction, alter forest nutrient cycling, alter tree metabolism, or change forest stress conditions by influencing insect pests, microbial pathogens, and by directly damaging foliar tissue. All but the latter of these impacts is extremely subtle, visibly asymptomatic, and detectable only by very careful forest monitoring.

The primary adverse response of a forest ecosystem to a sustained intermediate dose and Class II interaction would be reduced growth and consequently reduced biomass. Reduced essential element availability, decreased photosynthesis, altered carbon allocation, increased respiration, increased insect and disease stress, and decreased foliar tissue would all contribute to a reduction in tree growth rates and ultimately to lessen forest biomass. Alterations in the reproductive strategies of individual tree species or a differential response of these species to reduced nutrition, altered metabolism and pest stress, and to direct foliar injury may cause changes in competitive ability and may ultimately lead to alterations in tree succession and species composition. These relationships are outlined in Table 16-1. Useful reviews of Class II vegetative responses to air pollutants include Heck et al. (1977), Jensen et al. (1976), Weinstein and McCune (1979), Laurence and Weinstein (1981), and McLaughlin (1985).

Table 16-1. Interaction of Air Pollution and Temperate Forest Ecosystems under Conditions of Intermediate Air Contaminant Load, Designated Class II Interaction.

Forest soil and vegetation: Activity and response	Ecosystem consequence and impact
1. Forest tree reproduction, alteration or inhibition	1. Altered species composition

2. Forest nutrient cycling, alteration
    a. Reduced litter decomposition
    b. Increased plant leaching, soil leaching, and soil weathering
    c. Disturbance of microbial symbioses
	2. Reduced growth, less biomass

3. Forest metabolism
    a. Decreased photosynthesis
    b. Increased respiration
    c. Altered carbon allocation
	3. Reduced growth, less biomass

4. Forest stress, alteration
    a. Phytophagous insects, increased or decreased activity
    b. Microbial pathogens, increased or decreased activity
    c. Foliar damage increased by direct air pollution influence
	4. Altered ecosystem stress: Increased or decreased insect infestations; Increased or decreased disease epidemics; Reduced growth, less biomass, altered species composition

What is the evidence available to support the hypotheses that forest ecosystem exposure to intermediate doses of air pollutants result in reduced growth and biomass, altered species composition, and increased forest pest impact?

A. Forest Growth Reduction Caused by Air Pollution

The most fundamental characteristic of an ecosystem is its productivity. Forest productivity is high relative to other ecosystems and net productivity of 1200 dry g m^{-2} yr^{-1} for trees and shrubs together is quite typical for temperate forests (Whittaker, 1975).

1. Natural Regulation of Forest Growth

Forest productivity is strongly controlled by a large number of environmental factors, genetic determinants, system age, and stand dynamics. With regard to environmental regulators of growth, temperature coupled with water, nutrient, and light availability are recognized to be the most fundamental. Other variables, including carbon dioxide and insect and pathogen activity, are also extremely important (Hennessey et al., 1986) (Figure 16-1). Genetic determinants provide a range and boundaries for growth patterns. Age is very importantly related to forest growth. In even-age forests, it is possible to generalize three major stages of growth. Growth from seedling to pole size is proportional to age.

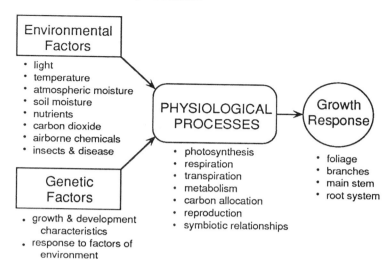

Figure 16-1. The physiological processes that determine tree growth are regulated by environmental and genetic factors.
Source: Woodman (1986).

Forest growth is highest from pole size to mature tree, where the leaf area index is highest and most stable. Eventually the combined influence of natural aging and tree size causes a reduction in the leaf area index and growth (Woodman 1986) (Figure 16-2). Stand structure and competition are additional variables exerting important regulation over tree growth. Substantial evidence suggests that multistoried, mixed species stands are more productive than single storied, single species stands (Kelty, 1984, Assmann 1970, Goff and Zedler, 1968, Whittaker, 1966).

Forest growth is complex in measurement as well as concept. Addition of woody tissue is the dominant feature of forest growth. The accumulation of woody biomass (living weight) represents gross photosynthetic production less respiratory losses. Individuals interested in the commercial value of forest systems will measure growth in terms of annual wood increment. The latter is calculated from measurements of tree bole diameter and height, and is modified by the amount of bole taper. Typically annual increment is expressed as volume of wood per unit of land area. It may also be expressed as weight of wood per unit of land area. Site index is another indicator of forest growth. It represents the average height of the dominant trees in a given forest system at some arbitrary age, typically 50 years. Temperate forest trees conveniently add a readily discernable layer of wood every year to their main stem. The annual increment of radial growth is easily measured by examining ring width (Nash et al., 1975, NSF, 1977, Heikkenen, 1984). Sensitive measurement of cambial growth may be obtained by the use of dendrometers. These devices quantify growth by mea-

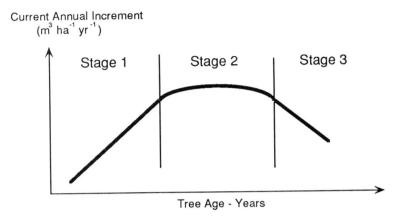

Figure 16-2. Tree age and growth in even-aged forests.
Source: Woodman (1986).

suring increases in stem circumference or single stem radii (Fritts, 1976; Kramer and Kozlowski, 1979). Measurement of basal area, or stem cross-sectional area in square centimeters at an arbitrary height above the ground, is yet another measure used to quantify forest tree growth. Regression equations are commonly used to estimate biomass by using diameter or diameter and height measurements (Tritton and Hornbeck, 1982).

In recognition of the variety of Class II interactions discussed, air quality may also influence forest productivity in numerous environments.

2. Sulfur Dioxide

The capacity of sulfur dioxide to reduce plant productivity is well established. The ability of this gas to reduce the yield of commercially important plants is of special concern to those investigating agricultural ecosystems (Davis, 1972; Jones et al., 1977; Sprugel et al., 1979). Chamber fumigations have provided evidence that sulfur dioxide can reduce the growth of forest seedlings (for example, Keller, 1978; Kress and Skelly, 1980) as reflected in height growth and radial wood increment (Figure 16–3). The most impressive evidence for sulfur dioxide suppression of forest growth stems from investigations carried out in natural forest environments exposed to elevated ambient sulfur dioxide from surrounding point sources.

Scheffer and Hedgcock (1955) included an investigation of growth impact in their study of smelter influence on adjacent coniferous ecosystems in Washington and Montana. Damage in the Deerlodge National Forest from the Washoe smelter located close to Anaconda, Montana was investigated in 1910 and 1911. A small number of sapling or pole-size trees were felled within the zone of visible forest damage, in the transition zone, and in areas not subject to

PICEA   ABIES

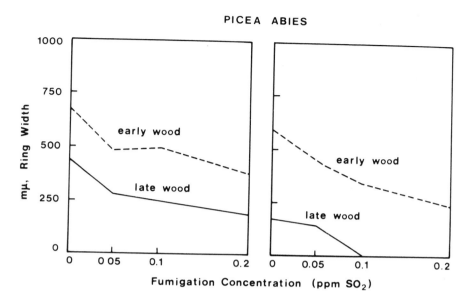

Figure 16-3. The influence of laboratory fumigation with sulfur dioxide on the average radial ring width of early wood and late wood of two spruce clones. Source: Keller (1980).

smelter damage. Radial diameter increments were determined for the sample trees. Species given primary attention included subalpine fir, lodgepole pine, and Douglas fir. Retardation of diameter growth of all three species was indicated for all years from 1892 through 1910 in the zone of visible injury. Growth suppression of transition zone trees was not well defined, but was suggestive of some degree of intermediate retardation. The authors concluded, however, that the differences in growth rates of the different zones surrounding the Washoe smelter were indicative, but not proof of, a sulfur dioxide effect. They were unable to separate the influence of the sulfur gas from differences due to natural forces that may have also influenced growth in the various zones. Conifer forests of the upper Columbia River Valley, Washington subject to sulfur dioxide from a smelter located at Trail, British Columbia were studied from 1928 through 1936. Height increment and current annual increase in length of the terminal shoot were employed to evaluate sulfur dioxide impact on ponderosa pine growth in forest zones exhibiting varying degrees of foliar symptoms. In zone 1, 60–100% of the foliar tissue exhibited sulfur dioxide injury symptoms, while zones 2 and 3 had 30–60% and 1–30%, respectively. Measured trees included saplings, height range 3–7 m (8–20 feet), and pole size trees, height range 8–17 m (25–50 feet). Height increments were taken during the fall by bending small trees and felling large ones. The results of these measurements are presented in Figure 16–4. The authors concluded that there was no indication of

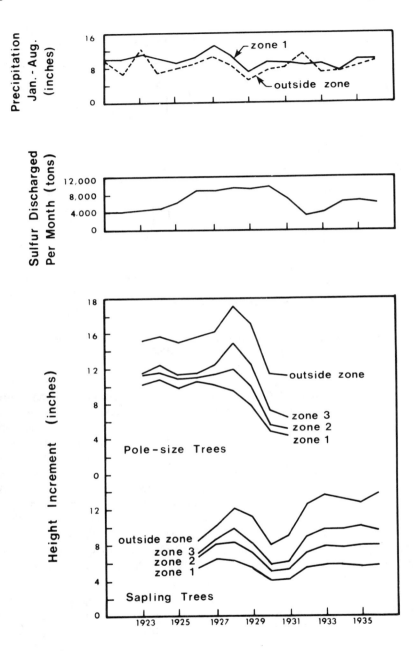

Figure 16-4. Annual height increment of ponderosa pine in the upper Columbia River Valley, Washington, subject to sulfur dioxide stress from a smelter in Trail, British Columbia.
Source: Scheffer and Hedgcock (1955)

sulfur dioxide influence on height growth in any zone prior to 1926. After 1926, however, height growth in zones 1, 2, and 3 showed generally less response to precipitation differences relative to the response outside the symptomatic zones. It was concluded that there was a definite retardation of height growth in 1928 resulting from 1927 injury (a 40% increase in sulfur dioxide discharge began in 1926) in zones 1 and 2, and approximately a year later in zone 3, and that the retardation was sustained in all three zones through the end of observations in 1936. Annual radial diameter growth was also determined for the ponderosa pines of the various zones (Figure 16-5). Diameter growth revealed initial retardation as early as 1921 in zone 1. Retardation in zones 2 and 3 began in 1924. All zones exhibited sustained suppression of diameter growth through the conclusion of the experiment in 1934. The average percentage reductions in diameter growth were 38, 24, and 17% for zones 1, 2, and 3, respectively. The authors judged that height growth curves and diameter growth curves were generally similar, except that the diameter measurements appeared to reveal the suppression sooner. It is extremely unfortunate that we do not know the ambient concentrations of sulfur dioxide in the atmospheres of the various zones along with the diurnal and seasonal fluctuations of these concentrations.

Samuel N. Linzon, Air Management Branch of the Ontario, Canada Department of Energy and Resources Management, conducted a 10 year (1953–1963) assessment of the Sudbury smelter district of Ontario on the growth of surrounding forests (Linzon, 1971). During the study period three large smelters were discharging approximately two million tons of sulfur dioxide to the atmosphere annually. The investigation was concentrated on eastern white pine growth because of its susceptibility to sulfur dioxide injury and because of its commercial importance. Forty-two permanent sample plots containing eastern white pine ranging in age from 65 to 85 years were established. Fume area plots were established at increasingly greater distances northeast of Sudbury in line with the prevailing wind, while control plots were located beyond the influence of smelter effluent. The degree of foliar injury was employed to segregate the fume area into inner, intermediate, and outer zones. Severe tree damage occurred up to 48 km (30 miles) northeast of Sudbury and the inner zone comprised an area of approximately 1865 km^2 (720 miles^2). The intermediate and outer zones contained 4144 km^2 (1600 miles^2) and 7770 km^2 (3000 miles^2), respectively. Radial increment cores were taken from 20 dominant living white pine trees on each sample plot. Annual radial growth for the 1940–1960 period was measured on each core. The results indicated a gradual decline in the growth of white pine in areas adjacent to the smelters, while a constant growth pattern was indicated in other areas. Height and diameter measurements of the pines were employed to construct a local volume table for each plot. Throughout the inner fume zone there was a net average annual loss in total volume of 0.03 m^3 (0.1 feet^3) per white pine tree in the 18–30 cm (7–12 inch) diameter class due to the combination of mortality and reduced growth. In the control plots each tree added 0.09 m^3 (0.30 feet^3) in total volume per year. The total reduction of volume per tree per year in the inner zone, therefore, was 0.12 m^3 (0.40 feet^3). Unfortunately the

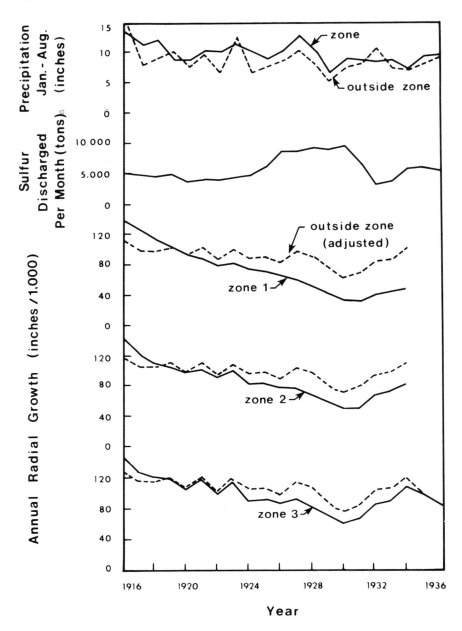

Figure 16-5. Annual radial increment of ponderosa pine in the upper Columbia River Valley, Washington, subject to sulfur dioxide stress from a smelter in Trail, British Columbia. Actual measurements (solid line) are compared with theoretical growth (dotted line) that could have occurred in the absence of sulfur dioxide injury. Source: Scheffer and Hedgcock (1955).

suppression of volume growth of stressed, but not killed, trees was not clear from this study. The relative significance of white pine weevil infestation, *Cronartium ribicola* infection, sulfur dioxide influence, and other stress factors in causing mortality was not clear. It is obvious, however, that the growth of eastern white pine is dramatically influenced over a very large area by the Sudbury smelter complex.

Stemple and Tryon (1973) conducted a unique investigation to assess the influence of sulfur dioxide and fly ash from coal burning railroad locomotives on surrounding forest vegetation. A deciduous forest ecosystem, dominated by white oak, and situated on a 267 m (800 foot) hill with a railroad at its base was examined in northern West Virginia. Sample plots were established in the "damage area" delineated by a layer of fly ash on the soil and a control area not influenced by the railroad. Measurements were made of site index (Figure 16-6), tree height, and annual radial increment. The site quality of the land adjacent to the railroad was drastically reduced. The authors judged that the coal tonnage hauled past the damaged area was directly correlated with the sulfur dioxide and other effluents impacting the area. Annual radial increment of the white oaks varied inversely with the annual coal tonnage moved through the study forest.

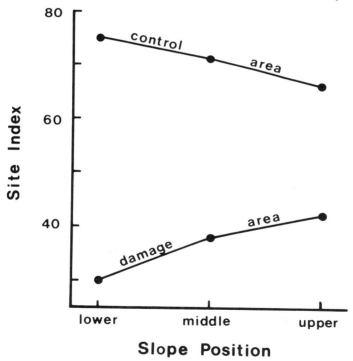

Figure 16-6. Site index for white oaks in Preston County, West Virginia, subject to effluent from coal-burning locomotives compared with an unimpacted control site. Source: Stemple and Tryon (1973).

John M. Skelly and colleagues of the Virginia Polytechnic Institute and State University, Blacksburg, Virginia have examined the growth of a variety of forest trees subject to effluent from the U.S. Army Radford Ammunition Plant in southwest Virginia. The Radford Arsenal initiated operations in 1939 and has had three significant production peaks that have corresponded with the major United States defense efforts in World War II, the Korean conflict, and the Vietnam conflict. The fact that the ammunition plant was situated in an isolated location, coupled with the realization that major effluent release was periodic and was associated with massive national defense efforts, presented the investigators with an opportunity to correlate surrounding forest growth with the quantity of air pollutants (sulfur dioxide and nitrogen oxides) released from the facility. Initial efforts to evaluate the influence of past pollutant levels on the growth of two important commercial tree species involved eastern white pine and yellow poplar (Stone and Skelly, 1974). Sample plots were established in the prevailing downwind direction. Annual ring widths of 43 eastern white pine and 50 yellow poplar, all in dominant or codominant crown classes, were determined to the nearest 0.05 mm with a dendrochronograph. A highly significant inverse relationship was found by linear regression analyses to exist between the fluctuating production levels of the arsenal and the annual ring widths of white pine and yellow poplar (Figure 16-7). Severely symptomatic 13-year-old eastern white pine in the vicinity of the arsenal have been shown to average only 66% of the height of trees lacking symptoms (Skelly et al., 1972). In an effort to better resolve the relationship in intensity of foliar symptoms of white pine to growth suppression, a study was conducted in a stand 1.6 km (1 mile) downwind of the closest arsenal emission source of nitrogen oxides. The trees of this stand ranged in age from 35 to over 135 years and exhibited foliar symptoms ranging from severe to asymptomatic. Analysis of regression correlations indicated no significant growth rate differences between symptom classes during pollution peaks. Growth of asymptomatic trees was judged to be reduced as much as that of injured trees during peak release episodes (Phillips et al., 1977a). The growth response of young loblolly pine has also been examined in the vicinity of the arsenal (Phillips et al., 1977b). Three stands of plantation loblolly ranging in age from 15 to 18 years and located 1.4–2.7 km (0.9–17 miles) northeast or northwest of the main power facility were examined. A significant inverse relationship was demonstrated between annual radial increment growth and arsenal production levels in two of the loblolly pine stands. In both of these stands additional analyses suggested theoretical reductions in diameter growth without the presence of visible injury.

Legge et al. (1977) have conducted field studies on the growth of lodgepole pine-jack pine hybrids, white spruce, and aspen subjected to sulfur dioxide and hydrogen sulfide from a natural gas processing plant in Whitecourt, Alberta, Canada. Growth of lodgepole pine-jack pine hybrids was quantified by examining basal area increments. Significant decreases in basal area increment for hybrid stands after 1965 were judged to be the result of unfavorable moisture conditions (excess water) rather than sulfur gas emissions.

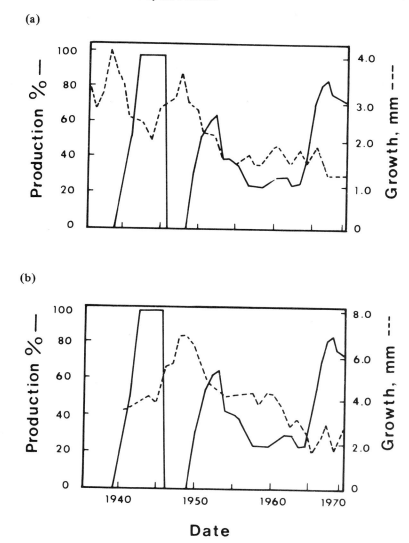

Figure 16-7. Influence of periodic production levels, and associated effluent containing sulfur dioxide and nitrogen oxides, from a military arsenal on the average annual ring width of 43 white pine trees (a) and 50 yellow poplar trees (b).
Source: Stone and Skelly (1974).

It is important to realize that all studies reviewed in this section have considered the amount of rainfall occurring over their study periods. Failure to do this would, of course, meaningfully reduce the confidence of experimental results, as precipitation exerts such a profound influence on tree growth. Perhaps the most disappointing aspect of the sulfur dioxide studies is infrequent presentation of dose information. Ambient concentrations and exposure durations were rarely monitored or reported. For historical studies it is clearly frequently impossible to provide this information. For contemporary investigations, however, inclusion of this information would greatly increase the utility of the data. Horntvedt (1970) has reported the results of a field study of the response of radial increment growth of spruce subjected to ambient sulfur dioxide. Reductions of 45, 27, and 18% in diameter growth over a 9-year period corresponded directly to average annual ambient sulfur dioxide levels of 543, 143, and 144 μg m^{-3}, respectively.

Keller (1985) provides a summary of sulfur dioxide and forest growth research. The evidence is clear, at sufficient exposure this gas can suppress growth and reduce forest tree vigor in the absence of any visible symptoms.

3. Acid Deposition

Initial delineation of the acid precipitation problem and early efforts to describe attendant environmental ramifications have come from Scandinavia (Bolin, 1971). It is not surprising, therefore, to realize that a considerable effort has been made in Sweden and Norway to evaluate the influence of acid rain on forest growth (Dahl and Skre, 1971; Sundberg, 1971).

Jonsson and Sundberg (1972) and Jonsson (1976) developed a model to examine the statistical correlation of Scotch pine and Norway spruce growth, as measured by annual ring widths for the period 1910–1965, with the intensity of increasing acid precipitation initiated in 1950 in southern Sweden (below 61°N latitude). Fifty percent of the forest growth in Sweden occurs in the latter area. The authors stratified their analysis of over 4000 trees by site classes and by region of differing susceptibility to acid rain stress. The latter classification was based on regional intensity of acid rain, chemistry of regional lakes and rivers, and distribution of soil types. The latter criterion was deemed especially important, and regions with base poor tills or sand deposits, and with aquatic resources of low pH and cation concentration, were judged to be particularly liable to stress from acid precipitation. The authors made a sincere effort to consider nonpollution factors that may have influenced forest growth. Special attention was given to differences in climatic factors and silvicultural practices in the study regions. The model predicted reduced forest growth in more susceptible regions of approximately 0.3–0.6% for the period 1951–1965 relative to less susceptible areas. The analysis did not enable the authors to conclude that acid rain was the cause of this reduced growth. Likewise acid precipitation was not eliminated as a cause and the analysis did not support any alternative explanations for the poorer growth observed. Jonsson and Sundberg concluded their study by presenting predictions of future Swedish forest growth given various scenarios of sulfur deposition.

Abrahamsen et al. (1976a,b) have examined trends in Norwegian forest growth and acid rain in a manner very similar to the Jonsson and Sundberg Swedish study. Comparisons of tree growth, as indicated by ring widths, were examined between regions presumed or known to have different inputs of acid precipitation and between sites of differing susceptibilities to acid stress due to soil characteristics. No consistent regional growth differences were observed. In fact, "somewhat better" development of pine was observed in one region (Sørlandet area) following 1950. Less productive sites, poor vegetation types, and shallow soils did not appear to be more sensitive to soil acidification. The authors concluded that no clear effects of acid precipitation on diameter growth of spruce and pine were detected by the regional tree-ring analyses.

Artificial applications of acid rain to Norway spruce and lodgepole pine seedlings have been conducted at the Sønsterud Forest Nursery in Norway (Tveite and Teigen, 1976). After 3 years of field applications no negative impacts of acid application were detected. Lodgepole pine exhibited a 20% stimulation of height growth after 3 years following application of 50 mm water per month at pH 4 and 3. A stimulation of height growth of approximately 15% was recorded for Norway spruce given a similar treatment. Laboratory applications of simulated acid precipitation have revealed no effect or a slightly stimulatory influence on the growth of seedlings of northeastern United States forest species (Wood and Bormnann, 1974, 1975). Working with 1-year-old yellow-poplar seedlings, on the other hand, Dochinger and Jensen (1985) observed that treatment with acid mist (pH 2.5) significantly reduced height, leaf dry weight, and new stem dry weight relative to mist treatment at pH 5.5.

In the United States the forest regions subject to the most acidic precipitation are located in the Northeast. In a comprehensive study of production and biomass of the northern hardwood forest conducted at the Hubbard Brook Experimental Forest in New Hampshire, Whittaker et al. (1974) observed a significant decline in growth from 1956–1960 to 1961–1965. Wood volume growth declined 17% between these two periods. Net ecosystem production was estimated as 350 g m^{-2} yr^{-1} above ground and 85 g m^{-2} yr^{-1} below ground for 1956–1960 and 238 and 52 g m^{-2} yr^{-1} for 1961–1965. A widespread drought occurred throughout the Northeast during 1961–1965, but examination of wood volume growth patterns, which for some trees could be traced for two centuries at Hubbard Brook, revealed no previous decrease similar to the precipitous decrease recorded in 1961–1965. The authors noted that the period of growth decrease was coincident with a period of increasing acidity in precipitation and inferred that this may have been responsible for the decrease in productivity.

Cogbill (1976, 1977) selected two United States mountainous, remote forest sites to assess the effect of acid precipitation on tree growth by tree-ring analysis. A northern hardwood forest site in the White Mountain National Forest, New Hampshire and a red spruce site in the Great Smoky Mountains National Park, Tennessee were studied. Average acidity in the precipitation in New Hampshire was pH 4.1 and in Tennessee was pH 4.4 at the time of the study. The initiation date of increased acidity of precipitation was unknown for both sites but was presumed to be prior to 1955 for New Hampshire and approximately 1955 for Tennessee. The tree-ring chronologies for the three northern hardwood species and the spruce are presented in Figure 16-8. No clear indication of regional decrease in tree growth was found. The variation due to climate

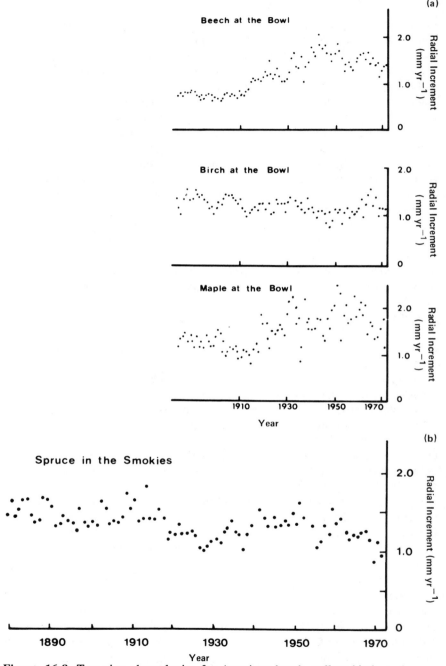

Figure 16-8. Tree ring chronologies for American beech, yellow birch, and sugar maple of the Bowl Natural Area, White Mountain National Forest, New Hampshire (a) and red spruce of the Great Smoky Mountains National Park, Tennessee (b). Source: Cogbill (1977).

showed no recent trend. Cogbill concluded that no correlation of forest growth and acid rain could be established for eastern North America.

Increment cores from pitch pine, shortleaf pine, and loblolly pine taken from the Pine Barren region of southern New Jersey were observed to indicate a decrease in growth rate from 1955 to 1979 (Johnson et al., 1981). In view of a strong statistical relationship between stream pH, an index of precipitation pH, and the tree growth rate decline, these authors suggested that acid deposition may have been the growth limiting factor.

LeBlanc et al. (1987a) have studied 47-52 year old conifer plantations (red pine, Scots pine, Norway spruce) growing at 250 m in Warrensburg, New York on the southeastern edge of the Adirondack Mountains. Some of there plantations existed on soils judged to have high risk of adverse impact from acid deposition (Chapter 9). Analyses of dendroecological data indicated that long-term variation in growth of sampled trees was related to stand dynamics and climatic factors, with little evidence for extended term impact of acid deposition. Species/site trees identified as potentially more susceptible to acidic deposition effects did exhibit synchronous growth decreases after 1960, but the beginning of decreased growth was coincident with documented climatic anomalies (drought and extreme winter temperatures) (LeBlanc et al., 1987b). In view of their inability to document decreased tree growth, unrelated to natural determinants of growth, in simple plantation forests located in one of the North American regions judged to be at high risk to acid deposition effects, LeBlanc et al. (1987c) concluded that it may be very difficult to document anomolous (pollution induced) decreases in mixed forests growing on more fertile sites in eastern North America. A comprehensive review of acid deposition influences on forest productivity (NCASI, 1983) concluded that evidence for damaging effects has not been provided. The evidence available to date linking high acid deposition and forest growth declines is generally correlative only, and frequently has not adequately evaluated alternative natural causes of growth impact.

4. Ozone

The ability of ozone to reduce the growth of agricultural plants has been appreciated for approximately 30 years (Todd and Garber, 1958) and has been adequately reviewed (Heck et al., 1988). The National Crop Loss Assessment Program (Heck et al., 1984) employed open-top field chambers in various United States locations to establish that ambient ozone causes yield reductions in soybean (10%), peanut (14-17%), turnip (7%), head lettuce (53-56%), and red kidney bean (2%) (Heck et al., 1982). Economic analyses have indicated that the benefits to society of moderate (25%) ozone reductions would be approximately $1.7 billion (Adams et al., 1985).

A large amount of evidence, generated from studies employing seedling-or sapling-size trees, also indicates that ozone can impact the growth of forest trees. Growth of plane trees in ambient greenhouse air in Washington, DC, was observed by Santamour (1969) to be only 75% of the height growth in filtered air.

Table 16-2. Height Growth of Loblolly Pine Seedlings at Termination of Exposure to 50 ppb Ozone, 140 ppb Sulfur Dioxide, and/or 100 ppb Nitrogen Dioxide for 6 hr day^{-1} for 28 Consecutive Days, Expressed as a Percent of the Control Trees.

Family	$O_3{}^a$	SO_2	NO_2	$O_3 + NO_2$	$O_3 + SO_2{}^b$	$O_3 + SO_2 + NO_2$
4-5 x 523	94	79^c	102	81	77^c	74^c
14-5 x 517	97	118	96	90	88^c	88^c

^{a}Average of three O_3 treatments.
^{b}Average of two $O_3 + SO_2$ treatments for 4-5 X 523.
^{c}Different from the control at $P = 0.05$ according to the Duncan's New Multiple Range Test.
Source: Kress et al. (1982).

Jensen (1973) observed that the growth of 1-year old sycamore seedlings was reduced by ozone doses of 300 ppb (588 μg m^{-3}) for 8 hr day^{-1} for 5 days week^{-1} for 5 months. Jensen (1982) also treated silver maple and eastern cottonwood seedlings with ozone at 0, 100, 200, or 300 ppb (0, 196, 392, or 588 μg m^{-3}) for 12 hrs day^{-1} for 60 days and found that relative growth, leaf-area expansion, and leaf-weight rates declined with increasing ozone exposure.

Kress et al. (1982a) found that low dose exposure of loblolly pine seedlings to ozone, nitrogen dioxide, and sulfur dioxide in combination can result in significant height growth suppression. All three pollutant concentrations employed were below the National Ambient Air Quality Standards (Table 16–2). Similar exposure of American sycamore to these three pollutants also resulted in significant growth suppression. Height growth was significantly suppressed by ozone alone in some cases (Kress et al., 1982b). In additional experiments using comparable exposures to ozone and nitrogen dioxide, Kress and Skelly (1982) observed that white ash and yellow poplar exhibited significant growth stimulations when exposed to ozone at 50 ppb (98 μg m^{-3}) and yellow poplar and Virginia pine were the only species that failed to show any significant growth response to ozone at 150 ppb (294 μg m^{-3}). Fumigation with ozone or sulfur dioxide alone did not significantly affect shoot height growth or seedling dry weight of yellow poplar, but in combination with each other and with nitrogen dioxide, a greater than additive response occurred (Mahoney et al. 1984). This led John Skelly and his colleagues to emphasize that gaseous pollutants may interact synergistically to reduce plant growth.

Hogsett et al. (1985) have studied the growth response of two varieties of slash pine seedlings to chronic ozone exposures. Emergent seedlings were exposed continuously to two daily peak exposure profiles of ozone having 7 hr (0900-1600) seasonal means of 104 and 76 ppb (204 and 149 μg m^{-3}) over a 112 day period. Destructive harvests at 7-day intervals over the exposure period were used to assess visible injury and to construct growth curves for stem diameter, plant height, top and root dry weight, and needle number and length. Visible injury was found to be slight, but all growth parameters decreased significantly

Table 16-3. Percent Reduction from Control Treatment on Growth of Each Slash Pine Variety Exposed to Low and High Ozone Profiles at the Final Harvest (112 days).

	Low ozone		High ozone	
	Var.	Var.	Var.	Var.
Parameter	*elliottii*	*densa*	*elliottii*	*densa*
Stem diameter	24	30	40	50
Plant height	22	30	33	41
Plant dry weight	21	19	50	48
Top dry weight	17	16	46	44
Root dry weight	38	34	67	68
Needles				
Number of primaries	7	9	19	21
Length of primaries	5	6	18	20
Number of secondaries	33	12	55	43
Length of secondaries	3	16	22	29

Source: Hogsett et al. (1985).

with time and ozone concentration. Root growth was the most severely impacted (Table 16–3).

The use of open-top chambers in field settings has allowed growth studies to evaluate the influence of ozone on larger seedlings and sapling trees. In the initial application of this exposure technique, Duchelle et al. (1982) employed open-top chambers to evaluate the effect of ambient ozone in the Shenandoah National Park in Virginia on the growth of eight planted forest species native to the Virginian Appalachian Mountains. Height growth was suppressed for all species at the end of the second growing season when grown in open plots (no chamber) and ambient chambers compared to those grown in chambers with charcoal-filtered air.

Wang et al. (1985) employed open-top chambers to study ambient ozone effects on trembling aspen in Millbrook, New York (approximately 110 km north of New York City). Over a three-year period, four clones representing a range of pollutant sensitivities were exposed to charcoal-filtered and ambient air. Ambient ozone significantly reduced (12–24%) above--ground dry-matter production and modified tree morphology; root/shoot ratios, and rates of leaf senescence. For two clones, biomass was reduced in the absence of visible foliar symptoms. Growth reductions were not significant for eastern cottonwood or black locust tested by similar procedures (Wang et al., 1986) (Table 16-4).

Due to its commercial importance in the southeastern United States, a variety of recent studies have examined loblolly pine response to ambient ozone. Shafer et al. (1987) planted seedlings of four full-sib families of loblolly in a field near Raleigh, North Carolina. Open top chambers were employed to expose the seedlings to ozone ranging from 0.5 to 1.96 times ambient ozone. Responses of stem height, stem diameter, biomass, and other characteristicies were quantified by regression. All relationships were linear for three families, but one family

Table 16-4. Results from Ambient-Open, Ambient-Chambered, and Filtered Chambered Treatments on Saplings of Three Tree Species, Millbrook, New York, 1984.

Species/char	Ambient-open	Ambient-chambered	Filtered-chambered
Hybrid poplar			
leaf weight, g	34.1^a	41.4^b	$50.9^{c}**$
stem weight, g	49.9^a	63.2^b	$78.6^{c}**$
total weight, abovegrd, g	82.0^a	103.2^b	$126.9^{c}**$
height, cm	189^a	208^b	$231.0^{c}**$
diameter, mm	12.4^a	13.7^b	14.6^b
no. of leaves per tree			
on main stem	40.4^a	41.8^a	$51.5^{b}**$
on lateral shoots	16.6^a	30.0^a	52.0^b
total no. of leaves	57.0^a	71.8^a	$103.5^{b}**$
number of lateral shoots	1.1^a	3.7^b	$7.9^{c}**$
Cottonwood			
leaf weight, g	107.6	93.5	113.9
stem weight, g	86.0	76.7	88.5
total weight, abovegrd, g	193.6	170.2	202.4
height, cm	162.0^a	175.0^{ab}	181.0^b
diameter, mm	18.6	17.1	17.8
no. of leaves per tree			
on main stem	31.8	29.7	33.6
on lateral shoots	183.0	175.0	192.0
total no. of leaves	214.0	205.0	226.0
number of lateral shoots	22.9^a	18.8^b	20.3^{ab}
Black locust			
stem + leaf weight, g	106.5	76.2	91.5
height, cm	94.1	113.0	123.0
diameter, mm	14.6	12.0	11.9
no. of leaves/tree			
(July 31, 84)	51.4^a	36.5^b	41.3^b

[a]All weights are oven-dry weights for means of all trees per treatment. Diameters were measured approximately 5 cm above the ground. Means with different letters are significantly different at $\alpha = 0.05$ using the Duncan's option in VA-SAS. Asterisks indicate differences significant at a = 0.01. Total weight above ground may not equal leaf + stem due to missing leaf data.
Source: Wang et al. (1986).

exhibited no significant growth response. Dose-response equations suggested a maximum growth suppression of 10% for ambient air compared to charcoal-filtered air. In a similar experiment, Adams et al. (1988) exposed loblolly pine seedlings, from five half-sib families, to ambient, subambient (0.6 X ambient), and elevated [ambient + 60 ppb (120 µg m^{-3})] ozone for one growing season in open-top chambers in the Tennessee Valley. Elevated ozone resulted in significantly reduced above ground volume and secondary needle biomass relative to seedlings grown in ambient air. Subambient ozone did not result in seedling size significantly different from ambient. Evidence was presented indicating that loblolly pine response to ozone is strongly regulated by genotype.

Kress et al. (1988) employed open-top chambers in North Carolina to examine the influence of both ozone and acid precipitation on loblolly pine growth. Exposure to elevated ozone (90 ppb, 176 µg m^{-3}) 12 hrs day^{-1} suppressed the diameter growth and foliar biomass of seedlings in one growing season treatment. There was no apparent effect of the acidic precipitation treatments.

In light of the evidence presented, it is clear that ozone, at concentrations common in numerous regions in North America, is capable of reducing the growth of some seedling and sapling size trees. In his comprehensive review of this topic, however, Pye (1988) cautions that extrapolation of information from managed environment studies with young trees to predictions concerning large trees in natural environments is difficult for numerous reasons. Differences in mature tree carbon allocation, canopy structure, competition, and canopy microclimate may all mediate mature tree response to ambient ozone.

5. Fluorides

It is apparent from research conducted on western United States conifers that elevated atmospheric fluoride can reduce the growth of trees contaminated from industrial sources. Treshow et al. (1967) recorded a reduction of up to 50% in the annual radial growth of Douglas fir subject to ambient fluoride. This suppression was noted irrespective of foliar symptoms. Extreme reductions in annual diameter growth of symptomatic ponderosa pine, presumably due to fluoride accumulation, have been recorded by Shaw et al. (1951) and Lynch (1951).

In his comprehensive study of lodgepole pine subjected to elevated fluorides from the Anaconda Aluminum facility in Columbia Falls, Montana (Chapter 13), Clinton F. Carlson included tree growth assessment in his observations. A 1973 investigation revealed that statistically significant growth losses attributed primarily to fluorides (secondarily to insects) were observed in 14 of 17 unmanaged stands for the period 1968–1973 in the vicinity of the aluminum plant (Carlson and Hammer, 1974a,b). Radial growth reductions, estimated from a prediction model, ranged from 0.005 mm (0.002 inch) to 1.8 mm (0.071 inch). This radial growth loss was equivalent to an average volume loss of 3.7 m^3 ha^{-1} yr^{-1} (57 board feet acre^{-1} yr^{-1}) for the measured stands. Extrapolation to the entire lodgepole pine ecosystem impacted by the industry suggested a total growth loss estimate of approximately 38,505 m^3 (8.5 million board feet) yr^{-1} (Carlson and Hammer, 1974a,b). Radial growth measurements were also made on Douglas fir and western white pine in the vicinity of the Anaconda plant. Ten-year periodic radial growth for two periods, 1958–1967 and 1968–1977, was measured on increment cores. Current 10-year periodic radial growth decreased dramatically with increasing foliar fluoride. Impact, defined as radial growth loss plus mortality, was calculated by Carlson (1978) to be 156,000 m^3 (5.5 million cubic feet) of usable wood for the period 1968–1977 on the 5360 ha (13,245 acres) influenced by the industry.

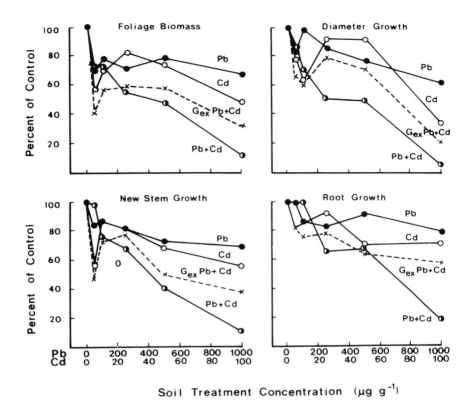

Figure 16-9. Growth parameters of sycamore seedlings grown in an agricultural soil amended with various combinations of lead and cadmium chloride. Treatments included lead alone (●), cadmium alone (○), or lead plus cadmium combined (◐). Values of expected growth for the combined treatment (Gex Pb + Cd) were calculated by multiplication of the reduction in growth due to separate heavy metal treatments and are indicated by dashed lines. (From Carlson and Bazzaz, 1977.)

6. Trace Metals

Most of the evidence implicating trace metal pollutants in tree growth suppression comes from studies involving seedling-size plants. Carlson and Bazzaz (1977) grew 2- to 3-year old seedlings of American sycamore in an agricultural soil treated with various amounts of lead and cadmium chloride. Various growth parameters were found to be synergistically affected by the lead-cadmium treatment (Figure 16–9). Diameter growth, foliage growth, and new stem growth of sycamore appeared to be synergistically affected by the combined lead plus cadmium treatment at lead/cadmium soil concentrations greater than 250/25 μg g^{-1}.

Lamoreaux and Chaney (1977) exposed first year seedlings of silver maple to cadmium concentrations they judged might be common in soils near industrial areas or in soils treated with sewage sludge or effluent. Seedlings were grown in sand amended with 0, 5, 10, or 20 μg g^{-1} cadmium chloride on a weight basis of rooting medium. Leaf, stem, and root dry weight were significantly reduced by all cadmium treatments.

Two- and 3-year-old seedlings of red maple, white pine, and Norway spruce were grown in sand and were irrigated with nutrient solution supplemented with either 0, 0.5, 1, 2, 4, 8, or 16 μg l^{-1} cadmium. All three species were exposed to 0, 6.25, 25, 50, 100, 200, and 400 μg l^{-1} zinc. Exposure of seedlings to these trace metals in an artificial soil was also examined. Highly significant correlations between a root growth index and cadmium and zinc concentrations in the nutrient solution suggested that increasing metal levels resulted in poorer root development. Seedlings in the cadmium- and zinc-amended soils developed symptoms similar to the nutrient solution study, except that injury was less severe at a given treatment level.

Kelly et al. (1979) grew first year seedlings of white pine, loblolly pine, yellow poplar, yellow birch, and choke cherry in a greenhouse environment in a natural forest soil collected in northwestern Indiana. Natural concentrations of trace metals in the soil were 0.6, 11.4, 2.0, and 20.6 μg g^{-1} cadmium, lead, copper, and zinc, respectively. The soil was amended with cadmium chloride to produce cadmium levels of 0, 15, and 100 μg g^{-1}. Shoot elongation and root and shoot dry weights were reduced by increasing levels of cadmium. Shoot height for all species was reduced by the 100 μg g^{-1} cadmium treatment, although yellow poplar and yellow birch heights after 17 weeks were not significantly influenced. Heights of white pine, loblolly pine, and choke cherry were not significantly different between the 0 and 15 μg g^{-1} treatments at the end of 17 weeks (Figure 16-10).

Baes and McLaughlin (1984) have employed trace element analysis of stem wood in an effort to correlate tree growth impact and air quality. Annual growth rings of short-leaf pine from the Great Smoky Mountains National Park showed suppressed growth and increased iron content between 1912 and 1963. During this period, smelting activity and large sulfur dioxide releases were occurring at Copperhill Tennessee, 88 km upwind. The highest trace metal concentrations detected in this study were in the range of 200–690 ppm aluminum, 0.47–7.5 ppm cadmium, 150–450 ppm manganese, and 27–120 ppm zinc. While these levels approach toxicity thresholds for certain agricultural crops, it is not clear what the impact may be on forest tree growth.

7. Dust

Bohne (1963) presented radial increment evidence suggesting that poplar tree growth was reduced by coarse particles from a cement plant 1.7 km (1 mile) distant. Heavy accumulations of cement kiln dust have been observed to reduce early growing season elongation of coniferous twigs and foliage (Darley, 1966).

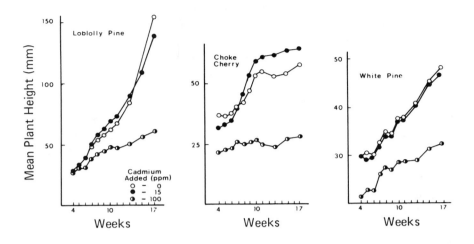

Figure 16-10. Height growth of loblolly pine, choke cherry, and white pine seedlings grown in a natural forest soil amended with various levels of cadmium.. Source: Kelly et al. (1979).

Terminal shoot growth of limestone dust contaminated eastern hemlock was shown to be reduced by Manning (1971).

Brandt and Rhoades (1973) measured annual ring widths to assess the influence of limestone dust on the growth of four forest tree species in southwestern Virginia growing in clean and dusty sites. A reduction in radial growth of at least 18% was exhibited by red maple, chestnut oak, and red oak. Growth of yellow poplar, on the other hand, was increased by 76%.

In general it can be concluded that coarse dusts may have an adverse effect on the growth of forest vegetation, mainly attributable to crust formation on foliage.

8. Contemporary Forest Growth Declines of Uncertain Cause

Recent efforts to review forest growth patterns over the past several decades have frequently revealed decreased growth rates. Climatic stresses, biotic stress agents, and aging are commonly responsible for these declines. In some cases, however, investigators argue that reduced air quality, as well as natural factors, may be exerting stress on forest productivity. It is not possible to provide a comprehensive review of forest growth studies in this volume. A few examples will be reviewed.

a. Red Spruce

The spruce-fir forest type exists on approximately 4.5 million ha in northern New England and New York (McCrerry et al., 1986). The red spruce and balsam fir components of this forest occur in five major stand types; spruce-fir swamps, spruce-fir flats, mixed spruce-hardwood, spruce-fir slopes, and old field spruce (Murphy 1917). At low elevation sites, red spruce stands are commonly even-aged, having followed disturbance associated with agricultural operations, forest harvest, or biotic agent epidemics. At high elevation sites, red spruce stands typically occur in complex mosaic patterns of uneven-aged condition, generally undisturbed by human cultural operations.

Evidence has been presented indicating increased mortality and morbidity or decreased radial or basal area growth increment for red spruce at low-and high-elevation sites in the Northeast (Baldwin, 1977; Johnson, 1983; Siccama et al., 1982; Weiss et al., 1984). While there is reasonable consensus that growth rates have declined over the past several decades across all size classes, age classes, and elevations, there is no agreement regarding cause. Numerous investigations emphasize changes in climate or stand dynamics (aging, competition) as the dominant regulators of growth. Basal area data from 3000 dominant or codominant red spruce from varied sites throughout the northeast revealed a consistent increase in annual growth from 1910 to 1920 to approximately 1960, with general decrease during the period 1960 to 1980, with early 1980 rates 13–40% below the 1960 peak (Hornbeck and Smith, 1985; Hornbeck et al., 1986). For low elevation spruce this growth decline was judged to be primarily due to normal forest aging (Federer and Hornbeck 1987). Examination of high elevation (900-1100 m) red spruce growth on Whiteface Mountain, New York indicated reduced growth beginning in the mid-1960s and continuing in the 1970s and 1980s (LeBlanc et al. 1986, LeBlanc et al., 1987, Raynal et al., 1987). These authors observed that the growth decreases documented were coincident with known climatic anomalies: the mid-1960s drought, which was the most severe (intensity and duration) over the past two centuries, and abnormally warm (1975–1976, 1980–1981, 1982–1983) and abnormally cold (1976–1977, 1977–1978, 1978–1979) winter temperatures. McLaughlin et al. (1987) analyzed increment cores from 1000 red spruce distributed in high elevation Appalachian forests in the eastern United States. These authors concluded that the observed growth decreases over the past 20–25 years were anamolous with respect to both climate change and forest aging. Federer et al. (1987) emphasize that red spruce growth decline appears unique relative to other species and that high elevation spruce response may be distinct in cause from low elevation factors.

b. Southern Pine

Periodic timber inventories conducted by the U.S. Department of Agriculture, Forest Service documented that net annual growth of pine in the Atlantic Coast States from Virginia to Florida has peaked and turned downward after many years

Table 16-5. Changes in Net Annual Growth of Southern Pine Growing Stocks on Timberland in the Southeast over the Most Recent Forest Inventory and Analysis Remeasurement Period.

Ownership class	Physiographic region	Net annual growth		
		From	To	Change
		--Million cubic feet--		
Public	Coastal plain	146	177	+31
	Piedmont & mountains	89	70	-19
	Total	235	247	+12
Forest industry (includes leased)	Coastal plain	446	683	+237
	Piedmont & mountains	176	182	+6
	Total	622	865	+243
Other private	Coastal plain	956	992	+36
	Piedmont & mountains	954	708	-246
	Total	1,910	1,700	-210
All owners	Coastal plain	1,548	1,852	+304
	Piedmont & mountains	1,219	960	-259
	Total	2,767	2,812	+45

Source: Sheffield et al. (1985).

of increase (Sheffield et al. 1985). Average annual radial growth of southern pine under 40 cm in diameter has declined by 30–50% over the past three decades. The most pronounced declines have been measured in pines growing on nonindustrial private forest land (Table 16-5).

As in the case of red spruce, numerous factors may be responsible for the decreased radial growth in southern pine. Important factors, as identified by Sheffield et al. (1985), include stand aging, increased stand densities, increased hardwood competition, drought, lowered water tables, land use history (i.e. loss of old-field sites), diseases, atmospheric deposition, and interactive effects of the preceding.

In his critiques of the southern pine decline, Lucier (1986, 1988) correctly indicates that the Forest Inventory and Analysis data cannot indicate the cause of decline. He further indicates that the reductions in radial and basal area growth are not equivalent to reductions in forest productivity. Clearly our understanding of the decline in net annual growth of southern pine is incomplete (Sheffield and Cost, 1987). It is also clear, however, that the impact of the decline on the southern forest industry is significant (Knight 1987) and that air quality may be potentially involved.

c. Other

A variety of additional United States studies have presented evidence for deciduous (McClenahen and Dochinger, 1985) and conifer (Mader and Adams, 1987) growth declines that are potentially associated with air quality. In addition, numerous studies in Europe have provided comparable evidence: Federal Republic of Germany (Schutt and Cowling, 1985, Kienast, 1982), Switzerland (Braun and Flückiger, 1987), and Finland (Hari et al., 1986).

B. Altered Succession and Species Composition

As a result of the considerable varietal and species variation in relative susceptibility to the various Class II interactions and the redundancy characteristic of most forest ecosystems, it is reasonable to suppose that differential tolerance to air pollution influence at the species level may be reflected in altered patterns of succession and species composition at the ecosystem level.

1. Autogenic and Allogenic Processes

Ecologists recognize two major types of processes that influence ecosystem succession. *Autogenic processes* are those resulting from biological factors within the system. In forest ecosystems autogenic processes would include site alterations caused by the vegetation, the influence of one plant species on another (Fisher, 1980), and the impact of native insect or disease microorganisms. *Allogenic processes*, on the other hand, are abiotic factors that influence succession from without the system. Geochemical and climatic forces are especially important examples of allogenic factors that influence forest ecosystems. Idealized ecosystem development characteristically is portrayed as an orderly change of biological progression, occurring in a more or less constant environment (Odum, 1969; Woodwell, 1974). It has been generally assumed that autogenic processes dominate allogenic processes in terrestrial ecosystem succession. This generalization, however, is quite inconsistent with data generated by recent imaginative studies with forest ecosystems. The importance of fire (an allogenic force) in influencing presettlement forest ecosystems in the North Central states of the United States has been substantial (Loucks, 1970; Frissell, 1973; Heinselman, 1973). The significance of wind stress (an allogenic force) has been suggested to exert substantial control over successional development of forest ecosystems in New England (Stephens, 1955, 1956; Raup, 1957; Henry and Swan, 1974). Forest management practices imposed by humans, for example, clear-cutting, may simulate the influence of natural allogenic forces on forest development and may interrupt progress toward a steady state condition (Bormann and Likens, 1979). Conversely, other forest management procedures, for example, fire control, may eliminate a controlling allogenic force and permit succession to proceed toward an unnatural steady state condition. Class II stresses imposed on forest ecosystems by air pollutants may be considered a twentieth century allogenic process of potential importance to forest ecosystem development. As in the case of clear-cutting, this force related to human activity might be expected to alter the attainment of steady-state conditions. Air pollution stress would appear to have certain unique qualities that may make it an allogenic influence of particular importance. Length of exposure to this force precludes evolutionary adjustment and its influence, in certain areas, may be quite continuous rather than cyclic, as are windstorms and fires. What is the evidence available to support the importance of air pollution as an allogenic force of significance in forest ecosystem development?

2. Air Pollution and Species Composition

In 1968, prior to sophisticated understanding of most Class II interactions, Treshow provided an excellent review of the impact of air contaminants on plant populations. Treshow's review, along with a variety of additional papers from the late 1960s, for example, Niklfeld (1970), Hajdúk and Ruzicka (1968), and Trautmann et al. (1970), have indicated alterations in successional pattern or species composition in forest ecosystems subject to air pollution exposure. Guderian and Kueppers (1980) reviewed plant community response to atmospheric pollution and developed critical research questions based on information available at that time.

Hayes and Skelly (1977) have monitored total oxidants and associated oxidant injury to eastern white pine in three rural Virginia sites between April 1975 and March 1976. Varieties of pine categorized as sensitive and intermediate to oxidant stress were judged to be under stress. The authors speculated that susceptible eastern white pine in the Blue Ridge and Southern Appalachian Mountains may be rendered less competitive by air pollution stress. Shifts in species composition away from white pine importance may be occurring in certain eastern regions. Brandt and Rhoades (1973), in their investigation of limestone dust impact in deciduous forests in southwestern Virginia, predicted changes in species composition resulting from dust influence. Dusty sites had reduced seedling and sapling density of red maple, chestnut oak, and red oak. This observation, along with documentation of reduced mean basal area and lateral growth of these trees, led the authors to suggest that yellow poplar, more resistant to stress caused by dust accumulation, would increase in importance in these hardwood stands.

Treshow and Stewart (1973) have conducted one of the few studies truly concerned with air pollution impact on an entire vegetative community. Portable fumigation chambers were placed over representative plants in intermountain grassland, oak, aspen, and conifer communities. Ozone fumigations were conducted to establish injury thresholds for 70 common plant species indigenous to these communities (Table 16–6). Generally, injury was evident at varying concentrations above 15 pphm (294 $\mu g\ m^{-3}$). Species that were found to be most sensitive to ozone in the grassland and aspen communities investigated included some dominants that were considered to be key to community integrity. The

Table 16-6. Injury Thresholds for 2-hr Field Fumigations with Ozone in Grassland, Oak, Aspen, and Conifer Intermountain Plant Communities.

Species	Injury threshold (pphm ozone)	
Grassland-oak community species		
Trees and shrubs		
Acer grandidentatum	over	40
Acer negundo	over	25
Artemesia tridentata		40
Mahonia repens	over	40
Potentilla fruticosa		30
Quercus gambelii		25
Toxicodendron radicans	over	30
Perennial forbs		
Achillea millefolium	over	30
Ambrosia psilostachya	over	40
Calochortus nuttallii	over	40
Cirsium arvense		40
Conium maculatum	over	25
Hedysarum boreale		15
Helianthus anuus	over	30
Medicago sativa		25
Rumex crispus		25
Urtica gracilis		30
Vicia americana	over	40
Grasses		
Bromus brizaeformis		30
Bromus tectorum		15
Poa pratensis		25
Aspen and conifer community species		
Trees and shrubs		
Abies concolor		25
Amelanchier alnifolia		20
Pachystima myrsinites	over	30
Populus tremuloides		15
Ribes hudsonianum		30
Rosa woodsii	over	30
Sambucus melanocarpa	over	25
Symphoricarpos vaccinioides		30
Perennial forbs		
Actaea arguta		25
Agastache urticifolia		20
Allium acuminatum		25
Angelica pinnata	under	25
Aster engelmanni		15
Carex siccata		30
Cichorium intybus		25

Table 16-6 (continued)

Species	Injury threshold (pphm ozone)	
Cirsium arvense	under	40
Epilobium angustifolium		30
Epilobium watsoni		30
Eriogonum heraclioides		30
Fragaria ovalis		30
Gentiana amarella	over	15
Geranium fremontii	under	25
Geranium richardsonii		15
Juncus sp.	over	25
Lathyrus lanzwertii	over	25
Lathyrus pauciflorus		25
Mertensia arizonica		30
Mimulus guttatus	over	25
Mimulus moschatus	under	40
Mitella stenopetala	over	30
Osmorhiza occidentalis		25
Phacelia heterophylla	under	25
Polemonium foliosissimum		30
Rudbeckia occidentalis		30
Saxifraga arguta	under	30
Senecio serra		15
Taraxacum officinale	over	25
Thalictrum fendleri	over	25
Veronica anagallis-aquatica		25
Vicia americana	over	25
Viola adunca	over	30
Annual forbs		
Chenopodium fremontii	under	25
Collomia linearis	under	25
Descurainia californica		25
Galium bifolium	over	30
Gayophytum racemosum		30
Polygonum douglasii	over	25
Grasses		
Agropyron caninum	over	25
Bromus carinatus	under	25

Source: Treshow and Stewart (1973).

most dramatic example was aspen itself. Single 2-hr exposure to 15 pphm ozone caused severe symptoms on 30% of the foliage exposed. White fir seedlings require aspen shade for optimal juvenile growth. The authors judged that significant aspen loss might restrict white fir development and alter forest succession. In a companion study, Harward and Treshow (1975) pursued their interest in evaluating ozone impact on aspen communities by evaluating the growth and reproductive response of 14 understudy species to ozone. Plants were fumigated in greenhouse chambers throughout their growing seasons. It was concluded from these fumigations that plant sensitivities varied sufficiently to make probable major shifts in composition in aspen communities following only a year or two of exposure to ozone above concentrations of 7–15 pphm (137–294 µg m^{-3}). The authors observed that comparable doses are widespread in the vicinity of urban areas and that widespread impacts on plant community stabilities may be common in nature.

The efforts of Michael Treshow and colleagues highlight the importance of examining shrub and herb strata when assessing the air pollution impact on forest ecosystems. Nyborg (1978) has made an interesting and related observation. While most commercially important forest trees develop well in soil with a pH as low as 3.5, a variety of forest shrubs exhibits a gradient in tolerance of low soil pH. In Alberta, Canada, Nyborg suggested that measurements on forest soils acidified by windblown elemental sulfur showed that when soil pH was 5 to 4, the number of species in the understory was reduced, when pH was 4 to 3 only a few species grew, and when the pH was less than 3 there was no undergrowth!

McClenahen (1978) has provided a most interesting study with quantitative data on the impact of polluted air on the various strata of a forest ecosystem. Forest vegetation was measured in seven stands on similar sites in a 50 km area of the upper Ohio River Velley. The stands were situated along a gradient of polluted air containing elevated concentrations of chloride, sulfur dioxide, fluoride, and perhaps other contaminants. Species richness (number of different species), evenness (dominance index — low values indicate dominance by one or a few species), and Shannon diversity index were typically reduced within the overstory, subcanopy, and herb strata near industrial sources of air contaminants. Increasing air pollutant exposure reduced canopy stem density, but abundance of vegetation in other strata tended to increase along the same gradient. The relative importance of sugar maple was greatly reduced in all strata with increasing pollutant dose, while yellow buckeye appeared tolerant of poor air quality. In the shrub layer the importance of spicebush increased with increasing pollutant exposure.

In southern California the predominant native shrubland vegetation consists of chaparral and coastal sage scrub. The former occupies upper elevations of the coastal mountains, extending into the North Coast ranges, east to central Arizona, and south to Baja California; while the latter occupies lower elevations on the coastal and interior sides of the coast ranges from San Francisco to Baja California. Westman (1979) applied standard plant ordination techniques to these shrub communities to examine the influence of air pollution. The reduced cover of native species of coastal sage scrub documented on some sites was statistically indicated to be caused by elevated atmospheric oxidants. Sites of high ambient oxidants were also characterized by declining species richness.

Winner and Bewley (1978) investigated the vascular plant and bryophyte synecological responses in a white spruce forest subjected to high sulfur dioxide exposure in central Alberta, Canada. Both understory components were found to decline in coverage as sulfur dioxide stress increased. Rosenberg et al. (1979) studied sulfur dioxide impact on forest composition upwind and downwind of a small coal-burning electric-generating station in central Pennsylvania. Species diversity and importance values of certain species were inversely related to distance from the source of emission. Diversity and importance of eastern white pine and black birch increased with distance, whereas the importance of white oak and red maple decreased. Impacts on species richness were estimated to extend 4.3 km downwind and 3.4 km upwind, while impacts on species diversity were estimated to extend 6.4 km downwind and 3.3 km upwind (Figures 16-11, 16-12).

Changes in population tolerance to pollution stress may impact fitness and influence species diversity. Berrang et al. (1986) applied ozone to red maple from natural populations in nine national parks and trembling aspen from natural populations in five national parks. Ozone tolerances of aspen populations from high and intermediate ozone exposure areas were greater than ones from low ozone exposure areas. Red maple populations from regions with high ozone exposure showed more tolerance than those from regions with low ozone exposure.

The influence of air pollution stress on succession and ecosystem species composition probably varies with the age and successional status of the forest. Harkov and Brennan (1979) have observed that most woody plants susceptible to ozone injury are generally early successional plant species. Most trees intermediate or tolerant of ozone stress are typically mid- or late-successional types. It is not unreasonable to propose, as Harkov and Brennan did, that late successional forest communities may be the most resistant to compositional change as a result of chronic air pollution exposure. Mature ecosystems are also typified by other characteristics that may increase their resistance to air pollution stress (Table 16–7). Low net production may reduce the potential importance of restrictions imposed by air contaminants on photosynthesis. Closed and slow nutrient cycling may make nutrient capital less liable to loss by air pollutant influence.

Generalizations are difficult, however, when assessing abiotic stress on forest community development (West et al., 1981, Zobel et al., 1976). Deletion of specific species from mixed plant communities does not necessarily lead to loss of production, stability, or recovery potential. The work of Allen and Forman (1976) suggests that numerous stresses, including sulfur dioxide exposure, that led to selective species removal, also allowed increased productivity of the remaining species.

C. Altered Forest Pest Influence

Native phytophagous forest insects and microorganisms that function as tree stress agents represent critically important autogenic influences on forest ecosystem structure and function. Interaction of these stresses with forest trees

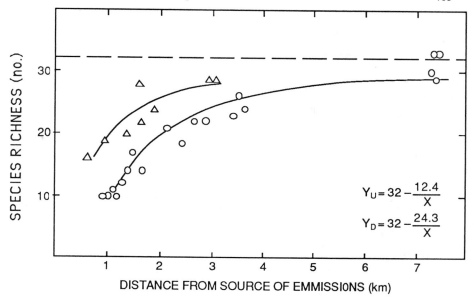

Figure 16-11. Effects of sulfur emissions from a Pennsylvania coal-burning power station on species richness in mixed-oak forests up- and downwind of source. Dashed line indicates asymptote for combined up- and downwind data.
Source: Rosenberg et al. (1979).

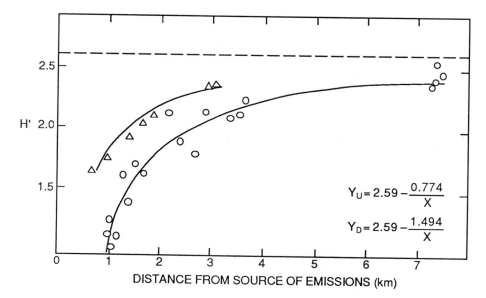

Figure 16-12. Effects of sulfur emissions from a Pennsylvania coal-burning power station on mean Shannon–Weiner Diversity Index (H') at different distances up- and downwind. Dashed line indicates the asymptote for combined up- and downwind data.
Source: Rosenberg et al. (1979).

Table 16-7. Characteristics of Ecosystem Development.

Ecosystem attributes	Developmental stages	Mature stages
Community energetics		
1. Gross production/community respiration (P/R ratio)	Greater or less than 1	Approaches 1
2. Gross production/standing crop biomass (P/B ratio)	High	Low
3. Biomass supported/unit energy flow (B/E ratio)	Low	High
4. Net community production (yield)	High	Low
5. Food chains	Linear, predominantly grazing	Weblike, predominantly detritus
Community structure		
6. Total organic matter	Small	Large
7. Inorganic nutrients	Extrabiotic	Intrabiotic
8. Species diversity— variety component	Low	High
9. Species diversity— equitability component	Low	High
10. Biochemical diversity	Low	High
11. Stratification and spatial heterogeneity (pattern diversity	Poorly organized	Well organized
Life history		
12. Niche specialization	Broad	Narrow
13. Size of organism	Small	Large
14. Life cycles	Short, simple	Long, complex
Nutrient cycling		
15. Mineral cycling	Open	Closed
16. Nutrient exchange rate, between organism and environment	Rapid	Slow
17. Role of detritus in nutrient regeneration	Unimportant	Important
Selection pressure		
18. Growth form	For rapid growth ("r-selection")	For feedback control ("K-selection")
19. Production	Quantity	Quality
Overall homeostasis		
20. Internal symbiosis	Undeveloped	Developed
21. Nutrient conservation	Poor	Good
22. Stability (resistance to external perturbations)	Poor	Good
23. Entropy	High	Low
24. Information	Low	High

Source: Odum (1969).

throughout their life cycles exerts powerful control on reproduction, stand density, plant distribution, and tree competition.

We have reviewed the considerable evidence concerning the influence of air pollutants on insects and disease microbes and the abnormal physiology they induce in individual tree hosts (Chapters 13 and 14). It is extraordinarily unfortunate, but we do not have substantial information on the interaction of air pollutants with insect pests and microbial pathogens at the population level! It is further true that our understanding of native insect and microbial influences on forest ecosystems over extended time intervals, in the absence of air pollutants, is not as sophisticated as it should be.

It is tempting to make some generalizations from the information that is available. It is undoubtedly true that the activities of some insects and some pathogens are increased, while others are decreased, in forest ecosystems subject to air contaminant stress. In the instance of oxidant induced predisposition of western conifers to bark beetle infestation, enhancement of this important insect in ponderosa pine populations does not appear to occur, as predisposed and infested trees do not appear to function as efficient brood trees for insect population build-up. In the case of infection by the ubiquitous *Armillaria* spp. fungi, it is judged that chronic stress imposed by air pollutants may enhance the significance of this fungus in certain forest tree populations, but not in others.

D. Case Study: Response of a Forest Ecosystem to Air Pollution

Our appreciation of the influence of Class II relationships resulting from the air pollution influence on forest ecosystems will not significantly advance unless and until comprehensive investigations of natural ecosystems under ambient air contaminant influence are planned and conducted. The most comprehensive North American effort to study forest ecosystem response to ambient atmospheric contaminants has been performed in southern California under the leadership of Paul R. Miller, Research Plant Pathologist, U.S.D.A. Forest Service, Pacific Southwest Forest and Range Experiment Station, Riverside, California, and was sponsored by the U.S. Environmental Protection Agency and the U.S.D.A. Forest Service. The southern California forest study and the ecosystem model developed to systematically investigate Class II air pollution impacts is thoroughly described in Taylor (1974), National Academy of Sciences (1977), and Miller (1977). In the author's judgment, it is essential to establish similar studies in eastern as well as other western forests.

1. San Bernardino Forest

The concentration of human activities, meteorology, and topography interact to produce elevated atmospheric concentrations of ozone, nitrogen oxides, and peroxyacetylnitrate in southwestern California. The adverse impact from these oxidants has been especially high in the coniferous forest of southern California. In

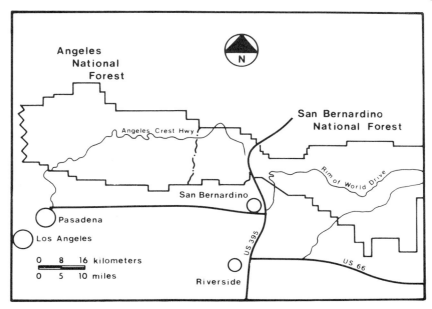

Figure 16-13. Location of the Los Angeles urban complex, an efficient producer of atmospheric oxidants, in relation to the San Bernardino and Angeles National National Forests.
Source: Wert et al. (1970).

Figure 16-14. Community-level interactions in a mixed conifer ecosystem. 1, Competition between woodpeckers and small mammals; 2, Climate control of oxidant concentration in different forest communities; 3, Effect of precipitation and temperature on soil moisture and soil temperature in different forest communities; 5, Effect of cone crop abundance on cone insect populations in different forest communities; 6, Effect of cone crop abundance on small-mammal populations in different forest communities; 7, Fruiting bodies of nonpathogenic fungi as food for small mammals in different forest communities; 8, Smog-caused mortality and morbidity in different forest communities; 9, Fruiting bodies of pathogens as food for small mammals in different forest communities; 10, Effect of temperature and evaporative stress on species composition in different forest communities; 11, Relationship between soil characteristics and pollution density of burrowing small mammals in different forest communities; 12, Relationship between soil microarthropods and plants; 13, Relationship between soil characteristics and microarthropod population; 14, Bark beetle mortality caused by natural enemies in different forest communities; 15, Effect of bark beetles on tree mortality and vigor in different forest communities; 16, Relationship between soil characteristics and forest community composition and growth; 17, Relationship between soil characteristics and species distribution and behavior of nonpathogenic fungi; 18, Relationship between soil characteristics and species distribution and behavior of pathogens; 19, Influence of forest community type on populations of natural enemies of bark beetles; 20, Woodpecker distribution and density in different forest communities; 21, Ef-fect of pathogens on tree vigor and mortality in different forest communities; 22, Relationship between non-pathogenic fungi and forest community composition and growth.
Source: Taylor (1974).

the San Bernardino National Forest (Figure 16-13), mortality of ponderosa and Jeffrey pines has been considerable over the past two decades. Since 1972, 12 investigators representing various research disciplines developed an integrated study to evaluate Class II interactions in the forests of the San Bernardino Mountains.

The San Bernardino Mountains are part of the Transverse Range in southern California. Their position east of the Los Angeles basin allows them to function as a barrier to the movement of marine air. The position and elevation of the San Bernardino Mountains facilitates the development of inversion layers in the Los Angeles basin. Air pollutants trapped beneath the inversion layer are drawn into the forest at higher elevations in the mountains by a "chimney effect", resulting from radiant heating of the south-facing slopes. As a result, a gradient in the concentration of oxidants exists across the coniferous forest in the mountains. Monitoring stations on the southwestern edge of the forest had average hourly ozone concentrations of 120 ppb (235 μg m^{-3}) from May to September during 1974–78. Monitoring stations to the north and east in the

Figure 16-14. Caption on preceding page.

forest are characterized by lesser concentrations. Camp Angeles, located 34 km southeast of the stations with the highest ozone concentrations, had average hourly ozone levels of 60 ppb (118 µg m^{-3}) (McBride et al., 1985).

In 1975 a systems simulation modeling process was initiated. Initially an inventory of ecosystem components and processes was performed. Organism, tree, community, and stand level interactions were summarized and have been presented graphically in Taylor (1974). Figure 16-14 presents the community level interactions identified. Only interactions judged to have especially important roles in regulating forest structure and function were selected for immediate study. The basic unit for modeling purposes was defined as the forest stand. Stands could be comprised of 10–200 trees on land areas from 100–25,000 m^2. Subsystems treated at the stand level included tree population dynamics, oxidant flux, canopy response, stand tree growth, stand moisture dynamics, tree seedling establishment, cone and seed production, litter production, litter decomposition, and small mammal population dynamics. Time resolution varied with subsystem and ranged from hours to several years.

Eighteen major study plots were selected along an east-to-west gradient of air pollutants (Figure 16-15). Monitoring stations were established along the pollution concentration gradients to provide accurate exposure information for the major study plots. Operation of this project for several years has produced an enormous amount of data. Data digestion and model refinement permits assessment of oxidant impact on southern California forests. Model predictions have been extended beyond the study ecosystem. This project was a success for the organization and perspective it has given to air-pollution–forest-ecosystem studies.

2. Forest Growth in San Bernardino

One of the few correlations of growth parameters of large trees growing under field conditions with ambient ozone levels has been provided by the comprehensive oxidant study conducted in the San Bernardino National Forest in California (Miller, 1977). Radial growth of ponderosa pine was compared for periods of low pollution (1910–1940) and high pollution (1941–1971) (Table 16-8). The average annual rainfall between these periods was 111 cm and 117 cm yr^{-1}, respectively. The 0.20 mm difference in average annual growth betwen the two periods was attributed to air pollution. Average 30-year-old trees grown in the two periods were estimated to have diameters of 30.5 cm (1910–1940) and 19.0 cm (1941–1971). An average 30-year-old tree grown in contemporary air was estimated to reach 7.0 m height and 19 cm diameter, and to be capable of producing one log 1.8 m long with a volume of 0.047 m^3. An average 30-year-old tree grown in the absence of oxidants was estimated to be 9.1 m in height, 30.5 cm in diameter, and produce one log 4.9 m long with a volume of 0.286 m^3 (Figure 16–16).

Figure 16-15. Air mass contaminated with photochemical oxidants moving into the forests of the San Bernardino Mountains, California.
Source: Prof. Joe R. McBride, University of California, Berkeley.

3. Forest Succession in San Bernardino

The forests of the San Bernardino Mountains in southern California have been subject to oxidant stress from the Los Angeles metropolitan complex for 30 years. In 1970, Cobb and Stark concluded that if air pollution from the Los

Table 16-8. Average Annual Radial Growth of 19 Ponderosa Pine Trees in Two Levels of Oxidant Air Pollutants in the San Bernardino National Forest, California.

High pollution		Low pollution	
Age[a] (years)	Average radial growth (cm) 1941-1971	Age[a] (years)	Average annual radial growth (cm) 1910-1940
20	0.20	60	0.52
21	0.33	55	0.49
29	0.22	55	0.61
22	0.33	57	0.34
25	0.30	64	0.40
35	0.23	63	0.55
27	0.29	60	0.44
28	0.31	65	0.46
35	0.26	60	0.75
22	0.43	71	0.67
39	0.21	63	0.71
35	0.34	71	0.65
29	0.37	66	0.78
33	0.37	63	0.53
35	0.34	60	0.33
35	0.37	70	0.38
36	0.35	61	0.32
36	0.33	62	0.37
34	0.36	59	0.37

Source: Miller (1977).
[a] Age at 1.4 m above ground in 1971.

Angeles basin continued to increase, there will be a conversion from well stocked forests dominated by ponderosa pine to poorly stocked stands of less susceptible tree species in the San Bernardino Mountains. Miller (1973) has provided a thorough discussion of this oxidant-induced forest-community change. Ponderosa pine is one of five major species of the "mixed conifer type" that covers wide areas of the western Sierra Nevada and the mountain ranges, including the San Bernardino Mountains, in southern California from 1000 to 2000 m (3000–6000 feet) elevation. Other species represented include sugar pine, white fir, incense cedar, and California black oak. The response of these five major tree species to oxidant air contaminants in the San Bernardino National Forest has been variable. Ponderosa pine exhibits the most severe foliar response to elevated ambient ozone. A 1969 aerial survey conducted by the U.S.D.A. Forest Service indicated 1.3 million ponderosa (or Jeffrey) pines on more than 405 km^2 (100,000 acres) were stressed to some degree. Mortality of ponderosa pine has been extensive. Actual death is typically attributed to bark-beetle infestation of air pollution stressed trees. White fir has suffered slight damage, but scattered trees have exhibited severe symptoms. Sugar pine, incense cedar, and black oak

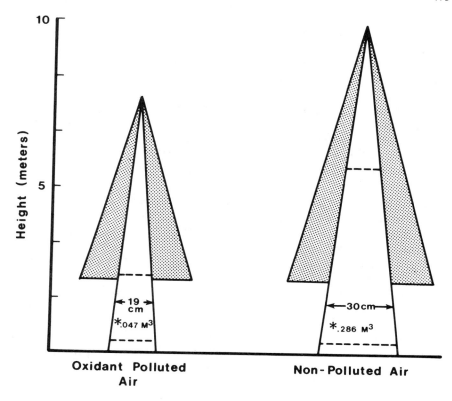

Figure 16-16. Calculated average growth of 30-year-old 15-cm San Bernardino National Forest, California, ponderosa pines in polluted and nonpolluted air based on radial growth samples from 1941–1971 and 1910–1940.. The asterisk indicates wood volumes in log with 15-cm top (min. merchantable diameter).
Source: Miller (1977).

have exhibited only slight foliar damage from oxidant exposure. A 233 ha (575 acre) study block was delineated in the northwest section of the San Bernardino National Forest in order to conduct an intensive inventory of vegetation present in various size classes and to evaluate the healthfulness of the forest. Ponderosa pines in the 30 cm (12 inch) diameter class or larger were more numerous than other species of comparable size in the study area. These pines were most abundant on the more exposed ridge crest sites of the sample area. Mortality of ponderosa pine ranged from 8 to 10% during 1968–1972. The loss of a dominant species in a forest ecosystem clearly exerts profound change in that system. Miller concluded from his investigation that the lower two thirds of the study area will probably shift to a greater proportion of white fir. It was judged that incense cedar will probably remain secondary to white fir. Sugar pine was presumed to be restricted by lesser competitive ability and dwarf mistletoe infection. The rate of composition change was deemed dependent on the rate of ponderosa pine mortality. The upper one third of the study area, characterized as more environmentally severe due to climatic and edaphic stress, supports less vigorous white fir growth. Following the loss of ponderosa pine in this area, sugar and incense cedar may assume greater importance. Miller judged, however, that natural regeneration of the latter species may be restricted in the more barren, dry sites characteristic of the upper ridge area. California black oak and shrub species may become more abundant in these disturbed areas. Additional and intensive research on forest composition in the San Bernardino National Forest has been reported (Miller, 1977). Tree population dynamics were examined on 18 permanent plots established in 1972 and 1973 and on 83 temporary plots established in 1974 to investigate forest development as a function of time since the most recent fire. Generally, the data still support the hypothesis that forest succession toward more tolerant species such as white fir and incense cedar occurs in the absence of fire. In the presence of fire, pine may be favored by seedbed preparation and elimination of competing species. These more recent studies suggest a larger number of forest subtypes may exist within the forest ecosystem than initially realized.

In 1982 and 1983, plots dominated by ponderosa pine were resurveyed. Plots were grouped on the basis of foliar symptoms into severe or slight injury groups. No significant difference in seedling establishment was observed in ponderosa pine on either severe-or slight-injury plots. Differences in regeneration of other species could not be correlated with foliar symptomology. Projection of stand age structures suggested the eventual dominance of incense cedar on severe injury plots in the mixed conifer forest (McBride et al., 1985).

The changes in forest composition caused by oxidants in this southern California forest have created a management concern, as well as ecological change, because the forest is intensively used as a recreational resource and the loss of ponderosa pine is judged to reduce the aesthetic qualities of the forest.

E. Summary

Perhaps the single most important characteristic of an ecosystem is its productivity. Compared to a variety of other terrestrial ecosystems, the productivity of forest ecosystems is relatively high. Productive forests are critically important, not only for the obvious relationship between wood volume and commercial products in managed forests, but also for the regulation and maintenance of quality for associated ecosystems, amenity functions, and general climatic and terrestrial stability. It is disconcerting to realize, therefore, that there is substantial and impressive evidence to indicate that two widespread air contaminants, sulfur dioxide and ozone, are capable of reducing forest growth. The more localized release of fluoride can also reduce the amount of forest biomass.

Evidence from a variety of studies examining forest growth in the vicinity of large point sources of sulfur dioxide has indicated significantly reduced growth. Generally, the correlation of growth impact with degree of foliar injury caused by sulfur dioxide is not high. Growth retardation occurs in the absence of any visible indication of stress. Most sulfur dioxide studies have accounted for precipitation influence on forest growth over the study periods.

Evidence for ozone suppression of forest growth has been provided by controlled environment, open-top chamber, branch-chamber, and field studies. As in the case of sulfur dioxide, ozone suppression of growth occurs without the development of visible symptoms. Unlike sulfur dioxide, however, ozone exposure sufficient to cause growth reductions is widespread over North American forests. Oxidant-related forest growth reduction may represent the most important relationship between air contamination and forest ecosystems.

Acid precipitation studies have not demonstrated a significant influence on forest growth. Investigations conducted in the most seriously impacted temperate forest regions in Scandinavia and the United States have not provided convincing evidence that acid rain either reduces or increases forest growth.

There are two important deficiencies in forest growth–air-pollution-stress research. The first relates to the paucity of growth studies in mature, natural forests. This forces extrapolation from studies generally using young trees and typically growing in managed environments. The second serious limitation relates to the inability to partition reduced growth to the various Class II interactions that may actually be responsible for it. For example, what percentage of reduced growth may be due to reduced nutrition, reduced photosynthesis, increased insects or disease, or increased foliar damage? Future investigations of forest growth, as impacted by air quality, must also include better accounts of growth influencing factors other than precipitation and air pollutants. Better awareness of additional climatic factors, stand dynamics, impacts of insect and disease influence, and management strategies must be indicated.

As forest deposition records become more extended, correlations of forest growth trends with regional air quality (McLaughlin et al., 1983, 1984; McLaughlin, 1985) will become more accurate. If air quality impacts can be successfully partitioned from other growth impacting variables, modeling (McLaughlin and Bräker, 1985) will allow quantification of forest growth patterns as influenced by atmospheric deposition.

There is increasing appreciation of the importance of allogenic forces on forest ecosystem succession. The significance of fire and wind stress on forest development is substantial in certain environments. It is concluded that air pollutant impact also exerts critically important control over forest succession and species composition. Long-term, continual stress tends to decrease the total foliar cover of vegetation, to decrease the species richness, and to increase the concentration of dominance by favoring tolerant species. The importance of air contaminants is probably most significant during early and midsuccessional stages of forest development. Again modeling efforts have great utility in evaluating successional impacts as in evaluating growth impacts of air pollutants. McLaughlin et al. (1978) applied FORET, designed to model successional dynamics of eastern deciduous forests, to study the long-term interaction of air pollution stress on forest community dynamics. Their results indicated that the response of individual trees in a forest stand may differ greatly from results predicted on the basis of responses determined in the absence of plant competition.

Very unfortunately we are unable, with the data available at this time, to evaluate the importance of insect and disease alterations resulting from air pollution interactions at the population level. This topic must be considered of highest priority for future research.

References

Abrahamsen, G., R. Horntvedt, and B. Tveite. 1976a. Impacts of acid precipitation on coniferous forest ecosystems. In L.S. Dochinger and T.A. Seliga, eds., 1st International Symp. Acid Precipitation and the Forest Ecosystem. U.S.D.A. Forest Service, Upper Darby, PA, pp. 991-1009.

Abrahamsen, G., K. Bjor, R. Horntvedt, and B. Tveite. 1976b. Effects of acid precipitation on coniferous forest. In F.H. Braekke, ed., Impact of Acid Precipitation on Forest and Freshwater Ecosystems in Norway. SNSF Project Research Report No. 6, Oslo, Norway, pp. 37-63.

Adams, R.M., S.A. Hamilton, and BA. McCarl. 1985. Assessment of the economic effects of ozone on U.S. agriculture. J. Air Pollu. Control Assoc. 35: 938-943.

Adams, M.B., J.M. Kelly, and N.T. Edwards. 1988. Growth of *Pinus taeda* L. seedlings varies with family and ozone exposure level. Water, Air, Soil Pollu. (in press.)

Allen, E.B. and R.T. Forman. 1976. Plant species removals and old-field community structure and stability. Ecology 57: 1233-1243.

Assmann, E. 1970. The Principles of Forest Yield Study. Pergamon Press, Oxford, 506 pp.

Baes, C.F. III and S.B. McLaughlin. 1984. Trace elements in tree rings: Evidence of recent and historical air pollution. Science 224: 494-497.

Baldwin, H. I. 1977. The induced timberline of Mount Monadnock N.H. Bull. Tor. Bot. Club 104: 324-333.

Berrang, P., D.F. Karnosky, R.A. Mickler, and J.P. Bennett. 1986. Population changes in eastern hardwoods caused by air pollution. Proc. 9th North Am. Forest Biol Workshop, Stillwater, OR, June 1986.

Bohne, H. 1963. Schädlihkeit von staub aus zementwerken für waldbestände. Allg. Forstz. 18: 107-111.

Bolin, B. (ed.). 1971. Air Pollution Across National Boundaries: The Impact on the Environment of Sulfur in Air and Precipitation. Report of the Swedish Preparatory Committee for the U.N. Conference on Human Environment. Stockholm, Sweden, 96 pp.

Bormann, F.H. and G.E. Likens. 1979. Catastrophic disturbance and the steady state in northern hardwood forests. Am. Sci. 67: 660-669.

Brandt, C.J. and R.W. Rhoades. 1973. Effects of limestone dust accumulation on lateral growth of forest trees. Environ. Pollu. 4: 207-213.

Braun, S. and W. Flückiger. 1987. Untersuchungen an gipfrieben von buche (*Fagus sylvatica* L.). Botanica Helvetica 97: 61-73.

Carlson, C.E. 1978. Fluoride induced impact in a coniferous forest near the Anaconda aluminum plant in northwestern Montana. Unpublished Ph.D. Thesis. Univ. Montana, Missoula, MN, 165 pp.

Carlson, R.W. and F.A. Bazzaz. 1977. Growth reduction in American sycamore (*Plantanus occidentalis* L.) caused by Pb-Cd interaction. Environ. Pollu. 12: 243-253.

Carlson, C.E. and W.P. Hammer. 1974a. Impact of fluorides and insects on radial growth of lodgepole pine near an aluminum smelter in northwestern Montana. U.S.D.A. Forest Service, Northern Region Rept. No. 74-75, 14 pp.

Carlson, C.E. and W.P. Hammer. 1974b. Impact of fluorides and insects on radial growth of lodgepole pine near an aluminum smelter in northwestern Montana. U.S.D.A. Forest Service, Northern Region Rept. No. 74-25, 14 pp.

Cobb, F.W. and R.W. Stark. 1970. Decline and mortality of smog-injured ponderosa pine. J. For. 68: 147-149.

Cogbill, C.V. 1976. The effect of acid precipitation on tree growth in eastern North America. In L. S. Dochinger and T.A. Seliga, eds., 1st Intl. Symp. Acid Precipitation and the Forest Ecosystem. U.S.D.A. Forest Service, Upper Darby, PA, pp. 1027-1032.

Cogbill, C.V. 1977. The effect of acid precipitation on tree growth in eastern North America. Water Air, Soil Pollu. 8: 89-93.

Dahl, E. and O. Skre. 1971. An investigation of the effect of acid precipitation on productivity. Kinferens om avsvavling, Stockholm, Nov. 11, 1969. Nordsforsk, Miljövardssedrtariatet, Pub. 1: 27-39.

Darley, E.F. 1966. Studies on the effect of cement-kiln dust on vegetation. J. Air Pollu. Control Assoc. 16: 145-150.

Davis, D.R. 1972. Sulfur dioxide fumigation of soybeans: Effects on yield. J. Air Pollu. Control Assoc. 22: 12.

Dochinger, L.S. and K.F. Jensen. 1985. Effect of acid mist and air pollutants on yellow-poplar seedling height and leaf growth. U.S.D.A. Forest Service Res. Paper No. NE-572, Northeastern Forest Esp. Sta. Broomall, PA, 4 pp.

Duchelle, S.F., J.M. Skelly, and B.I. Chevone. 1982. Oxidant effects on forest tree seedlings growth in the Appalachian Mountains. Water, Air, Soil Pollu. 18: 363-373.

Federer, C.A. and J.W. Hornbeck. 1987. Expected decrease in diameter growth of even-aged red spruce. Can. J. For. Res. 17: 266-269.

Federer, C.A., J.W. Hornbeck, and R.B. Smith. 1987. Regional dendrochronologies of red spruce and other species in New England. U.S.D.A. Forest Serv., Gen. Tech. Rep. Publica. No. 255, Northeastern Forest Exp. Sta., Broomall, PA.

Fisher, R.F. 1980. Allelopathy: A potential cause of regeneration failure. J. For. 78: 346-350.

Frissell, S.S. Jr. 1973. The importance of fire as a natural ecological factor in Itasca State Park, Minnesota. Quat. Res. 3: 397-407.

Fritts, H.C. 1976. Tree Rings and Climate. Academic Press, New York, 567 pp.

Goff, F.G. and P.H. Zedler. 1968. Structural gradient analysis of upland forests in the western Great Lakes area. Ecol. Monogr. 38: 65-86.

Guderian, R. and K. Kueppers. 1980. Response of plant communities to air pollution. In U.S.D.A. Forest Service Gen. Tech. Report No. PSW-43, Pacific Southwest Forest and Range Exp. Sta., Berkeley, CA, pp 187-199.

Hajdúk, J. and M. Ruzicka. 1968. Das studium der schäden an wildpflanzen und pflanzengessellschaften verusacht durch luftverunreinigung. In Air Pollution, Wageningen, Pudoc, pp. 183-192.

Hari, P., T. Raunemaa, and A. Hautojärvi. 1986. The effects on forest growth of air pollution from energy production. Atmos. Environ. 20: 129-137.

Harkov, R. and E. Brennen. 1979. An ecophysiological analysis of the response of trees to oxidant pollution. J. Air Pollu. Control Assoc. 29: 157-161.

Harward, M. and M. Treshow. 1975. Impact of ozone on the growth and reproduction of understory plants in the aspen zone of western U.S.A. Environ. Conserv. 2: 17-23.

Hayes, E.M. and J.M. Skelly. 1977. Transport of ozone from the Northeast U.S. into Virginia and its effect on eastern white pines. Plant Dis. Reptr. 61: 778-782.

Heck, W.W., W.W. Cure, J.O. Rawlings, L.J.Z. Zaragoza, A.S. Heagle, H.E. Heggestad, R.J. Kohut, L.W. Kress, and P.J. Temple. 1984. Assessing impacts of ozone on agricultural crops: I. Overview. J. Air Pollu. Control. Assoc. 34: 729-735.

Heck, W.W., A.S. Heagle, and E.B. Cowling. 1977. Air pollution: Impact on plants. In New Directions in Century Three: Strategies for Land and Water Use. 32nd Annual Meeting, Soil Conservation Soc. Am., Aug. 7–10, 1977, Richmond, VA, pp. 193-202.

Heck, W.W., O.C. Taylor, R. Adams, G. Bingham. J. Miller, E. Preston, and L. Weinstein. 1982. Assessment of crop loss from ozone. J. Air Pollu. Control Assoc. 32: 353-361.

Heck, W.W., O.C. Taylor, and D.T. Tingey. 1988. Response of crops to air pollutants. Environ. Pollu. 53: 1-473.

Heikkenen, H.J. 1984. Tree-ring patterns: A key-year technique for crossdating. J. For. 82: 302-305.

Heinselman, M.L. 1973. Fire in the virgin forests of the Boundary Waters Canoe Area, Minnesota. Quat. Res. 3: 329-383.

Hennessey, T.C., P.M. Dougherty, S.V. Kossuth, and J.D. Johnson, (eds.). 1986. Stress Physiology and Forest Productivity. Martinus Nighoff, New York, NY, 248 pp.

Henry, J.D. and J.M.A. Swan. 1974. Reconstructing forest history from live and dead plant material: An approach to the study of forest succession in southwest New Hampshire. Ecology 55: 772-783.

Hogsett, W.E., M. Plocher, V. Wildman, D.T. Tingey, and J.P. Bennett. 1985. Growth response of two varieties of slash pine seedlings to chronic ozone exposures. Can. J. Bot. 63: 2369-2376.

Hornbeck, J.W., and R.B. Smith. 1985. Documentation of red spruce growth decline. Can. J. For. Res. 15: 1199-1201.

Hornbeck, J.W., R.B. Smith, and C.A. Federer. 1986. Growth decline in red spruce and balsam fir relative to natural processes. Water Air Soil Pollu. 31: 425-430.

Hornvedt, R. 1970. SO_2 injury to forests. J. For. Utiliz. 78: 237-286.

Jensen, K.F. 1982. An analysis of the growth of silver maple and eastern cottonwood seedlings exposed to ozone. Can. J. For. Res. 12: 420-424.

Jensen, K.F. 1973. Response of nine forest tree species to chronic ozone fumigation. Plant Dis. Reptr. 57: 914-917.

Jensen, K.F., L.S. Dochinger, B.R. Roberts, and A.M. Townsend. 1976. Pollution responses. In R. Miksche, ed., Modern Methods in Forest Genetics. Springer–Verlag, New York, pp. 189-215.

Jones, H.C., N.L. Lacasse, W.S. Liggett, and F. Weatherford. 1977. Experimental Air Exclusion System for Field Studies of SO_2 Effects on Crop Productivity. U.S. Environmental Protection Agency, Publ. No. EPA-600/7-77-122, Washington, DC, 67 pp.

Johnson, A.H. 1983. Red spruce decline in the northeastern U.S.: Hypotheses regarding the role of acid rain. J. Air Pollu. Control Assoc. 33: 1049-1054.

Johnson, A.H., T.G. Siccama, D. Wang, R.S. Turner, and T.H. Barringer. 1981. Recent changes in patterns of tree growth rate in the New Jersey pinelands: A possible effect of acid rain. J. Environ. Qual. 10: 427-430.

Jonsson, B. 1976. Soil acidification by atmospheric pollution and forest growth. In L.S. Dochinger and T.A. Seliga, eds., Proc. 1st Intl. Symp. of Acid Precipitation and the Forest Ecosystem. U.S.D.A. Forest Service, Upper Darby, PA, pp. 837-842.

Jonsson, B. and R. Sundberg. 1972. Has the acidification by atmospheric pollution caused a growth reduction in Swedish forests? Royal College of Forestry, Res. Note No. 20, Stockholm, Sweden, 48 pp.

Keller, T. 1978. Wintertime atmospheric pollutants — Do they affect the performance of deciduous trees in the ensuing growing season? Environ. Pollu. 16: 243-247.

Keller, T. 1980. The effect of a continuous springtime fumigation with SO_2 and CO_2 uptake and structure of the annual ring in spruce. Can. J. For. Res. 10: 1-16.

Keller, T. 1985. SO_2 effects on tree growth. In W.E. Winner, H.A. Mooney, and R.A. Goldstein, eds., Sulfur Dioxide and Vegetation. Stanford Univ. Press, Stanford, CA, pp. 250-263.

Kell, J.M., G.R. Parker, and W.W. McFee. 1979. Heavy metal accumulation and growth of seedlings of five forest species as influenced by soil cadmium levels. J. Environ. Qual. 8: 361-364.

Kelty, M.J. 1984. The Development and Productivity of Hemlock-Hardwood Forests in Southern New England. Unpublished Ph.D. Thesis, Yale Univ., New Haven, CT, 206 pp.

Kienast, F. 1982. Jahrringanalytische untersuchungen in immussions-gefährdeten Waldschadengebieten des Wliser Rhotenalls. Geographica Helvetica 3: 143-148.

Knight, H.A. 1987. The pine decline. J. For. 85: 25-28.

Kramer, P.J. and T.T. Kozlowski. 1979. Physiology of Woody Plants. Academic Press, New York, 811 pp.

Kress, L.W. and J.M. Skelly. 1980. The interaction of O_3 hr, SO_2 hr, and NO_2 and its effects on the growth of two forest species. Pl. Dis. 64: 849-852.

Kress, L.W. and J.M. Skelly. 1982. Response of several eastern forest tree species to chronic doses of ozone and nitrogen dioxide. Pl. Dis. 66: 1149-1152.

Kress, L.W., J.M. Skelly and K.H. Hinkelmann. 1982a. Growth impact of O_3 hr, NO_2 hr, and/or SO_2 on Pinus taeda. Environ. Monitor. Assess. 1: 229-239.

Kress, L.W., J.M. Skelly, and K.H. Hinkelmann. 1982b. Growth impact of O_3 hr, NO_2 hr, and/or SO_2 on Plantanus occidentalis. Agric. Environ. 7: 265-274.

Kress, L.W., H.L. Allen, J.E. Mudano, and W.W. Heck. 1988. Response of loblolly pine to acidic precipitation and ozone. Paper No. 88-70.5. 81st Ann. Meeting, Air Pollu. Control Assoc., Dallas, TX, June 19–24, 1988, 11 pp.

Lamoreaux, R.J. and W.R. Chaney. 1977. Growth and water movements in silver maple seedlings affected by cadmium. J. Environ. Qual. 6: 201-205.

Laurence, J.A. and L.H. Weinstein. 1981. Effects of air pollutants on plant productivity. Ann. Rev. Phytopathol. 18: 257-271.

LeBlanc, D.C, D.J. Raynal, E.H. White, and E.H. Ketchledge. 1986. Characterization of historical growth patterns in declining red spruce trees. In G.C. Jacoby and J. W. Hornbeck, eds.,Proc. Intl. Symp. on Ecological Aspects of Tree-Ring Analysis. USDOE Conf. No. 8608144, Washington, DC.

LeBlanc, D.C, D.J. Raynal, and E.H. White. 1987a. Dendroecological analysis of acidic deposition effects on forest productivity. In T. C. Hutchinson and K.M. Meema, eds., Effects of Atmospheric Pollutants on Forests, Wetlands, and Agricultural Ecosystems. NATO ASI Series Vol. G16. Springer–Verlag, NY, pp. 291-306.

LeBlanc, D.C, D.J. Raynal, and E.H. White. 1987b. Acidic deposition and tree growth: I. The use of stem analysis to study historical growth patterns. J. Environ. Qual. 16: 325-333.

LeBlanc, D.C, D.J. Raynal, and E.H. White. 1987c. Acidic deposition and tree growth: II. Assessing the role of climate in recent growth declines. Environ. Qual. 16: 334-340.

LeBlanc, D.C, D.J. Raynal, E.H. White and E.H. Ketchledge. 1987. Comparative historical growth patterns of Adirondack conifers. In D.P. Lavinder, ed., Proc. IUFRO Symp., Tree Growth in a Changing Physical and Chemical Environment. Univ. British Columbia, Vancouver, BC.

Legge, A.H., D.R. Jaques, R.G. Amundson, and R.B. Walker. 1977. Field studies of pine, spruce, and aspen periodically subjected to sulfur gas emissions. Water Air Soil Pollu. 8: 105-129.

Linzon, S.N. 1971. Economic effects of sulfur dioxide on forest growth. J. Air Pollu. Control Assoc. 21: 81-86.

Loucks, O.L. 1970. Evolution of diversity, efficiency and community stability. Am. Zool. 10: 17-25.

Lucier, A.A. 1986. Summary and Interpretation of U.S.D.A. Forest Service Report on "Pine Growth Reductions in the Southeast." National Council of the Paper Industry for Air and Stream Improvement, Tech. Bull. No. 508, New York, 15 pp.

Lucier, A.A. 1988. Pine growth rate changes in the southeast. A summary of key issues for forest managers. South. J. Appl. For. (in press).

Lynch, D.F. 1951. Diameter growth of ponderosa pine in relation to the Spokane-blight problem. Northwest Sci. 25: 157-163.

Mader, D.L. and D.R. Adams. 1987. Growth of Massachusetts Forests in Relation to Potential Effects from Acid Deposition. Report to Mass. Department Environmental Protection. Boston, MA., February, 1987.

Mahoney, M.J., J.M. Skelly, B.I. Chevone, and L.D. Moore. 1984. Response of yellow poplar (*Liriodendron tulipifera* L.) seedling shoot growth to low concentration of O_3 hr, SO_2 hr, and NO_2 hr. Canad. J. For. Res. 14: 150-153.

Manning, W. 1971. Effects of limestone dust on leaf condition, foliar disease incidence and leaf surface microflora of native plants. Environ. Pollu. 2: 69-76.

McBride, J.R., P.R. Miller and R.D. Laven. 1985. Effects of oxidant air pollutants on forest succession in the mixed conifer forest of southern California. In Air Pollutants Effects on Forest Ecosystem Symp. Proc., Acid Rain Foundation, St. Paul, MN, May 8-9, 1985, pp. 157-167.

McClenahen, J.R. 1978. Community changes in a deciduous forest exposed to air pollution. Can. J. For. Res. 8: 432-438.

McClenahen, J.R., and L.S. Dochinger. 1985. Tree ring response of white oak to climate and air pollution near the Ohio river valley. J. Environ. Qual. 14: 274-280.

McCreery, L.R. M. Miller–Weeks, M.J. Weiss, and I. Millers. 1987. Cooperative survey of red spruce and balsam fir decline and mortality in New York, Vermont and New Hampshire — A progress report. Symp. Proc., IPM Symposium for Northern Forests, Madison, WI, March 1986, 13 pp.

McLaughlin, S.B. 1985. Effects of air pollution on forests. A critical review. J. Air Pollu. Control Assoc. 35: 512-534.

McLaughlin, S.B. and O.V. Bräken. 1985. Methods for evaluating and predicting forest growth responses to air pollution. Experientia 41: 310-319.

McLaughlin, S. B., T.J. Blasing, L.K. Mann, and D.N. Duvich. 1983. Effects of acid rain and gaseous pollutants on forest productivity: A regional scale approach. J. Air Pollu. Control Assoc. 33: 1042-1049.

McLaughlin. S.B., D.C. West, and T.J. Blasing. 1984. Measuring effects of air pollution stress on forest productivity. Tappi J. 67: 74-80.

McLaughlin, S., D.J. Downing, T.J. Blasing, E.R. Cook, and H.S. Adams. 1987. An analysis of climate and competition as contributors to decline of red spruce in high elevation Appalachian forests of the eastern United States. Oecologia 72: 487-501.

McLaughlin. S.B., DC West, H.H. Shugart, and D.S. Shriner. 1978. Air pollution effects on forest growth and succession: Applications of a mathematical model. Ann. Meeting, Air Pollu. Control Assoc. Houston, TX, June 25–30, 1978.

Miller, P.R. 1973. Oxidant-induced community change in a mixed conifer forst. In J. Naegle, ed., Air Pollution Damage to Vegetation, Adv. Chem. Series. No. 122, Am. Chem. Soc., Washington, DC, pp. 101-107.

Miller, P.R., (ed.). 1977. Photochemical Oxidant Air Pollutant Effects on a Mixed Conifer Forest Ecosystem. A Progress report. U.S. Environmental Protection Agency, Publ. No. EPA-600/3-77-104, Corvallis, OR, 338 pp.

Mitchell, C.D., and T.A. Fretz. 1977. Cadmium and zinc toxicity in white pine, red maple, and Norway spruce. J. Am. Soc. Hort. Sci. 102: 81-84.

Murphy, L.S. 1917. The red spruce — Its growth and management. U.S.D.A. Forest Service, Bull. No. 544. Washington, DC, 100 pp.

Nash, T.H., H.C. Fritts, and M.A. Stokes. 1975. A technique for examining non-climatic variation in widths of annual tree rings with special references to air pollution. Tree-Ring Bull. 35: 15-24.

National Academy of Sciences. 1977. Ozone and Other Photochemical Oxidants. Chapter 12. Ecosystems. NAS, Washington, D C, pp. 586-642.

National Council of the Paper Industry for Air and Stream Improvement. 1983. Acidic Deposition and its Effects on Forest Productivity — A Review of the Present State of Knowledge, Research Activities, and Information Needs — Second Progress Report. Tech. Bull. No. 392, New York, 104 pp.

National Science Foundation. 1977. Tales the tree rings tell. Mosaic 8: 2-9.

Nikfield, H. 1970. Pflanzensoziologische Beobachtungen in Rauchschadengebiet eines aluminiumwerkes. Zentbl. Ges. Forstw. 84: 318-329.

Nyborg, M. 1978. Sulfur pollution in soils. In J.O. Nriagu, ed., Sulfur in the Environment. Part II. Ecological Impacts. Wiley, New York, pp. 359-390.

Odum, E.D. 1969. The strategy of ecosystem development. Science 164: 262-270.

Phillips S.O., J.M. Skelly, and H.E. Burkhart. 1977a. Eastern white pine exhibits growth retardation by fluctuating air pollutant levels: Interaction of rainfall, age, and symptom expression. Phytopathology 67: 721-725.

Phillips, S.O., J.M. Skelly, and H.E. Burkhart. 1977b. Growth fluctuation of loblolly pine due to periodic pollution levels: Interaction of rainfall and age. Phytopathology 67: 716-728.

Pye, J.M. 1988. Impact of ozone on the growth and yield of trees: A review. J. Environ. Qual. (in press).

Raup, H.M. 1957. Vegetational adjustment to the instability of the site. In Proc. 6th Technical Meeting of the Intl. Union for the Protection of Nature, June 1956, Edinburgh, pp. 36-48.

Raynal, D.J., D.C. LeBlanc, B.T. Fitzgerald, E.H. Ketchledge and E.H. White. 1988. Historical growth patterns of red spruce and balsam fir at Whiteface Mountain, New York. Proc. In Effects of Atmospheric Pollutants on the Spruce-Fir Forests of the Eastern United States and Federal Republic of Germany. U.S.D.A. For. Ser. Gen. Tech. Rep. No. 255, NE-Northeastern Forest Esp. Sta.. Broomall, PA.

Rosenberg, C.R., R.J. Hutnik, and D.D. Davis. 1979. Forest composition at variances from a coal-burning power plant. Env. Pollu. 5: 307-317.

Santamour, F.S. Jr. 1969. Air Pollution Studies on *Plantanus* and American elm seedlings. Plant Dis. Reptr. 53: 482-485.

Shafer, S.R., A.S. Heagle, and D.M. Camberato. 1987. Effects of chronic doses of ozone on field-growth loblolly pine: Seedling responses in the first year. J. Air Pollu. Control. Assoc. 37: 1179-1184.

Scheffer, T.C. and G C. Hedgcock. 1955. Injury to Northwestern Forest Trees by Sulfur Dioxide from Smelters. U.S.D.A. Forest Service Tech. Bull. No. 1117, Washington, DC, 49 pp.

Sheffield, R. and N.D. Cost. 1987. Behind the decline. J. For. 85: 29-33.

Schutt, P. and E.B. Crowling. 1985. Waldsterben, a general decline of forests in central Europe: Symptoms, development, and possible causes. Plant Dis. 69: 548-558.

Shaw, C.G., G.W. Fischer, D.F. Adams, and D.W. Lynch. 1951. Fluorine injury to ponderosa pine: A summary. Northwest Sci. 15: 156.

Siccama, T.G., M. Bliss, and H.W. Vogelmann. 1982. Decline of red spruce in the Green Mountains of Vermont. Bull. Tor. Bot. Club 109: 162-168.

Skelly, J.M., L.D. Moore, and L.L. Stone. 1972. Symptom expression of eastern white pine located near a source of oxides of nitrogen and sulfur dioxide. Plant Dis. Reptr. 56: 3-6.

Sprugel, D.G., J.E. Miller, RN. Muller, H.J. Smith, and P.B. Xerikos. 1979. Effect of SO_2 fumigation on yield and seed quality in field-growth soybeans. Argonne National Laboratory Publ. No. ANL-ERC-72-22, Argonne, IL, 27 pp.

Stemple, R.B. and E.H. Tyron. 1973. Effect of coal smoke and resulting fly ash on site quality and radial increment of white oak. Castanea 38: 396-406.

Stephens, E.P. 1955. Research in the biological aspects of forest production. J. For. 53: 183-186.

Stephens, E.P. 1956. The uprooting of trees: A forest process. Soil Sci. Am. Proc. 20: 113-116.

Stone, L.L. and J.M. Skelly. 1974. The growth of two forest tree species adjacent to a periodic source of air pollution. Phytopathology 64: 773-778.

Sundberg, R. 1971. On the estimation of pollution-caused growth reduction in forest trees. The IASPS Symposium on Statistical Aspects of Pollution Problems. Boston, MA.

Symeonides, C. 1979. Tree-ring analysis for tracing the history of pollution. Application to a study in northern Sweden. J. Environ. Qual. 8: 482-486.

Taylor, O.C. 1974. Oxidant Air Pollutant Effects on a Western Coniferous Forest Ecosystem. Annual Progress Report 1973-1974. Statewide Air Pollution Research Center, Univ. California, Riverside, CA, 111 pp.

Todd, G.W. and M.J. Garber. 1958. Some effects of air pollutants on the growth and productivity of plants. Bot. Gaz. 120: 75-80.

Trautmann, W., A. Krause, and R. Wolff-Straub. 1970. Veränderungen der bodenvegetation in kiefernforsten als folge industrieller luftverunreinigungen im Raum Mannheim-Ludwigshafen. Schraftinr. Reihe Vegetationsk. 5: 193-207.

Treshow, M. 1968. The impact of air pollutants on plant populations. Phytopathology 58: 1108-1113.

Treshow, M. and D. Stewart. 1973. Ozone sensitivity of plants in natural communities. Biol. Conserva. 5: 209-214.

Treshow, M., F.K. Anderson, and F. Harner. 1967. Response of Douglas-fir to elevated atmospheric fluorides. For. Sci. 13: 114-120.

Tritton, L.M. and J.W. Hornbeck. 1982. Biomass Equations for Major Tree Species of the Northeast. U.S.D.A. Forest Service, Genl. Tech. Rep.. NE-69, Northeastern Forest Exp. Sta. Broomall, PA, 46 pp.

Tveite, N.B. and O. Teigen. 1976. Acidification experiments in conifer forest. 3. Tree Growth Studies. SNSE Project, Research Report No. 7, Oslo, Norway.

Wang, D., D.F. Karnosky, and F.H. Bormann. 1985. Effects of ambient ozone on the productivity of *Populus tremuloides* Michx. grown under field conditions. Can. J. For. Res. 16: 47-55.

Wang, D., F.H. Bormann, and D.F. Karnosky. 1986. Regional tree growth reductions due to ambient ozone: Evidence from field experiments. Environ. Sci. Technol. 20: 1122-1125.

Weinstein, L.H. and D.C. McCune. 1979. Air pollution stress. In H. Mussell and R. Staples, eds., Stress Physiology and Crop Plants. Wiley, New York, pp. 328-341.

Weiss, M.J., L.R. McCreery, I. Millers, M. Miller-Weeks, and J.T. O'Brien. 1984. Cooperative Survey of Red Spruce and Balsam Fir Decline and Mortality in New York, Vermont, and New Hampshire. U.S.D.A. Forest Service, Publ. No. NA-TP-11. Northeastern Forest Exp. Sta., Broomall, PA, 53 pp.

Wert, S. L., P.R. Miller, and R.N. Larsh. 1970. Color photos detect smog injury to forest trees. J. For. 68: 536-539.

West, DC, H.H Shugart, and DC Botkin. 1981. Forest Succession Concepts and Applications. Springer-Varlag, New York, 496 pp.

Westman, W.E. 1979. Oxidant effects on Californian coastal sage scrub. Science 205: 1001-1003.

Whittaker, R.H. 1975. Communities and Ecosystems. Macmillan, New York, 385 pp.

Whittaker, R.H. 1966. Forest dimensions and production in the Great Smoky Mountains. Ecology 47: 103-121.

Whittaker, R.H., R.H. Bormann, G.E. Likens, and T.G. Siccama. 1974. The Hubbard Brook Ecosystem Study: Forest biomass and production. Ecol. Mono. 44: 233-252.

Winner, W.E. and J.D. Bewley. 1978. Contrasts between brophyte and vascular plant synecological responses in an SO_2-stressed white spruce association in central Alberta. Oecologia 33: 311-325.

Wood, T. and F.H. Bormann. 1974. The effects of an artificial acid mist upon the growth of *Betula alleghaniensis* Br. Environ. Pollu. 7: 259-268.

Wood, T. and F.H. Bormann. 1975. Short-term effects of an artificial acid rain upon the growth and nutrient relations of *Pinus strobus* L. In L.S. Dochinger and T.A. Seliga, eds., Proc. 1st Intl. Symp. Acid Precipitation and the Forest Ecosystem. US.D.A. Forest Service, Upper Darby, PA, pp. 815-825.

Woodman, J.M. 1986. Mature tree response workshop report. Southern Commercial Forest Research Cooperative. Dec. 8-10, 1986. St. Louis, MO.

Woodwell, G.M. 1974. Success, succession and Adam Smith. BioScience 24: 81-87.

Zobel, D.B., A. McKee, G.M. Hawk, and C.T. Dyrness. 1976. Relationship of environment to composition, structure, and diversity of forest communities of the central western cascades of Oregon. Ecol. Mono. 46: 135-156.

SECTION III

FOREST ECOSYSTEMS ARE INFLUENCED BY AIR CONTAMINANTS IN A DRAMATIC MANNER — Class III INTERACTIONS

17
Forest Ecosystem Destruction: A Localized Response to Excessive Air Pollution

Under conditions of excessive exposure, that is, atypically high atmospheric concentrations of one or more contaminants for extended (or continuous) time periods, the impact on forest ecosystems may be very severe and dramatic. This response is designated a Class III interaction. The reaction of vegetation in this case is characterized by severe morbidity and mortality caused directly and unambiguously by air pollutants.

Atmospheric burdens of contaminants are generally of sufficient magnitude to cause Class III interactions *only* in those portions of forest ecosystems in the vicinity of major point sources of atmospheric contaminants. Stationary sources of primary importance include energy production facilities, for example, electric generating plants, gas purification plants; metal related industries, for example, copper, nickel, lead, zinc, or iron smelters; aluminum production plants; and a variety of other industries, for example, cement plants, chemical plants, and pulp mills. The forest area impacted by these facilities is typically confined to a zone of a few kilometers immediately surrounding the plant and for a distance of several kilometers in the downwind direction. The extent of the latter influence is quite variable and is primarily controlled by source strength of the effluent, local meteorology, regional topography, and susceptibility of the area vegetation.

Terrestrial ecosystems respond to gradients of natural environmental stress by predictable changes in structure. As environmental conditions become more damaging and restrictive, the size of dominant plants becomes smaller and a progression from systems dominated by trees to shrubs to grasses is typical. This pattern of structural change can readily be observed along temperature and wind gradients in mountainous regions, along atmospheric and soil salt gradients in maritime regions, and along moisture gradients in zones of differing annual precipitation input. It has been proposed that environmental stresses imposed on terrestrial ecosystems by anthropogenic activities may induce similar alterations in ecosystem structure (Curtis, 1956). Woodwell (1970) has developed this hypothesis and has indicated that evidence from studies dealing with the ecosystem effects of ionizing radiation, persistent pesticides, and of eutrophication is especially supportive. Woodwell generalized that a common reaction of forest ecosystems to environmental stresses is a "systematic dissection of strata layer by layer." Moving along a gradient of increasing stress; trees are eliminated first, then taller shrubs, then lower shrubs, then herbs, and finally bryophytes. Change in forest structure as described would be the most obvious but not the sole alteration resulting from stress. Associated with gross simplification would also be altered ecosystem functions, including reduced rate of energy fixation, reduced biomass, increased nutrient loss, and reduced animal populations. As a result of the considerable influence that intact forest ecosystems exert on surrounding geology and climate, gross simplification would also cause appreciable change in local erosion and sedimentation, hydrology, and meteorology.

Woodwell (1970) and Whittaker and Woodwell (1978) have indicated that air pollution stress results in forest ecosystem response comparable to other severe anthropogenic stresses. As indicated, Class III interactions are generally associated with industrial point sources of atmospheric contaminants. The two most important and pervasive of the latter are sulfur dioxide and fluoride. What is the evidence that these pollutants destroy and simplify forest ecosystems? A comprehensive review of this question has been provided by Miller and McBride (1975).

A. Sulfur Dioxide

Smelting of metal-bearing ores and combustion of high-sulfur fossil fuels for energy generation results in excessive production of sulfur dioxide. Since many of these facilities are located in forested regions, examples of Class III interactions are considerable.

1. Copper Basin, Tennessee

This area centered in Ducktown, Tennessee consists of approximately 243 km^2 (60,000 acres) and was originally covered with southern deciduous forest. Mining operations were initiated in the basin in 1850 and smelting operations were most active between 1890 and 1895. By 1910 gross forest simplification resulting from excessive sulfur dioxide had created three new vegetative zones surrounding Ducktown (Haywood, 1905; Hedgcock, 1914; Hursh, 1948). In a 27 km^2 (10.5 mile^2) area closest to the source vegetation was devastated and largely eliminated. All trees and shrubs were destroyed, and only a few, isolated islands of sedge grass occurred in the outer portions of this zone. A belt of grassland ecosystem, 68 km^2 (17,000 acres) in size, surrounded the barren zone. The principal grassland species was broomsedge. A transition zone of somewhat indefinite boundary and consisting of approximately 120 km² (30,000 acres) was located beyond the grassland. Few trees were located along the inner edge of the transition zone. Sassafras, red maple, sourwood, and post oak were common in the middle of the transition forest. The uninfluenced forest beyond the impact of the smelter consisted principally of mixed oaks, hickory, dogwood, sourwood, black tupelo, and some eastern white pine. The distance of vegetative impact extended 19–24 km (12–15 miles) to the north and approximately 16 km (10 miles) to the west of the smelter. Eastern white pine damage was recorded 32 km (20 miles) from the industry (Hursh, 1948).

Sheet and gully soil erosion has been excessive in the acutely damaged inner zone. Micrometeorological changes in the inner zone relative to the surrounding forest have been substantial; summer air temperature averages are 1–2°C higher, while the winter air temperature averages 0.3–1°C lower, the soil temperature is 11°C higher in the summer; the wind velocities are 5–15 times higher and rainfall is consistently lower (Hepting, 1971).

Reforestation of Copper Basin has not been rapid due to severe erosion, low soil nutrients and moisture, and high winds. Recent reforestation efforts have employed both pine and hardwood species. New techniques that have significantly increased the successful establishment of trees include use of specific ectomycorrhizae, broadcast applications of dried sewage sludge, and slit applications of sewage sludge and commercial fertilizer tablets (Berry, 1983).

2. Sudbury, Ontario

In the Sudbury area, a century of sulfur dioxide fumigation, copper and nickel particulate deposition, fire, soil erosion, and increased frost action have interacted to create 10,000 ha (25,000 acres) of barren land and 36,000 ha (89,000 acres) of stunted, open birch-maple woodland (Winterhalder, 1988). Three large nickel and copper smelters have historically discharged several thousand tons of sulfur dioxide daily into the surrounding atmosphere. Sulfur dioxide emissions from this area have approximated 10% of the North American sulfur dioxide total and 25% of the smelter total. Extensive simplification of the mixed boreal forest ecosystem surrounding this region has occurred primarily via the mortality of eastern white pine throughout a 1865 km² (720 square mile) area to the northeast of the Falconbridge, Copper Cliff, and Coniston smelters. This region has been extensively studied and reviewed by Gorham (1970), Linzon (1978), and Winterhalder (1984). Class III relationships exist with black spruce and balsam fir in addition to white pine. Acute impact on the latter species has been recorded in excess of 40 km (25 miles) from the source of sulfur dioxide. Red oak, red maple, and red-berried elder are more tolerant and may exist in disturbed forests as close as 1.6 km (1 mile) to the smelters. Morbidity and mortality of forest trees in the Sudbury region continued to spread at least until the construction of the world's tallest smokestack of 403 m (1250 feet) at Copper Cliff in 1972 and the closing of the Consiton smelter also in 1972.

Soil erosion has followed the destruction of surrounding forest ecosystems. Rainfall has been made highly acidic, commonly less than pH 3.0 in 1971. Elevated nickel and copper concentrations in soils have been recorded to distances of 50 km (31 miles). Eroded sediment has contaminated area lakes and water sources, with resulting long-distance transport of trace metal loads (Hutchinson and Whitby, 1976).

Since 1978, the Regional Municipality of Sudbury has been reclaiming areas of damaged land. Through 1984, over 2640 ha (6500 acres) have been planted with grass and 387,580 trees have been planted (Lautenback, 1985). Reclamation strategy involves manual application of limestone, fertilizer, and a grass-legume seed mixture to barren slopes. Within a few years, reclaimed sites are characterized by mixtures of native herbaceous and woody species (Winterhalder, 1987, 1988).

3. Wawa, Ontario

Forest destruction in the vicinity of an iron smelter in Wawa, northern Ontario, exhibits a more discrete pattern, one smelter relative to three, and a shorter history of impact resulting in less equilibration of vegetative response than in Sudbury. The Wawa smelter initiated operations in 1939 (significantly expanded in 1949) and has released as much as 100,000 tons of sulfur dioxide annually to the surrounding atmosphere. Vegetative impact is primarily confined to a strip northeast from the plant in the direction of the prevailing wind. Symptoms of sulfur dioxide damage may be observed for at least 32 km (20 miles) to the northeast. The mixed boreal forest in the Wawa area consists mainly of white spruce, black spruce, balsam fir, jack pine, white cedar, larch, and white pine in the dominant layer. Mountain maple and *Pyrus decora* are common in the understory.

Gordon and Gorham (1963) have systematically studied the response of the forest to sulfur dioxide from Wawa and have presented evidence consistent with Woodwell's pattern of forest ecosystem response to stress. These investigators established a series of vegetative sampling plots along a transect running from the smelter to 58 km (36 miles) northeast of Wawa. The variety of the flora declined acutely from approximately 20-40 species per 40 m² quadrat beyond 16 km (10 miles) of the facility to 0-1 species within 3 km (2 miles). The authors recognized four zones of destroyed or simplified forest ecosystem along the transect. In the zone of "very severe" impact, within 8 km (5 miles) of the source, both tree and continuous shrub layers were nonexistent. Some elder remained alive, but was symptomatic. In the zone of severe damage, extending from 16–19 km (10–12 miles) the forest canopy was still lacking. In the zone of considerable damage, 19–27 km (12–17 miles) from the source a discontinuous tree canopy was present. Tree mortality in this region was high. Only a few white birch and white spruce persisted. Tall shrubs were vigorous and *Pyrus* and mountain maple were common. Elder was present in large numbers in the understory. In the zone of moderate damage, from 27–37 km (17–23 miles), the

Table 17-1. Number of Species Recorded, Tree Symptoms, and Soluble Soil Sulfate at Increasing Distances from an Iron Smelter Along a Transect in the Direction of the Prevailing Wind.

Distance NE of smelter (miles)	Number ground flora species		Tree damage (aerial estimate)	Soluble sulfate in soil (meq 100 g^{-1} ignition loss)
	Quadrat 20 m × 2 m	Accessory		
0.56	0	2	Very severe	31
1.0	1	3	Very severe	24
1.2	1	2	Very severe	18
1.6	1	2	Very severe	16
2.1	2	1	Very severe	15.5
2.5	1	1	Very severe	14.2
3.2	3	2	Very severe	12.2
4.7	2	8	Very severe	8.2
5.8	5	10	Severe	6.5
6.4	9	6	Severe	5.0
6.6	15	5	Severe	6.9
7.3	8	14	Severe	3.9
9.5	15	23	Severe	4.1
10.8	26	14	Severe	4.8
11.7	20	16	Considerable	7.6
14.0	39	16	Considerable	6.1
15.7	27	15	Considerable	5.7
17.3	32	17	Moderate	4.2
19.2	27	10	Moderate	9.2
21.9	20	12	Moderate	9.0
25.5	36	16	No damage	9.4
27.9	27	19	No damage	9.1
30.1	27	17	No damage	8.0
36.1	27	10	No damage	4.8

Source: Gordon and Gorham (1963).

tree canopy was continuous but symptoms of stress were obvious. Table 17-1 presents a summary of ground flora species recorded at increasing distances from the smelter.

Gordon and Gorham concluded that the forest ecosystem was "peeled off in layers" as the smelter source was approached from the northeast. The tree stratum was intact at 37 km (23 miles) distance. It was discontinuous within 37 km and absent within 27 km (17 miles). The shrub stratum dominated from 27 km to 19 km (12 miles). Herbs were dominant from 19 km to 8 km (5 miles) and there was no continuous plant cover within 8 km. Erosion at Wawa was extensive in the devegetated area (Figure 17-1).

A large number of additional examples involving severe morbidity or mortality of forest species in the immediate vicinity of industrial point sources of sulfur containing contaminants may be cited (Table 17-2). Unfortunately, most

Table 17-2. Additional Examples of Boreal and Temperate Forest Ecosystems that Have Exhibited Class III Relationships with Sulfur Contaminants from Industrial Point or Area Sources.

Location	Source	Severely injured or killed species	Reference
Redding, California, USA	Smelter	Pine, oak	Haywood (1905,1910)
Anaconda, Montana, USA	Smelter	Douglas fir, lodgepole pine	Scheffer and Hedgcock (1955)
Missoula, Montana, USA	Pulp mill	Ponderosa pine, Douglas fir	Carlson (1974)
Superior, Arizona, USA	Smelter	Paloverde	Wood and Nash (1976)
Jackson, Mississippi USA	Nuclear electric generating facility (cooling tower SO_4^{2-})	Red oak, white pine, sassafras, white ash	Rochow (1978)
Colstrip, Montana, USA	Electric generating complex	Ponderosa pine	U.S.D.A. Forest Service (1978)
Trail, B.C., Canada	Smelter	Ponderosa pine, Douglas fir, western larch, lodgepole pine	Scheffer and Hedgcock (1955)
Anyox, B.C., Canada	Smelter	Western red cedar, western hemlock, Pacific silver fir, Sitka spruce	Errington and Thirgood (1971)
Yellowknife, NW Territories, Canada	Smelter	Black spruce, white spruce, paper birch, poplar, willow	Hocking et al. (1978)
Industrial Pennines, Great Britain	Mixed industrial	Scotch pine	Farrar et al. (1977)
Ruhr Valley, West Germany	Mixed industrial	Scotch pine	Knabe (1970)
Rouen, France	Mixed industrial	Scotch pine	Décourt (1977)

Figure 17-1. Forest destruction in the vicinity of a smelter operation in Wawa, Ontario, Canada. Years of effluent discharge (sulfur dioxide) has caused mortality in various forest strata in the downwind plume direction. At 30 km (18 miles) northeast of the smelter the white spruce, balsam fir, and white birch forest exhibits normal structure (a). As the smelter is approached in the plume zone, the various strata are sequentially destroyed. At 16 km (10 miles) from the smelter, the tree stratum is almost completely killed, with only scattered white spruce and white birch remaining

(c)

(d)

alive; the shrub layer is vigorous and consists of mountain maple, showy mountain ash and invading red-berried elder (b). Beginning at approximately 13 km (8 miles) from the smelter the tree stratum is completely destroyed and mortality in the shrub layer is extensive; climbing buckwheat is vigorous in the herb stratum (c). Within 6 km (4 miles) of the smelter almost all vegetation is destroyed and only scattered; symptomatic climbing buckwheat remains (d).
Source: Prof. Eville Gorham, University of Minnesota, Minneapolis.

reports do not include useful and accurate suggestions of the size of forest areas impacted. If these were available and could be totaled the sum would be impressive.

B. Fluorides

The release of particulate sodium and aluminum fluoride and gaseous carbon tetrafluoride and hydrogen fluoride from aluminum ore reduction facilities and phosphate fertilizer plants has resulted in Class III relationships in a variety of forest ecosystems surrounding these industries (Miller and McBride, 1975).

Two significant United States examples of fluoride damage are located in Columbia Falls, Montana and Franklin Park (Spokane), Washington. In Columbia Falls, the impact of the Anaconda Company aluminum reduction plant on the surrounding forest has been previously discussed (Chapter 12). Severe morbidity and mortality are concentrated in ponderosa pine and lodgepole pine on 8 km^2 (2000 acres) surrounding the plant. In Franklin Park, the Kaiser aluminum ore reduction operation has caused similar severe stress on ponderosa pine forests over a 130 km^2 (50 mile^2) area surrounding the plant. Because foliar fluoride analyses have been employed to document damage in both of these examples, confidence in assigning the destruction to atmospheric fluorides released by the plants is considerable (Carlson, 1972; Adams et al., 1952). Mortality of 81 ha (200 acres) of Douglas fir has been documented adjacent to a phosphate reduction facility in Georgetown Canyon, Idaho (Treshow et al., 1967). In their review, Miller and McBride (1975) indicated that additional examples of localized forest destruction caused by fluorides have occurred in the states of Washington, Oregon, and Montana.

Examples of Class III forest relationships with fluoride are numerous in Europe (Scurfield, 1960). Robak (1969) observed that none of the larger aluminum plants in Norway has been able to avoid some damage to the more sensitive conifers existing within 2 km of their smoke outlets. In some cases Scotch pine forests have been destroyed up to a distance of 10–13 km (6–8 miles) from fluoride sources. Gilbert (1975) has indicated that forest destruction by fluorides in Norway is consistent with the Woodwell hypothesis concerning relative susceptibility of the forest strata. In Hungary, an oak-pine forest has been destroyed to a distance of 800 km (875 yards) from an aluminum smelter (Keller, 1973). In Raushofen, Austria, approximately 800 ha (2000 acres) of mixed forest in the vicinity of an aluminum plant has been destroyed (Jung, 1968).

Hälgren and Nyman (1977) have investigated the influence of gaseous pollutants from an iron-sintering plant in Vitafors, Sweden on surrounding Scotch pine forests. The Vitafors industry releases significant amounts of both sulfur dioxide and hydrogen fluoride. Fluoride symptoms were not useful in distinguishing damage caused by sulfur and fluoride content. Foliar levels representative of emission free areas (baseline) were judged to be 0.07–0.12% sulfur and 10 μg g^{-1} fluoride. Elevated fluoride was detected to a distance of approximately 5 km (3 miles) from the industry. Foliar sulfur concentrations did not appear to be correlated with distance from the source. This study emphasizes the difficulty of making specific phytotoxicity judgements concerning industrial effluents, as these frequently contain more than one pollutant.

C. Other Pollutants

A variety of additional examples of Class III interactions could be cited involving industrial release of numerous other pollutants. Release of coarse dust, chlorine, ammonia, and other materials has resulted in local forest destruction. The significance of these incidents is not great, however, as they are usually infrequent in occurrence and are commonly highly localized.

The most important deficiency in our understanding of Class III relationships concerns oxidant influence on forest ecosystems. Oxides of nitrogen, ozone, and peroxyacetylnitrate are generated and released from area sources, frequently large urban complexes or regions of concentrated industrial-commercial-transportation activity, and as such are transported into surrounding forests in association with large air mass movements rather than in small, discrete plumes characteristic of point source industrial release of sulfur dioxide and fluoride. The oxidants are also rapidly metabolized by the biota and soils, and their detection in association with the vegetation is not possible, as they are neither persistent nor accumulative. As a result of these limitations, we have less than acceptable appreciation of the availability of ambient oxidants to cause acute morbidity and mortality in natural forests. The most convincing evidence for Class III oxidant involvement concerns the severe impact of ozone on ponderosa pine of the western montane forest along the western side of the Sierra Mountains in California (Chapter 15) (Miller and McBride, 1975). An approximate dose of more than 80 ppb ($157\mu g$ m^{-3}) for 12 hr has been judged to cause moderate to severe damage to ponderosa pine.

In the eastern United States considerable interest has been focused on eastern white pine varieties, as this species readily exhibits foliar symptoms in response to elevated oxidant concentrations. This species is a major component of four forest types and an associate in 14 other types, with a range extending over 28,350 km^2 (7 million acres) from the Lake States to the Appalachian Mountains (U.S.D.A. Forest Service, 1973). Documentation of field injury is available (Linzon, 1965; Berry and Ripperton, 1963; Berry, 1961; Skelly and Jonston, 1978). Fumigation experiments have suggested that other widespread eastern pine species, for example, Virginia pine and jack pine, may be even more susceptible to oxidant injury than white pine (Chapter 15). It has been suggested that if oxidant concentrations in eastern forests reach daily peaks in the range of 300–600 ppb (588–1176 μg m^{-3}) (cf., Table 2–10, Chapter 2), widespread impact on forest ecosystems with susceptible pine species may occur (Miller and McBridge, 1975). Experiments designed to explore this possibility remain a research priority.

Table 17-3. Interaction of Air Pollution and Temperate Forest Ecosystems under Conditions of High Air Contaminant Load, Designated Class III Interaction.

Forest soil and vegetation: Activity and response	Ecosystem consequence and impact
1. Severe morbility, excessive foliar damage	1. Dramatic change in species composition, reduced biomass, increased erodebility, nutrient attrition, altered microclimate and hydrology
2. Mortality	2. Forest simplification or destruction

D. Summary

Under conditions of high air pollution dose, forest trees may be severely injured or killed. This may result in severe purterbations to ecosystem structure and function (Table 17-3). Responses of forest trees and ecosystems to excessively damaging doses of air contaminants are designated Class III relationships. A very large number of examples of Class III relationships have been documented in North America and Europe. Typically these situations involve the impact of sulfur dioxide or fluoride released from industrial sources on surrounding forests. Generally an elliptical pattern of forest stress occurring in the prevailing downwind direction is recorded. The extent of the influence downwind and the area of forest impact is very variable and is controlled by emission strength, local meteorology, and topography and tree species susceptibility. The fact that many industries are located in valleys appears to exacerbate the damage by restricting plume expansion and reducing atmospheric mixing. In the western United States oxidants are imposing a Class III stress on portions of the western montane forests of southern California. A similar relationship for several eastern United States pine species and oxidants has been suggested but not clearly documented in the field.

In extreme Class III situations, irrespective of the specific pollutants, forest communities react, first, by losing sensitive species, second, by losing the tree stratum, and third, by maintaining cover in resistant shrubs and herbs widely recognized as seral or successional species. In less extreme Class III situations, the loss of sensitive species may be followed by maintenance of a tree stratum by resistant species. In this case, the forest consists of a smaller number of dominant plants and is characterized as being "simplified." The report of McClenahen (1978) provides a very nice example of air pollution induced forest simplification in the upper Ohio River Valley.

A fundamental question that has intrigued ecologists for a long time concerns the relationship between ecosystem simplification and stability (Holling, 1973). Are simplified ecosystems less stable? Ecosystems with complex structure can be maintained with relatively less energy. Ecosystem maturity is generally associated with greater complexity (Mergalef, 1963). May (1973) has reviewed a variety of theoretical models dealing with population stability in diverse biological communities. He concluded that there is no indication that increasing diversity and complexity occasion enhanced community stability. Further, as Langford and Buell (1969) have indicated, temperate forest ecosystems have traditionally been viewed as systems that naturally combine characteristics of low diversity with high stability. The loss of individual tree species from forest ecosystems by air pollution stress cannot, therefore, be judged to make the residual systems less stable, only more simple. The impact of species loss to a given ecosystem will be largely dependent on the overall significance of that species to the functions of the system.

Whittaker (1965) has indicated that natural communities are mixtures of unequally important species. The loss of an important species from a forest ecosystem will have greater impact on system function than a less important species. Productivity appeared, to Whittaker, to be the best predictor of relative species importance. When evaluating responses of forest ecosystems to Class III interactions, we should evaluate the consequences of the loss of particular species accordingly. By this criterion, the loss of ponderosa pine in some mixed conifer stands in California and the loss of white pine in eastern forests would be significant in certain forest types and insignificant in others. In addition to the importance of tree mortality for ecosystem considerations, species loss must also be evaluated in terms of management objectives imposed by people on the stressed ecosystem. If a severely impacted or destroyed species has commercial or aesthetic importance, its loss will be significant. In the case of gross forest destruction, regional impacts on nutrient cycling, soil stabilization, sedimentation and eutrophication of nearby aquatic systems, and climatic and hydrologic influences are important. Biomass reduction results in a corresponding reduction in the total inventory of nutrient elements held within a system and loss of the dominant vegetation destroys cycling pathways and mechanisms of nutrient conservation (Chapter 9). Research on the northern hardwood forest has clearly established that retention of nutrients within a forest ecosystem is dependent on constant and efficient cycling between the various components of the intrasystem cycle and that deforestation impairs this retention (Likens et al., 1977). Extensive nutrient loss can pollute downstream aquatic resources and can result in depauperization of a site and have long-term consequences with regard to future plant growth potential. Sundman et al. (1978) have reported significant failures in reforestation after clear-cutting of coniferous forests in northern Finland during recent decades. Increases in soil instability and erosion follow extensive mortality of dominant vegetation, particularly in regions with steep or unstable slopes. Increased erodibility was found to follow deforestation at the Hubbard Brook Experimental Forest in New Hampshire (Bormann et al., 1969). This mature forest ecosystem, when undisturbed, was little affected by erosion, with an average annual particulate matter export of 2.5 tons km^{-2} yr^{-2}. Deforestation and repression of growth for 3 years increased export to a maximum of 38 tons km^{-2} yr^{-1} (Bormann et al., 1974). Soil erosion has been extensive at both Copper Hill, Tennessee, and

Wawa, Ontario. In areas subject to fire stress, for example, southern California, an increase in the abundance of dead trees and increased shrub cover act to increase the fire hazard. Fire incidence has become more frequent in a coastal western hemlock forest severely stressed by sulfur dioxide from a copper smelter in Anyox, British Columbia (Errington and Thirgood, 1971).

Fortunately, efforts to reclaim and reforest numerous Class III sites are active. Significant advancements have been made regarding cultural practices that facilitate revegetation.

Class III interactions between forest ecosystems and air pollutants are the most dramatic of all responses. The forests areas involved in Class III interactions, however, are much less than in the Class I and II interactions.

References

Adams, D.F., D.J. Mayhew, R.M. Gnagy, E.P. Rickey, R.K. Koppe, and I.W. Allan. 1952. Atmospheric pollution in the ponderosa pine blight area, Spokane County, Washington. Ind. Eng. Chem. 44: 1356-1365.

Berry, C.R. 1961. White pine emergence tipburn, a physiogenic disturbance. U.S.D.A. Forest Service, Southeast Forest Exp. Sta. Paper No. 130, 8 pp.

Berry, C.R. 1983. Growth response of four hardwood tree species to spot fertilization by nutrient tablets in the Tennessee Copper Basin. Reclam. Rev. Res. 2: 167-175.

Berry, C.R., and L.A. Ripperton. 1963. Ozone, a possible cause of white pine emergence tipburn. Phytopathology 53: 552-557.

Bormann, F.H., G.E. Likens, and J.S. Eaton. 1969. Biotic regulation of particulate and solution losses from a forest ecosystem. BioScience 19: 600-610.

Bormann, F.H., G.E. Likens, T.G. Siccama, R.S. Pierce, and J.S. Eaton. 1974. The effect of deforestation on ecosystem export and the steady-state condition at Hubbard Brook. Ecol. Monogr. 44: 255-277.

Carlson, C.E. 1972. Monitoring fluoride pollution in Flathead National Forest and Glacier National Park. U.S.D.A. Forest Service, Div. State and Private Forestry, Missoula, MN, 25 pp.

Carlson, C.E. 1974. Sulfur damage to Douglas-fir near a pulp and paper mill in western Montana. U.S.D.A. Forest Service, Div. State and Private Forestry Publ. No. 74-13, 41 pp.

Curtis, J.T. 1956. The modification of mid-latitude grasslands and forests by man. In W.L. Thomas, ed., Man's Role in Changing the Face of the Earth. Univ. Chicago Press,. Chicago, IL, pp. 721-736.

Décourt, N. 1977. Premier inventair des effets de la pollution atmosphérique sur le massif forestier de Roumare. Biologie et Fôret, pp. 435-447.

Errington, J.C. and J.V. Thirgood. 1971. Search through old papers helps reconstruct recovery at Anyox from fume damage and forest fires. Northern Miner. Annu. Rev., pp. 72-75.

Farrar, J.F., J. Relton, and A.J. Rutter. 1977. Sulfur dioxide and the scarcity of *Pinus sylvestris* in the industrial Pennines. Environ. Pollu. 14: 63-68.

Gilbert, O.L. 1975. Effects of air pollution on landscape and land-use around Norwegian aluminum smelters. Environ. Pollu. 8: 113-121.

Gordon, A.G. and E. Gorham. 1963. Ecological aspects of air pollution from an iron-sintering plant at Wawa, Ontario. Can. J. Bot. 41: 1063-1078.

Gorham, E. 1970. Air pollution from metal smelters. Naturalist 21: 12-15, 20-25.

Hälgren, J.E. and B. Nyman. 1977. Observations on trees of Scots pine (*Pinus sil-vestris* L.) and lichens around a HF and SO_2 emission source. Stud. For. Suec. No. 137, 40 pp.

Haywood, J.K. 1905. Injury to vegetation by smelter fumes. U.S.D.A., Bur. Chem. Bull. No. 89, 23 pp.

Haywood, J.K. 1910. Injury to vegetation and animal life by smelter wastes. U.S.D.A., Bur Chem. Bull. No. 113, 63 pp.

Hedgcock, G.G. 1914. Injuries by smelter smoke in southeastern Tennessee. J. Wash. Acad. Sci. 4: 70-71.

Hepting, G.H. 1971. Air pollution and trees. In W.H. Matthews, F.E. Smith, and E.D. Goldberg, eds., Man's Impact on Terrestrial and Oceanic Ecosystems. MIT Press, Cambridge, MA, pp. 116-129.

Hocking, D., P. Kuchar, J.A. Plambeck, and R.A. Smith. 1978. The impact of gold smelter emissions on vegetation and soils of a sub-arctic forest-tundra transition ecosystem. J. Air Pollu. Control Assoc. 28: 133-137.

Holling, C.S. 1973. Resilience and stability of ecological systems. Annu. Rev. Ecol. System. 4: 1-23.

Hursh, C.R. 1948. Local climate in the copper basin of Tennessee as modified by re-moval of vegetation. U.S.D.A. Circular No. 774, 38 pp.

Hutchinson, T.C. and L.M. Whitby. 1976. The effects of acid rainfall and heavy metal particulates on a boreal forest ecosystem near the Sudbury smelting region of Canada. In L.S. Dochinger and T.A. Seliga, eds., Proc. 1st Intl. Symp. Acidic Precipitation and the Forest Ecosystem, U.S.D.A. Forest Service, Genl. Tech. Rep. No. NE-23, Upper Darby, PA, pp. 745-765.

Jung, E. 1968. Bestandsumwandlungen im rauchschadensgebiete von ranshofen. miedzynarodowej. Konj. Wplyw. Zanieczyszcen Powietrza Na. Lasy, 6th, Katowice, 1968, pp. 407-413.

Keller, T. 1973. Report on the IUFRO meeting "Air pollution effects on forests." Sopron, Hungary, Oct. 9–14, 1972. Eur. J. For. Pathol. 3: 56-60.

Knabe, W. 1970. Distribution of Scots pine forest and sulfur dioxide emissions in the Ruhr area. Staub. Reinhalt. Luft. 30: 43-47.

Langford, A.N. and M.F. Buell. 1969. Integration, identity and stability in the plant association. Adv. Ecol. Res. 6: 83-135.

Lautenbach, W.E. 1985. Land Reclamation Program 1978-1984. Regional Municipality of Sudbury. Sudbury, Ontario, Canada.

Likens, G.E., F.H. Bormann, R.S. Pierce, J.S. Eaton, and N.M. Johnson. 1977. Biogeochemistry of a Forested Ecosystem. Springer–Verlag, New York, 146 pp.

Linzon, S.N. 1965. Semimature-tissue needle blight of eastern white pine and local weather. Ont. Dep. Forestry, Res. Lab. Inform. Rept. No. O-X-1.

Linzon, S.N. 1978. Effects of airborne sulfur pollutants on plants. In J.O. Nriagu, ed., Sulfur in the Environment, Part II, Ecological Impacts. Wiley, New York, pp. 109-162.

Mergalef, R. 1963. On certain unifying principles in ecology. Am. Natur. 97: 357-374.

May, R.M. 1973. Stability and Complexity in Model Ecosystems. Princeton Univ. Press, Princeton, NJ, 235 pp.

McClenahen, J.R. 1978. Community changes in a deciduous forest exposed to air pollution. Can. J. For. Res. 8: 432-438.

Miller, P.R. and J.R. McBride. 1975. Effects of air pollutants on forests. In J.B. Mudd and T.T. Kozlowski, eds., Responses of Plants to Air Pollution. Academic Press, New York 195-235.

Robak, H. 1969. Aluminum plants and conifers in Norway. In Air Pollution Proc. 1st Eur.Congr. on the Influence of Air Pollution on Plants and Animals. Centre for Agric. Publish. and Documentation, Wageningen, The Netherlands, pp. 27-31.

Rochow, J.J. 1978. Measurements and vegetational impact of chemical drift from mechanical draft cooling towers. Environ. Sci. Technol. 12: 1379-1383.

Scheffer, T.C. and G.C. Hedgecock. 1955. Injury to northwestern forest trees by sulfur dioxide from smelters. U.S.D.A. Forest Service. Tech. Bull. No. 1117, 49 pp.

Scurfield, G. 1960. Air pollution and tree growth. For. Abstr. 21: 339-347, 517-528.

Skelly, J.M. and J.W. Johnston. 1978. A status report of the deterioration of eastern white pine due to oxidant air pollution in the Blue Mountains of Virginia, Proc. Amer. Phytopathol. Soc. 5: 398.

Sundman, V., V. Huhta, and S. Niemela. 1978. Biological changes in northern spruce forest soil after clear-cutting. Soil Biol. Biochem. 10: 393-397.

Treshow, M., F.K. Anderson, and F. Harner. 1967. Responses of Douglas fir to elevated atmospheric fluorides. For. Sci. 13: 114-120.

U.S.D.A. Forest Service. 1973. Silvicultural Systems for the Major Forest Types of the United States. Agr. Handbk. No. 445, U.S.D.A. Forest Service, Washington, DC, 114 pp.

U.S.D.A. Forest Service. 1978. Forest Insect and Disease Conditions in the United States — 1976. U.S.D.A. Forest Service, Washington, DC, 40 pp.

Whittaker, R.H. 1965. Dominance and diversity in land plant communities. Science 147: 250-260.

Whittaker, R.H. and G.M. Woodwell. 1978. Retrogression and coenocline distance. In R.H. Whittaker, ed., Ordination of Plant Communities. Dr. W. Junk, The Hague, The Netherlands, pp. 51-70.

Wood, C.W. and T.N. Nash. 1976. Copper smelter effluent effects on Sonoran desert vegetation. Ecology 57: 1311-1316.

Woodwell, G.M. 1970. Effects of pollution on the structure and physiology of ecosystems. Science 168: 429-433.

Winterhalder, K. 1977. Spatial heterogeneity in an industrially attenuated landscape before and after revegetation. In Landscape Ecology and Management Polyscience Publications, Montreal, Canada.

Winterhalder, K. 1984. Environmental degradation and rehabilitation in the Sudbury area. Laurentian University Review, Northern Ontario. Environmental Per-spectives 6: 15-47.

Winterhalder, K. 1988. Trigger factors initiating natural revegetation processes on barren, acid, metal-toxic soils near Sudbury, Ontario smelters. Proc. 1988 Annual Meeting of American Society for Surface Mining and Reclamation, April 17–22, 1988. Pittsburgh, PA.

18
Forest Dieback/Decline:
A Regional Response to Excessive
Air Pollution Exposure

Healthy forests are the rule and diseased forests the exception. Whenever forest trees become visibly symptomatic or die in large numbers and over widespread areas the phenomenon is dramatic and distressing to human managers and observers. The latter part of the twentieth century has seen numerous examples of forest dieback/decline in temperate and tropical latitudes. What is the role of atmospheric deposition in these forest dieback/decline phenomena?

A. Some Concepts of Forest Morbidity and Mortality

In recognition of the extraordinary diversity of forest tree species, habitats, and management strategies, it is not surprising to realize that tree morbidity and mortality patterns are also diverse. Despite their frequently extended term life cycles, trees are not immortal, and abnormal physiology and mortality of individual trees is commonly associated with one or more of the stress factors introduced in Chapter 1. The manner in which these stress factors influence trees, however, is very variable and is significantly influenced by the evolutionary history of the tree and the living or nonliving nature of the stress factor.

In the instance of tree species and environments with a short evolutionary history, i.e., the introduction of exotic biotic insects or pathogens or exportation of trees to locations beyond their natural ranges, the action of single biotic stress factors may result in direct and dramatic disease and death. North American examples of importance in this century include the chestnut blight disease of the American chestnut, Dutch-elm disease of the American elm, and white-pine blister rust disease of the eastern and western white pines. In these examples, it is surely true that several stresses are actually involved in mortality. It is appreciated, nevertheless, that the specific fungal disease agents involved in the above examples are the primary and dominant agents of death.

In contrast, in instances of tree species and environments with long evolutionary history, i.e., native tree species growing within their recognized range of distribution, single biotic stress factors rarely result in direct and dramatic disease and death. In this case, biotic stress agents are typically involved in highly interactive, sequential and extremely complicated patterns of impact. Not so with abiotic stress factors. The influence of volcanic eruption, wildfire, or windstorm can, and frequently does, cause acute and unambiguous morbidity and mortality in native forest systems.

As initially emphasized, most forests are healthy, and disease and death are the exception. While epidemics caused by exotic pathogens and insects, and destruction associated with fire and wind, are dramatic and recurrent, they are exceptional. Most disease and death in forest trees is the result of complex interactions of multiple factors (Franklin et al., 1987, Shigo, 1985) (Figure 18-1). Most disease and death in forest trees is subtle as only scattered individuals within stands are visibly symptomatic at any given time. Tree size and age are important variables in considerations of morbidity and mortality. In a simplified tree life table based on data for American beech in southeast Texas, Harcombe and Marks (1983) indicated the annual proportion dying varied as follows: seeds 0.90, seedlings 0.65, saplings 0.08, poles 0.06, and mature trees 0.02. Likewise, Kramer and Kozlowski (1979) have emphasized that large trees with complex crowns experience a "decrease in metabolism, gradual reduction in growth of vegetative and reproductive tissues, loss of apical dominance, increase in dead branches, slow wound healing, heartwood formation, increased susceptibility to injury from certain insects and diseases and are stressed from unfavorable environmental conditions." Excellent discussions of the relationships between age and size, and mortality have been provided by Franklin et al. (1987) and Peet and Christensen (1987).

B. Forest Dieback/Decline

One of the very most important forms of tree loss in temperate zone forests is a unique pattern of morbidity and mortality in which stand- or region-level stress symptoms or death occur. Forest dieback/decline occurs when a large proportion of a tree population exhibits visible symptoms of stress or unusual and consistent growth decreases or death over an area of many square kilometers. In contrast to single-tree impact, multiple-tree impact is dramatic, and its importance in long-term forest ecosystem development may be very great. The shorter term influence on forest management objectives is appreciated to be very severe in cases where the tree population exhibiting dieback/decline phenomenon is the focus of human management efforts. Useful discussions of the forest dieback/decline concept have been provided by Houston (1974, 1981), Kessler (1965), Manion (1981), Mueller–Dombois (1987), and Weaver (1965).

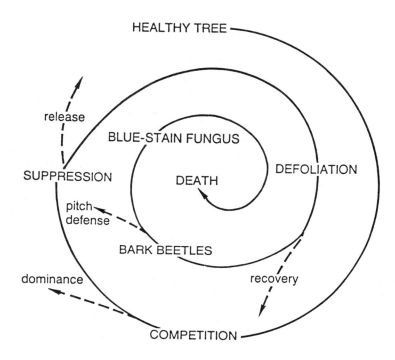

Figure 18-1. Factors involved in Douglas-fir mortality. In this example a healthy tree is suppressed by larger trees. It is predisposed to attack by defoliation. Partially defoliated, the weakened tree is attractive to bark beetles. Bark beetle infestation results in infection by the blue-stain fungus.
Source: Franklin et al. (1987).

1. Dieback/Decline Characteristics

Forest dieback/decline is irregular in distribution, discontinuous but recurrent in time, and the result of complex interactions of multiple stress factors (Smith, 1989). Nevertheless, these phenomena have sufficient common characteristics in various forest tree species to allow some generalization and synthesis of concept. Decline stress is more characteristic of mature forests than immature forests. Generally declines result from the sequential influence of multiple stress factors. Stress factors are both abiotic and biotic in nature. Abiotic stress factors common to numerous declines include drought and low- and high- temperature stress. Biotic agents of particular importance include defoliating insects, borers and bark beetles, root-infecting fungi, and canker inducing fungi. Typically declines are initiated by an abiotic stress, with mortality ultimately caused by a biotic stress agent. Quite commonly the abiotic stress responsible for decline initiation is the direct or indirect influence of change in some climatic parameter, for example, less than normal precipitation. Unfortunately our understanding of the interactive

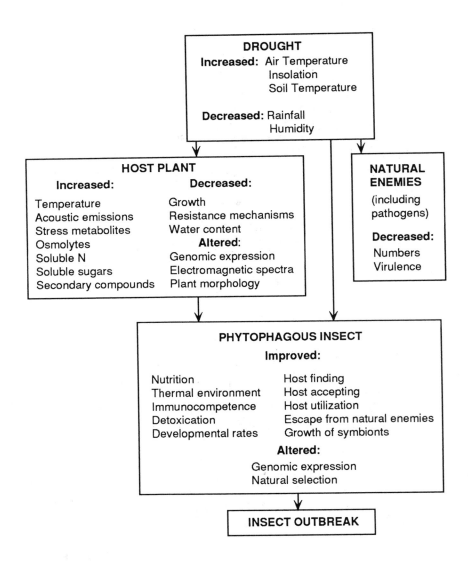

Figure 18-2. Hypothetical representation of how drought influences host plants, phytophagous insects, and their natural enemies to provoke insect outbreaks. Source: Mattson and Haack (1987).

and sequential nature of stresses of significance in forest dieback/declines is very poor. In the case of drought and insect herbivory, however, Mattson and Haack (1987) have presented an integrated hypothesis (Figure 18-2). Dieback/decline stress is typically characterized by a common progression of symptoms (Table 18-1).

Table 18-1. Symptoms Characteristic of Forest Trees Exhibiting Dieback/Decline Phenomenon. All Symptoms Are not Present in All Cases.

1. Reduced growth — reduced shoot elongation and diameter increment
2. Shorter internodes — tufted foliage
3. Fine root and mycorrhizal root destruction
4. Reduced food storage in roots
5. Premature development of foliar color in late summer/early fall
6. Reduced size of foliage and yellowing
7. Twig and branch dieback typically caused by weakly parasitic fungi
8. Outer crown dieback
9. Stimulation of stem and branch adventitious buds
10. Root infection by decay fungi, notably *Armillaria* species

Source: Manion (1981).

2. North American Forest Dieback/Decline

Presently dieback/decline stress is especially important in portions of the northern forest on sugar maple (McLaughlin et al., 1985; McIlveen et al., 1986; Parker and Houston, 1971; Wargo and Houston, 1974; Mitchell, 1987; Smith, 1987) and on American beech (Houston, 1975; Houston et al., 1979; Lacki, 1985). Very widespread decline and mortality of yellow birch occurred throughout New England, New York, and southeastern Canada from 1940 to 1965 (Greenridge, 1953; Hansbrough et al., 1950; Hawboldt, 1952). An important dieback/decline of white ash was most severe in the Northeast during 1950–1960 (Brandt, 1961; Hibben and Silverborg, 1978; Sinclair, 1965) and may again be significant (Castello et al., 1985).

The most widespread and serious declines over the past several decades in the central forest have been associated with members of the red oak group (scarlet, pin, red, and black oak) (Nichols, 1968; Skelly, 1974; Staley, 1965). The present distribution of oak decline is highly variable, but mortality can vary from 20% to 50% in severely impacted areas. Significant mortality has occurred in the George Washington and Jefferson National Forests in Virginia (Shriner, 1986). There is evidence that similar declines occurred during earlier periods in this century in the southern Appalachians (Balch, 1927). Dieback/declines of lesser importance that have occurred in the central and southern hardwood forests include dogwood decline (Hibben and Daughtrey, 1988; Walton, 1984), sweetgum decline (Miller and Gravatt, 1952), yellow-poplar decline (Toole and Huckenpahler, 1954), and magnolia decline (McCracken, 1985).

Conifers also exhibit wide-area dieback/decline stress. Important North American examples of the recent past include Alaska cedar (Shaw et al., 1985), western white pine (Leaphart and Copeland, 1957) and white spruce (Molnar and Silver, 1959) in the west; and shortleaf pine (Copeland and McAlpine, 1955), other southern pines (Sheffield et al., 1985; Chapter 16), and high elevation red spruce (Weiss et al., 1985; Section C this chapter) in the East.

Intensive investigations conducted on portions of these wide-area dieback/declines have produced proposals for the specific causal factors involved

Table 18-2. Important Forest Dieback/Decline Phenomena of the Past Century in North America.

Species	Initial stress	Contributing stress	Final stress
Yellow and white birch	warm summer temperatures, drought	foliage feeding insects	bronze birch borer, *Armillaria mellea* (root disease)
White ash	drought		*Cytophoma pruinosa* and *Fusicoccum* spp. (canker disease)
Sugar maple	drought	insect defoliation, harvesting, nutrient deficiency	*Armillaria* species (root disease) *Verticullium dahliae* (vascular disease)
American beech	beech scale *Cryptococcus fagisuga*	?	*Nectria* spp. canker fungi
Oak	drought, low temperatures spring	gypsy moth or oak leaf roller defoliation (insect)	*Armillaria mellea* (root disease), *Agrilus bilineatus* (insect borer)
Sweetgum	drought	nutrient stress	??
Western white pine	drought		*Armillaria mellea*, *Leptographium* spp. (root disease), *Europhium trinacriforme* (canker disease)
Shortleaf pine	drought, soil erosion, excessive soil moisture	nutrient stress, *Phytophthora* (root disease)	*Armallaria mella, Pythium* spp (root disease), southern pine beetle
Balsam fir	drought	eastern spruce budworm	*Armillaria mellea* (root disease)
White spruce	drought (?)	western spruce budworm (bud mortality)	*Aureobasidum pullulans* (twig, branch disease)
Red spruce (high elevation)	drought, climate warming	abiotic winter stresses	atmospheric deposition, biotic agents

(Table 18-2). Because of their apparent sequential and interactive nature, Manion (1981) proposed that stress agents involved in dieback/decline phenomena could be grouped into predisposing, inducing or inciting, and contributing factors. Judgements concerning the specific factors responsible for dieback/declines, however, have generally been based on observations made on very limited portions of areas impacted, have employed correlations to prove cause, have favored simplistic conclusions, have operated in the absence of any formal rules of proof, and frequently have been made in ignorance of comprehensive understanding of long-term forest ecosystem dynamics. As a result, the specific causes of only a handful of forest dieback/decline phenomena are known in detail. In most cases, including the majority of examples presented in Table 18-2, the causal nature of most North American dieback/declines remain incompletely understood!

3. European Forest Dieback/Decline

During the past decade, numerous forest species of Central Europe have exhibited dieback/decline symptoms. In selected regions, symptomology has been especially dramatic on silver fir, Norway spruce, Scots pine, European beech, and various oak species (particularly *Quercus robur* and *Q. petraea*).

In the Federal Republic of Germany, the third national survey of forest health, conducted from July to September 1985, revealed that 52% of the total forest land was exhibiting foliar symptoms of stress. By survey procedure, foliage loss was estimated as departure from a regional "healthy" (baseline) condition. Forests were recorded as healthy (damage category O) if they maintained 90% or more of their estimated normal foliar compliment. Stands with foliar losses from 11% to 25% percent were classified weakly damaged (category 1), from 26% to 60% were classified as moderately damaged (category 2), and > 60% were classified as heavily damaged (category 3). The 1985 inventory recorded that 33% of the symptomatic forests were classified in the lowest damage category (1), 19% in category (2), and 3% in the heavily damaged category (3) (BML, 1985).

The pattern of foliar symptoms in category (3) was generally scattered through the forest landscape and was not concentrated in large blocks. The forests of the southern part of the republic were more symptomatic than those of the north. In the south the most serious dieback/decline occurred in the eastern Bavarian Mountains along the border with Czechoslovakia, in the Black Forest, and in the Alps (Figure 18-3). Symptomology generally increased with tree age and exposure. Over previous survey periods, dieback/decline symptoms decreased in the area between the Danube and the foothills of the Alps, in Northrhine-Westfalia (area of high pollutant exposure), and in Scots pine forests. Symptoms appeared to intensify over previous surveys in the Alps and with oak species (BML, 1985; Rehfues, 1987).

Norway spruce grows throughout large areas of the FRG and is the most important forest species in terms of productivity and economic value. In the 1985 forest survey, 48% of the spruce were classified as healthy, 32% slightly damaged, 18% moderately damaged, and 2% severely damaged (Figure 18-4) (Rehfuess 1988). The pattern of spruce mortality has generally been individual trees or small groups of trees rather than over large areas. The most severe dieback/decline has been restricted to the German Mittelgebirge and to the Alps.

Rehfuess (1987, 1988) and others have concluded that the morbidity and mortality currently exhibited by Norway spruce results from variable etiology and is

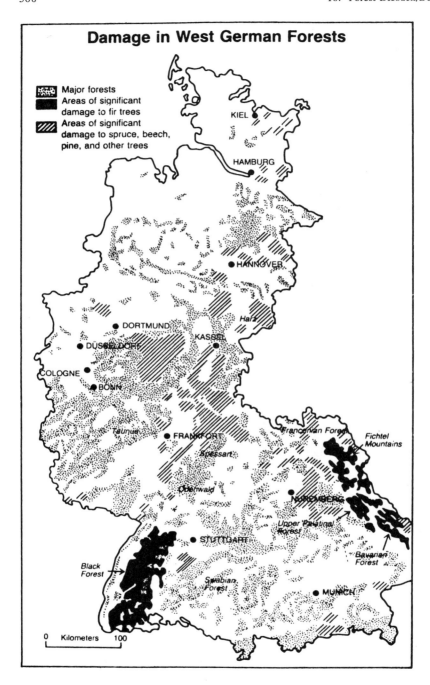

Figure 18-3. Forest damage in the Federal Republic of Germany. Source: Roberts (1983).

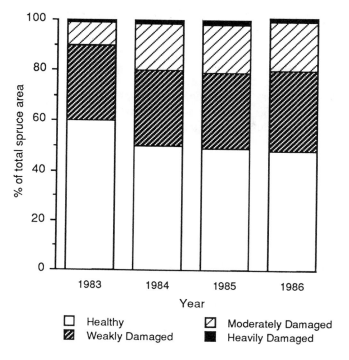

Figure 18-4. Inventory of Norway spruce health in the Federal Republic of Germany as revealed in national surveys of foliar symptomology.

due to region-specific stress factors. At high elevations with acidic bedrock, dieback/decline may be largely due to magnesium and calcium deficiency. Refuess (1987) has presented an integrated hypothesis to explain this deficiency. Oxidant exposure adversely impacts foliar wax and cell membrane integrity, which allows enhanced leaching of foliar magnesium and calcium by acidic cloud moisture. Root uptake of magnesium and calcium is restricted by high soil levels of exchangeable aluminum. Magnesium deficiency results in reduced photosynthesis, reduced growth, and reduced resistance to low temperature stress. In the Ore Mountains, along the border between Czechoslovakia and the German Democratic Republic, approximately 30,000 ha of spruce forests have been killed over the past decade. This mortality is judged to be primarily due to the combined impact of sulfur dioxide (30–34 ppb, 80–90 μg m^{-3} concentrations) and low temperature stress. At lower altitudes throughout southern Germany, foliar morbidity has been widely caused by needle cast fungi, including *Lophodermium piceae* and *Rhizosphaera kalkkoffii* . Frost stress in the late 1970s and early 1980s may have predisposed the foliage to fungal infection. Favorable precipitation and humidity during the subsequent growing seasons favored the development of the fungi. In higher altitudes of the calcareous Alps in

Southern Bavaria, spruce dieback/decline may be primarily due to drought and frost stress. Rehfuess observed that ozone or acid deposition stress may contribute to their decline. As in the case of North American dieback/declines, Rehfuess (1987) concludes that drought may be a significant initial and synchronizing stress in all contemporary Norway spruce declines.

In addition to Norway spruce, numerous additional forest dieback/decline phenomena are active in the Federal Republic of Germany. Silver fir exhibited foliar symptoms beginning in the mid-1970s. Presently severely damaged silver fir stands are concentrated in relatively small areas, especially at higher altitudes of the south German subalpine mountains. Deciduous declines, including European beech and oak species, have become more abundant in the mid-1980s. As in the case of Norway spruce, the causes of these dieback/declines are multiple and interactive (Prinz, 1987, 1988). It is likely that these dieback/declines do not result from a single, dominant stress, but more likely result from region specific multiple factors (Blank et al., 1988).

More comprehensive discussion of European forest dieback/declines is available in the reviews of: Bauer (1984), Krause (1987, 1988), Prinz et al. (1987), Guderian et al. (1987), and Blank et al. (1988).

C. Red Spruce Decline in North America

1. Introduction

Red spruce dieback/decline at high elevation locations along the Appalachian Mountain chain in eastern North America has been one of the most politically visible and intensively investigated wide-area forest declines of the 1980s (Figure 18-5).

Weiss et al. (1985) have indicated that extensive mortality of red spruce and balsam fir has occurred in the northeastern Untied States at numerous times during the last 200 years. The most important agents of mortality during this period have been fire, extreme wind events, and spruce beetle (*Dendroctonus rufipennis*), and/or spruce budworm (*Choristoneura fumiferana*) outbreaks. Griggs (1946) has indicated that stress observed in high elevation stands of red spruce and balsam fir in New Hampshire during the 1930s and 1940s was related to climatic change. During the 1960s mortality of red spruce and balsam fir was documented in numerous locations in northern New England (McLaughlin et al., 1987; Weiss et al., 1985). During the 1980s a variety of papers and reports have suggested an apparent increase in spruce-fir morbidity and mortality in the Northeast (for example, Friedland et al., 1984; Johnson and Siccama, 1984; Johnson et al., 1986; Scott et al., 1984 Siccama et al., 1982; Vogelmann et al., 1985). Red spruce occupies sites distributed over a large elevational range (~ 200–1200 m) in the northeastern region, and some evidence suggests greater morbidity and mortality at higher elevations. Weiss et al. (1985) assumed that approximately 10% of standing dead trees (or less) represent a normal (baseline) spruce-fir forest condition. One to three percent annual mortality has been recorded in recent in-

Figure 18-5. Red spruce mortality at 900 m on Camels Hump Mountain, Vermont.

ventories of commercial (low elevations) spruce-fir forests in New York, Vermont, and New Hampshire (McCrerry et al. 1987). Historically, estimates of annual mortality have ranged from 0.5 to 2% for low elevation, northern spruce-fir (Hertel et al., 1987). By contrast, if one assumes uniform annual decreases in recent surveys of high elevation (> 750 m) red spruce stem density or basal area reduction, current mortality appears to be 4-8 times historic low elevation rates; for example, 3.9% yr^{-1} 1965–1979 Camels Hump, Vermont (Siccama et al., 1982), 2.7% yr^{-1} 1965-1983 Camels Hump, Vermont (Vogelmann et al., 1985), 3.9% yr^{-1} 1964-82 Whiteface Mountain, New York (Scott et al., 1984), and 4% yr^{-1} 1982-1987 Whiteface Mountain, New York (Johnson, 1988). On an area basis, McCrerry et al. (1987) concluded that red spruce mortality is higher in New York and lower in New Hampshire and Vermont. The current increased mortality rate of high elevation red spruce in the northeastern region is not unprecedented in the past century (McLaughlin et al., 1987).

2. Symptoms of Red Spruce Dieback/Decline

The most prominent and consistently recorded symptom of the current red spruce morbidity is needle necrosis of the most recent foliage. Normal red spruce branch development has three buds elongated from each twig terminus each spring. Evans (1986) systematically observed hundreds of red spruce branch samples randomly selected from high elevation sites from New York, Tennessee, and North Carolina. Commonly needle necrosis was evident on one or two of the three first year twigs available. This necrosis, resulting in apical dieback and associated loss of foliar biomass and branch necrosis, has been generally described for stressed red spruce from a wide variety of sites in the East (Bruck et al., 1988; Johnson, 1988; McCrerry et al., 1987; McLaughlin et al., 1987; Mielke et al., 1986; Rock et al., 1986; Smith 1988). In addition to mortality, morbidity, and visible symptoms of stress, evidence previously discussed (Chapter 18) has been presented indicating decreased radial or basal area growth increment for red spruce at low- and high-elevation sites in the Northeast. Taken in aggregate, red spruce growth decrease, stress symptoms, morbidity, and mortality constitute "red spruce dieback/decline."

3. Red Spruce Stress Factors

A large number of stress factors are judged capable of contributing to red spruce dieback/decline. Competition and aging influence tree health in red spruce as they do in all forest species. In addition, evaluation of historic, regional-scale red spruce declines provides a list of potentially significant pathogens and insects (Table 18-3) and abiotic stress factors (Table 18-4).

4. Mt. Moosilauke Red Spruce Health Survey

The Forest Response Program established under the National Atmospheric Precipitation Assessment Program was a national research effort implemented to evaluate the impact of air pollutants on forest ecosystem health. In recognition of the apparent widespread nature of red spruce dieback/decline in high elevation forests of the East, and in recognition of the potential role of regional-scale air pollutants in this decline, the Spruce-fir Research Cooperative was established by the Forest Response Program. This program established six intensive research sites, which combined measurements of atmospheric deposition and meteorology with observations and assessments of red spruce and balsam fir health. The author's involvement with the Mt. Moosilauke site allows some observations for this mountain.

Mt. Moosilauke is the western-most peak of the New Hampshire White Mountains. The mountain covers an area of approximately 78 km^2, with a main axis approximately 9.6 km running in a north-south direction. It lies about 24 km east of the Connecticut River, 16 km southwest of the Presidential Range, and 8 km northwest of the Hubbard Brook Experimental Forest. Primary soil types include generally very well drained, shallow to thick haplorthods (spodosols) (low- to mid-elevation sites), and cryofolist (histosols), organic mats over rock (mid- to high-elevation sites). Vegetation zones of Mt. Moosilauke include the northern hardwood forest (with scattered red spruce) to approximately 760 m; boreal forest, red spruce-balsam fir association, 760–1280 m; balsam fir, 1280-1430 m; and alpine tundra above 1450 m. Red spruce has been a very important component of the forest of Mt. Moosilauke for approximately 2000 years. During the first half of the present century, however, large red spruce were

Table 18-3. Biotic stress agents judged potentially capable of significant roles in regional-scale decline of red spruce

Pathogens	Insects
1. Foliar pathogens	1. Spruce beetle
Lophodermium filiform *Chrysomyxa sp.*	*Dendroctonus rufipennis* *Polygraphus rufipennis*
2. Dwarf mistletoe	2. Spruce budworm
Arceuthobium pussillum	*Chroistoneura fumiferana*
3. Twig/branch canker pathogens	3. Swift (ghost) moth
Leucostoma (Cytospora/Valsa spp.) *Sirococcus conigenus*	*Pharmacis mustelinus* *(Hepialus gracilis)*
4. Stem decay pathogens	
Fomes pini *Fomitopsis pinicola* *Polyporus abietinus*	
5. Nematodes	
Sphaeronema sasseri *Pratylenchus* sp. *Crossonema* sp.	
6. Woody root pathogens	
Armillaria obsura *Scytinostroma galactinum* *Resinicium (Odontia) bicolor* *Polyporus balsameus* *Perenneporia subacida* *Coniophora puteana*	
7. Fine root pathogens	

Table 18-4. Abiotic Stress Factors Judged Potentially Capable of Significant Roles in Regional-Scale Decline of High-Elevation Red Spruce.

Stress	Reference
Temperature extremes	Conkey (1986), Friedland et al. (1984), Johnson (1988), Johnson et al. (1986), McLaughlin et al. (1987), Raynal et al. (1988), Spurr (1953), Wright (1955)
Moisture extremes	Curry and Church (1952), Johnson (1988, 1989), McLaughlin et al. (1987), Raynal et al. (1988), Reiners et al. (1984)
Wind	Harrington (1986), Rizzo (1986)
Soil conditions	Cronan and Shofield (1979), Cronan et al. (1978), Fahey and Lang (1975), Shortle and Smith (1988), Smith (1988b)
Atmospheric deposition	Johnson and Siccama (1984), Lovett et al., (1982), Mollitor and Raynal (1983), Smith (1958, 1988a), Vogelmann (1982), Weinstein et al. (1987), Bruck and Robarge (1988)

subjected to major disturbances including harvesting (up to approximately 1100 m) between 1890 and 1930, and a major hurricane in 1938.

During 1987 we conducted a health survey of red spruce on Mt. Moosilauke. A total of 189 dominant or codominant red spruce, averaging 10 m in height, were systematically inventoried for signs and symptoms on 19 permanent sample plots distributed over the mountain. A total of 46 trees were destructively sampled in the vicinity of permanent plots in order to closely observe foliage and root signs and symptoms. A total of 76 quantitative forest floor samples (organic soil layers Oi, Oe, and Oa) were collected in the vicinity of permanent plots for chemical and biological analyses. Forest floor samples were also collected from a low-elevation (250 m) red spruce-hemlock-northern hardwood stand at the Hubbard Brook Experimental Forest 8 km southeast of Mt. Moosilauke. The red spruce in this latter stand were vigorous and asymptomatic, and provided a reference perspective for Mt. Moosilauke observations.

We concluded from our survey that the status of red spruce health on Mt Moosilauke can presently be best described as being in a moderate decline status, beyond initial decline, but not in severe condition. This health status is less severe than other high elevation red spruce forests, e.g., Whiteface Mountain, New York (Johnson, 1989) and Mt. Mitchell, North Carolina (Bruck and Robarge, 1988) above 1000 m.

Mortality of standing red spruce on Moosilauke averaged 33%. Needle-loss and twig/branch necrosis was typically less than 50% of the full crown. The most consistent branch symptoms observed included necrosis (dieback), bunch, abrasion, break, broom and contortion (Figure 18-6). None of these symptoms is specific to cause, and a variety of abiotic and biotic stresses of well-appreciated importance to red spruce could be the agent responsible. We concluded that crown collisions during high-wind events may be especially important in this symptomology. The most consistent main stem symptoms observed included resinosis, branch base swellings, cankers, exposed roots, seams/cracks, stem curvatures, wounds, swelling, and breaks (Figure 18-7). Again none of these symptoms is specific to cause, and a variety of abiotic and biotic stresses were presumed operative. Numerous of the symptoms were consistent with *Phellinus pini* infection. While red spruce regeneration did not appear excessive,   seedlings observed frequently were vigorous and asymptomatic. Average needle lengths measured for the four previous growing seasons—1986, 1985, 1984, and 1983—did not exhibit significant differences by year, elevation, soil type, or crown position. Most needles on sampled branches were asymptomatic. The most frequent needle symptom observed on harvested branches was fleck (small chlorotic/necrotic spots). Internode lengths on harvested branches did not reveal significant differences over the 1986, 1985, and 1984 growing seasons. Internode length in 1983 was slightly less than in 1984 and 1985.

Signs and symptoms recorded on systematically exposed woody and fine roots were not frequent nor dramatic. Swift moth larvae and mycorrhizal root discoloration were not common. Forest floor pH measurements from all Moosilauke sites were not dramatically different than the asymptomatic red spruce stand at

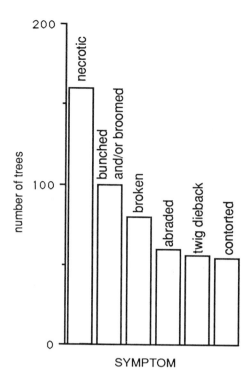

Figure 18-6. Frequency of branch symptoms recorded on Mt. Moosilauke permanent plots included in the 1987 pathology survey. Total trees systematically observed numbered 189.

the Hubbard Brook Experimental Forest. Bierman funnel extractions of forest floor samples consistently revealed high population densities of total nematodes (pathogenic species not distinguished), approximately 20,000–50,000 worms $100g^{-1}$ (dry wt) sample, but no consistent differences in Mt. Moosilauke sites relative to Hubbard Brook reference sites. Proton induced x-ray emission analyses of forest floor material indicated Moosilauke samples had higher concentrations than Hubbard Brook samples for the following elements: calcium, copper, manganese, nickel, phosphorus, sulfur, and zinc. Moosilauke samples had lower concentrations than Hubbard Brook samples for the following elements: aluminum, iron, and rubidium. Samples from both sites were comparable for chromium, lead, potassium, titanium, and vanadium (Figure 18-8).

5. Cause of Contemporary Red Spruce Decline

Red spruce dieback/decline has general characteristics consistent with other widespread, regional forest tree declines. Multiple stress factors are clearly in-

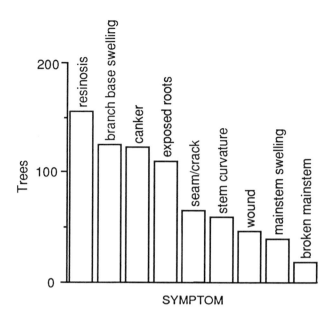

Figure 18-7. Frequency of main stem symptoms recorded on Mt. Moosilauke permanent plots included in the 1987 pathology survey. Total tree, systematically observed numbered 189.

volved in the present decline. Specific factors of importance vary from site to site over large elevational and latitudinal gradients. Most of the stress factors of paramount importance, climatic and weather extremes, and biotic agents, have been previously recognized as important in red spruce health. The specific role of atmospheric deposition has not been revealed by any unusual or unique symptom. The role of atmospheric deposition significance for red spruce dieback/decline cannot be fully assessed until controlled results from deposition monitoring are available for a longer period and until controlled environment studies more thoroughly explore potential mechanisms of impact. The most attractive hypotheses at present for air contamination involvement with red spruce dieback/decline include oxidant impact on foliar metabolism (Chapters 12 and 15), aluminum restriction of nutrient dynamics (Chapters 9 and 11), and global climate change (Chapter 19). With regard to the latter, Spurr (1953) has emphasized that the poor appearance and vigor of spruce in New England in the early 1950s was most probably a reflection of climatic warming. The latter author documented a mean annual temperature increase of 2.2°C (2.8°F) from 1817 to 1950 in New Haven, Connecticut. Long-term coupling of measurements of atmospheric deposition (exposure) and measurements of forest health are essential to assess the role of the former in the latter (Chapter 20).

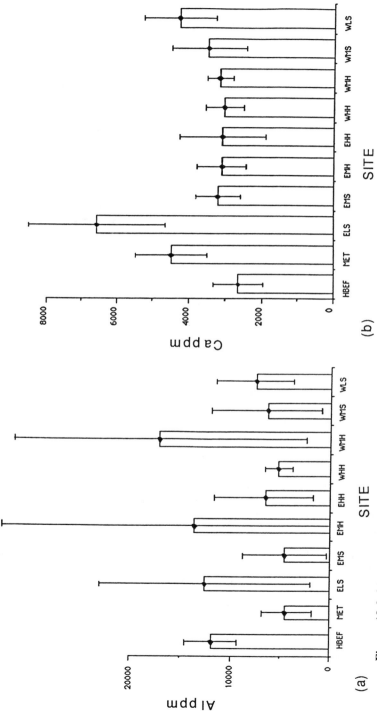

Figure 18-8. Mean and 95% confidence intervals of concentration of aluminum (a) and calcium (b) in the forest floor of a stand o hardwoods with asymptomatic canopy red spruce at 250 m in the Hubbard Experimental Forest, NH and the nine sample sites of survey of Mt. Moosilauke permanent vegetation plots. MET = meteorological station, ELS = east-side low elev spodosol, EMS elevation spodosol, EMH = east-side mid-elevation histosol, EMH = east-side mid-elevation histosol, WHH = west-side high e WMH = west-side mid-elevation histosol, WMS = west-side mid-elevation spodsol, WLS = west-side low-elevation spodosol. All si of six samples, except the HBEF site (n = 8). Analyses by proton induced by x-ray emission.

D. Summary

One of the most important characteristics of temperate zone forests are wide area stresses termed *dieback/decline phenomena*. These occur when a large proportion of a tree population exhibits visible symptoms of stress or consistent growth decreases, typically over a widespread area. The importance of dieback/decline stress in long-term forest ecosystem development may be very great. It may be the principal means for regeneration in some systems (Mueller–Dombois, 1983). The shorter term influence on forest management objectives is appreciated to be very severe in cases where the tree population exhibiting decline phenomena is the focus of management efforts. Current decline stress is especially important in North America in portions of the northern forest on sugar maple and on American beech. Very widespread decline and mortality of yellow birch occurred throughout New England, New York and southeastern Canada from 1940 to 1965. An important decline of white ash was most severe in the Northeast during the 1950-1960 period and may again be significant. Important North American conifer declines of recent or current importance include Alaska cedar, white pine, and white spruce in the West; and shortleaf pine, other southern pines, and currently high elevation red spruce in the East. Important Central European forest declines of recognized current importance include silver fir, Norway spruce, Scots pine, European beech, and oak species.

The dieback/decline stress is highly variable in distribution, discontinuous and recurrent in time, and the result of multiple interactive individual stress agents. These phenomena, nevertheless, exhibit several common characteristics in various forest tree species. Decline stress is more characteristic of mature forests than immature forests. Generally declines result from the sequential influence of multiple stress factors. Stress factors of recognized importance are both abiotic and biotic in nature. Abiotic stress factors common to numerous declines include drought and low- and high-temperature stress. Biotic agents of particular importance include defoliating insects, borers, and bark beetles, along with root-infecting and canker-inducing fungi. Typically declines are initiated by an abiotic stress, with mortality ultimately caused by a biotic stress agent. Quite commonly the abiotic stress responsible for decline initiation is the direct or indirect influence or change in some climatic parameter, for example, extended periods of less than normal precipitation. Decline stress is characterized by a common progression of symptoms that are not specific to cause.

In the Federal Republic of Germany, Norway spruce dieback/decline is hypothesized to result from a minimum of four sets of stress factors in various regions of the country. In North America, sugar maple dieback/decline has also been shown to result from various sets of stress factors in the United States and Canada. With regard to contemporary dieback/declines, it is probable that air pollutants are involved in some and not involved in others. Dieback/decline phenomena are frequently extended term, multiple factor, interactive, and subtle in initiation. They are importantly influenced by variations in climate, soil type, stand characteristics, and management strategies. As a result, successful diagnoses are most efficient when attempted at the stand or small area level. Diagnoses projected to regional or national scales are at risk of oversimplification and over-generalization. The best analyses of dieback/decline phenomena have been performed with adequate records of stand history and comprehensive appreciation of soil characteristics, climate variables, other abiotic and biotic stress factors, and atmospheric deposition.

With regard to air pollution involvement in forest dieback/decline, it is probable that air quality is not the sole or even primary cause, except in instances of excessive deposition (Class III situations). The more probable role of air contaminants is as a contributory stress in regions where climatic anomaly has initiated, and where other stresses have contributed, to the dieback/decline syndromes (Figure 18-9).

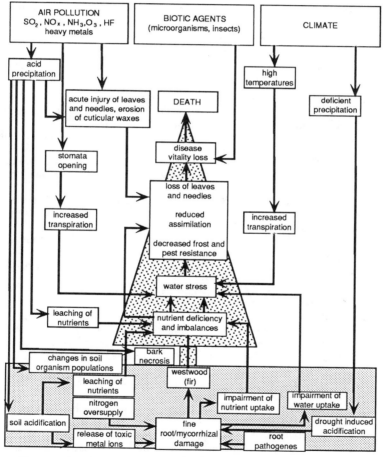

Figure 18-9. Potential interaction of multiple stresses in regional-scale forest dieback/decline phenomenon.
Source: Schäfer et al. (1988).

References

Bauer, F. 1984. Diagnosis and classification of new types of damage affecting forests. Commission of the European Communities. Publ.. No. DG VI-F3, Brussels, 20 pp.

Balch, R.E. 1927. Dying oaks in the southern Appalachians. Forest Worker. 3: 13.

Blank, L.W., T.M. Roberts, and R.A. Skeffington. 1988. New perspectives on forest decline. Nature 336- 27-30.

BML. 1985. Waldschadenserhebung 1985 Bundesminsterrum für Erährung. Landwirtschaft und Forsten, Bonn, FRG.

Brandt, R.W. 1961. Ash dieback in the northeast. U.S.D.A. Forest Service, Northeastern Forest Exp. Sta Paper No. 163, Upper Darby, PA, 8 pp.

Bruck, R.I. and W.P. Robarge. 1988. Change in forest structure in the boreal montane ecosystem of Mount Mitchell, North Carolina. Eur. J. For. Pathol. 18: 357-366.

Bruck, R.I., W. Robarge, S. Khorram, R. Bradow, W. Cure, S. Modena, J. Brockhaus, A. McDaniel, and P. Smithson. 1988. Observations of forest decline in the boreal montane ecosystem of Mt. Mitchell, North Carolina — An integrated forest response approach. Proceedings Effects of Atmospheric Pollutants on the Spruce-Fir Forests of Eastern United States and Federal Republic of Germany. U.S.D.A. Forest Service, Genl. Tech. Rep. No. 255, Northeastern Forest Exp. Sta., Broomall, PA.

Castello, J.D., S.B. Silverborg, and P.D. Manion. 1985. Intensification of ash decline in New York State from 1962 through 1980. Pl. Dis. 69: 243-246.

Conkey, L.E. 1986. Red spruce tree-ring widths and densities in Eastern North America as indicators of past climate. Quart. Res. 26: 232-243.

Copeland, O.L. and R.G. McAlpine. 1955. The interrelations of littleleaf, site index, soil, and ground cover in Piedmont shortleaf pine stands. Ecology 36: 635-641.

Cronan, C.S. and C.L. Schofield. 1979. Aluminum leaching response to acid precipitation: Effects on high-elevation watersheds in the Northeast. Science 204: 304-306.

Cronan, C.S., W.A. Reiners, R.C. Reynolds Jr., and G.E. Lang. 1978. Forest floor leaching: Contributions from mineral, organic, and carbonic acids in New Hampshire subalpine forests. Science 200: 309-311.

Curry, J.R. and T.W. Church. 1952. Observation of winter drying in conifers in the Adirondacks. J. Forest 50: 114-116.

Fahey, T.J. and G.E. Lang. Concrete frost along an elevational gradient in New Hampshire. Can. J. Forest Res. 5: 700-705.

Franklin, J.F., H.H. Shugart, and M.E. Harmon. 1987. Tree death as an ecological process. BioScience 37: 550-556.

Friedland, A.J., R. A. Gregory, L. Karenlampi and A.H. Johnson. 1984a. Winter damage to foliage as a factor in red spruce decline. Can. J. Forest Res. 14: 963-965.

Friedland, A.J., A.H. Johnson, and T.G. Siccama. 1984b. Trace metal content of the forest floor in the Green Mountains of Vermont: Spatial and temporal patterns. Water Air Soil Pollu. 21: 161-170.

Greenidge, K.N.H. 1953. Further studies of birch dieback in Nova Scotia. Can. J. Bot. 31: 548-559.

Griggs, R.F. 1946. The timberlines of North America and their interpretations. Ecology 27: 275-289.

Guderian, R., H-J. Ballach, A. Klumpp, G. Klumpp, K. Küppers, K. Vogels, and I. M. Willenberg. 1987. Reactions of Norway spruce (*Picea abies* (L). Karst) on air pollution in fumigation experiments and in damaged forest stands. In G. Hertel, ed., U.S.D.A. Forest Service, Genl. Tech. Bull No. 255, Northeastern Forest Exp. Sta., Broomall, PA.

Harcombe, P.A. and P. L. Marks. 1983. Five years of tree death in a Fagus-Magnolia forest, southeast Texas (USA). Oecologia 57: 49-54.

Harrington, T.C. 1986. Growth decline of wind-exposed red spruce and balsam fir in the White Mountains. Can. J. Forest Res. 16: 232-238.

Hawboldt, L.S. 1952. Climate and birch dieback. Dept. Lands and Forests, Bull. No. 6, Province of Nova Scotia, Canada, 37 pp.

Hertel, G.D., S.J. Zarnoch, T. Arre, C. Eager, V. Monhen, and S. Medlarz. 1987. Status of the spruce-fir cooperative research program. Proceedings, Air Pollu. Control Assoc., 80th Annual Meeting, New York, June 21–26, 1987, reprint number 87-34.2, 20 pp.

Hibben, C.R., and M.L. Daughtrey. 1988. Dogwood anthracnose in northeastern United States. Pl. Dis. 72: 199-203.

Hibben, C.R. and S. B. Silverborg. 1978. Severity and causes of ash dieback. J. Arboricul. 4: 274-279.

Hornsbrough, J.R., V.S. Jennsen, H.J. Macaloney, and R.W. Nash. 1950. Excessive birch mortality in the Northeast. Forest Pest Leaflet No. 52, Society of Amer. Foresters, Hillsboro, NH, 4 pp.

Houston, D.R. 1974. Diebacks and declines. Diseases initiated by stress, including defoliation. Arborists News 49: 73-76.

Houston, D.R. 1975. Beech bark disease. J. Forest 73: 660-663.

Houston, D.R. 1981. Stress triggered tree diseases. The diebacks and declines., U.S.D.A. Forest Service, Publ. No. NE-INF-41-81.Washington, DC, 36 pp.

Houston, D.R., E.J. Parker, R. Perrin, and K.J. Lang. 1979. Beech bark disease: A comparison of the disease in North America, Great Britain, France and Germany. Eur. J. Forest Pathol. 9: 199-211.

Johnson, A.H. 1988. Decline of red spruce in the high elevation forest of the northeastern U.S. World Resources Institute, Washington, DC.

Johnson, A.H. 1989. Decline of red spruce in the northern Appalachians: Determining if air pollution is an important factor. In G. M. Woodwell, ed., Proc. Markers of Air Pollution Exposure in Forests. National Acad. Science, Little Switzerland, NC, April 1988.

Johnson, A.H., and T.G. Siccama. 1984. Decline of red spruce in the northern Appalachians: Assessing the possible role of acid deposition.Tappi. J. 67: 68-72.

Johnson, A.H., A.J. Friedland, and J.G. Duschoff. 1986. Recent and historic red spruce mortality: Evidence of climatic influence. Water, Air, Soil Pollu. 30: 319-330.

Kramer, P. J. and T. T. Kozlowski. 1979. Physiology of Woody Plants. Academic Press, New York.

Krause, G.H.M. 1987. Forest decline and the role of air pollutants. In R. Perry, R. M. Harrison, J.N.B. Bell, and J.N. Lester, eds., Acid Rain: Scientific and Technical Advances. Selper Ltd., pp. 621-632.

Krause, G.H.M. 1988. Impact of air pollutants on above-ground plant parts of forest trees. CEC Symposium on Effects of Air Pollution on Terrestrial and Aquatic Ecosystems. Grenoble, May 18–22, 1987. D. Reichel Publ. Co., New york, NY.

Kressler, K.J. 1965. Dieback of managed, old-growth northern hardwoods in upper Michigan 1954–1964 — A case history. Pl. Dis. Rptr. 49: 483-486.

Lacki, M.J. 1985. Variation in radial growth of Amer. beech (*Fagus grandifolia* Ehrh.) at high elevations in the Great Smokey Mountains. Bull. Torr. Bot. Club 112: 398-402.

Leaphart, C.D. and O.L. Copeland. 1957. Root and soil relationships associated with the pole blight disease of western white pine. Soil Sci. Soc. Am. 21: 551-554.

Lovett, G.M., W.A. Reiners, and R.K. Olson. 1982. Cloud droplet deposition in fir forests: Hydrological and chemical inputs. Science 218: 1303-1304.

Manion, P.D. 1981. Tree Disease Concepts. Prentice-Hall, Englewood Cliffs, NJ, 399 pp.

Mattson, W.J. and R.A. Haack. 1987. The role of drought in outbreaks of plant-eating insects. BioScience 37: 110-118.

McCracken, F.I. 1985. Observations on the decline and death of southern magnolia. J. Aboricul. 11: 253-256.

McCreery, L.R., M. Miller–Weeks, M.J. Weiss, and I. Millers. 1987. Cooperative survey of red spruce and balsam fir decline and mortality in New York, Vermont and New Hampshire — A progress report. Symposium Proceedings, IPM Symposium for Northern Forests, Madison, WI, March 1986, 13 pp.

McIlveen, W.D., S.T. Rutherford, and S. N. Linzon. 1986. A historical perspective of sugar maple decline within Ontario and outside Canada. Ministry of the Environment Report No. ARB-141-86-Phyto, Ontario, Canada, 40 pp.

McLaughlin, D.L., S.N. Linzon, D.E. Dimma, and W.D. McIlveen. 1985. Sugar maple decline in Ontario. Ministry of the Environment. Report No. ARB-144-85-Phyto, Ontario, Canada, 18 pp.

McLaughlin, D.J. Downing, T.J. Blasing, E.R. Cook, and H.S. Adams. 1987. An analysis of climate and competition as contributors to decline of red spruce in high elevation Appalachian forests of the eastern United States. Oecologia 72: 487-501.

Mielke, M.E., D.G. Soctomah, M.A. Marsden, and W.M. Ciesla. 1986. Decline and mortality of red spruce in West Virginia. Report No. 86-4, U.S.D.A. Forest Service, Forest Pest Management, Ft. Collins, CO, 26 pp.

Miller, P.R. and G.F. Grawatt. 1952. The sweetgum blight. Pl. Dis. Rptr. 36: 247-252.

Mitchell, B. 1987. Air pollution and maple decline. Nexus (Atlantic Center for the Environment, Ipswich, MA) 9: 1-16.

Mollitor, A.V. and D.J. Raynal. 1983. Atmospheric deposition and ionic input in Adirondack forests. J. Air Pollu. Control Assoc. 33: 1032-1036.

Molnar, A.C. and G.T. Silver. 1959. Build-up of *Poiularia poulalans* (deBary) Berkhout within a severe spruce budworm infestation at Babine Lake, British Columbia. Forest Chron. 35: 227-231.

Mueller–Dombois, D. 1983. Canopy dieback and successional processes in Pacific forests. Pac. Sci. 37: 317-325.

Mueller–Dombois, D. 1987. Natural dieback in forests. BioScience 37: 575-583.

Nichols, J. 1968. Oak mortality in PA: A ten-year study. J. Forest 66: 681-694.

Parker, J. and D.R. Houston. 1971. Effects of repeated defoliation on root and root collar extractives of sugar maple trees. Forest Sci. 17: 91-95.

Peet, R.K. and N.L. Christensen. 1987. Competition and tree death. BioScience 37: 568-595.

Pring, B. 1987. Major hypotheses on forest damage causes in Europe and North America. Environment 12: 25-34.

Pring, B. 1988. Forest decline in the Federal Republic of Germany. U.S.D.A. Forest Service, Genl. Tech. Publ. No. 255, Northeastern Forest Exper. Sta., Broomall, PA.

Pring, B., G.H.M. Krause, and K.D. Jung. 1987. Development and causes of novel forest decline in Germany. NATO ASI Series, Vol. G16. T.E. Hutchinson and K.M. Meema, eds., Effects of Atmospheric Pollutants on Forests, Wetlands and Agricultural Ecosystems. Springer–Verlag, New York.

Raynal, D.J., D.C. LeBlanc, B.T. Fitzgerald, E.H. Ketchledge, and E.H. White. 1988. Historical growth patterns of red spruce and balsam fir at Whiteface Mountain, New York. Proceedings of Atmospheric Pollutants on the Spruce-Fir Forests of the Eastern United States and Federal Republic of Germany. U.S.D.A. Forest Service, Genl. Tech. Rep. No. NE-255, Northeastern Forest Exp. Sta., Broomall, PA.

Rehfuess, K.E. 1987. Perceptions on forest diseases in Central Europe. Forestry (UK): 60: 25-35.

Rehfuess, K.E. 1988. Personal Communication. Prof. Dr. Karl-Eugen Rehfuess, Lehrstuhl für Bodenkunde, Universität Müchen, Amalienstrasse 52, D800, Müchen 40, FRG.

Reiners, W.A., D.Y. Hollinger, and G.E. Lang. 1984. Temperature and evapotranspiration gradients of the White Mountains, New Hampshire, USA. Arc. Alp. Res. 16: 31-36.

Rizzo, D.M. 1986. Wind damage and root disease of red spruce and balsam fir in the subalpine zone of the White Mountains of New Hampshire. Unpublished Masters Thesis, Department of Plant Pathology, Univ. of New Hampshire, Durham, NH, 164 pp.

Roberts, L. 1983. Is acid deposition killing West German forests? BioScience 33: 302-305.

Rock, B.N., J.E. Vogelmann, D.L. Williams, A.F. Vogelmann, and T. Hoshizaki. 1986. Remote detection of forest damage. BioScience 36: 439-445.

Schäfer, H., H. Bossel, and K. Krieger. 1988. Modelling the responses of mature forest trees to air pollution. Geo J.17: 279-287.

Scott, J.T., T.G. Siccama, A.H. Johnson, and A.R. Breisch. 1984. Decline of red spruce in the Adirondacks, New York. Bull. Torr. Bot. Club 111: 438-444.

Shaw, C.G., A. Aglitis, T.H. Laurent, and P.E. Hennon. 1985. Decline and mortality of *Chamaecyparis mookatensis* in southeastern Alaska, a problem of duration but unknown cause. Pl. Dis. 69: 13-17.

Sheffield, R.M., N.D. Cost, W.A. Bechtold, and J. P. McClure. 1985. Pine growth reductions in the Southeast. U.S.D.A. Forest Service, Southeastern Forest Exp. Sta., Res. Bull. No. SE-83, Asheville, NC, 112 pp.

Shigo, A.L. 1985. Wounded forests, starving trees. J. Forest 83: 668-673.

Shortle, W.C. and K.T. Smith. 1988. Aluminum-induced calcium deficiency syndrome in declining red spruce. Science 240: 1017-1018.

Shriner, D.S. 1986. Eastern Hardwood Forest Response Research Cooperative. Cooperative Research Plan for FY 1986. U.S.D.A. Forest Service, U.S. Environmental Protection Agency, Broomall, PA, 44 pp.

Siccama, T.G., M. Bliss, and H.W. Vogelmann. 1982. Decline of red spruce in the Green Mountains of Vermont. Bull. Torr. Bot. Club 109: 162-168.

Sinclair, W.A. 1965. Comparisons of recent declines of white oak, oaks and sugar maple in northeastern woodlands. Cornell Plantations 20: 62-67.

Skelly, J.M. 1974. Growth loss of scarlet oak due to oak decline in Virginia. Pl. Dis. Reptr. 58: 396-399.

Smith, G.C. 1987. Preliminary report on maple decline in Massachusetts. Report presented to Department of Environmental Management. State of Massachusetts. Boston, MA, December, 1987.

Smith, W.H. 1985. Forest quality and air quality. J. Forest 83: 82-92.

Smith, W.H. 1988a. Effects of acid precipitation on forest ecosystems in North America. Chap. 5 in Vol. 4, D.C. Adriano and W. A. Salamons, eds., Advances in Environmental Science — Acid Precipitation, Springer–Verlag, New York, (in press).

Smith, W.H. 1988b. Red spruce rhizosphere dynamics: Spatial distribution of aluminum and zinc in the near-root soil zone. Forest Sci. (in press).

Smith, W.H. 1989. Forest health management for the future: An overview. Proc. Society of Amer. Foresters. National Convention, Rochester, NY, Oct. 18, 1988.

Spurr, S.H. 1953. The vegetational significance of recent temperature changes along the Atlantic seaboard. Am. J. Science 251: 682-688.

Staley, J.M. 1965. Decline and mortality of red and scarlet oaks. Forest Sci. 11: 2-17.

Toole, E.R. and B.J. Huckenpahler. 1954. Yellow-poplar dieback. Pl. Dis. Reptr. 38: 786-788.

Vogelmann, H.W. 1982. Catastrophe on Camels Hump. Nat. His. 91: 8-11.

Walton, G.S. 1984. Flowering dogwood decline due to drought disease, and cold winters. Fron. Pl. Sci. 37: 7-8.

Wargo, P.M. and D.R. Houston. 1974. Inspection of defoliated sugar maple trees by *Armillaria mella*. Phytopathology 64: 817-822.

Weaver, L.O. 1985. Diebacks and declines of hardwoods attributed to climatic changes — A review. Arborists News 30: 33-36.

Weinstein, L.H., R.J. Kohut, and J.S. Jacobson. 1987. Research at Boyce Thompson Institute on the Effects of Ozone and Acidic Precipitation on Red Spruce. Proceedings 80th Annual Meeting, Air Pollu. Control Assoc. New York, reprint no. 87-34.1, 20 pp.

Weiss, M.J., L.R. McCreery, I. Millers, J.T. O'Brien, and M. Miller–Weeks. 1985. Cooperative Survey of Red Spruce and Balsam Fir Decline and Mortality in New Hampshire, New York, and Vermont — 1984. Interim Report. U.S.D.A. Forest Service, Northeastern Area, P. O. Box 640, Durham, NH, 130 pp.

Wright, J.W. 1955. Species crossability in spruce in relation to distribution and taxonomy. Forest Sci. 1: 319-349.

Section IV

GLOBAL ATMOSPHERIC STRESS AND FOREST RISK ASSESSMENT

19
Alterations in Global Radiation Fluxes: Implications for Forest Health

Advances in global monitoring, climate modeling, and atmospheric and oceanic dynamics over the last 10 years have emphasized the need for biologists to become involved with ecosystem interaction with global-scale as well as local- and regional-scale air pollutants. This chapter will introduce potential interactions temperate forests may have with global-scale air contaminants. Global-scale pollutants, in contrast to local and regional contaminants, are released into the troposphere all around the earth by anthropogenic and natural sources, are capable of polluting the entire atmosphere of the earth, and may have their greatest significance to forest ecosystems by indirect consequences associated with alterations of radiation fluxes to and from the earth.

A. Global-Scale Air Contaminants

Global contaminants of special and/or potential significance to vegetative systems, due to their ability to regulate atmospheric radiation transfer, include very fine particles (dia < 2 μm) and a variety of trace gases.

1. Particulates

The principal aerosol found in the stratosphere consists of condensed sulfuric acid vapor (H_2SO_4) and water vapor. A dominant source of the former is presumed to be oxidation of surface-released carbonyl sulfide, although episodic volcanic injections introduce significant quantities of sulfur dioxide directly into the lower and mid stratosphere (Penner, 1988). The major types of tropospheric aerosol include soot, sulfate, maritime, crustal, and arctic haze. Arctic haze can be related to anthropogenic emissions (Wang et al., 1986).

The most common components of particulate material in the troposphere of North American urban and rural areas include sulfate and nitrate, lead halides, and elemental and organic carbon material. Other important components include material with a composition similar to the crust of the earth, such as fly ash, wind-blown soil dust, and large quantities of water (Hidy and Mueller, 1986). Particles are supplied by natural processes (wind erosion, sea spray, volcanic eruption, vegetation) and human activities (combustion, grinding). Particulate diameters range from less than 0.001 μm to giant particles in excess of 10 μm diameter (Chapter 2). Removal mechanisms from the atmosphere are quite efficient for coarse particles (diameter > 2 μm) and very fine particles (diameter < 0.1 μm). Certain fine particles (diameters approximately 0.1–2 μm) that escape coagulation and rainout and washout, however, may persist in the global atmosphere for extended periods and may have the potential to impact global albedo and atmospheric radiation dynamics (Charlson et al., 1987; Sagan et al., 1979; Thompson et al., 1984).

Particles may either heat or cool the surface of the earth, depending on their optical properties for both incident solar radiation and emitted thermal radiation.

2. Trace Gases

Trace gases commonly found in the troposphere currently number approximately 18 (Demerjian, 1986). Many, released at the surface of the earth, are persistent (stable) in the troposphere and migrate to the stratosphere. Several of these are of special interest as they appear to be accumulating (or decreasing) in some portion of the atmosphere, presumably due to a change in synthesis or removal mechanisms (Table 19-1). Noteworthy examples of increasing trace gases include carbon oxides, methane, nitrous oxide, and halocarbons (Cicerone, 1987; Penner et al., 1988).

a. Carbon Oxides

The literature on the global carbon budget and the significant role of carbon dioxide is extensive. Comprehensive and especially useful reviews include Bolin et al. (1979); Broecker et al. (1979); Clark (1982); Rotty and Reister (1987); U.S.D.O.E. (1985); and National Research Council (1983). Careful monitoring of carbon dioxide during the past three decades in Hawaii, Alaska, New York, Sweden, Austria, and the South Pole has firmly established that carbon dioxide is steadily increasing in the global atmosphere. It is estimated to have increased approximately 25% from 1800 to 1985 (U.S.D.O.E., 1985). The cause of this increase is thought to be the result of human activity. Nineteenth century land clearing for agriculture in the temperate latitudes was the initial contributor. Currently combustion of fossil fuels is the primary contributor. Tropical forest harvesting may also be making an important contemporary contribution

Table 19-1. Trace Gas Constituents of the Atmosphere with Estimated Rates of Annual Change and Importance for Global Radiation Flux.

Trace gas	Estimated annual increase	Importance for radiation flux and atmospheric chemistry
CO_2	0.4%	Absorbs infrared radiation; affects stratospheric O_3
CH_4	1.1%	Absorbs infrared radiation; affects tropospheric O_3 and OH; affects stratospheric H_2O and O_3
N_2O	0.2%	Absorbs infrared radiation; affects stratospheric O_3
$CFCl_3$	5.0%	Absorbs infrared radiation; affects stratospheric O_3
CF_2Cl_2	5.0%	Absorbs infrared radiation; affects stratospheric O_3
CH_3Cl_3	5.0%	Absorbs infrared radiation; affects stratospheric O_3
CH_3F_2Cl	12%	Absorbs infrared radiation; affects stratospheric O_3
$CFCl_2CF_2Cl$	10.0%	Absorbs infrared radiation; affects stratospheric O_3
Tropospheric O_3	0.8%[a]	Absorbs uv and infrared radiation
Stratospheric O_3	-0.3%[b]	Absorbs uv and infrared radiation
CO	≈1%	Involved in tropospheric O_3 and OH cycles
NO_x	?	Involved in O_3 and OH cycles and precursor of acidic nitrates
NMHC	?	Involved in tropospheric O_3 and OH cycles
$(CH_3)_2S$	?	Produces CNN which can alter cloud albedo; forms SO_2
OCS	?	Forms aerosol in stratosphere, which alters albedo
CS_2	?	May be major source of OCS in the troposphere
SO_2	?	Major precursor of acid rain
H_2S	?	Major natural source for SO_2
Tropospheric OH	-?	Scavenger for many atmospheric pollutants, including CH_4, CH_3, CCl_3, CH_3F_2Cl

[a]In lower troposphere
[b]In upper stratosphere
Source: Penner et al. (1988).

(Henderson–Sellers et al., 1988). The oceans, particularly the areas of open ocean, have been the most important sink for anthropogenic carbon dioxide. The current anthropogenic production of carbon dioxide from all sources is estimated to approximate 6×10^{15} g C yr^{-1} (WMO, 1985). The atmospheric carbon dioxide concentration has been estimated to have been approximately 290 ppm (5.2×10^5 µg yr^{-3}) in the middle of the nineteenth century. Today, the carbon dioxide concentration approximates 345 ppm (6.2×10^5 µg m^{-3}) and is increasing very

Table 19-2. Future Atmospheric Carbon Dioxide Concentrations for Three Carbon Emission Scenarios.

Carbon Emissions[a]	Year			
	2000	2025	2050	2075
	ppm			
High	380–400	480–530	760–840	1400–1550
Medium	380–400	440–480	540–600	670–760
Low	370–400	420–460	470–520	510–580

[a]The estimation of future carbon emissions varies with energy strategy, fuel types, and carbon cycle model applied, the authors presumed (greater than a 50–50 chance) that future emissions will be bounded by the high and low cases presented.
Source: Trabalka et al. (1986).

approximately 1.4 ppm (2.5×10^3 µg m^{-3}) per year. If the increasing rate continues, the carbon dioxide amount in the global atmosphere will double sometime in the latter half of the next century (Hansen, 1987; Holdgate et al., 1982; Trabalka et al., 1986) (Table 19-2). Carbon dioxide is the single most important atmospheric trace gas with regard to global climate change, as its radiative influence is second only to water vapor and its release has a dominant human component (Penner et al., 1988).

Carbon monoxide, despite the efficiency of the terrestrial soil sink and the atmospheric hydroxyl radical (OH) sink, also appears to be increasing slightly (Circerone, 1987; Sze, 1988). The present northern hemispheric average surface concentration is approximately 150 ppb (173 µg m^{-3}). Recent measurements indicate a global trend of about a one percent increase per year (Khalil and Rasmussen, 1988).

b. Methane

Methane (CH_4) is very stable in the troposphere, migrates to the stratosphere, and has an approximate atmospheric residence time of 10 years. Current remote-site sampling for methane and analysis of ice cores for methane both provide evidence for increases in this gas over the last century. The current rate of methane increase in the atmosphere is approximately 1–2% per year (Cicerone, 1987, Rowland, 1987). The current average abundance in the atmosphere is approximately 1.7 ppm (1.1×10^{-3} µg m^{-3}) in the northern hemisphere. The continuous monitoring record between 1979 and 1988 suggests an average annual increase around 1%, with significant interannual variability (Penner, 1988). The source of this methane increase is uncertain. Increased release or decreased sink removal of the gas are both possible. Natural release is primarily associated with anaerobic ecosystems due to the activity of methanogenic bacteria. Anthropogenic activities associated with increased methane production include increases in animal fermentation, rice culture, swamp creation, biomass burning, and oil and gas drilling. Ruminant animals, rice paddies and wetland emissions may be comparable and represent $7-10 \times 10^{13}$ g C yr^{-1} each. Biomass and fossil fuel burning may contribute an additional $3-4 \times 10^{13}$ g C yr^{-1} (Penner, 1988). The atmospheric sink for methane is provided by the OH radical.

c. Nitrous Oxide

The atmospheric residence time for nitrous oxide (N_2O) is in excess of 100 years. Analysis of ice core data and current monitoring indicate a statistically significant increase of approximately 0.50 ppb (0.9 μg m^{-3}) per year (Cicerone, 1987). The current global average concentration is 310 ppb (558 μg m^{-3}). Of the annual source strength ($\sim1.4 \times 10^{13}$ g N yr^{-1}), approximately one half is attributed to microbial nitrification processes in tropical and subtropical soils. An additional one third can be attributed to fossil fuel combustion (dominated by coal combustion) and biomass burning. The balance may largely come from temperate and boreal forests, and cultivated and fertilized soils (Denner et al., 1988). Emission rates of nitrous oxide from natural ecosystems appear to be higher than assumed previously by approximately 10 times (Bowden, 1986). Forest management activities, including whole-tree harvesting (Bowden and Bormann, 1986) and/or fertilization, may contribute to increased nitrous oxide release from temperate forest ecosystems (Chapter 3). The sinks for atmospheric nitrous oxide are not fully appreciated but are presumed to include photochemical destruction and reaction with atomatic oxygen to form molecular nitrogen and oxygen.

d. Halocarbons

Halocarbons include a wide variety of chemicals that combine carbon with a halogen. Important groups include chlorofluorocarbons (CFCs), chlorocarbons, and halons.

Chlorofluorocarbons were developed in the 1930s and have become widely used in aerosol spray cans, air conditioning, refrigeration, foam products (e.g., in cushions and insulating foams), solvents (e.g., electronics manufacture and dry cleaning), and in numerous miscellaneous uses. Measurements initiated in the late 1960s revealed certain of these chemicals accumulating in the atmosphere.

Important chlorofluorocarbons include CCl_2F_2, CCl_3F, CH_3Cl and CCl_4. CCl_2F_2 (CFC-12) and CCl_3F (CFC-11) are dominant in applications and account for over 80% of the current chlorofluorocarbon production globally. Mean annual increase rates in the atmosphere approximate 5% for both CFC-11 and CFC-12, with current concentrations of about 220 ppt for CFC-11 and 350 ppt for CFC-12 (Penner et al., 1988). The atmospheric lifetimes of CFC-11 and CFC-12 approximate 75 and 111 years respectively (WMO 1985). CH_3Cl results naturally from the decomposition of marine organisms. CCl_4 has both natural and anthropogenic sources (Lilian et al., 1975; Molina and Rowland, 1974; NAS, 1976; NAS, 1978; Singh et al., 1979). Despite a 1978 U.S. ban on the use of chlorofluorocarbons in aerosol sprays, more than 700,000 tons of halocarbons are released to the global atmosphere annually. Approximately 90% of the halocarbons released between 1955 and 1975 are still migrating through the troposphere on their way to the stratosphere. The atmospheric burden of these gases has increased 30–40% in recent years.

Chlorocarbons include methyl chloroform and carbon tetrachloride, and are used primarily as solvents and chemical intermediates. Methyl chloroform is primarily used as a general purpose solvent. Carbon tetrachloride is primarily used to make chlorofluorocarbons in the United States. Halons have been used in hand-held and total-flooding fire extinguishers since the 1970s. Annual production has been limited and emissions have been assumed to be only a small faction of production, based on the assumption that the halons remain inside fire extinguishers.

The 1987 Montreal Protocol will act to limit future emissions of CFC-11 and CFC-12 in particular. Despite this international agreement, however, future increases in the atmospheric abundance of halocarbons is anticipated.

B. Global Pollutant Regulation of Atmospheric Radiation Fluxes

The stratosphere, that portion of the earth's atmosphere extending from an altitude of about 15 to 50 kilometers, contains a diffuse layer of naturally occurring ozone that envelops the earth and screens out more than 99% of solar radiation at wavelengths shorter than 320 nanometers.

1. Natural Stratospheric Ozone Dynamics

Stratospheric ozone is formed by photolysis of oxygen at wavelengths shorter than 242 nanometers, followed by reaction of atomic oxygen with oxygen molecules on the surface of particulates. A variety of stratospheric mechanisms naturally act to destroy ozone. Approximately 50–70% of this destruction is caused by reaction with nitric oxide (NO). Nitric oxide is formed in the stratosphere by the interaction of nitrous oxide (N_2O, Table 19–1) with singlet oxygen. The catalytic cycle whereby nitric oxide destroys stratospheric ozone is as follows:

$$NO + O_3 \rightarrow NO_2 + O_2$$
$$O_3 + hv \rightarrow O_2 + O$$
$$NO_2 + O \rightarrow NO + O_2$$
$$\text{net: } 2\,O_3 + hv \rightarrow 3\,O_2$$

The formation and destruction of stratospheric ozone is normally in natural equilibrium so that the stratospheric ozone concentration remains constant (approximately 300 ppb, 5900 μg m^{-3}) (Hammond and Maugh, 1974).

2. Global Stratospheric Pollutants and Ozone Layer Integrity

Statospheric ozone can be decreased by a variety of catalysts produced by unusual natural events or anthropogenic activities. Important examples of these catalysts include oxides of nitrogen, chlorine, hydrogen, or bromine. Catalyst sources are varied and encompass unusual natural events, such as solar proton events and volcanic eruptions, and anthropogenic activities including nuclear explosions, supersonic aircraft, combustion, fertilizer use, forest harvesting, space shuttle rocket launchings, and halocarbon release (Cicerone, 1987).

Halocarbon pollutants may be particularly important in stratospheric ozone destruction. In summary, halocarbon molecules, for example, chlorofluoromethanes, released by various human activities slowly migrate through the troposphere. They pass through the tropopause and lower stratosphere and are decomposed (photolyzed by solar radiation — chiefly 175- to 220 nanometer wavelengths) in the mid to upper atmosphere. Free chlorine, resulting from this decomposition, can cause a rapid, catalytic destruction of ozone. One chlorine atom may destroy up to 100,000 ozone molecules in chain reaction sequence as follows (Cicerone, 1974):

$$Cl + O_3 \rightarrow C10 + O_2$$
$$C10 + O \rightarrow Cl + O_2$$
$$\overline{net\ O_3 + O \rightarrow O_2 + O_2}$$

In 1979, a stratospheric "ozone hole" was discovered over Antarctica. The cause of this hole is uncertain, although since the 1960s, there has been a 3.5 times increase in chlorine in the stratosphere over the Antarctic (Rowland, 1987). The Antarctic ozone hole phenomenon refers to springtime ozone column decreases of approximately 30–40 percent (Bowman, 1988). Evidence for decreases was provided initially by ground spectrophotometers (Farman et al., 1985) and subsequently by satellite observations (Stolarski et al., 1986). Reduced ozone amounts are naturally characteristic of the Antarctic in the spring (Farman et al., 1985; Dobson, 1966), but observations in recent years suggest ozone amounts much lower than long term averages.

Satellite ozone data from the Total Ozone Mapping Spectometer from 1979 through 1986 show that recent decreases in total ozone are global in distribution and are not confined to the Antarctic in the spring season. The decreases are largest in middle and high latitudes and occur in all seasons of the year (Bowman, 1988).

Initial estimates predicted an average global ozone column reduction between 1960 and 2030 of approximately 3% (NAS 1984) to 6.5% (Strodal and Isaksen, 1986) if chlorine emissions increased at 3% yr^{-1}. Frederick (1987) assumed halocarbon release rates fixed at 1980 levels and suggested an ultraviolet-B change at the surface of the earth of less than 10%, perhaps only 3% over the next several decades. Kerr (1987) has suggested possible average global depletion between 1955 and 2060 of approximately 16%, with even greater high latitude (> 60 degrees) springtime depletions (Strodal and Isakesen, 1986). A recent review of ground based and satellite ozone measurements concluded that there has been a decrease from 1978 to 1985 of 2.5% between 53°N and 53°S latitude, or approximately 0.35% yr^{-1} (Watson, 1988). Rodgers (1988) emphasized that photochemical models and measurements of atmospheric ozone trends have converged and are providing broad agreement globally. Model calculations all show maximum global depletion around 40 km, varying from 5% to 12% for different models. Of this, 4-9% is response to chlorofluorocarbons, and 1–3% is a response to reduced solar ultraviolet output approaching the solar minimum.

3. Greenhouse Effect and Global Climate Change

Climate is defined as the time-averaged value of meteorological quantities (Rind and Lebedeff, 1984). Over time, climate, like a forest, is characterized by change not constancy. In geological terms, the climate of the earth is most typically characterized by extended "moderate periods" with equable weather the year round, lack of icecaps, and generally warm seas. Humans evolved after the last "moderate period" and our development has been in a period of climatic revolution (Hepting, 1963). This period of revolutionary climatic alteration has been characterized by a complex of "cycles within cycles." Over the past 3000 years, for example, the general evidence suggests that the northeastern portion of North America has become cooler and more moist. Over the past several hundred years, however, and particularly during the first half of the present century, there is evidence for a moderating trend. There is general agreement that there has been a systematic fluctuation in recent global climate characterized by a net worldwide warming of approximately 0.5°C between the 1880s and the early 1940s. A model by Schneider and Mass (1975) computed that this warming trend in global temperature persisted to the early 1950s, was followed by temporary cooling through the early 1960s, and resumed warming into the mid 1970s. Warming has generally continued over the past two decades (Hansen and Lebedeff 1987) (Figure 19-1).

a. Regulation of Global Climate

The regulation of global climate is complex and incompletely appreciated. Numerous hypotheses have been proposed to explain the forces responsible for the variability of climate. The most plausible of these include: variations of the solar constant, changes in solar activity, passage of the solar system through an interstellar gas-dust cloud, variation in the velocity of the earth's rotation, gigantic surges of the Antarctic ice sheet, changes in the earth's orbital parameters, and alterations in the interactions between glaciers and oceans (Sergin, 1980). Climate may respond rapidly and dramatically to small changes in these independent variables (Bryson, 1974).

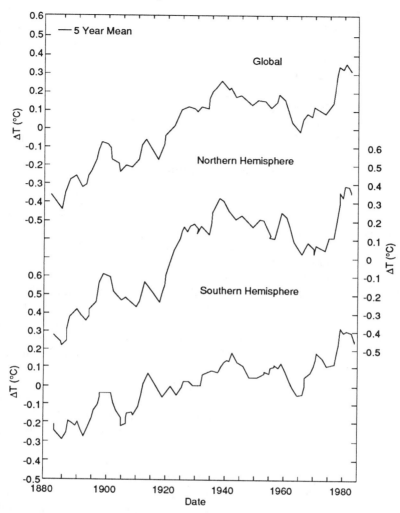

Figure 19-1. Global and hemispheric surface air temperature change estimated from meteorological station records. The 5-year running mean is the linear average for the 5 years centered on the plotted year.
Source: Hansen and Lebedeff (1987).

Added to this complexity and uncertainty is the suggestion that the activities of human beings, particularly land use activities (Sagan et al., 1979) and atmospheric contamination (Postel, 1986), are currently influencing global and regional climates. Numerous trace gases of the atmosphere, and all gases discussed in Section A2, have strong infrared absorption bands. As a result, these gases can have a significant effect on the thermal structure of the atmosphere because they absorb within the 7- to 14-μm atmospheric window, which transmits most of the thermal radiation from the surface of the earth and troposphere to space

(Wang et al., 1976). Presumably a primary result of more carbon dioxide, halo-carbon, and other trace gases in the atmosphere will be warming. While incoming solar radiation is not absorbed by carbon dioxide and these trace gases, portions of infrared radiation from the earth to space are. Over time, the earth could become warmer. While the forces controlling global temperature, are varied and complex, as suggested, the increase of 0.5°C since the mid-1800s is generally agreed to be at least partially caused by increased carbon dioxide. By 2000 it may increase an additional 0.5°C.

b. General Circulation Models

General circulation models, widely used to predict future climates, are numerical models of the earth-atmosphere system that solve the basic equations for atmospheric motion and provide boundary conditions of the earth and ocean (Hansen et al., 1983; Rind and Lebedeff, 1984). Quantities used to describe weather simulated by general circulation models include precipitation, evaporation, ground wetness, runoff, ground temperature, surface temperature, surface wind speed and direction, and heat flux. Several full-scale general circulation models are available for the United States. Some of the most widely used are those developed at the Geophysical Fluid Dynamics Laboratory, Princeton (GFDL), the National Center for Atmospheric Research, Boulder (NCAR), the Goddard Institute for Space Studies, New York (GISS), and Oregon State University, Corvallis (OSU).

General circulation models divide the landscape into a series of regions called *grid boxes*. Each grid box consists of a series of layers, which represent land, ocean, and layers of the atmosphere. Global weather is simulated using difference equations which detail the physics and dynamics of energy and material movement among the boxes. To evaluate the climatic change hypotheses, models are initially run with current atmospheric conditions. They are then run with altered atmospheric trace gas concentrations (for example, carbon dioxide) to simulate a new climate. The difference between these two equilibrium climate simulations provides an estimate of how the climate system may change (Rind and Lebedeff, 1984) (Figure 19-2). As each modeling group has formulated their model using different assumptions, all general circulation models are not in perfect agreement with regard to trace gas impact on global temperature and the hydrologic cycle (Figure 19-3). All models predict, however, warming and overall intensification of evaporation and precipitation in the warmer climate regimes of doubled atmospheric carbon dioxide.

c. Future Climate

Using the results of general circulation models, the National Academy of Sciences (1982a,b) has estimated a mean global average surface warming of $3 \pm 1.5°C$ ($5.5 \pm 2.7°$ F) in the next century. This amount of warming is far from trivial! The difference in average global temperature between the last glacial pe-

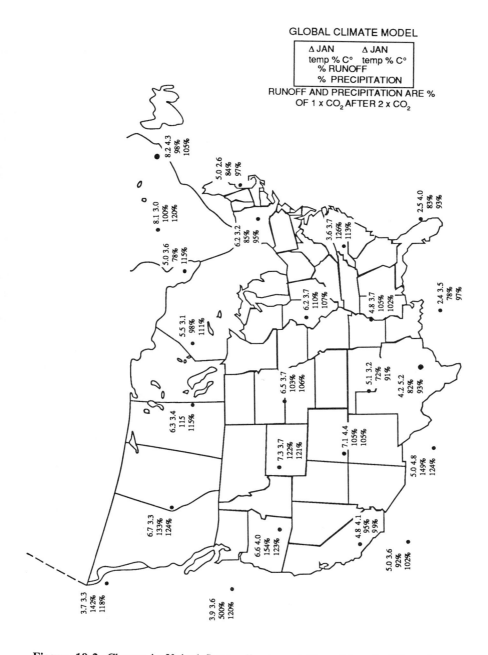

Figure 19-2. Change in United States climate variables with a doubling of atmospheric carbon dioxide, as estimated by the Goddard Institute of Space Studies general circulation model. Figures in the boxes represent the change in mean January temperature °C (left), change in mean June temperature °C (right), and percent runoff and precipitation of 1 X CO_2 after 2 X CO_2 for the grid points circled.
Source: U.S.E.P.A. (1988).

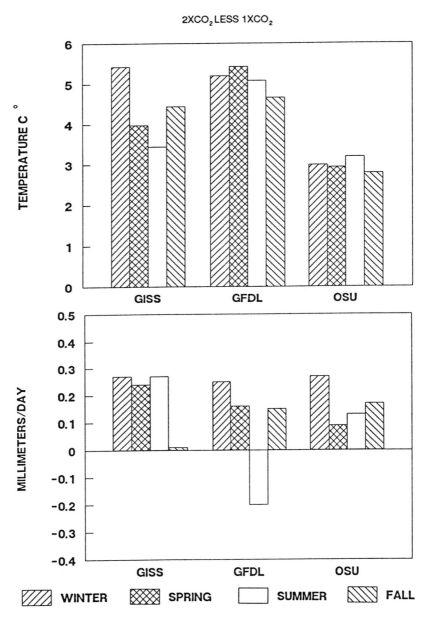

Figure 19-3. Seasonal average temperature (°C) and precipitation change for U.S. grid points as predicted by the Goddard Institute of Space Studies (GISS), Geographical Fluid Dynamics Laboratory (GFDL), and Oregon State University (OSU) general circulation models.
Source: U.S.E.P.A. (1988).

riod and the present (interglacial period) is approximately 4–5°C (Broecker, 1987). In 30 years, New York City could have the same annual temperature mean that Washington, D.C. presently has. In 30 years, New York City could annually average 35 days with temperatures above 32°C (90°F) rather than the current average of 14 days. In thirty years, the global mean temperature of the earth could be higher than it has been in the last several hundred thousand years (Hansen, 1987). Throughout most of the northern United States and Canada, the growing season could be increased by 20% or more (Figure 19-4). General circulation model predictions of the change in the hydrologic cycle with doubling of atmospheric carbon dioxide are less clear (Gleick, 1987a,b). While precipitation is generally estimated to increase, warming will intensify evaporation. The movement of water into and out of the soil-vegetation system has a large influence on the hydrologic cycle and is very variable regionally. Models have predicted that in the United States, portions of the Northwest could become wetter, portions of the Southern Great Plains could become somewhat drier, and portions of the Northeast could become considerably drier (Rind and Lebedeff, 1984).

C. Forest Ecosystem Impact of Increased Ultraviolet Radiation at the Surface of the Earth

Due to its biochemical action spectrum, ultraviolet-B radiation (280–320 nm) is toxic to biological systems. Organisms have evolved effective strategies to deal with this natural environmental stress. At a given latitude, the amount of ultraviolet-B radiation incident on the surface of the earth is presumed to vary within a range of acceptable plant tolerance. Recent evidence has suggested that global air pollutants, as well as natural forces, may regulate ultraviolet-B amounts received at ground level. If these forces act to increase ultraviolet-B exposure, forest tree health and forest ecosystem structure/function may be placed at risk.

1. Introduction

The amount of ultraviolet-B radiation, approximately 1.5% of total solar radiation reaching the outer atmosphere (Table 19-3), impacting the surface of the earth is regulated by latitude, elevation, season of the year, and a variety of atmospheric variables. Higher latitudes have reduced ultraviolet-B, while higher elevations have increased ultraviolet-B. Caldwell et al. (1980) have recorded that daily maximum total shortwave irradiance decreased by a factor of only 1.6 from the equator to the North Pole, whereas maximum ultraviolet-B irradiance decreased approximately tenfold.

Spatial regulation is largely determined by the mass of air through which radiation must pass. Higher latitudes have lower sun angles above the horizon and thereby filter ultraviolet-B through a larger air mass. High altitudes, on the other hand, provide a smaller air mass for ultraviolet-B transmission. Clouds, particu-

Table 19-3. Components of Solar Radiation Reaching the Outer Atmosphere of the Earth.

Wavelength	Percent
< 280 nm (UV-C)	0.5
280–320 nm (UV-B)	1.5
320–400 nm (UV-A)	6.3
Visible	38.9
Infrared	52.8
Solar constant	100.0

lates, and stratospheric ozone also contribute importantly to the natural atmospheric regulation of ultravioltet-B received. If stratospheric ozone is decreased 5–10%, as detailed in Section B1 of this chapter, greater amounts of ultraviolet-B will be received by vegetation on earth. A 10% reduction in ozone at 45 degrees north latitude would result in only a one percent increase in solar ultraviolet radiation (290–360 nm), but a 21 percent (DNA action spectra; Setlow 1974) to 28

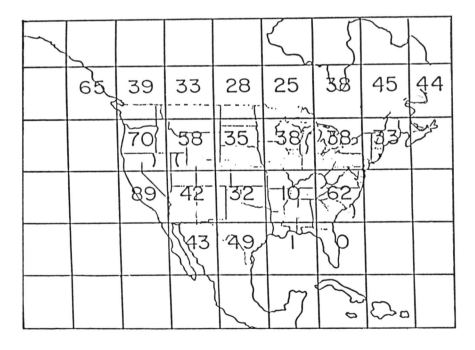

Figure 19-4. Increased length of the growing season (in days) for North America with doubled CO_2 as estimated by the Goddard Institute of Space Studies general circulation model. The growing season is defined as the time between the last spring frost and the first autumn frost.
Source: Rind and Lebedoff (1984).

percent (generalized plant action spectra; Caldwell, 1968) increase in biologically effective radiation.

Early plants evolved under conditions of much greater radiation than exists today. Ultraviolet-C (< 280 nm) is especially damaging to vegetation, and it is presumed that terrestrial plant evolution exploded during the Silurian, when UV-C radiation was presumably greatly diminished by formation of the ozone layer (Caldwell, 1979; Lowry et al., 1980).

2. Phytotoxicity of Ultraviolet-B

The effects of ultraviolet-B radiation on the biochemistry, physiology, morphology, and growth of plants has been studied extensively and is reviewed in Caldwell (1971, 1981), Klein (1978), Teramura (1983) and Hardy et al. (1988). Unfortunately most of the research conducted to date has been with herbaceous and agricultural plants, not woody species.

The targets of ultraviolet-B impact in plant tissues are adjacent pyrimadine bases within the DNA molecule. These bases may covalently bond to form dimers, thereby disrupting replication and protein synthesis. Even at current levels of ultraviolet-B radiation, plants are constrained from optimal growth, as indicated by the relatively rapid growth of shielded plants (Caldwell, 1968; Bogenreider and Klein, 1982; Gold and Caldwell, 1983; Basiouny, 1986). Enhanced ultraviolet-B has been shown to inhibit plant growth and productivity in nonwoody species (Tevini et al., 1981; Biggs and Webb, 1986; Murali and Teramura, 1985; Sisson and Caldwell, 1977; Dickson and Caldwell, 1978), as well as in tree species (Kossuth and Biggs, 1981). In general apparent photosynthesis is inhibited by ultraviolet-B radiation. The rate of dark respiration may be increased during and following ultraviolet-B exposure. Ultraviolet-B radiation may inhibit the transpiration rate of exposed plants, presumably due to stomatal damage and a subsequent increase in diffusion resistance.

Exposure to ultraviolet-B radiation may alter leaf morphology by affecting foliar size, thickness, and specific mass. Leaf thickness is often increased by ultraviolet-B radiation. Biomass accumulation and plant height growth are commonly inhibited by ultraviolet-B. Early life stages are frequently observed to be the most affected by ultraviolet-B exposure. Decreases in biomass productivity may in part be due to ultraviolet-B-induced decreases in photosynthesis, which have been documented for several species (Sisson and Caldwell, 1976, 1977; Van et al., 1976, 1977; Teramura et al., 1980, 1984; Sisson, 1981; Murali and Teramura, 1986; Brandle et al., 1977). Generalizations concerning radiation effects, however, must be treated with much caution, as experimental results are highly variable depending on the species and the age of the plant material used and on experimental conditions employed.

Differential susceptibility among sympatic species may in turn alter their competitive balance (Fox and Caldwell, 1978; Gold and Caldwell, 1983; Bogenreider and Klein, 1982).

3. Ultraviolet-B Stress and Forest Tree Species

Kossuth and Biggs (1981) exposed seedlings of seven coniferous species to ultraviolet-B radiation under growth chamber (low light) conditions. Significant reductions in biomass production, leaf area, and height growth resulted after 11 weeks in loblolly, slash, ponderosa, and lodgepole pine and in noble fir. A high-elevation Douglas-fir genotype did not exhibit these reductions. The study provided evidence for slight seedling susceptibility to realistic increases in ultraviolet-B. Roots tended to be more affected than shoots. Leaf area was lower for treated seedlings, except for noble fir and Douglas fir. Species from higher elevation sites tended to be least affected. In general, these tree seedlings were less sensitive to ultraviolet-B exposure than herbaceous species tested under similar conditions by the authors. A comparable study, but one using a better natural light simulation, was conducted by Sullivan and Teramura (1988). This investigation exposed seedlings of 10 conifers to ultraviolet-B in an unshaded greenhouse. Loblolly pine seedlings exhibited a 40% reduction in biomass and a 16% reduction in height following exposure to 19.1 kJ m^{-2} of biologically active ultraviolet-B. Height reductions were also recorded for lodgepole and red pine. Reduction of root biomass occurred in Scots pine. Fraser fir, white spruce, and white, pinyon, and black pine were not influenced.

Bogenrieder and Klein (1982) worked with deciduous species and found seedlings of European ash, European hornbean, European beech, and Norway maple to have significantly higher biomass when natural ultraviolet-B was excluded.

Semeniuk (1978) exposed six ornamental species to ultraviolet-B radiation ranging from 50% to 400% increases in biologically active radiation in a greenhouse. All six species — balsam fir, Norway and blue spruce, Chinese chestnut, red maple, and flowering dogwood — were insensitive (visual symptoms) to supplemental ultraviolet-B. Kaufman (1978), on the other hand, exposed Engelmann spruce and lodgepole pine seedlings to supplemental ultraviolet-B and ambient ultraviolet-B exclusion. He detected no adverse effects on growth, but did observe visual symptoms on treated seedlings during the following growing season.

Higher plants have evolved a variety of strategies to tolerate, avoid, or compensate for increased ultraviolet-B. Plants exclude much of the potentially injurious ultraviolet-B radiation from mesophyll tissue by absorbing it in the epidermis. In fact, epidermal ultraviolet-B absorptivity is well correlated with ultraviolet-B irradiance during plant development and growth (Robberecht & Caldwell, 1978, 1983; Robberecht et al., 1980). Flavonoids are especially efficient absorptive pigments. Plants also alter leaf morphology to increase ultraviolet-B reflectance. An effective avoidance strategy is growth in shaded habitats. Overstory plants may also reduce ultraviolet-B by orienting their leaf blades parallel to solar rays. Phenological adaptation to minimize radiation at the highest solar irradiance periods (June in mid latitudes) is another avoidance strategy. Photoreactivation and increased chlorophyll synthesis represent compensation strategies.

4. Ultraviolet-B Radiation and Community Dynamics

The possibility that community dynamics may be influenced by changes in ultraviolet-B radiation has been raised by a few studies that have investigated competitive and pest interactions.

Fox and Caldwell (1978) exposed three plant associations; agricultural crops and associated weeds, montane forage species, and disturbed-site weed associates to enhanced ultraviolet-B radiation under field conditions. This enhancement influenced competitive interactions in certain plant combinations. In some cases, evidence indicated that weed species gained competitive advantage (density and/or biomass) over crop species. In field studies with wheat and wild oats or goatgrass, shoot biomass was not affected by supplemental ultraviolet-B radiation, but the competitive ability of the species was altered (Gold and Caldwell, 1986).

A limited number of studies have also evaluated pest interactions as influenced by ultraviolet-B. Responses of insects associated with crop species appear variable. Aphid populations have been reduced by ultraviolet-B radiation, while spider mites were unaffected (Esser ,1980). Hayes (1978) has recorded developmental delays in pink bollworm and codling moth, but no influence on honeybees following ultraviolet-B treatment. Owens and Krizek (1978) recorded inhibition of fungal spore germination with increased exposure to ultraviolet-B. Carns et al. (1978) observed that pigmented spores of leaf infecting fungi were generally resistant to ultraviolet-B, but that increased ultraviolet-B appeared to decrease foliar infection in cucumber inoculated with spores of *Collectotrichum*.

Very unfortunately no studies were found dealing with the specific impacts of increased ultraviolet-B radiation on forest tree interactions other than Bogenrieder and Klein (1982). This latter study examined the competition of European ash and European hornbean, and European beech and Norway maple under conditions of filtered solar radiation. While the growth of each species was influenced the competitive balance did not appear to be. In the case of Norway maple, however, low-elevation varieties produced more biomass, had larger leaf area, and were taller when ultraviolet-B was reduced by filtration. Studies exploring forest pest interactions and altered ultraviolet-B flux are not available.

D. Forest Ecosystem Impact of Increased Carbon Dioxide and Altered Global Climate

The condition of the long-term climate is the most important regulator of forest ecosystem distribution. Climatic variables, principally temperature and moisture, establish the range of biotic components of ecosystems. Climate, in interaction with regional geology, determines the physical and chemical character of the soil substrate. In the short term, climate is variable and highly interactive with other forces that regulate the structure and function of forest ecosystems (Layser, 1980). In the instance of climate altered by changes in trace-gas input into the atmosphere, forest ecosystem response will integrate the impacts of altered carbon dioxide, temperature, and hydrologic cycles.

1. Carbon Dioxide

The bulk of existing evidence suggests that plants are generally limited in growth by current atmospheric levels of carbon dioxide (Strain, 1978). Increasing ambient carbon dioxide generally causes an initial increase in biomass (Wittwer, 1979). The two principal pathways of carbon fixation via photosynthesis, C_3 and C_4, respond differently but positively to carbon dioxide increase.

Under managed environmental growth conditions, a large body of information has been developed regarding plant response to carbon dioxide enrichment (Table 19–4). Comprehensive reviews of these responses are available (Allen, 1979; Brown, 1981; Dahlman et al., 1985; Kramer, 1981; Strain, 1978). Kramer (1981) has emphasized several important generalizations:

1. There are large differences in the amount of growth among plants of various species when subjected to high concentrations of carbon dioxide.
2. The response is greater in indeterminate plants such as cotton and soybean than in determinate species such as corn, sorghum, or tobacco.
3. It is greater in C_3 plants such as soybean and sunflower, than in C_4 plants such as corn and sorghum.
4. The largest growth response appears to occur in seedlings and decreases or ceases as plants grow older.
5. The partitioning of increased growth into various plant parts varies from plant to plant.

For the purposes of this chapter, it is important to realize that the vast majority of studies of plant response to elevated carbon dioxide has been obtained from investigations using agricultural species in artificial growth environments. Experiments with large woody plants, growing under natural conditions are almost nonexistent.

Table 19-4. Plant Response to Carbon Dioxide Enrichment.

1.	Photorespiration decreased
2.	Change in photosynthates produced
3.	Change in carbon allocation to organs
4.	Tolerance to atmospheric pollutants increased
5.	Root/shoot ratio increased
6.	Cytological changes observed
7.	Change in leaf senescence
8.	Increase in symbiotic nitrogen fixation

Source: Strain (1978).

2. Temperature and Moisture

Increasing trace gas injection into the atmosphere may elevate global mean temperature approximately 3–5°C in the next century. Some models predict temperature changes at the poles five times higher than at the equator (Manabe and Wetherald, 1975, 1980). In addition to the significance of this change for temperate zone forests, temperature differentials of this magnitude will alter precipitation patterns substantially. Generally climatologists agree that an intensification of hydrologic cycles is probable with temperature change, but predictions of alterations in precipitation patterns for specific regions are difficult and uncertain (Section B). Forest tree response to climate change can be predicted only with awareness of the magnitude of alterations in *both* temperature and moisture availability.

In a large number of managed environment studies (Dahlman et al., 1985), it has been demonstrated that increasing carbon dioxide concentrations decrease stomatal conductance and transpiration per leaf surface area. As a result, reduced precipitation available for plant uptake may be partially compensated by increased water use efficiency occasioned by increased carbon dioxide. Soil moisture is strongly controlled by temperature, on the other hand, and even in some areas receiving higher precipitation, total evapotranspiration increases more with warming, and as a result less soil moisture is available for tree growth. Waggoner (1984) has estimated that the projected change in weather by the year 2000, caused by increased atmospheric carbon dioxide, will cause moderate decreases of 1–12% in the yields of wheat, corn and soybeans in the American grain belt due to increased dryness. While agriculturists may be able to adopt new crop varieties to a drier climate, forests *cannot* be similarly manipulated.

3. Pest Interactions

The multiple and complex interactions between climate, biotic stress agents, and host trees are well beyond the scope of this book. It is critical to appreciate, however, that the circumstance of joining a susceptible forest tree with a virulent pathogen or vigorous insect in the same location will not necessarily produce disease. The climate must also be conducive. Climate, as expressed in temperature, moisture, air movement, and radiation in the environment of the forest is the primary regulator of tree health and growth. In a similar fashion, climate is the primary regulator of the population dynamics of microbial pathogens and insects. Long-term change in climatic trends, therefore, must be expected to have a profound influence on pest population dynamics and host tree interactions.

Evidence has been provided that a large number of diverse forest tree diseases have important climatic regulators. These diseases include rust, foliar, canker, vascular and root abnormalities.

Insect dynamics are also regulated by climatic variables. The indigenous eastern spruce budworm (*Choristoneura fumiferana*) exerts a profound influence on the successional dynamics of the spruce-fir forests of Maine and eastern Canada. Moisture stress is a primary regulator of budworm population dynamics. Regular budworm outbreaks have been recorded since the early 1700s. Budworm populations fluctuate from rare to excessive over time during nonoutbreak to outbreak episodes. When outbreaks develop there is widespread destruction of balsam fir in all mature stands. These outbreaks leave the less susceptible black and white spruces, the nonsusceptible white birch, and dense regeneration of both fir and spruce. In immature stands, insect damage is less and more fir survives. During nonoutbreak periods, young fir grow, together with spruce and birch, to form dense stands in which spruce and birch, in particular, are stressed by overcrowding. Succession proceeds to produce stands of mature and overmature trees with fir as the dominant species. The development of these overmature fir stands in association with a sequence of drought years are the predisposing conditions for budworm outbreak occurrence. Dry years appear to allow budworm populations to escape control by predators and parasites, to increase population densities, and in the presence of mature fir, to initiate outbreaks. Continued insect population increase eventually causes sufficient tree mortality to occasion a collapse of the population and a reduction back to low equilibrium density (Holling, 1973). In nonoutbreak periods, fir is favored in competition with spruce and birch. During outbreak periods, succession favors spruce and birch, as they are less susceptible to budworm influence. Without budworm impact, succession would eliminate spruce and birch. Fir persists because of its efficient regenerative capability, its rapid growth rate, and the irregular occurrence of climatic conditions favorable for budworm population expansion.

The geographic or host ranges of exotic microbial pathogens or insect pests may expand with climatic change. Previously innocuous endemic microbes or insects may be elevated to important pest status following climatic warming. Certain insects, for example, defoliators may need to consume greater foliar biomass to obtain nitrogen requirements due to the increase in the carbon/nitrogen ratio of the foliage resulting from carbon dioxide enrichment (Lincoln et al., 1986). Bark beetles are perhaps the most important insect pests of commercial forest species. Bark beetle population dynamics are closely associated with regional precipitation patterns.

4. Climate Change and Forest Response

In geologic time, plant species in disequilibrium with their climate have been extremely common. During the two million years of Quaternary time, during which the modern flora evolved, there were at least 16 glaciations lasting 50,000–100,000 years, with interglacial periods lasting 10,000–20,000 years. These interglacial periods were characterized by dynamic changing climates and dynamic changing, constantly adjusting forests. The present interglacial period (the Holocene epoch) has been characterized by numerous examples of forest tree distributions not in equilibrium with climate (Davis, 1981). The impressive average rates of Holocene range extensions in eastern North American tree species emphasize these disequilibria (Table 19-5). The global temperature change from the last glacial period to the present (~18,000 years) is very approximately 4°C (Kutzbach and Guetter 1986). It is possible, as we approach doubling of atmospheric carbon dioxide, that climatic warming may be quite rapid. It is possible that 4° C increase in global average temperature could be achieved in approximately 100 years. Such a rapid change would cause large numbers of tree species to be out of phase with their climate. Increased drought stress over widespread forest areas would be expected to initiate new rounds of progressive tree deterioration or dieback/decline phenomena (Chapter 18). Drought is the most common and important initiator of general forest tree decline. Competitive interactions would be changed and alterations in forest species composition would occur. Specific projections of change in forest ecosystems associated with climate alteration that are currently available are speculative and are based on one or more estimations of changes in temperature and precipitation (Slocum, 1985; Woodwell, 1986).

It might be expected that warming, especially with some drying, might cause an expansion of the central hardwood forest at the expense of the northern hardwood forest in the United States. White and red oak group species, hickories, and tulip poplar may be selectively favored over sugar maple, American beech, birches and associated conifers. Allen Solomon, Oak Ridge National Laboratory, has simulated the long-term response to climate for the eastern hardwood forest. His model predicted condensation of boreal species, such as spruce and hardwood, expansion in western Ontario and northwest Michigan, but no significant alteration in east-central Tennessee (Solomon, 1986). Frank Miller, Mississippi State University, has studied change in southern pine species with climate alterations. He has suggested that the range of loblolly pine will expand eastward and

Table 19-5. Average Rates of Holocene Tree Range Extensions in Eastern North America.

Species	Rate (m yr^{-1})
Jack/red pine	400
White pine	300-350
Oak	350
Spruce	250
Larch	250
Elm	250
Hemlock	200-250
Hickory	200-250
Balsam fir	200
Maple	200
Beech	200
Chestnut	100

Source: Davis (1981).

northward due to reduced precipitation and increased temperature west of the Mississippi River (Miller, 1983).

Jerry Leverenz, University of Washington, has examined the response of western conifers. His model suggests that ponderosa pine will increase in area and importance in California and in the Cascade Mountains in Oregon, and will decrease along the eastern slopes of the Rocky Mountains from Mexico to Canada. Douglas fir will maintain or expand its present range. Lodgepole pine will not change significantly. There may be local changes only for western hemlock and larch. Engelmann spruce will sustain a general reduction in acreage (Slocum 1985). Binkley and Larson (1987) have simulated the influence of climatic warming on the productivity of the managed northern hardwood forest and have concluded that an increase in productivity would occur at location 42°N 77°W, that forest management practices enhance the influence of climate change, and that relating global-scale change to local-scale forest effects is extremely difficult.

Gap models simulate forest dynamics by modeling the demographics of individual trees on a small model plot corresponding to the area of a forest gap. These models have been especially useful in projecting the dynamics of mixed-age, mixed-species forest stands (Botkin et al., 1982; Shugart and West, 1980; Shugart, 1984) and have been applied to climate change simulations. Urban and Shugart (1988) applied a gap model to examine the influence of climate change on the oak-pine forests of east Tennessee. Specific focus was on the influence of soil moisture availability and its regulation of species composition. In this area mesic soils are dominated by tulip tree, red maple, and basswood, intermediate soils by oak-hickory and other mixed hardwoods, and xeric soils by chestnut oak, black oak, and pine species. When the Goddard Institute of Space Studies climate-change scenario (Section 2b) was applied to the gap model, species compositions were shifted to the more xeric associations. Mesic soils shifted to intermediate sites, while intermediate soils shifted to the xeric condition. Botkin et al. (1988) applied a gap model to numerous general circulation model climate change scenarios for forests of the Great Lake States. Although precipitation increased in the climate models, evapotranspiration increased more and produced soils with less available soil moisture. In the Boundary Water Canoe Area, the boreal forest (white birch, balsam fir, white cedar, quaking aspen) was converted to northern hardwood forest (largely sugar maple) (Figure 19-5). On drier sites the climatic shifts were large enough to convert substantial forests to open woodlands, savannahs, or grasslands with small scattered trees. Pastor and Post (1988) have also employed gap models to assess northern forest response to climate change and have further included alterations in the nitrogen cycle. In general, on soils where there was no decrease in soil water availability associated with a doubled carbon dioxide concentration, the current mixed spruce-fir/northern hardwood forest was replaced by a more productive northern hardwood forest. In soils with decreased water availability, the spruce-fir/northern hardwood forest was replaced on some sites by a stunted pine-oak forest of much less biomass.

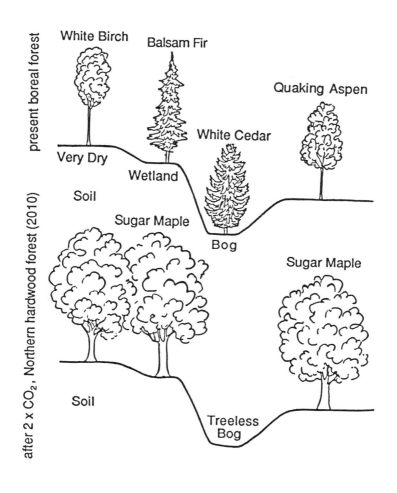

Figure 19-5. Predicted conversion during the next century of boreal forest to northern hardwood forest in Minnesota associated with climatic change resulting from a doubling of atmospheric carbon dioxide.
Source: Botkin et al. (1988).

Paleoecological evidence of past vegetative distributions can also be used to predict future vegetation patterns under specific climatic change regimes. Past climate-induced vegetation change can be reconstructed using time series of late Quaternary pollen data. These records indicate that ecological response surfaces can be used to model the equilibrium response of vegetation to climate change (Overpeck, 1988). Future vegetation patterns, simulated using three general circulation models of climate change at a doubling of current carbon dioxide levels, suggest that the effects of trace-gas-induced climate change will be significant in

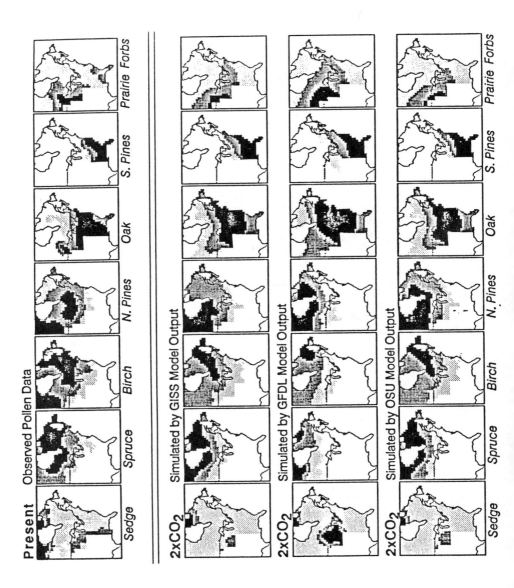

Figure 19-6. Percent North American vegetative patterns based on observed pollen data compared with vegetative patterns that would occur under climatic patterns with doubled atmospheric carbon dioxide as predicted by three general circulation models: Goddard Institute of Space Studies (GISS), Geographical Fluid Dynamics Laboratory (GFDL), and Oregon State University (OSU).
Source: U.S.E.P.A. (1988).

eastern North American forests (Figure 19-6). Large-scale warming and regional drying give rise to the most pronounced simulated changes (Overpeck and Bartlein, 1988; Shands and Hoffman, 1987).

E. Conclusions

Evidence currently available suggests that global-scale pollutants may have the potential, over the next century, of altering radiation fluxes to and from the earth. Potential consequences of these alterations include an increase in short-wave radiation incident on the earth and a reduction in long-wave radiation leaving the earth. Consequences of the former include receipt of greater amounts of ultraviolet-B radiation by the biota at the earth's surface and of the latter include global warming over the next century equivalent to that experienced by temperate vegetation over the last 18 millennia.

Halocarbon emissions to the global troposphere may be able to reduce natural ozone concentrations in the stratosphere and thereby increase the ultraviolet-B radiation received at the surface of the earth. Changes in atmospheric chemistry or physics that allow for increases in ultraviolet-B have a great implication for all elements of the biota. Ultraviolet-B radiation is well appreciated to be phytotoxic and able to influence plant growth and development at low levels of exposure.

Although there have been many published papers concerning the effects of ultraviolet-B radiation on vegetation, it is still difficult to accurately predict the effects of increased solar ultraviolet-B radiation on natural communities. This is due to several factors including: (a) many studies have employed unrealistically high ultraviolet-B irradiances; (b) many studies have involved very low photosynthetically active (400–700 nm) radiation during plant culture; (c) there are large interspecific and intraspecific differences in plant response to ultraviolet-B radiation; (d) little effort has been directed to nonagricultural, especially forest species; and (e) most ultraviolet-B stress research has concentrated on direct effects rather than indirect effects, e.g., pest/pathogen interactions.

Since it is possible that solar ultraviolet-B irradiance will increase in the next several decades due to stratospheric changes, it is critically important that the effects of ultraviolet-B radiation on plants be better understood. In particular, woody plants growing under field conditions must be exposed to realistic levels of ultraviolet-B radiation. Long-term and field studies are desperately needed to better understand ultraviolet-B radiation direct and indirect effects on forest trees and forest ecosystems.

The consequences of a warmer global climate, with even a very modest temperature increase, on the development of forest ecosystems will be profound. Warming, with increased carbon dioxide in the atmosphere, might enhance forest growth. Manabe and Stouffer (1980) have estimated that a doubling of atmospheric carbon dioxide would cause a 3°C warming at the U.S.-Canadian border, while Kellog (1977) has suggested that a rise of 1°C in mean summer temperature extends the growing season at the border by approximately 10 days. It is important to realize, however, that carbon dioxide may not be the most important limiting factor to the growth of forest trees. Limitations imposed by nitrogen availability or other nutrients obtained from the soil may be more important. Other changes associated with global warming may restrict forest growth. Physiological processes of plants — especially photosynthesis, transpiration, respiration and reproduction — are sensitive to temperature. With warming, respiration and decomposition may increase faster than photosynthesis. Transpiration and evaporation increases may enhance stress on some sites even if precipitation is increased. Reproduction may be altered by changes in the dynamics of pollinating insects, changes in flower, fruit or seed set, or changes in seedling production and survival. Relationships with important microbial and insect pests may be altered. New pest species may be created. Extended term change in temperature and available soil moisture will alter competitive relationships among principal tree species. Species placed at a competitive disadvantage will exhibit dieback/decline phenomena. Species placed at a competitive advantage will extend their vigor and range. The southern ranges of many forest species in the eastern United States could move several hundred to one thousand kilometers (600 miles) in a generally northward direction. The potential northern range of forest trees in the east could shift northward as much as 600–700 km (370–430 miles) over the next century. Slow migration rates of some species could limit movement to 100 km (60 miles). It could take centuries for eastern forests to fully migrate to potential northern distributions. If climate change is too rapid, some species may be eliminated (U.S.E.P.A., 1988).

The potential stresses imposed by regional-scale air pollutants (ozone, acid deposition) must be viewed against the background of uncertainty associated with the changes in radiation balances associated with global-scale air contaminants. If global warming does accelerate in the near term, and especially if expansive forest regions are subjected to reduced moisture availability, the expansion of forest tree decline will be very great. Such an occurrence could pale the impact of regional-scale pollutant deposition.

References

Allen, L.H. 1979. Potentials for carbon dioxide enrichment. In B. J. Barfield and J. F. Gerber, eds., Modification of the Aerial Environment of Crops. Am. Soc. Agricul. Engineers, St. Joseph, MI.

Binkley, C.S. and B.C. Larson. 1987. Simulated effects of climate warming on the productivity of managed northern hardwood forests. Climate change (submitted).

Bogenrieder, A. and R. Klein. 1982. Does solar UV influence the competitive relationship in higher plants? In J. Calkins, ed., The Role of Solar Ultraviolet Radiation in Marine Ecosystems. Plenum Press, New York, pp. 641-649.

Bolin, B., E.T. Degens, S. Kempe, and P. Ketner (eds.). 1979. The Global Carbon Cycle. Scope No. 13, Wiley-Interscience, 448 pp.

Botkin, D.B., J.F. Janak, and J.R. Wallis. 1972. Some ecological consequences of a computer model of forest growth. J. Ecol. 60: 849-873.

Botkin, D.B., R.A. Nisbet, and T. E. Reynales. 1988a. Effects of Climatic Change on Forests of the Great Lake States. U.S. Environmental Protection Agency, Report, March 15, 1988, Washington, DC.

Botkin, D.B., R A. Nisbet, and T.E. Reynales. 1988b. How soon will forests respond to CO_2 induced climatic change. Report to U.S. Environmental Protection Agency. March, 1988, Washington, DC.

Bowden, W. B. 1986. Gaseous nitrogen emissions from undisturbed terrestrial ecosystems: An assessment of their impacts on local and global nitrogen budgets. Biogeochemistry 2: 249-279.

Bowden, W.B. and F.H. Bormann. 1986. Transport and loss of nitrous oxide in soil water after forest clearcutting. Science 233: 867-869.

Bowman, K.P. 1988. Global trends in total ozone. Science 239: 48-50.

Broecker, W.S. 1987. Personal communication. Lemont–Doherty Laboratory, Columbia University, NY, Jan. 30, 1987.

Broecker, W.S., T. Takahaski, H.J. Simpson, and T.H. Peng. 1979. Fate of fossil fuel carbon dioxide and the global carbon budget. Science 206: 409-418.

Brown, S. (ed.). 1981. Global Dynamics of Biospheric Carbon. U.S. Dept. of Energy Conf. No. 8108131, Aug. 17, 1981, Bloomington, IN, 194 pp.

Bryson, R.A. 1974. A perspective on climatic change. Science 184: 753-760.

Caldwell, M.M. 1971. Solar UV irradiation and the growth and development of higher plants. Phytophysiology 6: 131-177.

Caldwell, M.M. 1968. Solar ultraviolet radiation as an ecological factor for alpine plants. Ecol. Monogr. 38: 243-268.

Caldwell, M. M. 1981. Plant response to solar ultraviolet radiation. In Encyclopedia of Plant Physiology. NS 12A: 170-197. Springer–Verlag, New York.

Caldwell, M.M., R. Robberecht, and W.D. Billings. 1980. A steep latitudinal gradient of solar ultraviolet-B radiation in the arctic-alpine life zone. Ecology 61: 600-611.

Carns, H.R., J.H. Graham, and S.J. Ravitz. 1978. Effects of UV-B radiation on selected leaf pathogen fungi and disease severity. UV-B Biological and Climatic Effects Research (BACER). Final Report. Report No. IAG-P6-0168. U.S. Environmental Protection Agency, Washington, DC.

Charlson, R.J., J.E. Lovelock, M.O. Andreae, and S. G. Warren. 1987. Oceanic phytoplankton, atmospheric sulfur cloud albedo and climate. Nature 326: 655-658.

Cicerone, R.J. 1987a. Changes in stratospheric ozone. Science 237: 35-42.

Cicerone, R.J. 1987b. Personal communication, Feb. 27, 1987. National Center for Atmospheric Research, Boulder, CO.

Cicerone, R.J., R.S. Stolarski, and S. Walters. 1974. Stratospheric ozone destruction by man-made chlorofluoromethanes. Science 185: 1165-1166.

Clark, W.C. (ed.). 1982. Carbon Dioxide Review. Oxford University Press, Oxford, UK, 469 pp.

Dahlman, R.C., B.R. Stain, and H. H. Rogers. 1985. Research on the response of vegetation to elevated atmospheric carbon dioxide. J. Environ. Qual 14: 1-8.

Davis, M.B. 1981. Quaternary history and the stability of forest communities. In D. C. West, H. H. Shugart, and D. B. Botkin, eds., Forest Succession Concepts and Application. Springer–Verlag, New York, p. 132-153.

Demerjian, K.L. 1986. Atmospheric chemistry of ozone and nitrogen oxides. In A.H. Legge and S.V. Krupa, eds., Air Pollutants and Their Effects on the Terrestrial Ecosystem. John Whiley and Sons, New York, NY.

Dobson, G.M.B. 1966. Dynamics of stratospheric ozone. Q. J. R. Meteorol. Soc. 92: 549-552.

Esser, G. 1980. Einfluss einer mach schadstoffimmission vermehsten einstrahlung von UV-B-licht auf kulturpfeanzen, 2. Versuchsjahr, Bericht Battelle Institut E.V. Frankfurt, BF-R-63, 984-I.

Farman, J.C., B.G. Fardniner, and J.D. Shanklin. 1985. Large losses of total ozone in Antarctica several seasonal $C10_x/No_x$ interaction. Nature 315: 207-210.

Fox, F.M. and M.M. Caldwell. 1978. Competitive interaction in plant populations exposed to supplementary ultraviolet-B radiation. Oecologia 36: 173-190.

Gleich, P.H. 1987a. Regional hydrolic consequences of increases in atmospheric CO_2 and other trace gases. Climatic Change 10: 137-161.

Gleich, P.H. 1987b. Global climatic changes and regional hydrology: Impacts and responses. In The Influence of Climate Change and Climatic Variability on the Hydrolic Regime and Water Resources. IAHS Publ. No. 168, Vancouver Symposium, August, 1987.

Gold, W.G. and M.M. Caldwell. 1983. The effects of ultraviolet-B radiation on plant competition in terrestrial ecosystems. Physiol. Plant. 58: 435-444.

Hammond, A.L., and T.H. Maugh. 1974. Stratospheric pollution: Multiple threats to earth's ozone. Science 186: 335-338.

Hansen, J. 1987. Personal communication, Goddard Instit. Space Studies, NASA, Columbia University, New York, Feb. 6, 1987.

Hansen, J., G. Russell, D. Rind, P. Stone, A. Lacr, S. Lebedeff, R. Reudy, and L. Travis. 1983. Efficient three-dimensional global models for climate studies: Models I and II. Mon. Wea. Rev. III: 609-662.

Henderson–Sellers, A., R.E. Dickinson, and M.F. Wilson. 1988. Tropical deforestation: Important processes for climate models. Climatic Change 13: 43-67.

Hardy, J., R. Worrest, J. Bailey, J. Chapman, and S. Holman. 1988. Ecological Effects of Ultraviolet-B Radiation. U.S.E.P.A. Research Plan 1989-1993. U.S. Environmental Protection Agency, Corvallis, OR.

Hayes, D.K. 1978. Influence of broad-band UV-B on physiology and behavior of beneficial and harmful insects. UV-B Biological and Climatic Effects Research (BACER). Final Report. Report No. IAG-D6-0168. U.S. Environmental Protection Agency, Washington, DC.

Hepting, G.H. 1963. Climate and forest disease. Annu. Rev. Phytopathol. 1: 31-50.

Hidy, G.M. and P.K. Mueller. 1986. The sulfur oxide-particulate matter complex. In A.H. Legge and S.V. Krupa, eds., Air Pollutants and Their Effects on the Terrestrial Ecosystem. John Wiley and Sons, New York, pp. 51-104.

Holdgate, M.W., M. Kassa, and G.F. White. 1982. World Environmental Trends between 1972 and 1982. Environ. Conserv. 9: 11-29.

Holling, C.S. 1973. Resilience and stability of ecological systems. Annu. Rev. Ecol. System. 4: 1-23.

Kaufman, M.R. 1978. The effects of ultra-violet radiation in Englemann spruce and lodgepole pine seedlings. In Research Report on the Impacts of Ultraviolet-B Radiation on Biological Systems: A Study Related to Stratospheric Ozone depletion. U.S. Environmental Protection Agency, Corvallis, OR.

Kellogg, W.W. 1977. Effects of human activities on global climate. Worth Meteorological Organ, Tech Note No. 156, Geneva.

Kerr, R.A. 1987. Halocarbons linked to ozone hole. Science 236: 1182.

Khalil, M., A.K. Rasmussen, and R. A. Rasmussen. 1988. Carbon monoxide in the earth's atmosphere: Indications of a global increase. Nature 332: 242-245.

Klein, R.M. 1978. Plants and near ultraviolet radiation. Bot. Rev. 44: 1-127.

Kossuth, S.V. and R.H. Biggs. 1981. Ultraviolet-B radiation effects on early seedling growth of Pinaceae species. Can. J. For. Res. 11: 243-248.

Kramer, P.J. 1981. Carbon dioxide concentration, photosynthesis and dry matter production. BioScience 31: 29-33.

Kutzbach, J.E. and P.T. Guetter. 1986. The influence of changing orbital parameters and surface founding conditions on climate simulations for the past 18,000 years. J. Atmos. Sci. 43: 1726-1759.

Lal, M., S.K. Dube, P.C. Linha, and A.K. Gain. 1986. Potential climatic consequences of increasing anthropogenic constituents in the atmosphere. Atmos. Environ. 20: 639-642.

Layser, E.F. 1980. Forestry and climatic change. J. For. 78: 678-682.

Lillian, D., H.B. Singh, A. Appleby, L. Lobbam R. Arnts, R. Gumpert, R. Hague, J. Toomey, J. Kazozis, M. Antell, D. Hansen, and B. Scott. 1975. Atmospheric fates of halogenated compounds. Environ. Sci. Technol. 9: 1042-1048.

Lowry, B., D. Lee, and C. Hebant. 1980. The origin of land plants. A new look at an old problem. Taxon 29: 183-197.

Manabe, S. and R.J. Stouffer. 1980. Sensitivity of a global climate model to an increase of CO_2 concentration in the atmosphere. J. Geophys. Res. 85: 5529-5554.

Manabe, S., and W.T. Wetherald. 1975. The effects of doubling the CO_2 concentration on the climate of a general circulation model. J. Atmos. Sci. 32: 3-15.

Manabe, S. and W.T. Wetherald. 1980. On the distribution of climate changes resulting from an increase in CO_2 content of the atmosphere. J. Atmos. Sci. 37: 99-118.

Miller, W.F. 1983. Water — A Resource in Demand. Proc. Symposium on Future Climate and Potential Impacts on Natural Resource Management. Southern Cooperative Series, Bull. No. 288, Mississippi Agri. and Forestry Exp. Sta., Mississippi State Univ., Mississippi State, MS, 34 pp.

Molina, M. and F. Rowland. 1974. Stratospheric sink for chlorofluormethanes. Nature 249: 810-811.

National Academy of Sciences. 1976. Halocarbons: Effects on Stratospheric Ozone. NAS, Washington, DC, 352 pp.

National Academy of Sciences. 1978. Chloroform, Carbon Tetrachloride and Other Halomethanes: An Environmental Assessment. NAS, Washington, DC, 294 pp.

National Academy of Sciences. 1984. Causes and Effects of Changes in Stratospheric Ozone: Update 1983. National Academy Press, Washington, DC.

National Research Council. 1983. Changing Climate: Report of the Carbon Dioxide Assessment Committee. U.S. National Research Council, National Academy Press, Washington, DC.

Overpeck, J. T. 1988. Modeling the transient response of vegetation to climatic change: A paleoecological time series perspective. Bull. Ecol. Soc. Am. (Supplement) 69: 251.

Overpeck, J.T. and P.J. Bartleim. 1988. Assessing the Response of Vegetation to Future Climate Change: Ecological Response Surfaces and Paleoecological Model Validation. U. S. Environmental Protection Agency. Report, March 15, 1988, Washington, DC.

Owens, O.V.H. and D.T. Krizek. 1978. Multiple effects of UV-B irradiation on fungal spore germination. UV-B Biological and Climatic Effects Research (BACER). Final Report. Report No. IAG-D6-0168, U.S. Environmental Protection Agency, Washington, DC.

Pastor, J. and W.M. Post. 1988. Response of northern forest to CO_2-induced climatic change: Dependence on soil water and nitrogen availabilities. Nature (in press).

Penner, J.E., P.S. Connell, D.J. Wuebbles, and C.C. Covey. 1988. Climate Change and its Interactions with Air Chemistry: Perspectives and Research Needs. U.S.E.P.A. Project Report No. 280, Atmospheric Science, Research Triangle Park, NC, 90 pp.

Postel, S. 1986. Altering the Earth's Chemistry: Assessing the Risks. Worldwatch Paper No. 71, Worldwatch Institute, Washington, DC, 66 pp.

Rind, D. and S. Lebedeff. 1984. Potential climatic impacts of increasing atmospheric CO_2 with emphasis on water availability and hydrology in the United States. U.S. Environmental Protection Agency, Washington, DC.

Robberecht, R. and M.M. Caldwell. 1978. Leaf epidermal transmittance of ultraviolet radiation and its implications for plant sensitivity to ultraviolet-radiation induced injury. Oecologia 32: 277-287.

Robberecht, R. and M.M. Caldwell. 1983. Protective mechanisms and acclimation to solar ultraviolet-B in *Oenothera stricta*. Plant, Cell Environ. 6: 477-485.

Robberecht, R.,M. M. Caldwell, and W. D. Billings. 1980. Leaf ultraviolet optical properties along a latitudinal gradient in the arctic-alpine life zone. Ecology 61: 612-619.

Rodgers, C. 1988. Global ozone trends reassessed. Nature 332: 201.

Rotty, R.M. and D.B. Reister. 1986. Use of energy scenarios in addressing the CO_2 question. J. Air Pollu. Control Assoc. 36: 1111-1115.

Rowland, F.S. 1987. Personal communication, April 3, 1987. Chemistry Dept., Univ. California, Irvine, CA.

Sagan, C., O.B. Toon, and J.B. Pollack. 1979. Anthropogenic albedo changes and the earth's climate. Science 206: 1363-1368.

Schneider, S.H. and C. Mass. 1975. Variation in global climate during the twentieth century. Science 190: 741-743.

Semenink, P. 1978. Biological effects of ultraviolet radiation on plant growth and development in florist and nursery crops. In Research Report on the Impacts of Ultraviolet-B Radiation on Biological Systems: A Study Related to Stratospheric Ozone Depletion. U.S. Environmental Protection Agency, Corwallis, OR.

Sergin, V.Y. 1980. Origin and mechanism of large-scale climatic oscillations. Science 209: 147-1483.

Setlow, R.B. 1974. The wavelengths in sunlight effective in producing skin cancer: A theoretical analysis. Proc. Natl. Acad. Sci. USA 71: 3363-3366.

Shands, W.E. and J.S. Hoffman. 1987. The Greenhouse Effect, Climate Change, and U.S. Forests. The Conservation Foundation, Washington, DC, 304 pp.

Shugart, H.H. 1984. A Theory of Forest Dynamics. Springer–Verlag, New York.

Shugart, H.H. and D.C. West. 1980. Forest succession models. BioScience 30: 308-313.

Singh, H.B., L.J. Salas, H. Schiegeishi, and E. Scribner. 1979. Atmospheric halocarbons, hydrocarbons, and sulfur hexafluoride: Global distributions, sources and sinks. Science 203: 899-903.

Slocum. R.W. 1985. Major climate changes likely, say scientists. J. For. 83: 325-327.

Solomon, A.M. 1986. Transient response of forests to carbon dioxide-induced climate change: Simulation modeling experiments in eastern North America. Oecologia 68: 567-579.

Strain, B.R. (ed.). 1978. Report of the Workshop on Anticipated Plant Response to Global Carbon Dioxide Enrichment, August, 4–5, 1977. Dept. of Botany, Duke University, Durham, NC.

Strodal, F. and I.S.A. Isaksen. 1986. Ozone perturbations due to increases in N_2O, CH_4, and chlorocarbons: Two-dimensional time-dependent calculations. In J.G. Titus, ed., Effects of Changes in Stratospheric Ozone and Global Climate. Vol. I. Overview. U.S.E.P.A. and U.N. Environmental Program, pp. 83-119.

Sullivan, J.H. and A.H. Teraumura. 1988. The effects of ultraviolet-B radiation on seedling growth in the Pinaceae. Am. J. Bot. 75: 225-230.

Sze, N.D. 1977. Anthropogenic CO emissions: Implications for the atmospheric CO-OH-CH_4 cycle. Science 195: 673-674.

Teramura, A.A. 1983. Effects of ultraviolet-B radiation on the growth and yield of crop plants. Physiol. Plant 58: 415-427.

Thompson, S.L., V.V. Aleksandrov, G.L. Stenchikov, S.H. Schneider, C. Covey, and R.M. Chervin. 1984. Global climatic consequences of nuclear war: Simulations with three dimensional models. Ambio 13: 236-243.

Trabalka, J.R., J.A. Edmonds, J.M. Reilly, R.H. Gardner, and D.E. Reichle. 1986. Atmospheric CO_2 projections with globally averaged carbon cycle models. In J.R.

Trabalka and D.E. Reichle, eds., The Changing Carbon Cycle: A Global Analysis. Springer–Verlag, New York, pp. 534-560.

Urban, D.L. and H. H. Shugart. 1988. Forest Response to Climatic Variability. A Simulation Study for Southeastern Forests. U.S. Environmental Protection Agency. Report, March 15, 1988, Washington, DC.

U.S. Environmental Protection Agency. 1988a. The Potential Effects of Global Climate Change on the United States. Vol. I. Regional Studies. Vol. II. National Studies. Report to Congress, Oct. 1988. Office of Research and Development. U.S.E.P.A., Washington, DC.

U.S. Environmental Protection Agency. 1988B. Workshop on Effects of Global Warming. April 6–7 1988. Washington, DC.

U.S. Department of Energy. 1985. Atmospheric Carbon Dioxide and the Global Carbon Cycle. D.O.E. Publ. No. DOE/ER 0239., Washington, DC, 315 pp.

Waggoner, P.E. 1984. Agriculture and carbon dioxide. Am. Scient. 72: 179-184.

Wang, W.C., Y.L. Yung, A.A. Lacis, T. Mo, and J.E. Hansen. 1976. Greenhouse effects due to man-made perturbations of trace gases. Science 194: 685-690.

Wang, W.C., D.J. Wuebbles, W.M. Washington, R.G. Isaacs, and G. Molnar. 1986. Trace gases and other potential perturbations to global climate. Rev. Geophys. 24: 110-140.

Watson, R. 1988. Ozone Trends. Executive Summary. Panel Report. NASA, Washington, DC.

Wittwer, S.H. 1979. Future technological advances in agriculture and their impact on the regulatory environment. BioScience 29: 603-610.

Woodwell, G.M. 1986. Forests and climate: Surprises in store. Oceanus 29: 71-75.

World Meteorological Organization. 1985. Atmospheric ozone. Global Ozone Research and Monitoring Project. Report No. 16, WMO, Geneva.

20

Forest Quality and Air Quality: Forest Health Risk and Future Needs

Areas of temperate forests are currently experiencing significant perturbation from air pollution. The influence of air contaminants on specific forests is variable and is dependent on exposure (dose), hazard (importance of susceptible species in exposed forests and other site specific variables), management objectives, and the perspective of interest, i.e., individual tree, stand, population, community, or ecosystem. Before we consider the importance of specific air pollutants and specific forest systems, it is essential to have an overview of general forest health principles.

A. Fundamentals of Forest Health

Forest systems have enormous variability. Forests may differ in soil type, climate, aspect, elevation, species composition, and age. Forests may be uneven aged, evenaged, all aged, or overmature. Forests may be reproduced by seed, by coppice, or by planting. Some forests have their structure completely shaped by natural forces, others may be influenced by human forces as well as natural forces, while other forests may be completely artificial in design and establishment. Forest trees may be arrayed in a continuum of human management efforts ranging from no management (areas of slow growth or low tree density) to very intensive management (forest tree nurseries, seed orchards, arboreta, urban/suburban forests). All of this variability makes generalizations about forest health very difficult to formulate, but also makes them essential to attempt. The following 10 fundamentals of forest health attempt to summarize some of the most important concepts of forest stress.

1. Forests are Generally Healthy and Significant Tree Disease and Injury is Typically Confined to Scattered Individuals, Small Groups or Stands.

While we have a very long list of forest health problems, disease remains the exception rather than the rule. Temperate zone forests combine the characteristics of relatively low diversity with relatively high stability (Langford and Buell, 1969). Tree mortality rates vary greatly with The stage of forest development, they are high in seedling and vegetative closure stages, but medium to low in mature and old-age stages (Franklin et al., 1987).

2. Stress Factors that Influence Forest Health are Numerous, Varied, Recurrent, and Highly Interactive.

Prof. John S. Boyce, Yale University, defined tree disease in 1913 as "any disturbance of the normal function of a tree and this may be caused by fungi, mistletoes, insects, impoverished condition of the soil, poisonous gases in the air, and numerous other agencies" (Boyce, 1913). Over the last three quarters of a century since Boyce's comprehensive definition of tree disease, we have added vast amounts of information to our understanding of forest health, but not to his list of stress factors.

Factors capable of causing injury, disease and mortality in forest systems are termed *stresses*. Stress factors may be biotic or abiotic, natural or unnatural. Stresses recognized to have widespread and general importance in forest systems include climatic, pathogenic, entomologic, anthropogenic, wildlife, fire, and stand dynamic elements. Trees are large and long lived. Tree health integrates the influences of all stresses acting concurrently, sequentially, and interactively.

3. Stresses Play Important Natural Roles in Forest Ecosystem Structure and Function.

Forest ecosystems are dynamic and not static. Over time they are characterized by variability not constancy. There is substantial evidence to support the conclusion that numerous forest stress factors play fundamental roles in forest form and process (Smith 1986). Even forest insects and pathogens that cause massive destruction in the short term, may play essential and beneficial roles in long term forest change. These roles involve regulation of species competition, species composition and succession, primary production, and biogeochemical cycling (Dinoor and Esched, 1984; Huffaker, 1974; Mattson and Addy,1975; Seastedt and Crossley, 1984).

In natural forest ecosystems, disturbance or perturbation caused by environmental or biotic forces may be necessary to maintain maximum diversity and productivity. The natural tendency in eastern United States forest ecosystems toward periodic perturbation at intervals of 50–200 years recycles the system and maintains a periodic wave of peak diversity and a corresponding wave of peak primary production (Loucks, 1970).

4. Specific Stress Factors of Importance Vary with Age, Location, Species Composition, *and* with Management Objectives.

Stress factor significance exhibits great temporal and spatial variation. Numerous stresses have greatest importance at young tree ages, while others are most significant at old tree ages. Stresses vary with soil type, aspect, elevation, and microclimate. White-pine blister rust is an important disease in northern Wisconsin and northern New England, but is less significant in southern Wisconsin and southern New England (Marty, 1966, Van Arsdel et al., 1961) because of climate.

Management objectives define the relative importance of specific stress factors. Unlike agricultural ecosystems, forest ecosystems do not have "absolute stresses," only "relative stresses." The significance of a particular stress is relative to the management goals and objectives imposed on the forest system or forest tree by the human manager. Microbial decay of stem wood, for example, could result in the loss of substantial wood volume in a forest being managed for sawlogs, other solid wood products, pulp or chips. This same decay process would be almost completely without significance if the same forest was being managed for watershed protection. It could even be of value in a forest being managed, at least in part, to favor cavity nesting birds. Complete or partial defoliation of deciduous trees caused by foliar pathogens or insects, in a single year, could cause significant economic loss in a forest being managed as a family campground. A single year of defoliation in this same forest, if it was being managed for timber production, would be largely without importance. Root disease that predisposed scattered trees to windthrow would be inconsequential to the manager of a wilderness area, while of central concern to the manager of a recreational forest.

5. The Influence of Individual Stresses is Very Difficult to Partition from other Stresses Unless Catastrophic in Character or Unique in Appearance.

Ambiguous tree disease is the rule and unambiguous tree disease the exception. Unambiguous stresses with catastrophic impact include volcanic eruption such as Mt. St. Helens; the October 12, 1962 windstorm in the Pacific Northwest, hurricane Camille (August 17, 1969) in the Gulf States, the 1938 (September 21, 1938) hurricane in New England, and the 1988 wildland fires in Wyoming and California. Equally unambiguous and unique are wide-area defoliations by tussock moths, eastern and western spruce budworms, and gypsy moths; and wide-area mortality caused by chestnut blight disease, Dutch elm disease, fusiform rust or white pine blister rust. While these stresses are dramatic, they are not the norm of forest health dynamics. More typically tree morbidity and mortality is the result of complex interactions of multiple stress factors (Franklin et al., 1987a; Waring, 1987; Shigo, 1985) and the relative importance of specific stresses is ambiguous.

6. Dramatic, Wide Area, Multispecies Stresses are Unusual and Exceptional. Dramatic, Wide Area, Single Species Stresses (Most Dieback/Decline Phenomena) Are Common Historically, Are Potentially Recurrent, and Invariably Caused by Multiple Stress Interactions.

Wide area, multispecies mortality results from numerous natural (volcanic, wind, fire) stresses and anthropogenic (point source air pollutants, land draining/flooding, harvesting) stresses, but is typically restricted in area and limited in recurrence. Wide area single species dieback/decline phenomena, on the other hand, are constantly occurring somewhere in the temperate zone, are likely to be recurrent in a single place over time, and are invariably caused by multiple stresses. Dieback/decline stress is more characteristic of mature forests than immature forests. Generally declines result from the sequential influence of abiotic and biotic stress factors (Manion, 1981). Abiotic stress factors common to numerous declines include drought and low-and high-temperature stress. Biotic agents of particular importance include defoliating insects, borers and bark beetles, root infecting fungi, and canker inducing fungi. Typically declines are initiated by an abiotic stress, with mortality ultimately caused by a biotic stress factor.

7. Historically, Most Regional-Scale Forest Dieback/Decline Phenomena Have Had Suggested, but Unproven Etiologies.

A close examination of the literature related to North American dieback/decline phenomena of the past century reveals more speculation than evidence relative to cause and effect of decline phenomena. We have no formally accepted "rules of proof" for causality in dieback/decline diseases. The most conclusive documentations of causes associated with dieback/declines have been those concerned with specific stands of limited regions (Houston, 1974; Sinclair, 1964).

The study of cancer in humans provides a useful analogy with the study of dieback/decline phenomena in forests. Cancer is recognized as a multistage process and numerous cancers involve multiple stress interactions. Confident assessment of the specific factors involved in certain cancers, and the relative importance of these factors, are not possible because of the current limited capabilities of the science of toxicology (Wilkinson, 1987). Confident assessment of the specific factors involved in wide-area forest dieback/decline phenomena likewise exceed the current limited capabilities of forest science. Future advances in both human health and forest health sciences will facilitate our understanding of etiologies in both cases.

8. Throughout the Temperate Zone, Over the Last Century, the Most Consistent and Dominant Initiator and/or Contributor to Forest Dieback/Decline Has Been Drought.

Less than normal (adequate) water supply has been associated with almost all wide-area dieback/decline phenomena (Chapter 18). The quality of the evidence supporting this association is highly variable. Major environmental factors associated with restricted water supply to trees include: a) near exhaustion of available soil water, b) cold temperatures, c) anerobic conditions, d) impervious soils, e) high potential evaporation, and f) inadequate nutrition (Waring and Schlesinger, 1985). Forests subjected to chronic drought typically have trees with an abnormally large component of their carbon allocated to roots (Waring, 1983). Sustained drought halts photosynthesis, depletes carbohydrate reserves, attenuates defensive compound synthesis, and ultimately reduces canopy mass (Pook 1984, Waring, 1987). Evidence linking enhanced pathogen (Schoeneweiss 1986) and insect (Mattson and Haack, 1987) stress with drought-stressed plants has been presented.

9. A Role for Regional Scale Rir Pollutants in Forest Dieback/Decline Phenomena Is Indicated, a Role for Global-Scale Pollutants, if Global Warming Occurs, Is Certain.

Regional-scale air pollutants (most importantly, oxidants) exhibit direct and interactive effects on forest ecosystems in portions of the temperate zone. The integration of stresses imposed by regional-scale pollution is causing growth reductions in some forest species and is involved in dieback/decline phenomena in susceptible tree species at ambient levels in specific regions (Smith, 1984, 1985).

Global-scale air pollutants (carbon dioxide, nitrous oxide, methane, halocarbons) have the potential over the next century to cause global warming equivalent to that experienced by temperate vegetation over the last 18 millennia (Smith, 1987). Climatic warming will alter components of the hydrologic cycle. Those areas receiving less precipitation will become drier. Some areas will receive more precipitation, but enhanced drying due to intensification of evapotranspiration caused by warming may still induce water deficits in forest vegetation. Waggoner (1984) has estimated that the projected change in weather by the year 2000, caused by climatic warming, will cause moderate decreases of 1–12% in the yields of wheat, corn, and soybeans in the American grain belt due to increased dryness. While agriculturists may be able to adopt new crop varieties to drier climates, forests cannot be similarly manipulated. Increased drought stress over widespread forest areas would be expected to initiate new rounds of progressive tree health deterioration or dieback/decline phenomena. It is possible, as we approach doubling of atmospheric carbon dioxide, that climatic warming may be quite rapid. It is possible that a 3.5°C increase in global average could be achieved in approximately 60 years. Such a rapid change could cause large numbers of tree species to be "out-of-phase" with their climate (cf., Raup 1937).

10. High-Risk Forest-Health Situations for Managed Forests Have Generally Fallen, and Will Continue to Fall, into One of Four Classes.

Risk is a measure of the probability that an adverse effect will occur (Wilkinson, 1987). Our experience with forest tree management suggests four situations that have the potential to significantly increase forest health risk. The four classes of health risk for managed forest include: exotic agent risk, exotic tree risk, cultural risk, and climate risk (Table 20-1).

Exotic agent stress results when the activities of human beings transfer biotic agents of stress from one continent or country to another, or when human activities expose trees to novel physical or chemical stress factors. Introduced biotic disease agents may be released from population controls imposed by their natural predators, parasites and competitors. In the absence of coevolution, tree defense mechanisms may not have evolved. Classic examples of very damaging exotic biotic pests include the causal agents of chestnut blight, Dutch elm disease, white pine blister rust, and beech bark disease; and numerous insects including gypsy moth, European elm bark beetle, European pine shoot moth, beech scale, red pine scale, and hemlock scales and adelgid. Exotic abiotic stresses are numerous and varied, with air pollution and soil structural and chemical changes representing important examples.

Exotic tree stress results when the activities of human beings introduce forest trees to new environments. While some introductions have had outstanding extended term success records — Monterey pine in New Zealand (Smith, 1980) and Douglas fir and Sitka spruce in Europe (Hermann, 1987) — many others have had less distinguished records. Examples of the latter include: United States pines in Brazil (Hodges and May, 1972) and Chile (Ciesla, 1988), eucalyptus in southern California (Schriven et al., 1986), Monterey and maritime pines in Hawaii (Bega et al., 1978), Douglas fir (Read and Sprackling, 1976) and eastern white pine (Sprackling and Read, 1976) in Nebraska, Rocky Mountain juniper in

Table 20-1. Classes of Significant Health Risks for Managed Forests.

I.	Exotic Agent
	A. Biotic (pathogen, arthropod)
	B. Abiotic (chemical, physical)
II.	Exotic Tree
III.	Cultural
IV.	Climate

Kansas (Tisserat and Rossman, 1988), and sugar maple and flowering dogwood in urban environments. Tree introductions risk exposing trees to unfavorable environmental conditions and to biotic pests to which they have not evolved defensive strategies.

Cultural stress results when silvicultural, arborcultural or nursery practices (e.g., fertilization, pesticide use, monocultures, gene-pool reduction, spacing, thinning, fire, pruning, and site preparation) act to directly or indirectly exacerbate pest problems. Intensive forest management practices have certainly acted to reduce or eliminate the damage caused by a certain insect, pathogen, and abiotic stresses. Shorter rotations have eliminated numerous pest problems associated with mature and overmature forests. Soil fumigation has eliminated the short-term risk of seedling root damage caused by nematodes and damping-off fungi in nurseries. Proper species and planting site selection has minimized risks associated with wind, ice, and snow damage, and from numerous rust and root diseases. Fertilization has reduced the risk of nutrient stress and low vitality, which acts to predispose trees to certain biotic stress agents that favor trees with low vigor. Unfortunately, however, numerous forest tree stress and pest relationships have been intensified by contemporary forest cultural practices (Shea, 1970). The conversion of fusiform rust from a minor disease of southern pine to the most widespread, destructive, and economically significant disease of southern pine (Miller and Schmidt, 1987) is a classic example. In awareness of the obligate parasitic nature of the pathogen (*Cronartium quercuum* f. sp. *fusiforme*), it is not surprising that almost everything employed to enhance pine growth — site selection (Schmidt et al. 1988), site preparation (Miller, 1972), fertilization (Rowan, 1977, Blair and Cowling, 1974), planting on former agricultural land (Miller, 1972), — also enhances disease development. Root diseases have also been increased by cultural practices. Management practices imposed on second-growth stands of Douglas fir increase *Phellinus weirii* and the severity of laminated root rot (Thies, 1984). Thinning in southern pine increases damage caused by annosus root rot by facilitating the spread of *Heterobasidion annosum* (Anderson and Mistretta, 1982). Insect damage has frequently been increased by cultural activities. More frequent, extensive, and severe outbreaks of eastern spruce budworm have been associated with forest management practices in eastern Canada (Blais, 1983).

Climate stress results not only from the direct influence of climatic variables on tree health, but also from the regulation that climate exerts on biotic pest populations and on tree disease processes. The activities and population dynamics of many of the most serious microbial pathogens (Heptig, 1963) and insect species (Wallner, 1987) are importantly influenced by climatic parameters. Over time, climate, like a forest, is characterized by change not constancy. The natural regulation of climate is complex and is incompletely appreciated. Numerous hypotheses have been proposed to explain the forces responsible for natural climate variability. The most plausible of these include: variations of the solar constant, changes in solar activity, passage of the solar system through an interstellar gas-dust cloud, variation in the velocity of the earth's rotation, gigantic surges of the Antarctic ice sheet, change in the earth's orbital parameters, and alterations in the interactions between glaciers and oceans (Sergin 1980). In addition, alterations of local (Landsberg, 1970), regional (Sagan et al., 1979), and global (Lal et al., 1986) climate by the activities of human beings is not insignificant relative to natural regulators.

A careful examination of the future trends of these four classes of stress risk for managed forests is not encouraging. Without exception, all classes of risk appear to exhibit increasing trends. Exotic agent stress introductions may be stable if we assume that a thorough mixing of the world's biota has been achieved. The present intensity of transfer and travel would suggest that maybe it has not (Wilson, 1987; Gibbs and Brasier, 1988). Recent introductions of the longhorn borer (*Photacantha semipunctata*) onto eucalyptus in southern California (Scriven et al., 1986), pine wood nematode (*Bursaphelenchus xylophilus*) onto hard pines in China (Mamiya, 1987), and woolly adelgid (*Adeges tsugae*) onto Eastern hemlock in Connecticut (McClure, 1987) give us little support for complacency. In addition, numerous chemical and physical alterations of the environment, resulting from human activities, continue to expose trees to "exotic" stress forces (Tattar, 1978). This is particularly dramatic in the instance of air quality and forest health (this book). With regard to Class II, exotic tree risks, we see very active planting of exotic tree species, especially in developing countries. Aforestation and reforestation of hundreds of thousands of hectares with exotic conifers is active in South America, Africa, and China. Caribbean pine (*Pinus caribaea*), over the past several decades, has been moved from its restricted natural range in central America, the Bahamas, and Cuba and introduced into over 40 tropical and subtropical countries (Lamb, 1973). This species is attractive because of its growth rate and wood qualities, but not because of our confidence in its pest resistance nor long-term tolerance of environmental stresses in latitudinal and elevational sites widely different from its natural range. Cultural practice stress, Class III, is increasing in both developed and developing countries. In developed nations, e.g., the United States, the trend toward increasing dependence on wood and fiber from plantation forests rather than natural forests is clear. Southern pine management represents only a slightly modified "agricultural crop system" (Miller and Schmidt, 1987). As previously emphasized intensive forest culture clearly solves some pest problems but it also creates new ones. In developing countries, intensive culture is typically superimposed on exotic trees planted in closely spaced monocultures. The risks of stress disaster are very high. Agriculture and forestry have always been at the mercy of serious damage caused by extreme climatic events. Climate risks continue and are intensified by the increasing appreciation of human perturbation of the atmosphere and the prospect of global warming over the next century that is equivalent to that experienced over the last 18 thousand years.

These fundamentals of forest health reveal disturbing trends in forest health risk. Clearly this risk is complex in character and management. Both regional-scale and global-scale air contaminants are concluded to be of importance in contemporary forest health risk assessments. This book has attempted to review the evidence provided to support this latter conclusion. What are the pollutants of special importance and what are the forest systems at particular risk?

B. Relative Importance of Specific Air Pollutants to Forest Ecosystems

A very large number of pollutants have been identified in this volume to influence forest tree health. The most important of these are contaminants most widely distributed and those with the greatest potential to adversely impact forest systems in the near or long term. In order of importance, the following pollutants are judged to be of primary concern.

1. Photochemical Oxidants

Ozone is concluded to be the single most important air pollutant adversely impacting temperate zone forest tree health. A large number of Class II interactions have been demonstrated for ozone. This phytotoxic gas can adversely influence the metabolism of individual forest trees by impacting photosynthesis, respiration, carbon allocation, reproduction, and pest and pathogen interactions. Depending on species susceptibilities and forest composition, ozone can influence productivity, nutrient cycling, and inter-specific competition at the ecosystem level. Ozone is interactive with other air pollutants and stresses in forest systems. Ozone exposures are maximum during the day and during the growing season when trees are physiologically most active and are able to absorb this phytotoxic gas. Additional oxidants, for example, various adelhydes, peroxyacetylnitrates, and hydrogen peroxide are also common in the troposphere in numerous locations, but are not thought to as important as ozone for tree health.

During the growing season, a very large proportion of both eastern and western forest systems is subject to elevated ozone exposure (Figure 20-1). There is substantial evidence that photochemical oxidants, or their precursors, can be transported over considerable distances from their areas of origin. Montane forest ecosystems may be subject to elevated and sustained ozone exposures due to a lack of nitrogen oxide scavengers.

The hydrocarbons released from forest foliage may be important in the formation of ozone in rural atmospheres. The soil and vegetation of forest ecosystems may also remove substantial quantities of ozone from polluted environments. These Class I interactions, however, are judged to be secondary in importance to the extemely large number of Class II interactions that have been demonstrated for ozone.

While photochemical oxidants have not, and will not, cause dramatic localized forest destruction, as in the case of sulfur dioxide and fluorides, the large forest areas subject to Class II interactions with this gas make this pollutant of enormous importance to those concerned with forest ecosystem health. Scientific consensus stresses the primary importance of ozone for agricultural and forest ecosystem stress (Abelson, 1987).

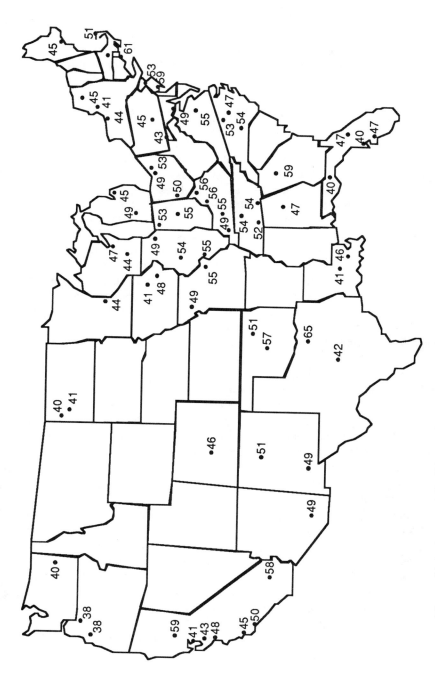

Figure 20-1. Average daily 7 hr maximum ozone concentration (ppb) in rural areas during the growing season. Source: National Acid Precipitation Assessment Program (1987).

In the United States, the trend of urban ozone pollution has been relatively stable or very slightly decreased from 1977 through 1986. With the exception of Minneapolis, however, no major urban area in the country is in attainment of the National Ambient Air Quality Standard (Chapter 2) for ozone (U.S.E.P.A., 1988a). The precursors of ozone synthesis, nitrogen oxides and volatile organic compounds, are multiple source and difficult and expensive to manage. As a result, dramatic near-term reductions in tropospheric ozone concentrations are not anticipated.

2. Carbon Oxides and Assorted Trace Gas Pollutants

Carbon oxides and assorted trace gas pollutants are concluded to be the most important group of air pollutants with potential to adversely impact forest ecosystems due to their potential to alter radiation fluxes to and from the earth. Carbon dioxide, carbon monoxide, methane, nitrous oxide, and at least five halocarbons (Chapter 19) exhibit increasing concentration trends in the troposphere, where they function to restrict infrared radiation loss from the earth. This restriction has the clear potential to increase the surface temperature of the earth. In association with warming, near term climate change may also involve perturbations to the hydrologic cycle and intensification of severe storm and wind events. Any or all of these changes in regional climate regimes have enormous potential to directly and indirectly change forest ecosystems. Initial change could include significant increases in the area and number of tree species exhibiting dieback/decline symptomology. Eventual change could encompass alterations of competitive interaction among forest species, alterations of insect and pathogen interactions, and eventually migrations and alterations of species composition in portions of the temperate and boreal forest zones.

In addition to the combustion and industrial sources of carbon oxides and other trace gases, forest manipulations by humans may also be responsible for increased atmospheric loading of radioactively active trace gases. Forest harvesting and burning in the tropical latitudes of Brazil, Africa, Indonesia and the Philippines may reduce sink function and may be important sources of carbon dioxide. Soil disturbance resulting from human activities may result in increased source strength of carbon monoxide, nitrous oxide, and methane.

It is not likely that the increasing concentration trends of these trace gas components of the atmosphere will be stopped or reversed in the near future. The direct effects of these gases on radiative forcing is clear. The climatic feedbacks associated with changes in radiation flux, however, are unclear. As a result, predictions of the magnitude and rate of global climate change are uncertain. General circulation models, while in general agreement regarding the effects of climate alteration on a global basis, cannot be regarded as reliable indicators of regional-scale climate changes resulting from global warming. As a result, specific impacts on forests of particular regions cannot be confidently assessed at this time.

In addition to climate modification, halocarbons eventually reaching the stratosphere have the additional potential of stratospheric ozone destruction with the associated increase in ultraviolet-B transmission through the atmosphere to earth. This increase, due to ultraviolet-B biological molecule action spectra, has the potential to directly and indirectly influence the health of trees as well as other plants.

3. Nitrogen Oxides and Secondary Products

Atmospheric pollutants containing nitrogen are ranked third in significance to forest ecosystems due to their important roles in secondary pollutant synthesis, acidification of precipitation, and difficulty of source management.

The deposition of gas-phase nitrogen compounds to forest systems may be relatively high (some locations) in the form of nitric acid, but is generally presumed low in the case of nitric oxide and nitrogen dioxide. The deposition of these gases (or oxidized products) to forest systems may constitute a Class I fertilization relationship. Nitrogen oxides are not judged to be directly involved in important Class II interactions.

Deposition of nitrogen in the form of nitrate in precipitation represents an important nitrogen-pollutant input to the northern and central hardwood forest (Figure 20–2). While event variation is very high, cloud water concentrations of nitrate exceed rainwater concentrations. Montane forests of eastern North America are subject to significant cloud water deposition and, therefore, are subject to especially high nitrate deposition. Even at high deposition rates, however, evidence provided has not clearly demonstrated an adverse, near-term impact on forest systems. The influence of associated hydrogen ion input to the forest will be discussed in the next section.

The most important role of nitrogen pollutants emitted to the atmosphere involves the role of nitrogen oxides in oxidant synthesis.

Data for the period from 1977 to 1986 suggests that annual average ambient levels of nitrogen dioxide have decreased slightly (approximately 14%) (U.S.E.P.A., 1988a). Major increases in nitrogen oxide emissions are possible over the next decade primarily due to the utility sector shift to coal. Coal-fired boilers release three to six times as much nitrogen oxides as do oil- and gas-fueled boilers. In addition, incinerators (municipal solid waste, tire, wood) run at high temperatures to minimize the release of toxics (e.g., chlorinated dioxins) and therefore release more nitrogen oxides than lower temperature combustion processes.

4. Sulfur Oxides and Secondary Products

Historically Class III impacts of sulfur dioxide deposition on forest systems have been dramatic. In portions of the forested temperate zone, e.g,. eastern Europe, these impacts remain very significant. In other portions, e.g., western Europe and North America, Class III impacts of sulfur dioxide have been largely elimi-

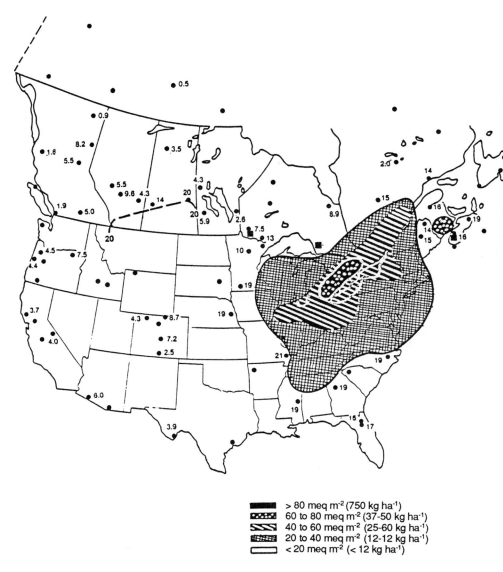

> 80 meq m^{-2} (750 kg ha^{-1})
60 to 80 meq m^{-2} (37-50 kg ha^{-1})
40 to 60 meq m^{-2} (25-60 kg ha^{-1})
20 to 40 meq m^{-2} (12-12 kg ha^{-1})
< 20 meq m^{-2} (< 12 kg ha^{-1})

Figure 20-2. Deposition of nitrate ions (millequivalents m^{-2}) compiled from various Canadian and American monitoring networks in 1980. One meq m^{-2} equals 0.62 kg ha^{-1}.
Source: Work Group 1 (1983).

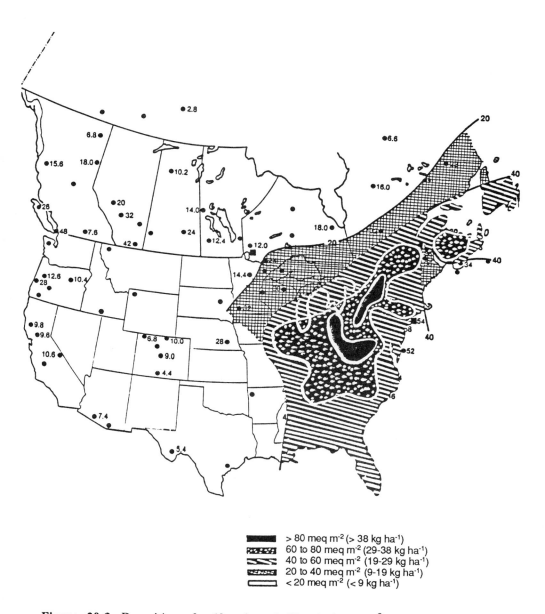

Figure 20-3. Deposition of sulfate ions (millequivalents m⁻²) compiled from various Canadian and America monitoring networks in 1980. One meg m⁻² equals 0.48 kg ha⁻¹.
Source: Work Group 1 (1983).

nated. In addition, approximately 35% of the sulfur dioxide produced in the United States leaves the continent, whereas almost all of the nitrogen oxides are converted to secondary pollutants and deposited in the United States (Abelson, 1987).

Unfortunately, however, atmospheric sulfur pollutants resulting from the transformation of sulfur dioxide to sulfuric acid and amonium sulfate represent contaminants of uncertain, but potential, long-term significance to forest ecosystems. In part due to this uncertainty and in part due to the restriction of deposition to eastern North American forest systems (Figure 20-3), and in part due to the high probability of source reduction, sulfur pollutants are ranked fourth with regard to forest impact. The efficacy of sulfur source reduction has in fact been detected at remote forest monitors. During the 20-year period, 1964–1983, decreases in the concentration and deposition of sulfate in precipitation at the Hubbard Brook Experimental Forest, White Mountain National Forest, New Hampshire, have been attributed to decreases in sulfur dioxide emissions from upwind sources (National research Council, 1983; Gschwandtner et. al., 1985). Significant decreasing trends in sulfate concentrations have also been recorded at 5 of 8 National Atmospheric Deposition Program monitoring sites in the Northeast and Midwest since their initiation in 1978 (Schertz and Hirsch, 1986). Hedin et al. (1987) have examined the relationship between decreased sulfate input to the Hubbard Brook Experimental Forest and precipitation acidity. Because precipitation acidity at Hubbard Brook is regulated by other strong acid anions (i.e., nitrate), as well as sulfate and also basic materials that act to neutralize strong acids (Driscoll et al. 1989), a stoichiometric correlation between extended term trends of sulfate and hydrogen ion input to the forest were not found. Nevertheless, the decreased input of sulfate at Hubbard Brook has resulted in large decreases in both the concentration and deposition of hydrogen ions to this forest (Hedin et al., 1987).

The wet deposition of sulfuric acid to forests increases the input of hydrogen ions to forest systems. Most North American forest ecosystems receive less than 1 kg ha^{-1} annually (Figure 20–4). At this deposition rate, short-term adverse impacts have not been detected. Greater uncertainty, however, is associated with extended-term influences. Over time, hydrogen ion input may influence the aluminum dynamics in forest soils so that this ion assumes prominence as a competitor for nutrient uptake in trees or is directly toxic to tree roots. In soils with low sulfate adsorption potential, nutrient cation loss in association with sulfate anions may degrade site quality over an extended time. Fortunately, forest soils at special risk to most of these influences do not represent the full region subject to elevated sulfuric acid deposition (Figure 20-5).

In forest soils deficient in sulfur, the input of sulfur pollutants could represent a Class I fertilization effect. Sulfur deficient soils, however, are of limited distribution.

During the period 1977–1986 in the United States, ambient concentrations of sulfur dioxide were reduced 37%. This reduction resulted largely from installation of sulfur dioxide emission collectors, use of low sulfur combustion strate-

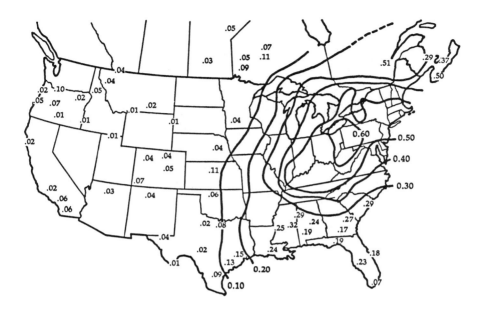

Figure 20-4. Annual deposition of hydrogen ion (H+) in precipitation during 1985. Source: National Acid Precipitation Assessment Program (1987).

gies, or the use of low sulfur fossil fuels in large combustors. It is highly likely that additional sulfur reduction strategies will be implemented in the near term. This is particularly important as utility and industrial sectors make a major shift in fuel source from oil to coal.

5. Particulates

Particulates represent an extremely heterogeneous group of atmospheric pollutants and generalizations are difficult. It is clear, however, that forest ecosystems exhibit important source and sink relationships with particulate pollutants. In 1987, the U.S. Environmental Protection Agency published new standards based on particulate matter smaller than 10 microns (PM10). Small particles have special significance for human health because of increased inhalability and retention, but also for forest ecosystems due to long distance transport, preferential contamination with toxic metals, and efficiency of vegetative capture via impaction. In addition, forest burning may be an extremely important source of PM10 particles. Approximately 149,000 ha (376,000 acres) are burned annually in Oregon, releasing approximately 97,000 tons of PM10. Forest burning rep-

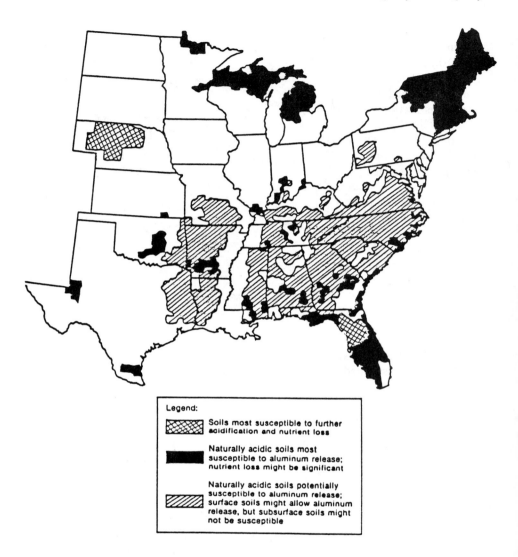

Figure 20-5. Soil sensitivity to acid deposition in the eastern United States. Source: Office of Technology Assessment (1984).

resents 84% of these emissions, with agricultural field burning providing the remainder (U.S.E.P.A., 1988a).

The pollen and fungal spores produced in forested areas have enormous medical and economic significance to a very large segment of the population of the temperate zone. The large surface-to-volume ratio that forest trees present to the atmosphere makes this vegetative form a relatively efficient particle interceptor.

Increased appreciation of the capability that forested regions have for air filtration in urban and industrial areas is needed. This is especially true with regard to PM10 particles.

With reference to the Class II influence on forest ecosystems, nitrate-, sulfate-, and trace-metal-containing particles are of particular interest. As emphasized in previous sections, evidence does not support dramatic, short-term impacts of these pollutants on forests. The ability of heavy metal particles to interfere with nutrient cycling in the extended term, however, is an important research topic.

The importance of particulate pollutant dynamics in global climate change is recognized as potentially important but of uncertain direction!

The U.S. Environmental Protection Agency (1988a) has suggested a 23% decrease in ambient particulate levels from 1977 to 1986. It is probable, however, that much of this reduction has been associated with particulates larger than 10 microns in size. Particulate collection strategies are more efficient for larger particles. PM10 particles may be reduced, unchanged, or increased over this same period. In 1985, 290 United States counties were nonattainment for the National Ambient Air Quality particulate standard (Figure 20-6). In portions of numerous western states, this nonattainment is in part due to natural wind-blown dust.

Over the next decade significant increases in fine particulate emissions may occur. The most important source of new contamination will be increased coal combustion. If efficient stack emission control devices are employed, and if chemical and physical coal cleaning procedures are developed and implemented, the increases associated with coal utilization can be moderated. It is important to realize, however, that volatile trace metals and those metals associated with particles with diameters less than 1 micron are not effectively removed by electrostatic precipitators. As previously indicated, increased coal and other combustion processes may release increased quantities of sulfur dioxide and especially nitrogen oxides. The secondary formation of sulfate and nitrate aerosols in industrial and power-generating facility plumes will further contribute to the atmospheric particulate load.

6. Hydrocarbons

Aside from pollen, hydrocarbons are the most important air pollutants generated by forest ecosystems. Increased appreciation of the role of these hydrocarbons in photochemical oxidant formation is needed. Current evidence indicates an uncertain role for forest hydrocarbons in atmospheric oxidant chemistry and in associated adverse impact on human and forest vegetation health.

Anthropogenic hydrocarbons, on the other hand, clearly function as primary precursors in the generation of oxidant pollutants in major urban and rural areas with roles of primary significance to surrounding forest ecosystems. As indicated in the ozone and nitrogen oxide discussions, these oxidants have been implicated in a variety of Class II and III relationships.

In the United States, hydrocarbon emissions have been decreased by the use of oxidation catalysts that burn hydrocarbons emitted from motor vehicle engines.

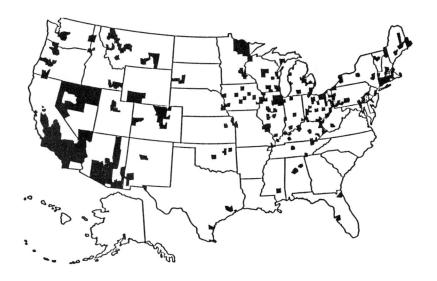

Figure 20-6. Counties in total or partial non attainment for particulates in 1985.
Source: U.S. Environmental Protection Agency (1985).

Despite continued increases in automotive use, there has been a net reduction in
hydrocarbon release. Vehicle turnover should continue to reduce hydrocarbon
emissions. Decreased uncontrolled burning of solid waste has also contributed to
reduced hydrocarbon emission. This downward trend is expected to continue.
Hydrocarbon emissions from industrial sources may increase if facilities are expanded or if leaks and spills occur that result in major evaporative losses.

C. Relative Importance of Air Pollutants
for Specific Forest Regions: United States
Summary of Risk

Temperate forests have historically been subjected to major change resulting
from the activities of human beings. For centuries the major influence was
gross destruction for agricultural, fuel, or other wood-product purposes. In the
present century the reduced need for agricultural land and increased forest management have reduced the adverse impact on forests in temperate latitudes.
Human activities of primary contemporary importance to forest structure and
function have included the introduction of exotic arthropod and microbial tree
pests into forest systems lacking evolutionary exposure to these destructive
agents, enhancement of native and natural stresses by cultural practices, and the
creation of artificial forests of one or a few commercially important species. In

the past three decades we have accumulated evidence to indicate that an additional major anthropogenic modifier of temperate forest ecosystem development is air pollution.

The interactions between air pollution and forest systems have been described in this book as falling into one of three classes. Due to the enormous complexities of atmospheric chemistry and transport, and forest ecosystem structure and function, these classes of interaction may not occur as discreet entities in time or space. It is perhaps more accurate to view the interactions of forests and air pollutants as a continuum of responses that vary greatly depending on the specific forest and the specific air contaminant of interest. The value of the three class perspective, however, is that it facilitates the appreciation of a comprehensive relationship and permits generalizations concerning ecosystem response as well as individual species reaction.

Class II interactions are judged to be the most significant as they are capable of inducing major alterations in the patterns and processes of forest ecosystems, are frequently subtle and insidious in character, and are widespread in occurrence. Forest ecosystems are able to act as sources of air contaminants and as sinks for air pollutants. Class I interactions follow Class II relationships in importance. Forests do have important roles in global major element cycles. Gross forest destruction is currently concentrated in the tropical zones, and it is concluded that destruction of tropical forests will greatly exceed the importance of changes in temperate forests relative to global nutrient cycling and climate dynamics over the next decade. Temperate forests are clearly efficient producers of a variety of air contaminants, most notably pollen, particulate and volatile hydrocarbons, and combustion products associated with forest fires and wood burning. These can be of extreme local and regional importance. Forests have generated these contaminants for a long period, however, and it is presumed that evolutionary adjustments in the biota have been made. The capability of forests to serve as sinks or repositories for air pollutants is very significant. The improvement of local air quality by large forests in urban or industrial environments is important. The regional and continental importance of forest soils as sinks for gaseous pollutants and trace metals is even more important. The Class I concept infers no adverse impact on forest ecosystems, however, and it is judged that the threshold of transition between Class I and Class II interactions has been crossed for several pollutants in numerous areas throughout the temperate zone. Class III interactions involving severe impact of air pollutants on forests are dramatic, but are judged relatively unimportant because of their extremely localized nature and decreased occurrence.

Risk assessment involves determination of the probability of the occurrence of an adverse impact. With specific regard to the influence of atmospheric deposition on a particular forest ecosystem, risk assessment involves a determination of air pollution exposure (section B) and a forest vulnerability (hazard) evaluation. The latter evaluation encompasses multiple factors, including susceptibility of woody species to air pollution impact, species diversity, other stresses concurrently impacting the system, management objectives, and others.

Regional and local differences in air quality are very great due to differences in source strength, climate, and topography. These differences, combined with variable forest ecosystem vulnerabilities, result in the differential significance of specific pollutants and the differential importance of Class I, II, and III interactions in various areas of the temperate zone. A very brief summary of the air pollution impact on the forests of the coterminous United States is presented by examining the influence of major contaminants in five major eastern and western forest systems.

1. Eastern United States

The major forest types of the eastern United States are dominated by deciduous species and are conveniently divided into the northern, central, and southern hardwood regions (Figure 20-7). Conifers have considerable significance in portions of these regions, with spruce and balsam fir being especially important in northern New England; white, red, and jack pine especially important in the Lake States; and longleaf, loblolly, and slash pine especially important in the Southeast.

a. Boreal Forest

The true boreal forest is not widespread in the coterminous United States. The northern portions of the northern hardwood forest, for example, in portions of Maine, have an aspect of the boreal forest due to the large amount of white spruce, balsam fir, and coniferous swamps and bogs. This forest does, of course, occupy a position of prominence in Canada, and a similar boreal forest covers large areas of Eurasia. The climate of this forest is extreme and the growing season short. The boreal forest has the lowest species diversity and lowest net primary productivity of any major North American forest type.

The spruce-fir boreal forest association is distributed, at high elevations, along the entire Appalachian Mountain chain in the eastern United States. This forest is characterized by short growing seasons, harsh climatic conditions, poor soil quality, and high pollutant loadings via cloud and dry deposition. As detailed in Chapter 18, atmospheric pollution is concluded to play a role in red spruce decline in portions of this montane forest. Pollutants of most importance are ozone (growth impact) and acid deposition (aluminum availability). The high-elevation boreal forest is concluded to be at high risk for long-term adverse impact from atmospheric deposition for a variety of reasons. Pollutant exposure is high and additional stresses are numerous. The impact of major climate changes will be especially dramatic in this forest, especially if droughts become more abundant or more severe.

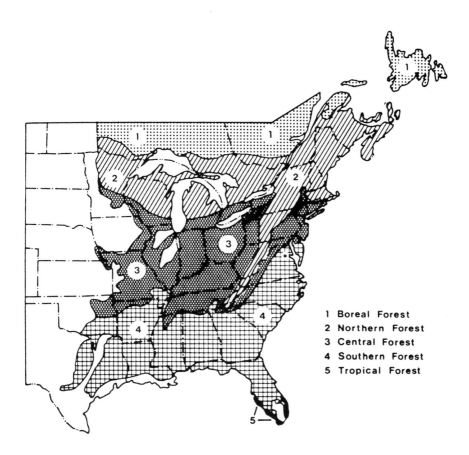

Figure 20-7. Major forest ecosystems of the eastern United States.
Source: Society of American Foresters (1954).

b. Northern Forest

The northern hardwood forest is dominated by yellow birch, American beech, and sugar maple. Coniferous species of great importance in selected areas include eastern hemlock and white, red, and jack pines. The maple-beech-birch forest covers over 14 million ha (35 million acres) in the New England, Middle Atlantic, and Lake States regions (Figure 20-7). Sugar maple is the most widely distributed of the three hardwood species. In recent history major, wide-area dieback/decline phenomena have characterized northern hardwoods. Yellow birch dieback/decline occurred throughout New England, New York, and southeastern Canada from 1940 to 1965 and was due in part to soil warming and drought. Beech bark disease, caused interactively by beech scale and the canker inducing *Nectria* fungus, is currently causing widespread beech morbidity and mortality. Over the past four decades, sugar maple has undergone at least eight distinguishable regional dieback/declines in the United States and southeastern Canada. Stresses concluded to be important in these maple declines have included insect defoliation, late spring frost, and drought (Hepting, 1971, Houston, 1981).

The northern hardwood forest is also subject to significant air pollutant exposure. Ozone exposure is particularly high during the growing season in numerous locations in this forest, especially in selected southern portions of the distribution. Midwestern sources of sulfur dioxide contribute sulfates and acid deposition to extensive areas of the northern hardwood forest. The Ohio River basin alone, with scores of large coal-burning power plants and numerous industrial boilers, generates polluted air masses that are regularly transported, depending on the wind, northeastward to the northern hardwood forests of Pennsylvania, New York, and New England and northwestward to the northern hardwood forests of Wisconsin and Minnesota, or due north into Ontario (Gavin et al., 1978).

Class II interactions with ozone may be quite widespread in the northern forest. Ozone and acid deposition are surely not the sole cause of all contemporary sugar maple dieback/decline situations. It is probable, however, that selected maple declines have important air contaminant involvement in specific locations. The pollutant with greatest stress potential is concluded to be ozone. The role of long-term deposition of strong mineral acids and heavy metals remains uncertain. If global climate warming continues, northward expansion of the northern hardwood forest is anticipated, along with increased dieback/decline phenomena in southern portions of its range.

c. Central Forest

The central forest is the largest of the eastern forests and covers approximately 46.5 million ha (115 million acres). It is dominated by various species of oak and hickory, with yellow poplar an important associate in various portions of the distribution. The forest is characterized by a favorable climate, extended growing season, and relatively high species diversity and productivity. The forest is extremely variable, however, and specific composition varies depending on local site conditions.

Contemporary stresses influencing the central forest are numerous. Insect defoliation is of chronic significance in the oak-hickory forest, with gypsy moth the most significant contemporary pest. During the past several decades, oak dieback/decline has increased in numerous, especially mountainous, sections of the forest. As in the case of sugar maple decline, insect defoliation, late spring frosts, fungal infection, and drought are important stresses in specific areas of oak decline.

Large portions of the central forest are characterized by high air pollution exposure. Central and eastern regions are subject to periodic high ozone events. In the case of selected central forest species, it is concluded that Class II interactions with ozone may be common. Specific involvement of oxidants in oak decline is uncertain. Short-term impacts from acid deposition are not demonstrated and long-term impacts are uncertain.

Oak decline will increase with significant climate change. Warming with drying will increase dieback/decline phenomena, especially in red, white, and scarlet oaks. Warming with unchanged or increased precipitation will favor more mesic communities to the disadvantage of oak.

d. Southern Forest

This extensive area is conveniently divided into two regions: an oak-pine area in the northern portion and a longleaf-loblolly-slash pine region in the southern (coastal) portion. The northern region is dominated by a variety of oaks and hickories, and conifers are restricted to the poorer soils and drier sites. While long-leaf pine forests dominate the landscape of the southern portion, the region is a mosaic of vegetative types with variable mixtures of pine and deciduous species. The southern portion is also characterized by extensive establishment of slash and loblolly plantations for commercial production. Major contemporary stress factors of potential importance in southern forests include fire, wind, southern pine beetle, and fusiform rust disease in plantations.

The air pollution risk to southern pine resources is concluded to be high for several reasons. Ozone exposure is high throughout much of the 84 million ha (208 million acre) region. Recent analysis of U.S. Forest Service Forest Inventory and Assessment data revealed a 16–20% reduction of radial growth in some natural pine stands in Georgia, South Carolina, and North Carolina. While all or most of this decrease may be due to stand aging, increased stand density, increased hardwood competition, loss of old-field sites, or climate change, it is additionally possible that oxidant pollution may also be involved. Selected soil types throughout the region are at an elevated risk with regard to long-term acid deposition influence. If climate warming occurs, widespread dieback/decline is possible in the southern forest due to drying. Higher temperatures and drier soils may make it impossible for most species to regenerate naturally. Natural forest stands could be converted to grassland. Commercial forests could flourish but only with increased management intensity, possibly including irrigation. Narrowness of the commercial forest gene-pool may increase the risk over extended periods of time.

e. Tropical Forest

This forest is restricted to the margins of the southern Florida peninsula and to the keys of south Florida. Certain Florida counties on the eastern side of the peninsula are characterized by high levels of oxidant and particulate pollutants. The significance of these contaminants to the mahogany, fig, and mangrove species of the tropical forest is not clear.

2. Western United States

The major forest types of the western United States are dominated by coniferous species and are conveniently divided into five major forest systems (Figure 20-8). While air pollutant exposure levels are generally lower in the west relative to the east (Figures 20-1 through 20-4), extreme variations in meteorology and topography, along with important point sources of emissions make atmospheric deposition highly variable. In addition, western conifers exhibit highly variable vulnerability to stress caused by the most important regional-scale pollutant, ozone (Table 20–2).

a. Western Montane Forest

The extensive western montane forest is largely characterized by ponderosa, lodgepole, and western white pines along with western larch.

Numerous Class III relationships have existed or exist in this forest in the immediate vicinity of various metal smelters and power generating facilities. Sulfur dioxide and fluoride emissions are primarily involved in these situations. Elevated sulfate inputs do occur in forests in the eastern portion of the distribution. In southern California this forest type is subject to excessive Class II interactions with oxidants (Chapter 16). Additional areas of Class I and II interactions are presumed to be associated with other major industrial and urban areas scattered throughout the range of this forest. Ozone is the most important regional scale pollutant. Increased drying associated with global-scale climate modification could have a severe impact on the western montane forest.

Table 20-2. Susceptibility of Western Conifers to Ozone Stress.

Very sensitive	Sensitive	Intermediate	Tolerant
Jeffrey pine/ Coulter pine hybrids Western white pine	Coulter pine Jeffrey pine Ponderosa pine Knobcone pine	White fir Douglas-fir California red fir	California cedar Giant sequoia Sugar pine

Source: Miller et al. (1983).

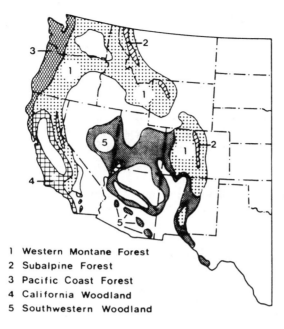

1 Western Montane Forest
2 Subalpine Forest
3 Pacific Coast Forest
4 California Woodland
5 Southwestern Woodland

Figure 20-8. Major forest ecosystems of the western United States. Source: Miller and McBride (1975).

b. Subalpine Forest

The subalpine forest consists largely of spruce-fir communities. Engelmann spruce and subalpine fir are the dominants throughout most of the range of this limited system. The research dealing with air pollution impact on this forest is limited, and the potential for Class I and II relationships is not clear. The presumption is one of minimal impact.

c. Pacific Coast Forest

The most common species in this forest is Douglas fir, which is seral and forms pure, even-aged stands following fire disturbance. Dominant climax species include western hemlock, Sitka spruce, and western red cedar.

This forest is subject to major air mass migrations from the Pacific Ocean and thereby generally has relatively clean air. Specific point sources and/or urban areas, however, can significantly increase local exposure to sulfur dioxide, ozone, and acid deposition. In Washington, Everett, Seattle and Tacoma are significant sources of ozone precursors. Lowland Douglas-fir forests in the Puget Sound area may be subject to elevated oxidant exposures and exhibit subtle Class II responses. Precipitation in the Puget Sound area is slightly acidic (range 4.5–4.9, winter 1986). Douglas fir growing on the Olympic Peninsula are exposed to low air pollution loads. Rainfall averages pH 5.2 in the Olympic National Park. Urban fog, for example, that associated with Seattle, may , on the other hand, be very acidic.

d. California Woodland

Dominant coniferous species of the California woodland include sugar and ponderosa pines. This forest is subject to extensive urban and suburban development, and certain areas are characterized by high doses of oxidant and particulate pollution. Oxidant levels are greatest in the central and southern portions of the Sierra Nevada, near cities such as Los Angeles, Bakersfield, and Fresno. Class I and II interactions are important in portions of this forest.

e. Southwestern Woodland

This relatively large forest system is generally characterized by relatively low-density pinyon pine and juniper species. Isolated Class III interactions occur in association with industrial sites. Research on potential Class I and II interactions has been minimal but is justified, particularly in regard to the elevated sulfur dioxide and particulate loads that have characterized southern Arizona atmospheres. Increased coal combustion in the Southwestern woodland would increase the need for additional research.

D. Future Needs

Adequate understanding, regulation, and management of the stresses imposed on forest ecosystems by air pollution will require increased efforts in monitoring, research, and prediction. Class I and II interactions imposed on forest systems by both regional- and global-scale pollutants will be effectively addressed only through an integrated effort involving all three of these elements.

1. Monitoring

Monitoring strategies are essential for determining changes or trends in ecological systems and in environmental parameters influencing these systems. These strategies involve repeated measurements over time of selected physical, chemical, or biological variables. Monitoring information can be of critical importance to hypothesis formulation, hypothesis testing, ecological prediction and ecological system risk assessment.

Initial or routine carbon dioxide monitoring revealed an increase in the concentration of this gas and has further established its rate of global atmospheric increase. Documentation of acidified precipitation in North America was revealed in the routine precipitation chemistry measurements initiated at the Hubbard Brook Experimental Forest, White Mountain National Forest, New

Hampshire, in the early 1960s. Earlier documentation in Europe had been provided by similar routine monitoring of precipitation in Scandinavia.

As important as monitoring efforts are, they have historically been characterized by numerous limitations and deficiencies. Some of the most important include the following:

1. Monitoring programs are frequently deficient in scientific need, clear definition, and long-term justification.
2. Monitoring efforts commonly do not recognize relevant temporal and spatial dimensions.
3. Data sets from uncorrelated or poorly administered monitoring programs may not be compatible nor comparable.
4. Monitoring programs are inherently costly, and maintaining continuity of the effort in mission, motivation, manpower, and money is a sizable challenge. Feasibility, utility, and scientific validity must be carefully evaluated, along with expense and manpower requirements, to ensure a successful strategy.
5. Monitoring efforts directed at documenting change in biological systems frequently cannot distinguish changes induced by natural forces from changes induced by anthropogenic forces.
6. There is no centralized collection, storage, maintenance, digestion, or summarization of monitoring information.

Monitoring efforts, addressing these historic deficiencies, must be continued, initiated, and sustained.

a. Monitoring Sites

Ideally monitoring sites should be located in forest locations where extended-term records of one or more important parameters already exist. This is particularly true in protected (long-term security of monitoring capability) and discipline integrated research locations such as the National Science Foundation Long-Term Ecological Research sites (Brennerman and Blinn, 1987; Callahan, 1984) (Figure 20–9), Department of Energy National Environmental Research Parks (Trabalka 1985), Man and the Biosphere Program Biosphere Reserves (Man and the Biosphere, 1984), or in similar locations (U.S.E.P.A. 1988b, c, d).

Where it is appropriate or necessary to establish a new forest site for long-term monitoring, a large amount of relevant site-specific information must be detailed. The minimal perspective required was suggested by the Eastern Hardwoods Research Cooperative, National Acid Precipitation Assessment Program (U.S. Forest Service 1985) (Table 20–3).

b. Meteorological and Atmospheric Deposition Monitoring

Abiotic climate stresses represent forces of *extreme* significance to forest health (Table 20-4). In addition, the exposure of forest trees to air contaminants is reg-

Figure 20-9. Location of the 15 Long-Term Ecological Research Sites supported by the National Science Foundation.
Source: Brenneman et al. (1987b).

Table 20-3. Site Variables Important for Initial Site Characterization.

PHYSIOGRAPHY	SOILS
slope	profile description
aspect	% large fragments, fines
elevation	thickness
latitude/longitude	texture analysis
pertinent microrelief	% base saturation
	total N
	available P
	extractable cations
PARENT MATERIAL	% cations
type	sulfur
depth to bedrock	
composition	TREES
	species
	DBH
VEGETATION	crown class (D, C, I, S)
forest type/association	tree condition
basal area	crown density
stand	crown ratio
species	mortality
stems area^{-1}	symptoms and signs
site index	
species composition	HISTORICAL DATA
age of canopy dominants	all information
age structure by stratum	qualitative or quantitative
estimate of crown closure	historical vegetation type,
(by spherical densitometry)	Kuchler's potential type
diameter distribution	Bailey's ecoregion
stocking	disturbance history
understory	climatic records, past 10 years

Table 20-4. Climatic/Meteorologic Stress Factors of Significance to Temperate Forest Health.

temperature extremes
humidity extremes
high winds
hail, ice, snow
lightning

ulated by meteorological variables. As a result, comprehensive meteorologic data must be collected at forest monitoring sites (cf., Karl and Quayle, 1988). Meteorologic parameters measured by the Mountain Cloud Chemistry Project, National Acid Precipitation Assessment Program are representative of the comprehensive information required (Table 20-5).

Measurement of numerous air contaminants are also essential for comprehensive assessment. Again the pollutants measured in dry (Table 20-6) and wet (Table 20-7) deposition by the Mountain Cloud Chemistry Project represent core

Table 20-5. Meteorological Measurements Recorded Hourly from May through October by the Mountain Cloud Chemistry Project, National Acid Precipitation Assessment Program.

Wind speed	Wind speed maximum
Wind direction	Wind speed minimum
Ambient temperature	Wind speed standard deviation
Relative humidity	Wind vector magnitude
Solar radiation	Wind direction standard deviation
Barometric pressure	Cloud detector
Precipitation total	

Table 20-6. Dry Deposition Monitored by the Mountain Cloud Chemistry Project, National Acid Precipitation Assessment Program.

Gases (hourly)	Gases (weekly)	Particulates (weekly)
ozone	ammonia	ammonium
nitric oxide	nitric acid	sulfate
nitrogen dioxide	nitrous acid	nitrate
nitrogen oxides		
sulfur dioxide		
hydrogen peroxide		

Table 20-7. Wet Deposition Monitored by the Mountain Cloud Chemistry Project, National Acid Precipitation Assessment Program.

pH	nitrate ion
liquid water content	sulfate ion
potassium ion	sodium ion
calcium ion	ammonium ion
magnesium ion	hydrogen peroxide (aqueous)
chloride ion	conductivity

components of a complete program. Wet deposition measurements should monitor rain, snow, fog, and/or cloud water as appropriate. In addition, routine determination of selected trace metals, including aluminum, lead, cadmium, zinc, and vanadium, should be included in both dry and wet deposition analyses. Screens for organic acids would be desirable additions to wet deposition measurements.

In addition to dry and wet inputs to the forest from the atmosphere, determination of additional chemical fluxes within the forest are highly desirable. These additional fluxes importantly include throughfall, stemflow, litterfall, forest floor leaching, and accumulation by the dominant vegetation. Specific analyses to determine these flux rates have been included in the Electric Power Research Institute's, Integrated Forest Study on Effects of Atmospheric Deposition

Table 20-8. Chemical Analyses Conducted to Establish Forest Ecosystem Flux Rates (kg ha^{-1} yr^{-1}) for the Integrated Forest Study on Effects of Atmospheric Deposition Supported by the Electric Power Research Institute.

Substrate	Analyses
Litterfall	Biomass, N, P, K, Ca, Mg, S, SO_4^{2-}
Throughfall	
Bulk, wet only	H^+, SO_4^{2-}, NO_3^-, total N, NH_4^+ Ca_2^+, K^+, Mg^{2+}, Na^+, Al_3^+, Cl^-, HCO_3^-, orthophosphate
Stemflow	H^+, SO_4^{2-}, NO_3^-, total N, NH_4^+ Ca_2^+, K^+, Mg^{2+}, Na^+, Al_3^+, Cl^-, HCO_3^-, orthophosphate
Leaching	H^+, SO_4^{2-}, NO_3^-,
Forest floor	total N, NH_4^+ Ca_2^+, K^+, Mg^{2+}, Na^+, Al_3^+, Cl^-, HCO_3^-, orthophosphate
Vegetation increment	
Overstory	Organic matter, N, S
foliage	SO_4^{2-}S, Ca, K, Mg, P
branch	
bole	
roots	

Source: Pitelka (1985).

(Petelka, 1985) and are presented in Table 20-8. Figure 20-10 presents a schematic graphic of the sampling sites of this integrated study.

Wet deposition in the United States is currently measured by the National Trends Network, National Acid Precipitation Assessment Program. Analysis of the 150 station National Trends Network has demonstrated that the network is capable of detecting statistically significant annual changes of 20–30% in hydrogen ion, sulfate, or nitrate concentrations in the eastern United States in regions with dimensions of four degrees of longitude and latitude (approximately 160,000 km^2) (National Acid Precipitation Assessment Program, 1988). The U.S. Environmental Protection Agency has established the National Dry Deposition Network. This network of approximately 100 sites across the continental United States will monitor the dry deposition of ozone, sulfur dioxide, particulate sulfate and nitrate, and nitric acid. In addition, wind speed, wind direction, temperature, relative humidity, solar radiation, precipitation, and delta temperature will be recorded. Ozone concentrations and meteorological parameters will be monitored continuously. Deposition of other pollutants will be inte-

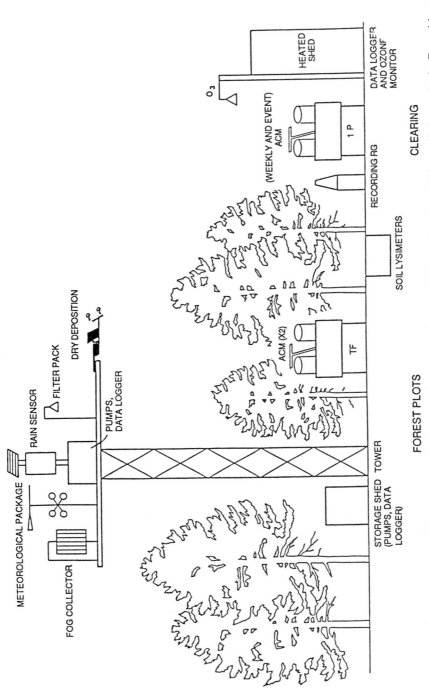

Figure 20-10. Intensive monitoring facilities of the study sites of the Integrated Forest Study on Effects of Atmospheric Deposition supported by the Electric Power Research Institute. TF = throughfall, IP = incident precipitation, ACM = Aerochem Metrics, Inc. sampler, RG = rain gauge. Source: Oak Ridge National Laboratory.

grated over 7-day daytime and nighttime sampling periods (Figure 20-11) (Hodges, 1988).

c. Forest Health Monitoring

In human health management, physicians have developed several routine procedures to monitor general wellness. These include body temperature, blood pressure and chemistry, heart rate, and weight. We must monitor comparable tree parameters to track forest health. A comprehensive discussion of this important topic is well beyond the scope of this book.

1. Long-Term Natural System Monitoring

Excellent reviews have been provided that examine the importance and challenge of long-term natural system monitoring efforts (Botkin, 1977; Callahan, 1984; Hinds, 1984; Strayer, et. al., 1986; Synnott, 1977). These reviews emphasize that extended-term monitoring programs must be biologically relevant, statistically credible, and cost effective. They must also have competent and secure leadership and be well planned and executed. Monitoring of ecosystem-level processes involves special challenges (Hinds, 1984). Monitoring the health of specific forest tree species is more straightforward and is our specific focus. Forest scientists have developed a variety of standard measurements that are useful in tree health assessments. Recent research efforts have proposed a large number of new potentially useful indices (Table 20-9).

2. Conventional Indices of Forest Tree Health

The general correlation of tree age and size (height, diameter) with site conditions is the most universal index of forest tree health. In the temperate zone, forest trees produce a new layer of stemwood each growing season, and the amount of current annual increment integrates the stresses imposed by abiotic and biotic factors and reflects tree health (Phipps, 1985; NCASI, 1987). The addition of wood density determinations to annual ring width measurements allows additional refinement of the assessment of environmental influences on tree health (Conkey, 1979, 1984a,b). Additional new developments in dendrochronology continue to advance the utility of tree ring information for atmospheric deposition studies (Van Deusen, 1988).

Cross-sectional area of tree stems, basal area, is an important conventional index of tree health. Normally, trees that occupy dominant positions in the canopy exhibit three phases of basal-area change: (a) a steady increase up to maturity, (b) a leveling off with maturity, and (c) a decline with older-mature and over-mature ages classes (Hornbeck et al. 1986). Decreases in basal area during (a) and (b) can reflect poor health.

Inventories of species specific symptoms (Ciesla and Hildebrandt, 1986; Muir and Armentano, 1985) and signs (Weiss et al., 1985) represent very useful gen-

Figure 20-11. National Dry Deposition Network monitoring station at the Hubbard Brook Experimental Forest, White Mountain National Forest, New Hampshire. The station consists of a 10 m meteorological tower on the left and a 10 m air quality tower on the right. The latter tower supports a Teflon filtration system for the quantification of atmospheric concentrations of sulfate, nitrate, nitric acid, and sulfur dioxide. Ozone is measured with an ultraviolet photometer.

Table 20-9. Forest Tree Health Indices.

Conventional	New
Diameter	Leaf area
Height	Leaf persistence
Annual increment (ring width)	Leaf chemistry
Baral area	Root area
Age	Root persistence
Signs/symptoms	Root chemistry
	Soil chemistry
	Stream chemistry

eral tree health indicies. Monitoring symptom development on particularly sensitive vegetation has been a long-established procedure (Heck, 1966; Manning and Feder, 1980; Steubing and Jäger, 1982; Winner and Bewley, 1978; Skye, 1979). Biomonitoring systems have been evaluated in forest ecosystems (Bennett and Stolte, 1985; Bennett, 1987).

3. New Indices of Forest Tree Health

To compliment traditional tree health indices, a variety of new indices have been proposed.

Yield or productivity of plants is largely the integral of photosynthesis in leaves. As a result, leaf area and temporal duration of leaf area represent a health index of substantial potential (Waggoner and Berger, 1987). Waring (1983, 1985) has proposed that the amount of leaf area supported by a deciduous forest increases to a maximum value as environmental conditions improve and decreases as conditions become less favorable. A large number of direct and indirect procedures are available to quantify forest leaf area. Throughout the growing season, canopy leaf area can be estimated by recording the amount of radiation in specific wavelengths absorbed or transmitted by leaves (Assar et al., 1984, Running et al., 1986, Waring et al., 1986). Overstory leaf area can also be assessed from periodic photographs with a fish-eye lens taken beneath the canopy. In addition, overstory leaf area can be estimated from collection of litterfall (Hinds, 1984), by linear correlation with the cross-sectional area of sapwood in tree stems (Marchand, 1984; Waring et al., 1981), and from awareness of light extinction through the canopy (Marshall and Waring 1986). Tree growth efficiency, as determined by stemwood production per unit of foliage area, has been proposed as a "universal" tree health index (Waring 1983). This health index records wood productivity in relation to the productive capacity of the tree. Since stemwood production is of low carbon allocation priority relative to the production of buds, shoots, or fine roots, it is judged to be a sensitive index of the relative ability of trees to mobilize carbohydrates (Christiansen et al., 1987).

Root biomass, particularly fine root biomass, may represent a particularly accurate forest tree health index. Unfortunately, however, direct measurement is difficult to impossible. Indirect measurement, however, may facilitate future utility (Marshall and Waring, 1985).

Chemical change indicators hold substantial promise for future utility in assessments of forest tree health impact associated with air pollutants (Muir and Armentano, 1985; Woodwell, 1989). Chemicals proposed as useful in monitoring strategies are extremely diverse and include nutrients, for example, sulfur (Huttunen et al., 1985; Legge et al., 1988), nitrogen (Jones and Coleman, 1989) and carbon (Luxmoore, 1989; McLaughlin 1989); stable isotope ratios of hydrogen, carbon, oxygen, nitrogen, and sulfur (Waring, 1987b); enzymes, for example, superoxide dismutase, peroxidase, and catalase, and nonenzyme defense compounds, for example, vitamin E, carotenes, glutathionl, and ascorbic acid (Richardson et al., 1989); and numerous others including starch, chlorophyll, and hormones (Muir and Armentano, 1985). Wessman et al. (1988) have employed images obtained via the Airborne Imaging Spectrometer, an experimental high-spectral resolution imaging satellite sensor developed by NASA, to estimate the lignin concentration of whole forest canopies in Wisconsin. Foliar lignin content is a primary rate-limiting factor of forest litter decomposition. The strong correlation between canopy lignin concentration and nitrogen availability (via nitrogen mineralization) in seven forest ecosystems on Blackhawk Island, Wisconsin suggested to Wessman et al. that canopy lignin may serve as an index for site nitrogen status.

As forest condition is a major, frequently primary, regulator of soil and associated stream chemistry, monitoring schemes directed to these associated resources may prove valuable in forest health monitoring. Extended term trends in soil pH, organic matter, cation exchange capacity, base saturation and extractable cations are intimately linked to forest health. The integrated forest watershed studies at Hubbard Brook in New Hampshire (Likens et al., 1977) and Coweeta in North Carolina (Swank et al., 1981) have dramatically documented the relationship between forest condition and streamwater chemistry.

4. Remote Sensing of Forest Health

Remote detection of forest stress has been widely studied for more than 25 years (Murtha, 1969 a,b; Weber and Polcyn ,1972). Larsh et al. (1970) employed aerial photography to assess air pollution damage to ponderosa pine in California. In recent years, enormous advancements in computer and satellite technology have revolutionized remote detection strategies (Greegor ,1986 Rock et al., 1986; Roller and Colwell ,1986).

Contemporary satellite systems collect reflected electromagnetic energy in the visible and infrared wavelengths from forest surfaces. Significant advancements have been made in spectral sensitivity, spatial resolution, and repeat viewing time. Digital data format has enhanced data processing and integration with other data sets. The Landsat multispectral scanner was the first satellite orbiting multispectral scanner. The second generation of Landsat satellites have supplemented the multispectral scanner with the Thematic Mapper, have improved spatial resolution from approximately 60 X 80 m to 30 X 30 m, and have a temporal cycle of 14 days.

The usefulness of satellite scanning for forest health depends on stress induction of alterations in reflected energy from forest trees. The greatest utility of remote scanning is in separating groups of stressed trees from healthy trees. Separation of stressed from healthy tree populations is most efficient with uniform tree size, low species diversity, uniform crown coverage, and low-relief terrain. Forest stands with variable structure, multiple species, crown openings, and location in sites with steep terrain represent systems more difficult to scan with present technology.

Rock et al. (1989) have documented a gradient of montane conifer stress in northeastern United States using Landsat Thematic Mapper data. The gradient described for red spruce health suggested most severe stress in the Adirondack Mountains of New York, moderate stress in the Green Mountains of Vermont, and least stress in the White Mountains of New Hampshire. While contemporary techniques do not allow the determination of the specific cause or causes of the west to east red spruce health gradient, the authors are hopeful that coupling satellite spectral measurements with ground-based spectral measurements and advances in spectral sensors will ultimately allow this resolution (Rock et al., 1986, 1989).

Future advancements in satellite measurements may even allow estimations of rate changes in fundamental plant processes such as photosynthesis, transpiration, maintenance respiration, and litter decomposition; as well as changes in phytochemistry such as altered starch or lignin amounts (Waring et al., 1986; Wessman et al., 1988).

While significant challenges remain to be addressed, the utility of remote sensing in air pollution–forest health studies is advancing rapidly and has significant potential for future monitoring strategies (Electric Power Research Institute 1988). Unless and until we get more serious and sophisticated about periodic assessment, we will not know whether the health of forest systems is improving, stable, or declining.

2. Research

Over the past several decades scientists have demonstrated that local-, regional-, and global-scale air pollutants are capable, in selected locations in selected forest types, of causing significant reductions in tree productivity and shifts in species composition, of interacting with other forest tree stress factors, and of altering forest ecosystem processes, including energy flow and biogeochemical cycling (Figure 20-12).

This demonstration requires that considerations of forest health be given prominent consideration in deliberations concerning national and international clean air laws, regulations, and treaties; strategic energy planning and facilities design; industrial, commercial, and transportation planning and siting, and natural science research funding.

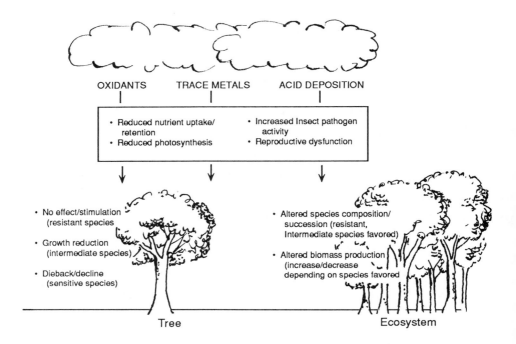

Figure 20-12. Forest tree and ecosystem response to regional-scale air pollution exposure.

Scientists can rarely resist the temptation to call for additional research (Kozlowski, 1980) and the author finds he is no exception. We must fully recognize that the applied sciences of forest health management are directly dependent on our basic understanding of tree and forest form and function. We desperately need greater fundamental understanding of tree biochemistry and physiology. We further need more appreciation of basic forest ecology and ecosystem dynamics. Research programs dedicated to specific stress factors must have both field and laboratory components. Studies of mechanisms of disease obtained in managed environments of the laboratory, growth chamber, and greenhouse must be combined with studies conducted in the field in order to obtain a comprehensive perspective. Efforts in controlled and natural environments must receive equivalent support and encouragement. Research information must be synthesized and integrated.

Forest health research is multidisciplinary research. Forest health integrates all of the stresses influencing trees over time. A large number of core disciples effectively contribute to our understanding of forest health: atmospheric sciences, soil science, fire science, biochemistry, meteorology, climatology, physiology, ecology, entomology, and pathology.

We must continue to strive to develop more effective communication between forest scientists and forest managers, decision makers, and society at large. Advancements in our understanding of forest stress dynamics, with no implementation of improved environmental management strategies, do nothing to reduce forest health risks. We further have a responsibility to work toward improving science competence at all levels of society. The great ignorance of science, even in a developed nation such as the United States (Hively, 1988), restricts the ability of society to evaluate and support the need for educational and research efforts.

a. Continuance and Establishment of Comprehensive Air Pollution–Forest Ecosystem Studies

A very high research priority is reserved for the continuance and establishment of comprehensive investigations to systematically examine Class II interactions in forest ecosystems located in those portions of the temperate zone particularly subject to air pollution exposure. These investigations should include analysis of air contaminant influence on soil metabolism and structure, nutrient cycling, tree reproduction, photosynthesis and respiration, important arthropod species and microbial pathogens, foliar symptoms of important vegetation in all forest strata, and a careful examination of forest productivity and alterations in successional trends and species dominance. These studies must be of extended term. They will require the participation of numerous scientific disciplines, minimally including pathology, entomology, meteorology, soil science, soil microbiology, ecology, and systems analysis. Continuous meteorological and air quality monitoring will be required. Air pollutants measured should include all wet- and dry-deposited pollutants previously identified (Section D:1:b.). The objective of these comprehensive studies will be to clarify and quantify various Class I and II interactions. The ecosystems will be evaluated for their ability to resist (inertia) and respond (resilience) to disturbance from air pollution stress. Model development for the various interactions will allow future projections, given various air quality scenarios, and will allow extrapolation of findings to other ecosystems.

Integrated and extended-term forest studies that have been initiated and supported by the National Science Foundation, National Acid Precipitation Assessment Program, U.S.D.A. Forest Service, U.S. Environmental Protection Agency, U.S. Department of Energy, the Electric Power Research Institute and others must be sustained. New integrated forest studies should be initiated in forest systems presently understudied. Minimal intensive study site distribution for regional-scale pollutants should include the northeastern montane forest system, the northern and central hardwood forest systems, the southern commercial forest system, and, in the west, the Pacific coast forest and the California woodland (Figures 20-7,20-8). With reference to global-scale pollutant potentials for forest ecosystems, integrated research programs are justified in any forest system with important recreational, wilderness, watershed, or commercial significance.

b. Specific Research Priorities

Several topics have especially high research priority for both the comprehensive ecosystem studies and investigations conducted in the field, greenhouse, or laboratory. It is important to realize, as in the case of epidemological investigations concerning air pollution impact on human health, field experimentation must be supported by a strong laboratory (controlled environment) research program. Particularly important research areas include the following:

1. Exposure-response information on visible (symptomatic) tree response to air pollutants, with experiments appropriately designed to accommodate and consider the influence of genetic factors, environmental factors, and interactions of air contaminants.
2. Development of accurate, relatively simple, and reproducible methodologies to identify and inventory visible (symptomatic) injury in the field employing ground, satellite, or other remote strategies.
3. Exposure-response information on invisible (asymptomatic) tree response, including an evaluation of the ability of air pollution exposure to influence tree reproduction, photosynthesis, respiration, carbon allocation, nutrient dynamics, growth, and competitive ability, along with the associated ecosystem dynamics of productivity, succession, and species composition.
4. Analysis of the ability of air pollution stress to interact (additively, synergistically, competitively) and to predispose, aggravate or reduce other tree stresses caused by arthropod, pathogen, edaphic, climatic, or human management strategies.
5. Determine the biochemical and physiological bases of air pollution stress on forest vegetation.
6. Determine the stand-, population-, community- and, ecosystem-level consequences of air pollution stress on forest systems.
7. Determine the ability of forest vegetation and forest soils to act as a sink and source for atmospheric contaminants.
8. Develop models of air pollution–forest-system interactions to facilitate research, integrate results, extrapolate information, and permit prediction.
9. Determine the direct and indirect effects of increased ultraviolet-B radiation on forest trees and the interaction of these effects with other stress factors of forest vegetation.
10. Evaluate the influence of increased carbon dioxide availability, warming, growing season extension, and increased and reduced soil moisture on tree health and the interaction of these changes with other forest stress factors.

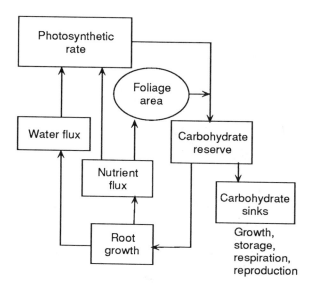

Figure 20-13. Conceptual model of carbon allocation in trees.
Source: Ford (1982).

c. Modeling and Prediction

Models are essential for the integration of research findings, hypothesis formulation, extrapolation of findings to other systems, and, very importantly, to future
prediction. Simple conceptual models of complicated tree processes (Figure 20-
13) or ecosystem processes (Figure 20-14) are particularly useful in hypothesis
formulation and integration of research results. Descriptive and process models
are required for effective extrapolation and prediction.

Pastor and Post (1986) have formulated a forest ecosystem model with ex-
plicit feedbacks between light, water, and nitrogen availabilities and their resul-
tant effects on productivity and species composition. Mäkelä et al. (1987) have
provided a simple dynamic model for the sensitivity and risk of forests exposed
over an extended term to sulfur deposition.

Schäfer et al. (1988a) have provided a review of model applications to air-pol-
lution-forest system interactions. Systems analysis and simulation models have
great utility for describing the interactive dynamics of pollutant exposure and
forest tree and stand development (Bossel and Schäfer, 1988; Krieger et al., 1988;
Schäfer et al., 1988b). Nance et al. (1988) have developed a flexible model that
permits simulation of some of the possible effects of atmospheric deposition on
the growth and development of loblolly pine stands. Burkhart et al. (1988) have
provided a model to perform sensitivity analyses of stand-level productivity from
hypothesized air pollution effects on individual tree diameter, height, and/or
crown development.

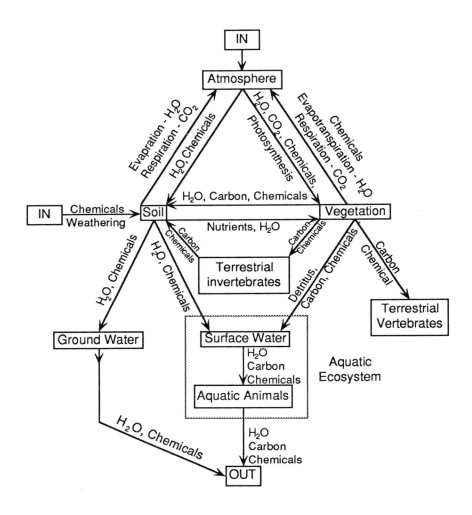

Figure 20-14. Conceptual forest ecosystem model. Arrows represent the movement of carbon (energy), water, and various chemicals between components. Source: U. S. Environmental Protection Agency (1988b).

E. Conclusion

The health and welfare of all societies are intimately linked to the health and welfare of their associated natural and artificial ecosystems. During the twentieth century we have made enormous advancements in our understanding of forests and their values to us. We have learned to appreciate, manage, and manipulate forests. We have created new forests and have damaged others. We have recognized many of the strengths and weaknesses of our production, stewardship, and regulatory policies.

The recognition that we are capable of stressing forests in so many ways has sobered our sense of accomplishment derived from the sophisticated appreciation of forest form and function, the ability to increase forest productivity, and the strategies for preserving forests in perpetuity and that we have developed.

Research indicates that regional air pollution is one of the significant contemporary anthropogenic stresses imposed on some temperate forest ecosystems. Gradual and subtle change in forest metabolism and composition over wide areas of the temperate zone over extended time, rather than dramatic destruction of forests in the immediate vicinity of point sources over short periods, must be recognized as the primary consequence of regional air pollution stress. Global air pollution, with its associated capability to cause rapid climate change, has the potential to dramatically alter forest ecosystems in the next century. The integrity, productivity, and value of forest systems is intimately linked to air quality. Failure to give careful consideration to forest resources in societal considerations of energy technologies and in management and regulation of air resources is unthinkable.

References

Abelson, P.H. 1987. Ozone and acid rain. Science 238: 141.

Anagostakis, S.L. and J.Kranz. 1987. Population dynamics of *Cryphonectria parasitica* in a mixed-hardwood forest in Connecticut. Phytopathology 77: 751-754.

Anderson, R.L and P.A. Mistretta. 1982. Management Strategies for Reducing Losses Caused by Fusiform Rust, Annosus Root Rot, and Littleleaf Disease. U.S.D.A. Forest Service, Agricul. Handbook No. 597, Asheville, NC, 30 pp.

Asrar, G., M. Fuchs, E.T. Hanemasu, and J.L. Hatfield. 1984. Estimating absorbed photosynthetic radiation and leaf area index from spectral reflectance of wheat. Agron. J. 76: 300-306.

Basiouny, F.M. 1986. Sensitivity of corn, oats, peanuts, rice, rye, sorghum, soybean and tobacco to UV-B radiation under growth chamber conditions. J. Argon. Crop Sci. 157: 31-35.

Bega, R.V., R.S. Smith Jr. A.P. Martinez, and C.J. Davis. 1978. Severe damage to *Pinus radiata* and *P. pinaster* by *Diplodia pinea* and *Lophodermium* spp. on Molokai and Lanai in Hawaii. Pl. Dis. Rptr. 62: 329-331.

Bennett, J.P. 1987. Establishment and Assessment of White Pine Biomonitoring Plots in the Great Smoky Mountains National Park. Publ. No. SP-86-41, National Park Service, Air Quality Division, Denver, CO, 31 pp.

Bennett, J.P. and K.W. Stolte. 1985. Using Vegetation Biomonitors to Assess Air Pollution Injury in National Parks. Milkweed Survey. Nat. Res. Rep. No. 85-1, National Park Service, Air Quality Division, Denver, CO, 18 pp.

Biggs, R.H. and P.G. Webb, 1986. Effects of enhanced ultraviolet-B radiation on yield and disease indicence and severity for wheat under field conditions. In R.C. Worrest and M.M. Caldwell, eds., Stratospheric ozone reduction, solar ultraviolet radiation and plant life. N.A.T.O. A.S.I. Series, Vol. G8. Springer–Verlag, Berlin.

Blair, R.L. and E.B. Cowling. 1974. Effects of fertilization, site, and vertical position on the susceptibility of loblolly pine seedlings to fusiform rust. Phytopathology 64: 761-762.

Blais, J.R. 1983. Trends in the frequency, extent, and severity of spruce budworm outbreaks in eastern Canada. Can. J. For. Res. 13: 539-547.

Bogenreider, A. and R. Klein. 1982. Does solar UV influence the competitive relationship in higher plants? In J. Calkins, ed., The role of solar ultraviolet radiation in marine ecosystems. Plenum Press, New York.

Bosel, H. and H. Schäfer. 1988. Eco-physiological dynamic simulation model of tree growth, carbon and nitrogen dynamics. International Union of Forestry Research Organizations. Forest Simulation Symptoms Conference. Univ. of California, Berkeley, CA, Nov. 2-5, 1988.

Botkin, D.B. 1977. Long-term Ecological Measurements. Conference Report. Woods Hole, MA, March 16-18, 1977. National Science Foundation, Washington, DC.

Boyce, J.S. 1913. Forest sanitation. The Tahoe 2: 14-17.

Brandle, J.R., W.F. Campbell, W.B. Sisson, and M.M. Caldwell. 1977. Net photosynthesis, electron transport capacity, and ultrastructure of *Pisum sativum* L. exposed to ultraviolet-B radiation. Plant Physiol. 60: 165-169.

Brenneman, J. and T. Blinn. 1987a. Long-term Ecological Research in the United States. A Network of Research Sites. Forest Science Dept., Oregon State Univ., Corvallis, OR, 39 pp.

Brenneman, J., J.F. Franklin, and J.J. Magnuson, (eds.). 1987b. LTER Network News (fall 1987). Forest Science Dept., Oregon State Univ., Corvallis, OR.

Burkhart, H.E., R.L. Amateis, and C.D. Webb. 1988. A model for sensitivity analyses of possible air pollution impacts on stand productivity. IUFRO Forest Simulation Systems Conference, Berkeley, CA, Nov. 2-5, 1988.

Caldwell, M.M. 1968. Solar ultraviolet radiation as an ecological factor for alpine plants. Ecol. Monogr. 38: 243-268.

Callahan, J.T. 1984. Long-term ecological research. BioScience 34: 363-367.

Christiansen, E., R.H. Waring, and A.A. Berryman. 1987. Resistance of conifers to bark beetle attack: Searching for general relationships. For. Ecol. Manage 22: 89-106.

Ciesla, W.M. 1988. Pine bark beetles: A new pest management challenge for Chilean foresters. J. For. 86: 27-31.

Ciesla, W.M and G. Hildebrandt. 1986. Forest decline inventory methods in West Germany: Opportunities for Application in North American Forests. Forest Pest Management, Report No. 86-3, U.S.D.A. Forest Service, Ft. Collins, CO, 31 pp.

Conkey, L.E. 1984. Dendrochronology and forest productivity: Red spruce wood density and ring width in Maine. In M.M. Harris and A.M. Spearing, eds., Research in Forest Productivity, Use, and Pest Control. U.S.D.A. Forest Service, Northeastern Forest Exp. Sta., Genl. Tech. Rep. No. NE-90, Broomall, PA, pp. 69-75.

Conkey, L.E. 1979. Response of tree-ring density to climate in Maine, USA. Tree-Ring Bull. 39: 29-38.

Conkey, L.E. 1984. X-ray densitometry: Wood density as a measure of forest productivity and disturbance. In D.D. Davis, ed., Air Pollution and Productivity of the Forest. Izaak Walton League of America, Washington, DC, pp. 287-296.

Dickson, J.G. and M.M. Caldwell. 1978. Leaf development of Rumex patientia L. (Polygonaceae) exposed to UV irradiance (280–320 nm). Am. J. Bot. 65: 857-863.

Dinoor, A. and N. Eshed. 1984. The role and importance of pathogens in natural plant communities. Annu. Rev. Phytopath. 22: 443-466.

Driscoll, C.T., G.E. Likens, L.O. Hedin, J.S. Eaton, and F.H. Bormann. 1989. Changes in the chemistry of surface waters. Environ. Sci. Technol. 23: 137-143.

Electric Power Research Institute. 1988. Proceedings: Seminar on Remote Sensing of Forest Decline Attributed to Air Pollution. March 11–12, 1987, Luxenburg, Austria. EPRI, Palo Alto, CA.

Fox, F.M. and M.M. Caldwell 1978. Competitive interactions in plant populations exposed to supplementary ultraviolet-B radiation. Oecologia 36: 173-190.

Franklin, J.F., H.H. Shugart, and M.E. Harmon. 1987. Tree death as an ecological process. BioScience 37: 550-556.

Galvin, P.J., P.J. Samsun, P.E. Coffey, and D. Romano. 1978. Transport of sulfate to New York State. Environ. Sci. Technol. 12:580-584.

Gibbs, J.N. and C.M. Brasier. 1988. Transatlantic perspective on exotic plant pathogens. Pl. Dis. 72: 373.

Gold, W.G. and M.M. Caldwell. 1983. The effects of ultraviolet-B radiation on plant competition in terrestrial ecosystems. Physiol. Plant. 58: 435-444.

Greegor, D.H. 1986. Ecology from space. BioScience 36: 429-432.

Gschwandtner, G., K.C. Gschwandtner, and K. Eldridge. 1985. Historic Emissions of Sulfur and Nitrogen Oxides in the Untied States from 1900 to 1980. Publ. No. EPA-600-57-85-009. U.S. Environmental Protection Agency, Washington, DC.

Heck, W.W. 1966. The use of plants as indicators of air pollution. Air Water Pollut. Intl. J. 10: 99-111.

Hedin, L.O., G.E. Likens, and F.H. Bormann. 1987. Decrease in precipitation acidity resulting from decreased $SO_4{}^{2-}$ concentration. Nature 325: 244-246.

Hepting, G.H. 1963. Climate and forest diseases. Annu. Rev. Phytopathol. 1: 31-50.

Hepting, G.H. 1971. Diseases of Forest and Shade Trees of the United States. U.S.D.A. Forest Service, Agr. Handbook No. 386, U.S. Government Printing Office, Washington, DC, 658 pp.

Hermann, R.K. 1987. North American tree species in Europe. J. For. 85 (12): 27-32.

Hinds, W.T. 1984. Towards monitoring of long-term trends in terrestrial ecosystems. Environ. Conserv. 11: 11-18.

Hively, W. 1988. Science observer: How much science does the public understand? Am. Scient. 76: 439-444.

Hodges, C.S. and L.C. May. 1972. A root disease of pine, Araucaria, and eucalyptus in Brazil caused by a new species of *Cylindrocladium*. Phytopathology 62: 898-901.

Hodges, M.G. 1988. Personal Communication. Atmospheric Studies, Hunter Environmental Services, Gainesville, FL

Hornbeck, J.W., R.B. Smith, and C.A. Federer. 1986. Growth decline in red spruce and balsam fir relative to natural processes. Water Air Soil Pollu. 31: 425-430.

Houston, D.R. 1974. Dieback and declines: Diseases initiated by stress, including defoliation. Arborists News 49: 73-76.

Houston, D.R. 1981. Stress Triggered Tree Diseases. The Diebacks and Declines, U.S.D.A. Forest Service Publ. No. NE-INF-41-81. Northeastern Forest Exp. Sta., Broomall, PA, 33 pp.

Huffaker, C.B. 1974. Some implications of plant-arthropod and higher-level, arthropod-arthropod food links. Environ. Entomol. 3: 1-9.

Huttunen, S., K. Laine, and H. Torvela. 1985. Seasonal sulphur contents of pine needles as indices of air pollution. Annu. Bot. Fennici 22: 343-359.

Jones, C.G. and J.S. Coleman. 1989. Biochemical indicators of air pollution effects in trees: Unambiguous signals based on secondary metabolites and nitrogen in fast-growing species? In G.M. Woodwell, ed., Markers of Air Pollution Effects in Forests. National Academy Press, Washington, DC.

Karl, T.R. and R.G. Quayle. 1988. Climate change in fact and in theory: Are we collecting the facts? Climatic Change 13: 5-17.

Kelman, A. 1987. The status of support for forest pathology research. Pl. Dis. 71: 387.

Kossuth, S.V. and R.H. Biggs. 1981. Ultraviolet-B radiation effects on early seedling growth of Pinaceae seedlings. Can. J. For. Res. 11: 243-248.

Kozlowski, T.T. 1980. Impacts of air pollution on forest ecosystems. BioScience 30: 88-93.

Krieger, H., H. Schäfer, and H. Bossel. 1988. Modelling and simulation of spruce stand dynamics under different stress regimes introducing the competition-classes-concept. Proc. 3rd Intl. Symposium on Systems Analysis and Simulation. Berlin, G.D.R., Sept. 12–16, 1988.

Lal, M., S.K. Dube, P.C. Sinka, and A.K. Jain. 1986. Potential climatic consequences of increasing anthropogenic consituents in the atmosphere. Atmos. Environ. 20: 639-642.

Lamb, A.F.A. 1973. Fast growing timber trees of the lowland tropics. *Pinus caribaea*. Vol. I. Comm. For. Inst., Univ., Oxford, UK, 254 pp.

Landsberg, H.E. 1970. Man-made climatic changes. Science 170: 1265-1274.

Langford, A.N. and M.F. Buell. 1969. Integration, identity and stability in the plant association. Adv. Ecol. Res. 6: 83: 135.

Larsh, R.N., P.R. Miller and S.L. West 1970. Aerial photography to detect and evaluate air pollution damaged ponderosa pine. J. Air Pollu. Control Assoc. 20: 289-292.

Legge, A.H., J.C. Bogner, and S.V. Krupa. 1988. Foliar sulphur species in pine: A new indicator of a forest ecosystem under air pollution stress. Environ. Pollu. 14: 159-162.

Likens, G.E., F.H. Bormann, R.S. Pierce, J.S. Eaton, and N.M. Johnson. 1977. Biogeochemistry of a Forested Ecosystem. Springer–Verlag, New York, 146 pp.

Loucks, O.L. 1970. Evolution of diversity, efficiency and community stability. Am. Zool. 10: 17-25.

Luxmoore, R.J. 1989. Nutrient use efficiency as a marker of air pollution and natural stress effects in forests. In G.M. Woodwell, ed., Markers of Air Pollution Effects in Forests. National Academy Press, Washington, DC.

Mäkelä, A., J. Materna, and W. Schöpp. 1987. Direct Effects of Sulfur on Forests in Europe — A Regional Model of Risk. Working Paper No. WP-87-57. International Institute for Applied Systems Analysis, Luxenburg, Austria, 38 pp.

Mamiya, Y. 1987. Origin of the pine wood nematode and its distribution outside the United States. In M.J. Wingfield, ed., Pathogenicity of the Pine Wood Nematode. American Phytopathological Press, St. Paul, MN, pp. 59-65.

Man and the Biosphere. 1984. Action Plan for Biosphere Reserves. Nature and Resources (UNESCO) 20: 1-12.

Manion, P.D. 1981. Tree Disease Concepts. Prentice-Hall, Englewood Cliffs, NJ, pp. 324-288.

Manning, W.J. and W.A. Feder. 1980. Biomonitoring Air Pollutants with Plants. Applied Science Publishers, London, UK., 141 pp.

Marchand, P.J. 1984. Sapwood area as an estimator of foliage biomass and projected leaf area for *Abies balsamea* and *Picea rubens*. Can. J. For. Res. 14: 85-87

Marshall, J.D. and R.H. Waring. 1985. Predicting fine root production and turnover by monitoring root starch and soil temperature. Canad. J. For. Res. 15: 791-800.

Marshall, J.D. and R.H. Waring. 1986. Comparative methods of estimating leaf area in old-growth Douglas-fir. Ecology 67: 975-979.

Marty, R. 1966. Economic guides for blister-rust control in the East. U.S.D.A. Forest Service, Northeastern Forest Exp. Sta., Res. Paper No. NE-45, Broomall, PA, 14 pp.

Mattson, W. J. and N.D. Addy. 1975. Phytophagous insects as regulators of forest primary production. Science 190: 515-521.

Mattson, W.J. and R.A. Haack. 1987. The role of drought in outbreaks of plant-eating insects. BioScience 37: 110-118.

McClure, M. 1987. Hemlock woolly adelgid may also attack spruce. Fron. Pl. Sci. 39: 7-8.

McLaughlin, S.B. 1989. Carbon allocation as an indicator of pollutant impacts on forest trees. In G.M. Woodwell, ed., Markers of Air Pollution Effects in Forests. National Academy Press, Washington, DC.

Miller, P.R. and J.R. McBride. 1975. Effects of air pollutants on forests. In J.B. Mudd and T.T. Kozlowski, eds, Responses of Plants to Air Pollution. Academic Press, New York, pp. 195-235.

Miller, P.R., G.J. Longbotham, and C.R. Longbotham. 1983. Sensitivity of selected western conifers to ozone. Pl. Dis. 67: 1113-1115.

Miller, T. 1972. Fusiform rust in planted slash pines: Influence of site preparation and spacing. For. Sci. 18: 70-75.

Miller, T. and R.A. Schmidt. 1987. A new approach to forest pest management research in the south. Pl. Dis. 71: 204-207.

Muir, P.S. and T.B. Armentano. 1985. Developing Procedures for Evaluation of Visible Air Pollution Effects in Broad-Leaved Species: Problem Definition, Protocol, and Research Needs. Workshop Proceedings, Butler Univ., Indianapolis, IN, April 3–4, 1985, 16 pp.

Murali, N.S. and A.H. Teramura. 1985. Effects of ultraviolet irradiance on soybean: VI. Influence of phosphorus nutrition on growth and flavonoid content. Physiol. Plant. 63: 413-416.

Murali, N.S. and A.H. Teramura. 1986. Effectiveness of UV-B radiation on the growth and physiology of field-grown soybean modified by water stress. Photochem. Photobiol. 44: 215-219.

Murtha, P.A. 1969a. Aerial Photographic Interpretation of Forest Damage: A Discussion of Concepts. Information Report No. FMR-X-18, Forest Management Institute, Canada Dept. Fisheries and Forestry, ottawa, Canada 17 pp.

Murtha, P.A. 1969b. Aerial Photographic Interpertation of Forest Damage: An Annotated Bibliography. Information Report No. FMR-X-16, Forest Management Institute, Canada Dept. Fisheries and Forestry, Ottawa, Canada, 76 pp.

Nance, W.L., J.E. Grissom, C.D. Nelson, H.E. Burkhart, and C.D. Webb. 1988. Simulating the effect of atmospheric deposition on loblolly pine stands. Fifth Biennial Southern Silvicultural Research Conference, Memphis, TN, Nov 1–3, 1988.

National Acid Precipitation Assessment Program. 1987. Interim Assessment, The Causes and Effects of Acid Deposition. Vol. IV. Washington, DC.

National Acid Precipitation Assessment Program. 1988. Annual Report 1987. NAPAP, Washington, DC, 76 pp.

National Council of the Paper Industry for Air and Stream Improvement. 1987. Tree Rings and Forest Mensuration: How Can They Document Trends in Forest Health and Productivity? Tech. Bull. No. 523, NCASI, New York, 66 pp.

National Research Council 1983. Acid Deposition: Atmospheric Processes in Eastern North America. National Academy Press, Washington, DC.

Oak, S.W. and F.H. Tainter. 1988. Risk prediction of loblolly pine decline on little-leaf disease sites in South Carolina. Pl. Dis. 72: 289-293.

Office of Technology Assessment. 1984. Acid Rain and Transported Air Pollutants: Implications for Public Policy. U.S. Government Printing Office, Washington, DC, 323 pp.

Pastor, J. and W.M. Post. 1986. Influence of climate, soil moisture, and succession on forest carbon and nitrogen cycles. Biogeochemistry 2: 3-27.

Pitelka, L.F. 1985. Project Summary. Integrated Forest Study on Effects of Atmospheric Deposition. Oak Ridge National Laboratory, Oak Ridge, TN, 35 pp.

Pjhipps, R.L. 1985. Collecting, preparing, crossdating, and measuring tree increment cores. Water-Resources Investigations Report No. 85-4148. U.S.
Geological Survey, Reston, VA, 48 pp.

Pook, E.W. 1984. Canopy dynamics of *Eucalyptus maculata* Hook. III. Effects of
drought. Aust. J. Bot. 32: 405-413.

Raup, H.M. 1937. Recent changes of climate and vegetation in southern New
England and adjacent New York. J. Arnold Arbor. 18: 79-117.

Read, R.A. and J.A. Sprackling. 1976. Douglas-fir in Eastern Negraska: A
Provenance Study. U.S.D.A. Forest Service, Rocky Mt. For. and Range Exp. Sta.,
Res. Paper RM-178, Ft. Collins, CO, 10 pp.

Richardson, C.J., R.T. Di Giulio, and N.E. Tandy. 1989. Free-radical mediated processes as markers of air pollution stress in trees. In G M. Woodwell, ed., Markers
of Air Pollution Effects in Forests. National Academy Press, Washington, DC.

Rock, B.N., J.E. Vogelmann, D.L. Williams, A.F. Vogelmann, and T. Hoshizaki.
1986. Remote detection of forest damage. BioScience 36: 439-445.

Rock, B.N., J.E. Vogelmann, and NJ. Defeo. 1989. The use of remote sensing for the
study of air pollution effects in forets. In G.M. Woodwell, ed., Markers of Air
Pollution Effects in Forests. National Academy Press, Washington, DC.

Roller, N.E.G. and J.E. Colwell. 1986. Coarse-resolution satellite data for ecological
surveys. BioScience 36: 468-475.

Rowan, S.J. 1977. Fertilizer-induced changes in susceptibility to fusiform rust vary
among families of slash and loblolly pine. Phytopathology 67: 1280-1284.

Royer, M.H. and W.M. Dowler. 1988. A World Plant Pathogen Database. Pl. Dis. 72:
284-288.

Running, W.W., D.L. Peterson, M.A. Spanner, and K.B. Teuber. 1986. Remote sensing of coniferous forest leaf area. Ecology 67: 273-276.

Sagan, C., O.B. Toon, and J.B. Pollack. 1979. Anthropogenic albedo changes and
the earth's climate. Science 206: 1363-1368.

Schäfer, H., H. Bossel, and H. Krieger. 1988a. Modelling the responses of mature
forest trees to air pollution. Geojournal 17:279-287.

Schäfer, H., H. Krieger, and H. Bossel. 1988b. Modelling air pollution effects on
plants, particularly on forest growth — A review. Conference on Air Pollution in
Europe: Environmental Effects, Control Strategies and Policy Options.
Stockholm, Sweden. September 26–30, 1988, pp. 1-10.

Schertz, T.L and R.M. Hirsch. 1986. Trend Analysis of Weekly Acid Rain Data —
1978-83. Water Resources Investigation Report 85-4211. U.S. Geological
Survey, Washington, DC.

Schmidt, R.A., T. Miller, R.C. Holley, R.P. Belanger, and J.E. Allen. 1988.
Relation of site factors to fusiform rust incidence in young slash and loblolly pine
plantations in the coastal plain of Florida and Georgia. Pl. Dis. 72: 710-714.

Schoweneweiss, D.F. 1986. Water stress predisposition to disease: An overview. In
P.G. Ayres, ed., Water, Fungi and Plants. Cambridge Univ. Press, New York, pp.
157-174.

Scriven, G.T., E.L. Reeves, and R.F. Luck. 1986. Beetle from Australia threatens eucalyptus. Calif. Agric. (July/Aug): 4-6.

Seastedt, T.R. and D.A. Crossley Jr. 1984. The influence of arthropods on ecosystems. BioScience 34: 157-161.

Sergin, V.Y. 1980. Origin and mechanism of large-scale climatic oscillations.
Science 209: 1477-1483.

Shea, K.R. 1970. Disease and insect activity in relation to intensive culture of
forests. Unasylva 19: 109-118.

Shigo, A.L. 1985. Wounded forests, starving trees. J. For. 83: 668-673.

Sinclair, W.A. 1964. Comparisons of recent declines of white ash, oaks, and sugar
maple in northeastern woodlands. Cornell Publications 21: 62-67.

Sisson, W.B. 1981. Photosynthesis, growth and ultraviolet irradiance absorbance of *Curcurbita pepo* L. leaves exposed to ultraviolet-B radiation (280–315 nm). Plant Physiol. 67: 120-124.

Sisson, W.B. and M.M. Caldwell. 1976. Photosynthesis, dark respiration and growth of *Rumex patientia* L. exposed to ultraviolet irradiance (280-315 nanometers) simulating a reduced atmospheric ozone column. Plant Physiol. 58: 563-568.

Sisson, W.B. and M.M. Caldwell. 1977. Atmospheric ozone depletion: Reduction of photosynthesis and growth of a sensitive higher plant exposed to enhanced UV-B radiation. J. Exp. Bot. 28: 691-705.

Skye, E. 1979. Lichens as biological indicators of air pollution. Annu. Rev. Phytopathol. 17: 325-341.

Smith, P.C. 1980. California conifers thrive in New Zealand. Califor. Agric. (Aug-Sept): 4-6.

Smith, W.H. 1981. Air Pollution and Forests. Springer–Verlag, New York, 397 pp.

Smith, W.H. 1984. Ecosystem pathology: A new perspective for phytopathology. For. Ecol. Mamt. 9: 193-219.

Smith, W.H. 1985. Forest quality and air quality. J. For. 83: 82-92.

Smith, W.H. 1986. Role of phytophagous insects, microbial, and other pathogens in forest ecosystem structure and function. In B.L. Bedford, ed., Modification of Plant-Pest Interactions by Air Pollutants. Ecosystems Research Center, Publ. No. ERC-117, Cornell Univ., Ithaca, New York, pp. 34-52.

Smith, W. H. 1987. Future of the hardwood forest: Some problems with declines and air quality. In R.L. Hay, F.W. Woods, and H. DeSelm, eds., Proc. Central Hardwood Forest Conference VI. Dept. of Forestry, Wildlife and Fisheries, Univ. of Tennessee, Knoxville, TN, pp. 3-13.

Society of American Foresters. 1954. Forest Cover Types of North America. Soc. Am. For., Washington, DC, 67 pp.

Sprackling, J.A. and R.A. Read. 1976. Eastern White Pine in Eastern Nebraska: A Provenance Study of Southern Appalachian Origins. U.S.D.A. Forest Service. Rocky Mt. Forest Range Exp. Sta., Res. Paper RM-179, Ft. Collins, CO, 8 pp.

Steubing, L. and H. J. Jäger, (eds.). 1982. Monitoring of Air Pollutants by Plants. Dr. W. Junk Publishers, The Hague, Netherlands.

Strayer, D., J.S. Glitzenstein, C.G. Jones, J. Kolasa, G.E. Likens, M.J. McDonnell, G. G. Parker, and S.T.A. Pickett. 1986. Long-Term Ecological Studies: An Illustrated Account of Their Design, Operation, and Importance to Ecology. Occasional Publ. No. 2., Institute of Ecosystem Studies, Millbrook, New York. 38 pp.

Swank, W.T., J.B. Waide, D.A. Grossley Jr., and R.L. Todd. 1981. Insect defoliation enhances nitrate export from forest ecosystems. Oecologia 51: 297-299.

Swanson, F.J. and J.F. Franklin. 1988. The Long-Term Ecological Research Program. EOS 69: 34, 36.

Synnott, T.J. 1977. Monitoring Tropical Forests: A Review with Special Reference to Africa. Monitoring and Assessment Research Centre, Technical E Report No 5. International Council of Scientific Unions. Chelsea College, University of London, London, UK, 45 pp.

Tattar, T.A. 1978. Diseases of Shade Trees. Academic Press, New York, 361 pp.

Teramura, A.H., R.H. Biggs, and S. Kossuth. 1980. Effects of ultraviolet-B irradiances on soybeans. II. Interaction between ultraviolet-B and photosynthetically active radiation on net photosynthesis, dark respiration, and transpiration. Plant Physiol. 65: 483-488.

Teramura, A.H., M.C. Perry, J. Lydon, M.S. McIntosh, and E.G. Summers. 1984. Effects of ultraviolet-B radiation on plants during a mild water stress. III. Effects on photosynthetic recovery and growth in soybean. Plant Physiol. 60: 484-492.

Tevini, M., W. Iwanzik, and U. Thoma. 1981. Some effects of enhanced UV-B irradiation on the growth and compositon of plants. Planta 153: 388-394.

Thies, W.G. 1984. Laminated root rot: The quest for control. J. For. 82: 345-356.

Tisserat, N.A. and A.Y. Rossman. 1988. A canker disease of Rocky Mountain juniper caused by *Botryophaeria* stevensii. Pl. Dis. 72: 699-701.

Trabalka, J.R. (ed.). 1985. Atmospheric Carbon Dioxide and the Global Carbon Cycle. Publ. No. DOE-ER-0239. U.S. Department of Energy, Washington, DC, 315 pp.

U.S. Environmental Protection Agency. 1985. Maps Depicting Non-Attainment Areas Pursuant to Section 107 of the Clean Air Act. U.S.E.P.A., Washington, DC.

U.S. Environmental Protection Agency. 1988a. Environmental Progress and Challenges: EPA's Update. Publ. No. EPA-230-07-88-033. U.S.E.P.A., Washington, DC, 140 pp.

U.S. Environmental Protection Agency. 1988b. Future Risk: Research Strategies for the 1990's Appendix C. Strategies for Ecological Effects Research. U.S.E.P.A. Science Advisory Board. Publ. No. SAB-EC 88-040. Washington, DC.

U.S. Environmental Protection Agency. 1988c. Research Initiative on Forested Ecosystems. Office of Research and Development. U.S.E.P.A. Washington, DC, 66 pp.

U.S. Environmental Protection Agency. 1988d. Synthesis and Integration of the Forest Response Program: The Path from Projects to Major Program Outputs. Synthesis and Integration Report No. 14. U.S.E.P.A. Environmental Research Laboratory, Corvallis, OR, 136 pp.

U.S. Forest Service. 1985. Research Plan for Eastern Hardwoods Research Cooperative. Northeastern Forest Exp. Sta., Broomall, PA.

Van Arsdel, E.P., A. J. Riker, R.F. Kouba, V.E. Suomi, and R.A. Bryson. 1961. The climatic distribution of blister rust on white pine in Wisconsin. U.S.D.A. Forest Service, Lake States Forest Exp. Sta., Station Pap. No. 87, St. Paul, MN, 34 pp.

Van Deusen, P.C. 1988. Analyses of Great Smoky Mountain Red Spruce Tree Ring Data. U.S.D.A. Forest Service. Southern Forest Exp. Sta., Genl. Tech. Report No. SO-69. New Orleans, LA, 67pp.

Van, T.K., L.A. Garrard, and S.H. West. 1976. Effects of UV-B radiation on net photosynthesis of some crop plants. Crop Sci. 16: 715-710.

Van, T.K., L.A. Garrard, and S.H. West. 1977. Effects of 298-nm radiation on net photosynthetic reactions of leaf discs and chloroplast preparations of some crop species. Environ. Exp. Bot. 17: 107-112.

Waggoner, P.E. and R.D. Berger. 1987. Defoliation, disease and growth. Phytopathology 77: 393-398.

Wallner, W.E. 1987. Factors affecting insect population dynamics: Differences between outbreak and non-outbreak species. Annu. Rev. Entomol. 32: 317-340.

Waring, R.H. 1983. Estimating forest growth and efficiency in relation to canopy leaf area index. Adv. Ecol. Res. 13: 327-354.

Waring, R.H. 1985. Imbalanced ecosystems: Assessments and consequences. For. Ecol. Mamt. 12: 93-112.

Waring, R.H. 1987a. Characteristics of trees predisposed to die. BioScience 37: 569-574.

Waring, R.H. 1987b. Distinguishing pollution from climatic effects by the analysis of stable isotope ratios in the cellulose of annual growth rings. In S.H. Bicknell, ed., California Forest Response Program Planning Conference. Proceedings Feb. 22–24, 1987 Conference. College of Natural Resources, Humboldt State University, Arcata, CA, pp. 90-98.

Waring, R.H. and W.H. Schlesinger. 1985. Forest Ecosystems: Concepts and Management. Academic Press, New York, 340 pp.

Waring, R.H., K. Newman, and J. Bell. 1981. Efficiency of tree crowns and stem-wood production at different canopy leaf densities. Forestry 54: 129-136.

Waring, R.H., J.D. Aber, J.M. Melillo, and B. Moore III. 1986. Precursors of change in terrestrial ecosystems. BioScience 36: 433-438.

Weber, F.P. and F.C. Polcyn. 1972. Remote sensing to detect stress in forests. Photogrammetric Engin. 38: 163-175.

Weiss, M.J., L.R. McCreery, I. Millers, J.T. O'Brien, and M. Miller–Weeks. 1985. Cooperative Survey of Red Spruce and Balsam Fir Decline and Mortality in New Hampshire, New York, and Vermont — 1984. Forest Pest Management. Interim Report, U.S.D.A. Forest Service, Northeastern Area, Durham, NH., 130 pp.

Wessman, C.A., J.D. Alber, D.L. Peterson, and J.M. Mehillo. 1988. Remote sensing of canopy chemistry and nitrogen cycling in temperate forest ecosystems. Nature 335: 154-156.

Wilkinson, C.F. 1987. The science and politics of pesticides. In G.J. Marco, R.M. Hollingworth, and W. Durham, eds., Silent Spring Revisited. Am. Chem. Soc., Washington, DC, pp. 25-46.

Wilson, C.L. 1987. Exotic plant pathogens — Who's responsible? Pl. Dis. 71: 863.

Winner, W.E. and J.D. Bewley. 1978. Terrestrial masses as bioindicators of SO_2 pollution stress. Oecologia 35:221-230.

Woodwell, G. (ed.). 1989. Markers of Air Pollution Effects on Forests. National Research Council Workshop Proceedings. April 25–27, Asheville, NC, National Academy Press, Washington, DC.

Work Group 1. 1983. Impact Assessment. Work Group 1. United States-Canada Memorandum of Intent on Transboundary Air Pollution. Final Report. Washington, DC.

Index